W0042260

K.P. Ghiggino (Ed.)

Progress in Pacific Polymer Science 3

Proceedings of the Third Pacific Polymer Conference
Gold Coast, Queensland, December 13–17, 1993

Springer-Verlag
Berlin Heidelberg New York
London Paris Tokyo
Hong Kong Barcelona Budapest

Dr. Ken P. Ghiggino
School of Chemistry
University of Melbourne
3052 Parkville, Victoria
Australia

ISBN-13:978-3-642-78761-4 e-ISBN-13:978-3-642-78759-1
DOI: 10.1007/978-3-642-78759-1

Cip data applied for

Softcover reprint of the hardcover 1st edition 1994

Typesetting: Camera-ready by authors
SPIN:10127464 52/3020-5 4 3 2 1 0 - Printed on acid -free paper

Progress in Pacific Polymer Science 3

THE PACIFIC POLYMER FEDERATION

The Pacific Polymer Federation (PPF) is an Association of societies in the Pacific region, which have at least part of their activities devoted to polymer science and technology.

The Mission of the PPF is to advance and benefit polymer science and technology by encouraging and facilitating interaction between polymer societies in the Pacific region and the exchange of scientific knowledge through visits by polymer scientists and participation in polymer conferences.

The PPF was founded in October 1987 by the Division of Polymer Chemistry of the American Chemical Society, the Society of Polymer Science, Japan and the Polymer Division of the Royal Australian Chemical Institute.

There are now 14 member societies of the PPF, comprising, in addition to the founding societies: The Society of Polymer Science of Korea (1988), the Macromolecular Science and Engineering Division of the Chemical Institute of Canada (1988), the Polymer and Industrial Section of the Malaysian Institute of Chemistry (1988), the Polymer Division of the Chinese Chemical Society of the People's Republic of China (1989), the Polymer Group of the New Zealand Institute of Chemistry (1989), the Division of High Polymer Physics of the American Physical Society (1990), the Plastics and Rubber Institute of Singapore (1990), the Polymer Society, Taipei (1993), the Division of Polymers and Organic Materials of the Mexican Academy of Materials Sciences (1993), the Division of Polymer Science of the Chemical Society of Vietnam (1993) and the Indonesian Polymer Association (1993). An application for membership is pending from the Polymer Society of Thailand.

The policies for the operation and development of the PPF are determined by the Council, which meets once per year. The current Officers of the PPF are: President: Joe Salamone (USA), Vice-President: Chung Yup Kim (Korea), Deputy Vice-President: Renyuan Qian (China), Secretary-Treasurer: Jim McGrath (USA). A PPF Newsletter is published annually.

PACIFIC POLYMER CONFERENCES

The main activity of the PPF is to hold a Pacific Polymer Conference every two years, which is organized by the host member society. The 1st Pacific Polymer Conference was held on 12-16 December 1989 in Maui (Hawaii, USA) and the 2nd Pacific Polymer Conference was held at Lake Biwa, Otsu, near Kyoto (Japan) on 26-29 November 1991.

The 3rd Pacific Polymer Conference, PPC-3, was held on 13-17 December 1993 at the Gold Coast, Queensland, Australia and organized by the Polymer Division of the Royal Australian Chemical Institute. There were 528 participants, including 117 students) from 23 countries, with the highest numbers coming from Australia (277), Japan (121), USA (40) and Korea (30). The scientific program comprised 50 Invited Lectures, 250 contributed Oral Presentations and 150 Posters, with up to 9 concurrent sessions. There were also Plenary Forums on "The Future of Polymer

Science and Technology in the Pacific Region" and "Polymer Materials and the Environment".

The 4th Pacific Polymer Conference will be held in Kauai (Hawaii, USA) in December 1995, the 5th Pacific Polymer Conference in Korea in 1997 and the 6th Pacific Polymer Conference in China in 1999.

PROGRESS IN PACIFIC POLYMER SCIENCE

Invited Lectures from each Pacific Polymer Conference have been published by Springer-Verlag as a hard-cover book of about 350 pages with the title: "Progress in Pacific Polymer Science", and provides a permanent record of the current status of major topics in polymer science. The 3rd Pacific Polymer Conference continues this tradition with this publication of "Progress in Pacific Polymer Science 3".

ACKNOWLEDGEMENTS

The 3rd Pacific Polymer Conference was organized by a Committee of 30 Australian polymer scientists, assisted by an International Advisory Committee representing all of the member societies of the PPF and an Honorary Advisory Committee. Particular mention should be made of the contributions of Peter Pomery (Secretary), David Hill (Treasurer, Program and Preprints), Graeme George (Program and Preprints), Bruce Guise (Promotion) and Ken Ghiggino (Editor of Progress in Pacific Polymer Science 3).

Jim O'Donnell
Chairman of the Organizing Committee for PPC-3
President of the Pacific Polymer Federation 1993
31 January 1994

PPF COUNCIL
(Photo taken at PPC-3 Banquet)

Back row (l to r)
Sung Chul Kim (Korea), Akihiro Abe (Japan), Chung Yup Kim (Korea), David Hill (Australia), Hardy Chan (Singapore), Koichi Hatada (Japan), Chee-Cheong Ho (Malaysia)

Front row (l to r)
Mrs. S. C. Kim (Korea), Renyuan Qian (China), Noer Surdia (Indonesia), Almeria Natansohn[a] (Canada), Jim O'Donnell (Australia), Supawan Tantayanon[b] (Thailand), Joe Salamone (USA), Neil Edmonds (NZ)

Absent
Jim McGrath (USA), Phil Pincus[c] (USA), Takeshi Ogawa (Mexico), Leo-Wang Chen (Taiwan), Dang van Luyen (Vietnam)

[a] for Basil Favis, [b] observer status at PPC-3, [c] for Hyuk Yu

Contents

Kinetics of Phase Separation of Polymer Blends
T. Nose 1

Cell Adhesion and Protein Adsorption on Gradient Surfaces
H.B. Lee, J.H. Lee 13

Rheological Properties of Blends: Facts and Challenges
P.J. Carreau, M. Bousmina, A. Ajji 25

Hydroboration Polymerization
Y. Chujo 41

Stereospecific Polymerization of Styrene
M. Kuramoto 53

Polymer Design for Minimally Adhesive Surfaces
K.J. Wynne, T. Ho, R.A. Nissan, X. Chen,
J.A. Gardella, Jr. 63

Some Recent Developments in the Aminoxyl Radical Trapping Technique
S.E. Bottle, W.K. Busfield, I.D. Grice, K. Heiland,
I.D. Jenkins, W. Meutermans, M. Monteiro 85

The Surface Excess Structure in 1,4-Polyisoprene-Polyacetylene Copolymer Solutions
J.W. White, J.C.H. Kerr, J. Penfold, P.M. Saville,
R.K. Thomas 99

Helix-Sense-Selective Polymerization of Methacrylates and Conformational Characteristics of the Polymers
Y. Okamoto, T. Nakano 111

Spectroscopic Studies of Conducting Polymers on Oxide Catalysts
G.J. Millar, G.F. McCann, G.A. Bowmaker, R.P. Cooney 123

Synthesis of Poly(p-Epoxystyrene-co-N-Phenylmaleimide-co-styrene)
B. Pulpoka, P.C. Ternai, F. Schue, S. Tantayanon 137

Production, Modifications and Applications of Chitosan in Vietnam
D. Van Luyen 147

A Novel Polymer Film that Controls Light Transmission
M. Honda, S. Hozumi, S. Kitayama 159

The γ-Initiated Degradation of Polypropylene: Stabilization by Phenols
D.J. Carlsson, S. Falicki, J.M. Cooke, D.J. Gosciniak 171

Construction of Polypeptide Tertiary Structure by the Template-Assisted Synthesis
Y. Imanishi, S. Kimura, T. Tsuchimoto 181

A New Family of Water-Soluble Polymers: Polymers of N-Substituted Acrylamide Derivatives
S. Fujishige....................................191

The Measurement and Meaning of Rate Coefficients in Free-Radical Polymerizations
J.P.A. Heuts, P.A. Clay, D.I. Christie, M.C. Piton, J. Hutovic, S.H. Kable, R.G. Gilbert....................................203

Structure -Reactivity Relationships in Acyclic Diene Metathesis (ADMET) Polymerization
K.B. Wagener....................................217

In-Situ Spectroscopy of Redox Reaction in PPY
D.Y. Kim, D.H. Kim, J.Y. Lee, C.Y. Kim....................................225

Disubstituted Poly(bithiophene)s: Syntheses and Applications
H.S.O. Chan, S.C. Ng, H.H. Huang, S.H. Seow....................................237

Fluorescence Studies of Polymer Diffusion Across Interface
M.A. Winnik, H.-B. Kim, S. Kawaguchi....................................247

Recent Studies on β-Crystalline Form of Isotactic Polypropylene
G. Shi....................................259

Controlling Sequence Distribution in Copolyesters
C.W. Potter, J.C. Lim, G. Serpe, J. Economy....................................271

Performance on Demand - A New Era in Polymer Science (A Case Study using Conducting Polymers)
G.G. Wallace....................................283

Reversible Optical Storage in Azo Polymers
A. Natansohn, P. Rochon....................................295

Highly Electrically Conductive Polymer Composites and Blends
T.J. Kang, Y. Miyaki, J.H. Han, T. Motobe, Y.E. Whang, S. Miyata....................................307

Single-Chain Single Crystal
H. Bu, E. Chen, S. Xu....................................313

Polymers in New Water Treatment Technologies
B.A. Bolto....................................325

Novel Application of Natural Rubber Latex as Natural Flocculant for Tin Tailings Slurry
C.C. Ho, K.C. Lee, E.B. Yeap....................................337

Liquid Natural Rubber: Preparation and Application
I. Abdullah....................................351

A Nuclear Magnetic Resonance Study of the Microstructure of Poly(Vinyl Alcohol)
S. Amiya....................................367

Enhancing Reinforcement Effects of Polymers in Asphalt: An Applied Approach to Solid Waste Disposal
W.H. Daly, Z. Qiu, J. Youngblood, I. Negulescu.............381

Rigid-Rod Polyimides with n-Alkoxymethyl Side Chains
S.-B. Park, Y.-J. Choi, W.-C. Zin, J.C. Jung..............401

New Developments in Perfluoroelastomers: Chemistry, Properties and Applications
A.L. Logothetis..411

Oriental Lacquer
Otto Vogl, Jan Bartus, Meifang Qin, John D. Mitchell.......423

Author Index...433
Subject Index..435

Kinetics of Phase Separation of Polymer Blends

Takuhei NOSE
Department of Polymer Chemistry, Tokyo Institute of Technology
2-12-1 Ookayama, Meguro-ku, Tokyo 152 JAPAN

INTRODUCTION

The phase diagram, especially coexistence curve, can provide us an indispensable information in studying the phase separation kinetics since it gives the stability and the fractional phase volume of the system that are most important factors controlling the phase separation processes. In spite of its importance, there are a very few studies because of experimental difficulties. In this study, we have overcome the difficulties and determined precise coexistence curves of polymer blends.[1-4] Taking advantage of the established coexistence curves, we have studies kinetics of early and late stages of phase separation. As to the early stage, time-evolution of concentration fluctuations in the *metastable* region has been studied using time-resolved synchrotron small-angle x-ray scattering.[5] For the late stage, the crossover from co-continuous to droplet structure has been investigated, which is primarily governed by the fractional phase volume.[6] Furthermore, using the information obtained here, the morphological structure induced by two-step temperature jump has also been investigated.[7-9] Through these studies, we will demonstrate importance of the coexistence curve in studying phase-separation kinetics.

EXPERIMENTAL

Materials and Samples

Polystyrenes(PS) were commercial ones with narrow molecular weight distributions. Poly(2-chlorostyrene)s(P2ClS) were radically polymerized and fractionated to be of narrow molecular-weight distribution. Characteristics of PS and P2ClS used and their sample codes are listed in Table 1. Di-n-butyl phthalate (DBP) , which is thermodynamically good for both of PS and P2ClS, was added to the blends for adjusting the phase boundary and the phase separation to appropriate range of temperature and rate, respectively. Blend systems used and their codes are shown in Table 2, and the sample code Y-PSXX, specifying the blend system Y with PS volume composition XX % in the total polymers, will be used.

K.P. Ghiggino (Ed.)
Progress in Pacific Polymer Science 3

Experimental Methods

To determine the coexistence curve, two methods using a specially designed centrifuge[10] and a scanning electron microscope(SEM) with image analysis, respectively, were employed. Volume ratios of separated phases for at least two different initial compositions enabled us to determine compositions of coexisting phases under the quasi-binary approximation.

Relatively early stage of phase separation was investigated with light scattering and synchrotron small-angle x-ray scattering(SAXS). SAXS experiments were carried out at the SUNY X3A2 Beamline, National Synchrotron Light Source, Brookheaven National Laboratory, using a modified Kratky collimator.[11]

Morphological structure of the fractured surface of phase-separated blends was observed by SEM, JEOL JSM-T220, using backscattered electrons. The fractured surface observed was flat, and PS-rich and P2ClS-rich phases gave a good contrast of SEM images without staining.

Table 1 Characteristics of materials

samples	codes	$M_w / 10^4$	M_w / M_n
PS	PS35K	3.5	<1.06
	PS37K	3.7	<1.06
	PS50K	5.0	<1.06
P2ClS	P2ClS90K	9.03	1.10
	P2ClS109K	10.9	1.09
	P2ClS118K	11.8	1.11
	P2ClS233K	23.3	1.19

Table 2 Blended Samples

system codes	combinations of PS and P2ClS	DBP content/ wt%
A	PS35K/P2ClS109K	0
B	PS50K/P2ClS90K	6.9
C	PS37K/P2ClS223K	22.6
D	PS50K/P2ClS118K	9.7

Table 3 Characteristic parameters of coexistence curves of the polymer blends

system	T_c	ϕ_c	exponent	amplitude		symmetry factor	
	°C		β	a^+	b^+	a^-	b^-
A	137	0.57	0.5	0.733	−0.230	−0.338	0.130
B	124.8	0.54_1	0.39	0.686	--	−0.734	--
C	134.3	0.52_4	0.33	0.667	−0.266	−0.370	0.0762

A,C: $\varepsilon=1-T/Tc$ ($\varepsilon<0.3$), B:$\varepsilon=1-Tc/T$ ($\varepsilon<0.1$)(put $b^+=b^-=0$)

RESULTS AND DISCUSSION

Coexistence Curves

Figure 1 shows coexistence curves of the blends A, B, and C with fixed different amounts of DBP. The composition ϕ_{PS} taken as the horizontal axis is the volume fraction of PS in the total polymer, i.e., $\phi_{PS}=[PS]/([PS]+[P2ClS])$ with [X] being the volume fraction of X in the blend.

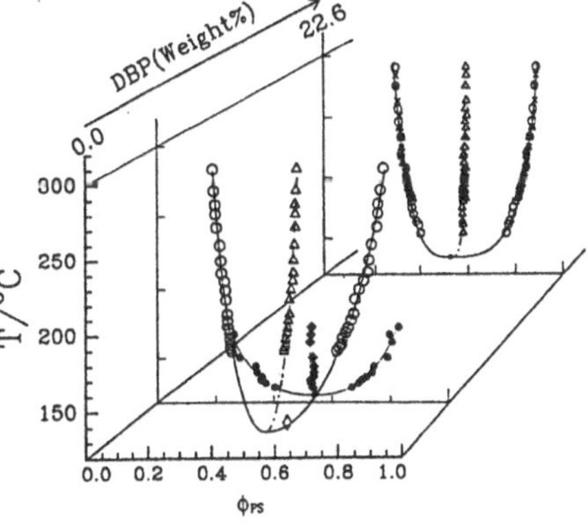

Figure 1. Coexistence curves of (a) System A (b) System B, and (c) System C blends, along with their diameters ($\equiv(\phi^++\phi^-)/2$).

According to the Wegner expansion,[12] the measured coexistence curves were analyzed by the expressions

$$\phi^+ = \phi_c + a^+\varepsilon^\beta + b^+\varepsilon^{\beta+\Delta} \quad (1)$$

$$\phi^- = \phi_c + a^-\varepsilon^\beta + b^-\varepsilon^{\beta+\Delta} \quad (2)$$

$$\phi^+ - \phi'' = A\_\varepsilon^\beta + B\_\varepsilon^{\beta+\Delta} \quad (3)$$

Here, ϕ^+ and $\phi^-$are compositions ϕ_{PS} in the PS-concentrated and PS-dilute coexistence phases, respectively. The values of exponent β and the coefficients were determined by the least-squares fitting of eqs 1-3 with the fixed value of Δ = 0.5. The obtained values with critical temperature T_C and critical composition ϕ_C are summarized in Table 3. The exponents β suggest the presence of the crossover[13, 14] from the mean-field to the Ising behavior with increasing the content of diluent DBP. Coexistence curve of the blend A without DBP exhibited the mean-field type critical exponent, while the blend C with DBP of more than 20 wt% has the Ising type critical exponent. The blend B with an intermediate DBP content shows an intermediate value of β.

In spite of the deviation from the mean-field type behavior, the coexistence curve of the blend B was well described by the Flory-Huggins theory under the quasi-binary approximation with the interaction parameter χ that was consistent with those estimated by the molecular weight dependence of cloud points.

Concentration Fluctuations In The Metastable Region

To study the growth of concentration fluctuation in the metastable region, static and time-resolved SAXS were measured for the blend A without DBP, A-PS64.

In the framework of the mean-field *linear* theory, Binder[15, 16] derived the time-dependent structure factor S(q,t) of a binary polymer blend as

$$S(q, t) = S_x(q) + [S(q, 0) - S_x(q)]\exp(-2\tau_q^{-1}t) \qquad (4)$$

where $S_x(q)$ is the structure factor in the limit of $t \to \infty$, i.e., in the one-phase region, $S_x(q)$ constitutes the equilibrium structure factor, and in the two-phase region, it corresponds to the virtual structure factor. The relaxation time τ_q of concentration fluctuation is given with a Fourier transform of the Onsager coefficient $\Lambda(q)$ as $\tau_q=S_x(q)/q^2\Lambda(q)$. The sign of relaxation time τ_q is determined by that of the structure factor $S_x(q)$, i.e., $S_x(q)$ determines the stability of the concentration fluctuation with the wavenumber q. If $S_x(q) > 0$, then the concentration fluctuation decays by a small thermal perturbation, while if $S_x(q) < 0$, then the fluctuation grows. Under the condition of weak asymmetry of polymeric indices,

$$\tau_q^{-1} \equiv (6/\tau_R)q^2R_m^2(1 - (\chi/\chi_s)f_D(q^2R_m^2)) \qquad (5)$$

with

$$f_D(q^2R_m^2) = (2/q^2R_m^2)[1 - (1 - \exp(-q^2R_m^2))/q^2R_m^2] \qquad (6)$$

The relaxation time τ_R specifies the time required for a chain molecule to diffuse over a length comparable to its own size, R_m is a mean radius of gyration, f_D is the Debye function for a Gaussian chain, and χ_s is the Flory-Huggins interaction parameter χ at the mean-field spinodal temperature T_s(m.f.).

The structure factor $S_x(q)$ was obtained from the mean-field static susceptibilities measured in the one-phase region, using the random phase approximation[14] and the assumption of continuous temperature change of χ-parameter over the one- and two-phase regions. The results are represented in Figure 2. Figure 3 represents the location of studied temperature range in the phase diagram. The composition ϕ_{PS} of the blend is off-critical, so that the system should change from stable to unstable state passing through metastable state with increasing temperature. The time-dependent structure factors were measured as a function of temperature near the phase boundary.

Figure 4 represents time-evolution of the normalized structure factor $\Delta S(q,t)/\Delta S(q,0)$ in the metastable(a) and unstable(b) regions, where $\Delta S(q,t)=$

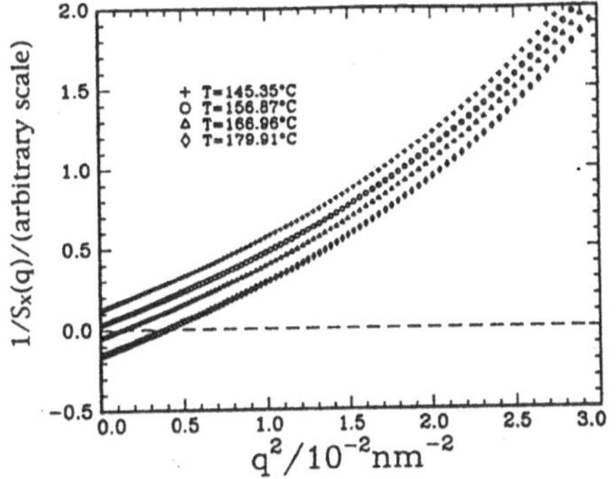

Figure 2 Structure factor $S_X(q)$ as a function of temperature for the blend A-PS64. Mean-field spinodal temperature T_S(m.f.) = 160.48°C

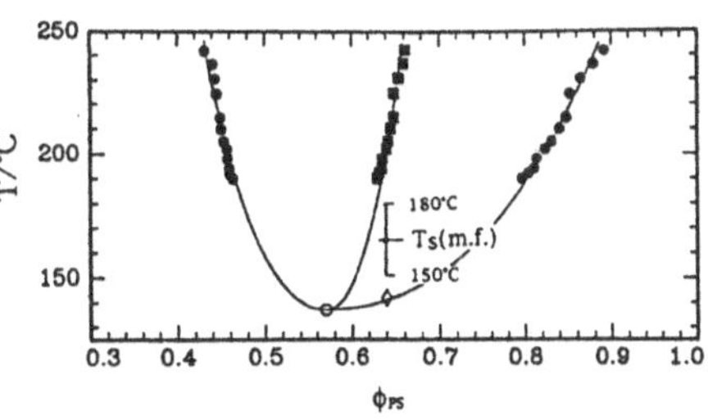

Figure 3 Coexistence curve of the blend system A. A vertical solid line indicates experimental region for phase-separation studies. T_s(m.f.)=160.48°C.

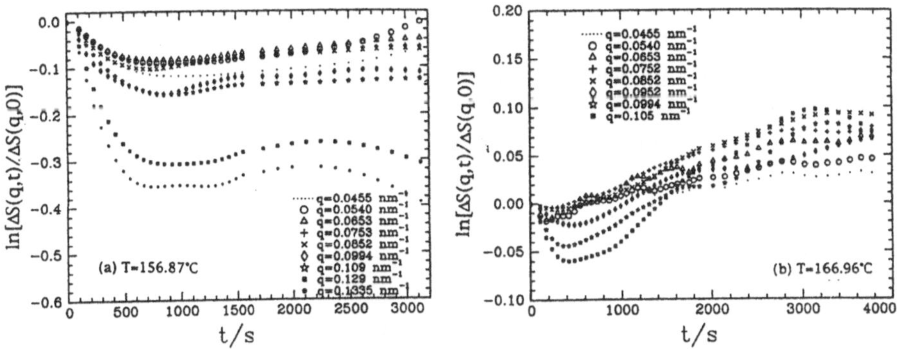

Figure 4 Time-evolution of the normalized structure factor in metastable(a) and unstable(b) of the blend A-PS64.

$S(q,t) - S_X(q)$. In the metastable region ($S_X(0)>0$) and in the unstable region ($S_X(0)<0$), but close to T_S(m.f.), $\Delta S(q,t)$ even at small wavenumbers q decreased initially, and then $\Delta S(q,t)$ increased, eventually leading to phase separation. When the temperature was higher enough than T_S(m.f.), $\Delta S(q,t)$ increased from the beginning. According to eq 4, the initial slope of the plots in Figure 4 was evaluated to determine values of t_q, which plotted against q^2 in Figure 5. Calculated curves of eqs 5 and 6 with an adjustable parameter of τ_R could reproduced the q-dependence of τ_q, as shown in Figure 5. The obtained relaxation time τ_R of a chain molecule was of the order of 100 s, and its temperature-dependence followed the WLF equation with the constants estimated from

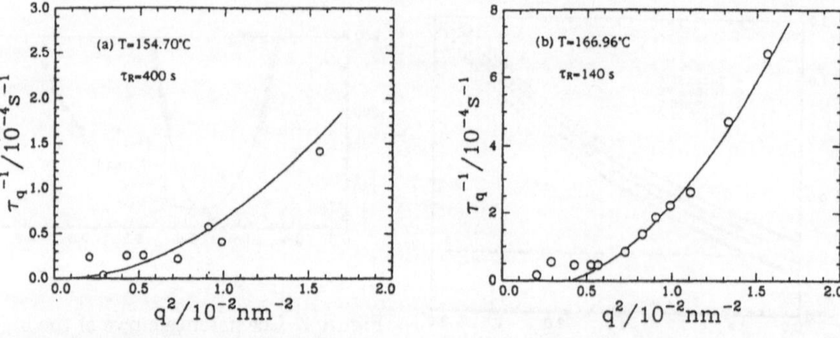

Figure 5 q-dependence of relaxation time τ_q at different temperatures[(a),metastable; (b),unstable] in the blend A-PS64. Solid curves are calculated curves by the Binder theory of eq 5 with eq 6.

viscoelastic properties. In conclusion, the initial kinetics could be described by the linearized theory of Binder.

Morphological Structure And Phase-Separation Kinetics At The Crossover Region

Microphase structure and growth mechanism of polymer blends in the late stage of phase separation are dominantly controlled by the volume fraction v of minority phase. At the critical condition v = 0.5 the co-continuous structure is formed, while at the off-critical condition v << 0.5 the droplet structure has to appear. Here, more explicit relation of the morphological structure to the fractional phase volume will be investigated using System B. In particular, our attention has been focused on morphological structure in crossover region from the critical to the off-critical.

In the late sage, the following scaling law for the growth of phase-separated domain structure is expected to be satisfied: The characteristic length scale L of the structure obeys the power law

$$L = q_m^{-1} = t^{\alpha} \tag{7}$$

which can be observed in Fourier space as the peak position q_m of structure factor S(q,t) or be evaluated in real space by image analysis of SEM photographs. The exponent α reflects morphology and growth mechanism.

Critical region and off-critical region: Figure 6 represents phase-volume fraction dependence of morphological structure for the blend B at 165°C. Location of phase-separation condition were indicated in Figure 7 (points 1,2, and 3). The dark area and the bright area in the positive pictures of SEM are PS-rich and P2ClS-reich phase, respectively, and the fraction v denotes that of PS-rich phase. The structure at v= 0.45 was co-continuous, growing with the exponent $\alpha = 0.92$. The exponent supports Siggia's growth mechanism[17] for fluid systems of continuous domain with curved interface. On the other hand, the droplet structure

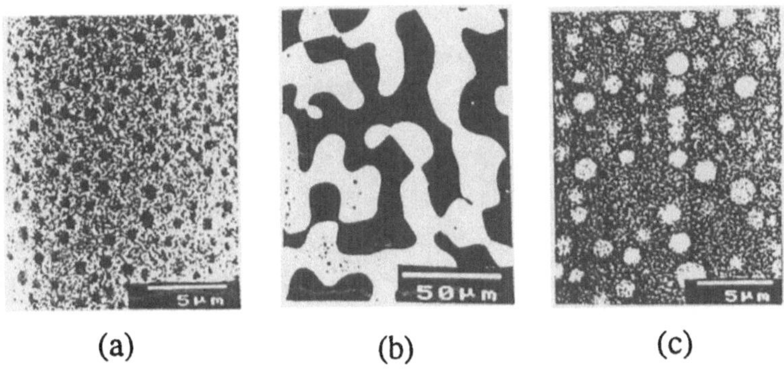

Figure 6. SEM pictures in the regions of droplet and co-continuous structures for System B blends at 165°C. (a)B-PS32:t=30.5h;v=0.12, (b)B-PS50:t=3.7h; v=0.46, and (c)B-PS70:t=30.5h;v=0.85

appeared at v=0.12 and 0.85. The growth rate was very low compared with that of the co-continuous structure, with the exponent α being around 0.3. The growth is considered to be driven by vaporization-condensation process of the Ostwald ripening.[18] Theoretically expected morphological structure and growth mechanism at asymptotic conditions, the critical and off-critical ones.

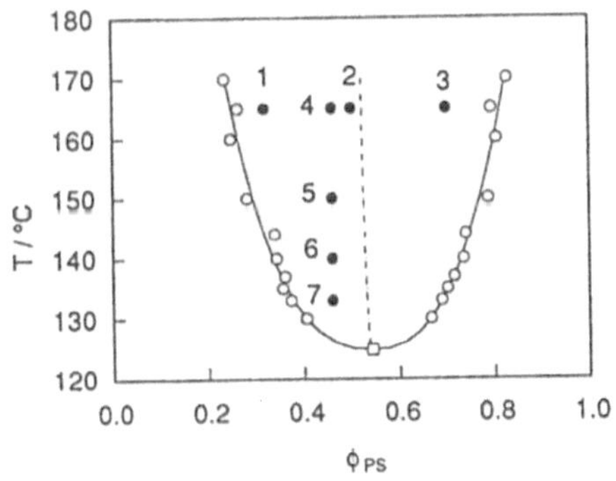

Figure 7. Coexistence curve of System B with location of experimental conditions (points1 - 7) for morphological studies.

<u>Crossover region:</u> Figure 8 shows change in morphological structure with decreasing phase-volume fraction v. In order to change v, the temperature was decreased at the fixed PS-composition, using B-PS46, instead of decreasing the PS-composition at a fixed temperature, for experimental simplicity, which is illustrated in Figure 7(points 4 → 5 → 6 → 7). The crossover from co-continuous to droplet structure was found between the phase-volume fractions 0.35 and 0.27. At v=0.29, the blend showed a heterogeneous structure with the co-continuous and droplet structures coexisting. Characteristic feature of the heterogeneity are that the two structures form different regions without being mingled uniformly, and that characteristic size of the droplet structure is much smaller than that of the co-continuous structure. Growth rate of the co-continuous structure was larger than that of the droplet structure, which results in enhancement of the heterogeneity, especially the size difference, with laps of time, as shown in Figure 9.

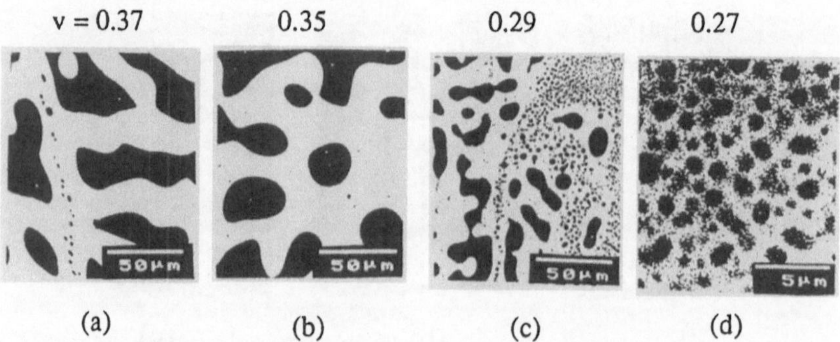

Figure 8. Morphological structure in the blend B-PS46 as the phase volume v decreases with decreasing temperature, crossing the crossover region. See Figure 7 as for experimental conditions.(points 4-7)

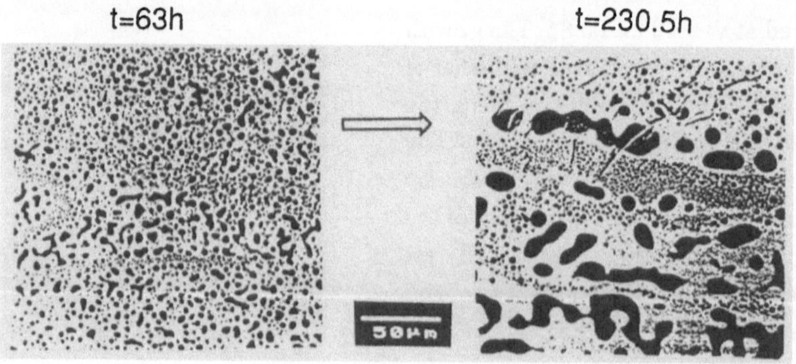

Figure 9 Structure change with laps of time in the crossover region of B-PS46 at 140°C.

Mechanism of the heterogeneous structure formation is considered to be as illustrated in Figure 10. In an early and intermediate stages of phase separation, modulated, essentially co-continuous structure produced by growth of symmetrical concentration fluctuations around the initial concentration extends uniformly over the whole system.(a) With further proceeding of the phase separation, some parts of the modulated structure of the minority PS phase break into droplets, and consequently the residual continuous structure becomes of *finite* size, since v is not large enough to keep the co-continuous structure stable.(b) As the growth rate of the continuous structure is much greater than that of the droplets, the time-evolution of the whole structure has to be induced and governed by growth of the continuous structure. The modulated continuous structure has larger spacing (or larger mesh size of the network) than that it would have in case of no breaking.(b-1) Because of finite size of the continuous structure, it shrinks with thickening of the "tube" of the continuous-structure network by Siggia's mechanism.(c-1) As the

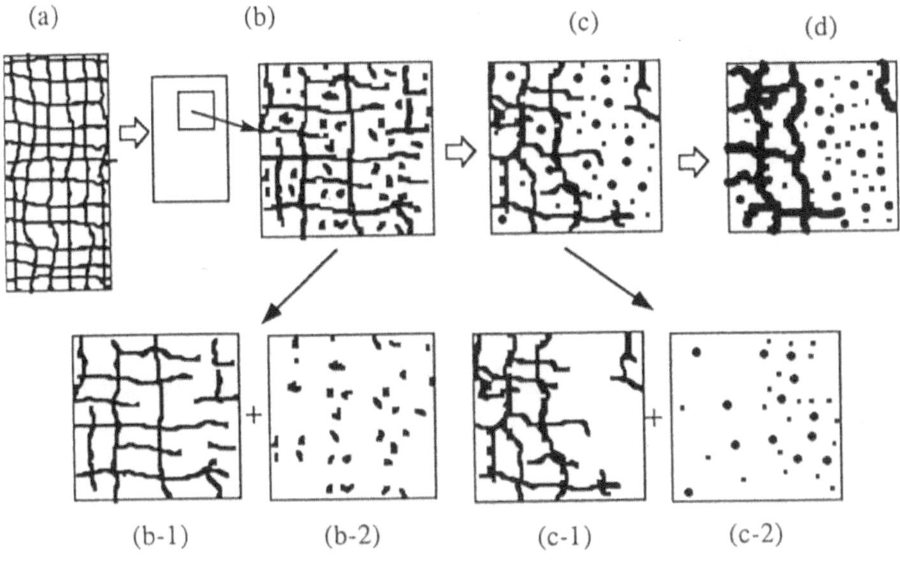

Figure 10 Schematic presentation of growth mechanism of the heterogeneous structure in the crossover region from co-continuous to droplet structure.

result, the majority phase including droplets is flowing out to exterior of the continuous structure(c-2) , and the droplets accumulate at the boundary of the continuous regions to form a droplet region.(d) The peak position q_m of scattered light intensity profile followed the power law of eq 7 with the exponent $\alpha = 0.91$ over a wide range of time including the stage were the heterogeneous structure was observable by SEM (Figure 9(a)), as shown in Figure 11. The exponent supports Siggia' mechanism, i.e., that the growth of co-continuous structure with flow in the "tube" is the dominant process in the whole-structure growth over earlier less-heterogeneous and later more-heterogeneous structures. The presence of droplet structure could give little effects on the scattering function probably because the scattered light from the larger structure has the stronger intensity.

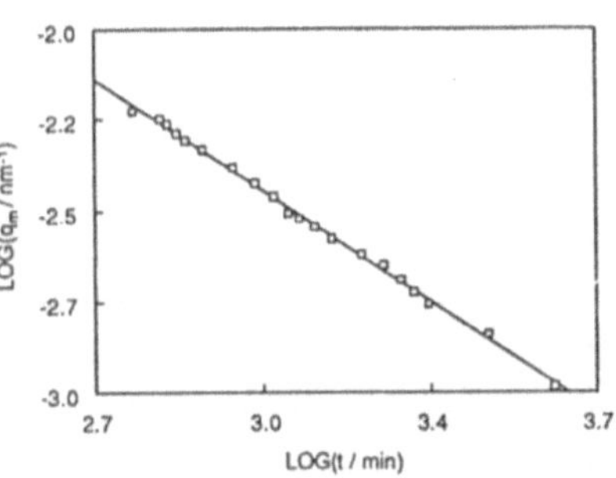

Figure 11 Time-evolution of characteristic wavenumber q_m in the crossover region for B-PS46 at 140°C

Phase Separation by Two-step Temperature Jumps

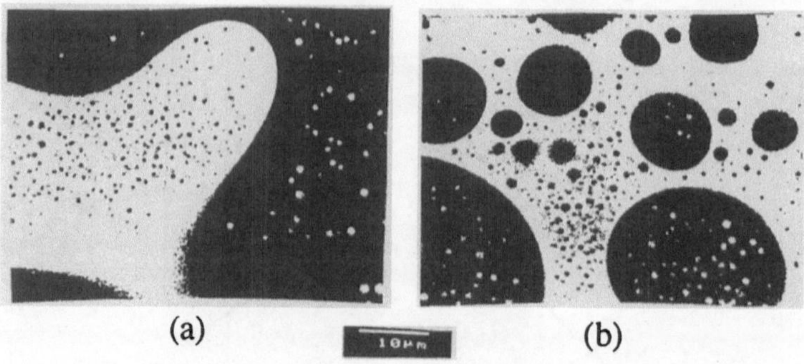

Figure 12 Domain structures induced by two steps temperature jumps at (a)critical and (b)off-critical compositions. (a) D-PS60, 135°C;37h → 155°C;11h, (b)D-PS50, 130°C;70h → 155°C;24h

Two-step temperature jumps, i.e., deep-quench jumps after formation of microdomain structure by a jump of shallow quench, can produce the so-called salami-type structure, large domain structure with small droplets inside. If the large domain has the co-continuous structure formed by the first jump of critical condition, the obtained structure has to be droplet-in-continuous-domain structure, while the first jump is made in off-critical condition, the structure should be droplet-in-droplet structure. In fact, we could produce the two types of the salami structure using blends of B-PS55, D-PS60 and D-PS50, of which examples are shown in Figure 12.

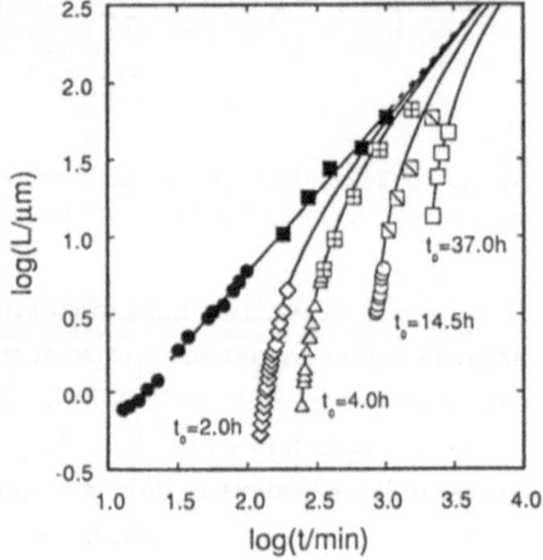

Figure 13. Growth of large co-continuous domain size L after the second temperature jump for Sample D-PS60 near the critical composition. 1st step T=135°C kept for the time t_0, 2nd step:T=155°C, filled points: the single step(i.e., t_0=0)

After the second jump, the large domain kept growing with laps of time, while droplets appear inside, grew in size first, showed no appreciable change for a while, and then started to decrease their number. To examine the growth rate of the large domain, we proposed the following equation[8] for the growth of characteristic length L as a function of the time t, which would have to be satisfied if the growth rate was not affected by the system history.

$$L=K(t+t_x-t_0)^{\alpha} \tag{8}$$

where t is taken to be such that t=0 when the first jump is made, and t_0 is the time when the second jump is performed. K and t_x are related to the length L_0 at $t=t_0$ by

$$L_0=Kt_x^{\alpha} \qquad (9)$$

Here, the scaling law of eq 7 was assumed. Figure 13 shows the growth of L as a function of t_0 for the co-continuous large domain of D-PS60. Experimental results agree with calculated curves of eq 7 when t_0 is small. However, as t_0 becomes larger, the observed values deviate towards smaller L from the calculated curves at longer times. Therefore, the growth rate of the co-continuous domain size after the second jump was almost the same but slightly lower than that of the single-step jump owing to the different thermal history, when t_0 is large.

References:

1 Kwak KD, Okada M, Nose T (1991) Polymer 32:864

2 Kwak KD, Okada M, Chiba T, Nose T (1992) Macromolecules 25:7204

3 Chu B, Wu D, Linliu K, Ying Q-C, Nose T, Okada M (1992) Phys Rev Lett 68:3184

4 Chu B, Ying Q, Linliu K, Xie P, Gao T, Li Y, Nose T, Okada M (1992) Macromolecules 25:7382

5 Ying Q, Chu B, Wu G, Linliu K, Gao T, Nose T, Okada M (1993) Macromolecules 26:5890

6 Okada M, Kwak KD, Chiba T, Nose T (1993) Macromolecules 26:6681

7 Okada M, Kwak KD, Nose T (1992) Polym J 24:215

8 Kwak KD, Okada M, Chiba T, Nose T (1993) Macromolecules 26:4047

9 Tao J, Okada M, Nose T (1993) Rept on Preog in Polym Phys Jpn 36:197

10 Chu B, Linliu K, Xie P, Ying Q, Wang Z, Shook J (1991) Rev Sci Instrum 62:2252

11 Chu B, Wu D, Wu C (1987) Rev Sci Instrum 58:5317

12 Wegner FJ (1972) Phys Rev B5:4529

13 de Gennes PG (1977) J Phys (Paris) Lett 38L:441

14 de Gennes PG (1979) Scaling Concepts in Polymer Physics. Cornell University Press, Ithaca, NY, and London, Chapter 4

15 Binder K (1983) J Chem Phys 79:6387

16 Binder K (1987) Colloid Polym Sci 265:273

17 Siggia ED (1979) Phys Rev A 20:595

18 Lifshitz IM, Slyozov VV (1961) J Phys Chem Solids 19:35

Cell Adhesion and Protein Adsorption on Gradient Surfaces

Hai Bang Lee and Jin Ho Lee*

Biomaterials Laboratory, Korea Research Institute of Chemical Technology,
P. O. Box 9, Daedeog Danji, Taejeon 305-606, Korea
*Department of Macromolecular Science, Han Nam University,
133 Ojeong Dong, Daedeog Ku, Taejeon 300-791, Korea

Abstract: A new method for preparing a wettability gradient on polymer surfaces was developed. Low density polyethylene sheets were treated in air with corona from a knife-type electrode whose power gradually increases along the sample length. The polymer surfaces oxidized gradually with the increasing power and the wettability gradient was created on the surfaces as evidenced by the measurement of water contact angles, Fourier-transform infrared spectroscopy in the attenuated total reflectance mode, and electron spectroscopy for chemical analysis. The wettability gradient surfaces prepared were used to investigate the interactions of model protein and cells in terms of the surface hydrophilicity/hydrophobicity of polymeric materials.

INTRODUCTION

Detailed analysis of surface-induced changes of adhered cells or adsorbed proteins requires a series of well-characterized surfaces, each having a constant chemical or physical property. It is often necessary a large number of time-consuming experiments, by which each surface is prepared and characterized separately. This is a laborious procedure with the strong possibility of a methodological error. We tried another approach, in which a surface property is changed on one sample in a gradient along the sample length (i. e., "gradient surface"). Such a gradient surface is of particular interest for basic studies of the interactions between biological species and surfaces since the effect of a selected property can be examined in a single experiment on one surface.

The preparation of surfaces with a wettability gradient was first described by Elwing et al. (1). A gradient of methyl groups was formed by diffusion of dimethyl dichloro silane through xylene onto flat hydrophilic silicone dioxide surfaces. Elwing and co-workers (1-7) and Hlady and co-workers (8-10) have used the wettability gradient surfaces to investigate surface hydrophilicity-induced changes of adsorbed proteins. The wettability gradient surfaces prepared by the above methods seem to be useful as basic research tools, but have a limitation; they can be applied to only hydrophilic inorganic substrates such as silicon, silica, quartz, or glass.

A method for preparing a wettability gradient on various polymer surfaces was developed by our group (11-13) and by Pitt et al. (14,15). The wettability gradient was produced via radio-frequency plasma discharge treatment by exposing the polymer sheets continuously to plasma.

K.P. Ghiggino (Ed.)
Progress in Pacific Polymer Science 3

The polymer surfaces oxidized gradually along the sample length with increasing plasma exposure time and thus the wettability gradient was created.

Recently, we developed a new method for preparing a wettability gradient on polymer surfaces using corona discharge treatment (16). The wettability gradient was produced by treating the polymer sheets with corona from a knife-type electrode whose power was changed gradually along the sample length. The polymer surfaces oxidized gradually with the increasing power and the wettability gradient was created on the sample surfaces. This method of preparing a wettability gradient on polymer surfaces is simpler and more practical than the plasma treatment method because the samples are discharged in air at atmospheric pressure in this method, whereas they are discharged under vacuum in the plasma treatment method.

We also prepared specific functional group (-COOH, -$CONH_2$, -OH group, etc.) gradients on polymer surfaces by the above corona treatment, followed by the graft copolymerization and subsequent substitution reactions (17). The functional group gradient surfaces prepared can be used to systematically investigate the interactions of biological or other species in terms of the surface functional groups and their density of polymeric materials.

In this study, the wettability gradient surfaces prepared on low density polyethylene (PE) by the corona discharge treatment was used to investigate cell and protein interactions continuously related to the surface wettability of polymeric materials.

PREPARATION AND CHARACTERIZATION OF WETTABILITY GRADIENT SURFACES

An additive-free PE sheets, 250-300 μm thick (Hanyang Chemical Co., Korea), were cut into 5.0 x 7.0 cm pieces and ultrasonically cleaned twice in ethanol for 30 min. The PE sheets were treated with a radio-frequency corona discharge apparatus made by our laboratory for the preparation of gradient surfaces (Fig. 1). The PE sheet was placed on the sample bed and dry air

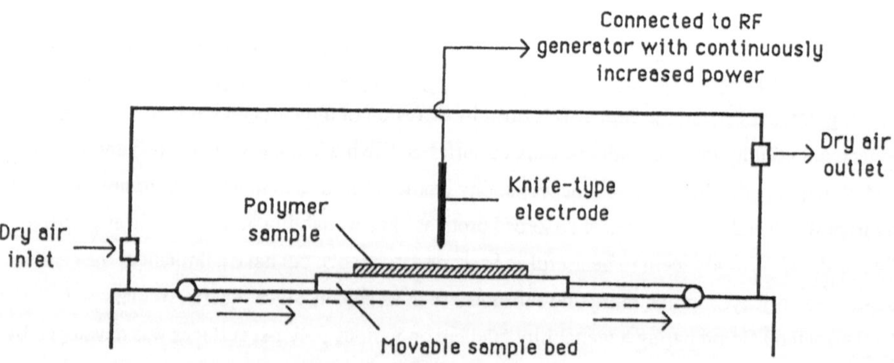

Fig. 1. Schematic diagram showing the corona discharge apparatus for the preparation of wettability gradient surfaces.

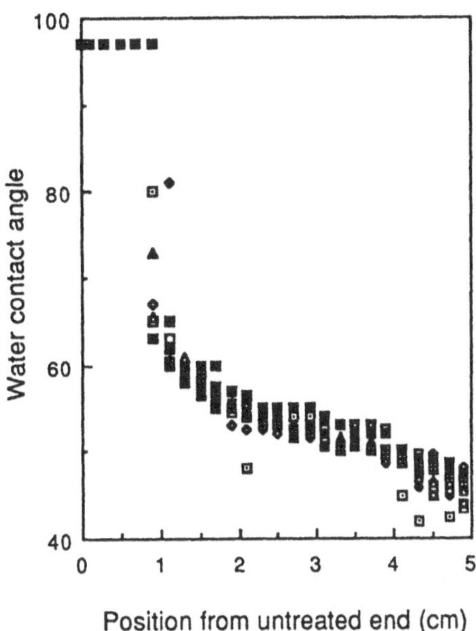

Fig. 2. Changes in water contact angle of corona-treated PE surface along the sample length. Sample numbers, n = 11.

was purged through the apparatus at a flow rate of 20 liter/min. The knife-type electrode was 1.0 mm away from the sample surface. At the same time that the sample bed was translated at a constant speed of 1.0 cm/sec, the corona was discharged from the knife-type electrode onto the sample with gradually increasing power (from 10 to 35 watt at 100 kHz). The sample sheet (5.0 x 5.0 cm) was treated for 5 sec. By this treatment, the sample surface was continuously exposed to the corona with increasing power.

The corona-treated PE surfaces were characterized by the measurement of water contact angle. The water contact angle, an indicator of the wettability of surfaces, was measured at room temperature using a contact angle goniometer (Model 100-0, Rame-Hart, Inc., USA). Drops of purified water (3 μl) were deposited onto the corona-treated surface along the sample length. The corona-treated PE surfaces did not show any visible changes, but the water contact angles of the surfaces gradually decreased along the sample length with increasing corona power (Fig. 2). The decrease in the contact angles (and thus the increase in wettability) along the sample length is probably due to the oxygen-based polar functionalities incorporated on the surface by the corona treatment. Fig. 3 shows the possible mechanism by which the oxidation that occurs on a polymer surface such as PE is effected by the corona treatment. The corona discharge treatment of the polymer surface may produce carbon radicals from the hydrophobic backbone, followed by the formation of unstable hydroperoxides through rapid binding with oxygen in air, and then the

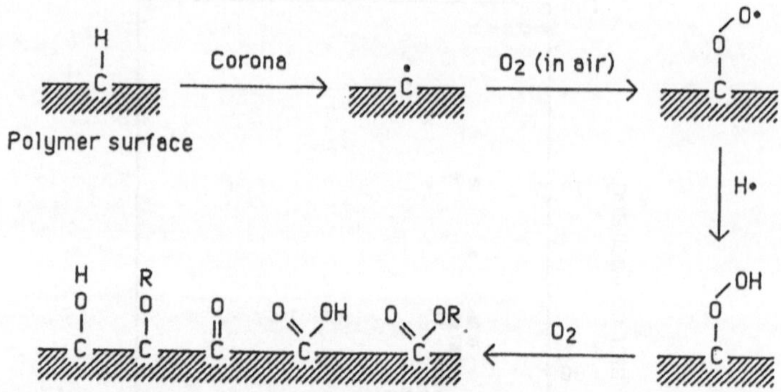

Fig. 3. Possible mechanism of the formation of oxygen-based functionalities on a polymer surface by corona discharge treatment.

decomposition of the hydroperoxides to produce various oxygen-based functionalities (hydroxyl group, ether, ketone, aldehyde, carboxylic acid, carboxylic ester, etc.) by reaction with additional oxygen (18-20). It seems that the oxygen-based functional groups produced on the PE surfaces increased with the increase in corona power and this contributed to the formation of the wettability gradient.

The changes in oxygen-based functional groups produced on the corona-treated PE surface along the gradient were investigated by Fourier-transform infrared spectroscopy in the attenuated total reflectance mode (FTIR-ATR). The FTIR-ATR spectra were obtained using a Digilab FTS 80 (Bio-Rad, USA) spectrophotometer equipped with KRS-5 reflection element (incidence angle, 45°). The corona-treated sheet was cut into five sections (1.0 x 5.0 cm each) in the direction perpendicular to the gradient. Each section was then cut in half (0.5 x 5.0 cm) and held against both faces of the reflection element. Fig. 4 shows FTIR-ATR spectra of the corona-treated PE gradient surfaces. The absorption band at ~1730 cm^{-1} corresponds to a C = O stretching band derived from carbonyl groups such as ketone, aldehyde, carboxylic acid, or ester. From the spectra, we can see that carbonyl groups were produced on the PE surfaces by corona treatment and gradually increased along the sample length with increasing corona power.

The changes in functional groups produced on the corona-treated PE surface along the gradient were further analyzed by electron spectroscopy for chemical analysis (ESCA) (ESCALAB MK II, V. G. Scientific Co., UK) equipped with Al $K\alpha$ radiation source at 1487 eV and 300 watt power at the anode. Fig. 5 shows ESCA carbon 1S core level spectra of the corona-treated PE gradient surface. The hydrophobic side of the gradient surface (section 1) showed an alkyl carbon peak at a binding energy, 285 eV. The corona-treated sections showed new peaks at higher binding energies, indicating the formation of carbon-oxygen functionalities, as labeled in Fig.5. The oxygen-based functional groups increased gradually along the sample length with increasing corona power.

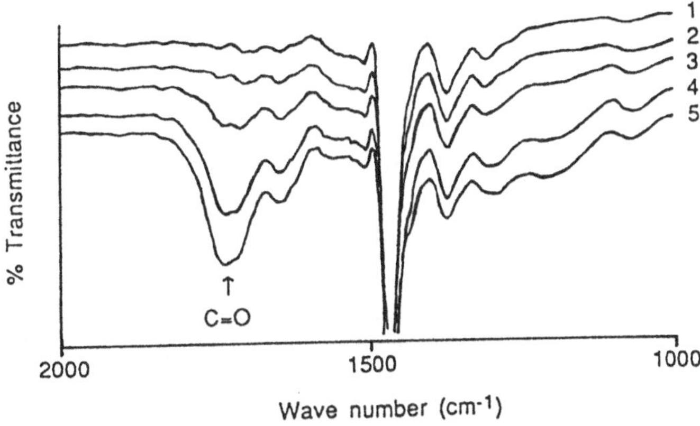

Fig. 4. FTIR-ATR spectra of corona-treated PE surfaces along the gradient. Numbers labeled on the spectra (1 to 5) represent the sections from the untreated hydrophobic end of the gradient surface.

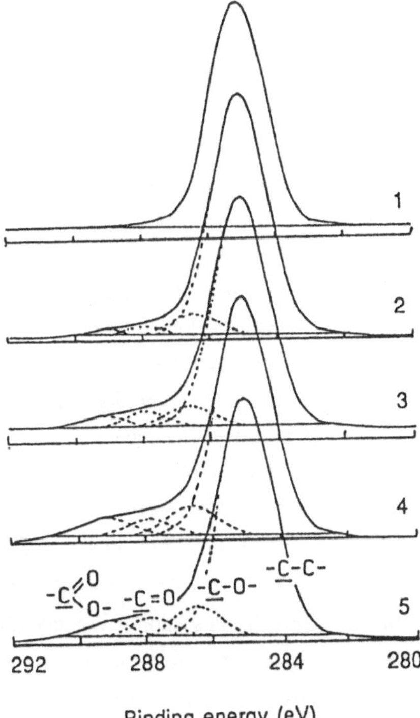

Fig. 5. ESCA carbon 1S core level spectra of corona-treated PE surfaces along the gradient. Numbers labeled on the spectra (1 to 5) represent the sections from the untreated hydrophobic end of the gradient surface.

CELL ADHESION AND GROWTH

The corona-treated PE surfaces were used to study the effect of polymer surface wettability on cell adhesion and growth. Chinese hamster ovary (CHO) cells (CHO-KI-BH_4, Oak Ridge National Laboratory, USA) were used as a model system because they exist as reasonably stable single cells and are not unreasonably fastidious in terms of culture requirements (21). They are grown in monolayer with fast generation time (about 12 hr).

The CHO cells routinely cultured in tissue culture polystyrene flasks (Corning, USA) at 37 ^oC under 5 % CO_2 atmosphere were harvested after the treatment with trypsin (0.05 % trypsin/0.02 % EDTA (Gibco Laboratories, USA)). The cells (4 x 10^4 cells/cm^2) were seeded to the corona-treated wettability gradient surfaces mounted in similar test chambers to those described by van Wachem et al. (22,23). The culture medium used was Ham's F-12 nutrient mixture (Gibco Laboratories) containing 5 % newborn calf serum, 100 units/ml penicillin, and 100 μg/ml streptomycin. After 2 hr culture, the cell adhesion to the sections of the gradient surface was determined by counting the number of detached cells with an electronic cell counter (Coulter Model ZM, USA). The results were expressed as the percentage of cells seeded to the surfaces. The cell growth on the wettability gradient surfaces was carried out at intervals of 24 and 48 hr in a same way to the cell adhesion experiment. The culture medium was changed once after 24 hr. The cell growth to the surfaces was expressed in terms of the number of cells attached per cm^2. Further detailed procedures for the cell culture were described in a previous paper (24).

The cells adhered and grown on the wettability gradient surfaces were also examined by a scanning electron microscopy (SEM, JSM-840A, Jeol Co., Japan). For the SEM observation, the cells attached on the gradient surfaces were fixed with 2.5 % glutaraldehyde in phosphate buffered saline (PBS, pH 7.2-7.3) for 15 min, followed by the dehydration in ethanol graded series, 50, 60, 70, 80, 90, and 100 %, for 10 min each. After drying at room temperature, the cell-attached surfaces were cut into five sections along the wettability gradient, gold deposited in vacuum, and examined by SEM with a tilt angle of 45°.

As seen in Fig. 6, the cells were adhered more onto the sections with moderate hydrophilicity of the wettability gradient surface. The maximum adhesion of the cells appeared at around a water contact angle of 50-55°. SEM observation also verified that the cells are more spread onto the sections with moderate hydrophilicity (Fig. 7). The cells were not spread well on the section 1 (water contact angle, about 98°) after 24 hr culture. The section 3 (contact angle, about 55°) shows better spreading of the cells than the section 2 (contact angle, about 62°) or section 5 (contact angle, about 48°). The cells on section 3 were completely spread and flatten after 24 hr culture. Cells attached on surfaces are spread only when they are compatible on the surfaces. It seems that surface wettability plays an important role for cell adhesion and spreading. Fig. 8 shows the number of CHO cells grown on the wettability gradient surface after 24 and 48 hr culture. After 48 hr culture the cells were grown more than twice compared to those after 24 hr culture. The cells were grown better on the hydrophilic sections than the hydrophobic ones. The maximum growth of the cells were appeared on the sections with moderate hydrophilicity at

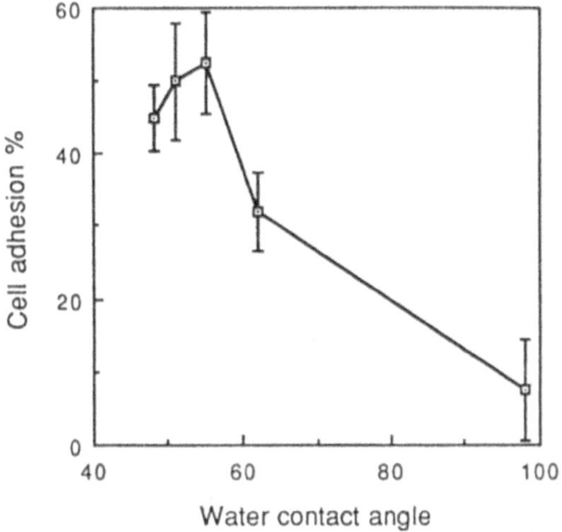

Fig. 6. CHO cell adhesion on corona-treated PE surface (Number of seeded cells, 4 x 10^4/cm^2; culture time, 2 hr). n = 3.

around a water contact angle of 50-55°. This trend was similar to that of cell adhesion after 2 hr culture as seen in Fig. 6. The fact that cells are more adhered, spread, and grown on the moderately wettable surfaces was also observed by others (22,23,25) as they cultured endothelial cells, platelets, or fibroblast onto various polymer substrates with different surface wettabilities. Preferential adsorption of some cell attachment proteins like fibronectin and vitronectin from culture medium onto the moderate wettable surfaces may be a possible reason for better cell adhesion, spreading, and growth.

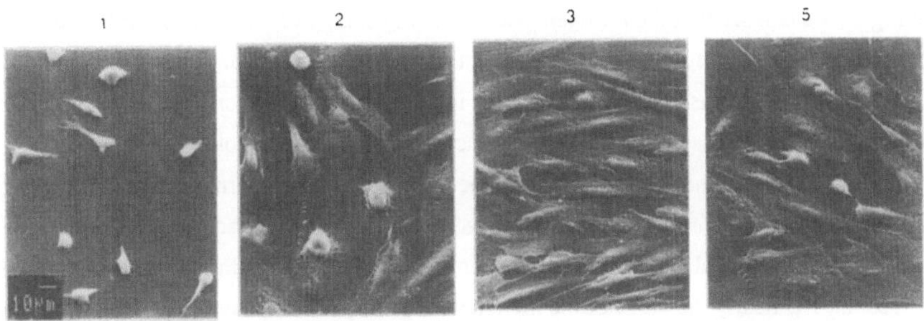

Fig. 7. SEM pictures of CHO cells grown on corona-treated PE surface along the wettability gradient after 24 hr culture (Original magnification, x 500). Numbers labeled (1 to 5) represent the sections from the untreated hydrophobic end of the gradient surface.

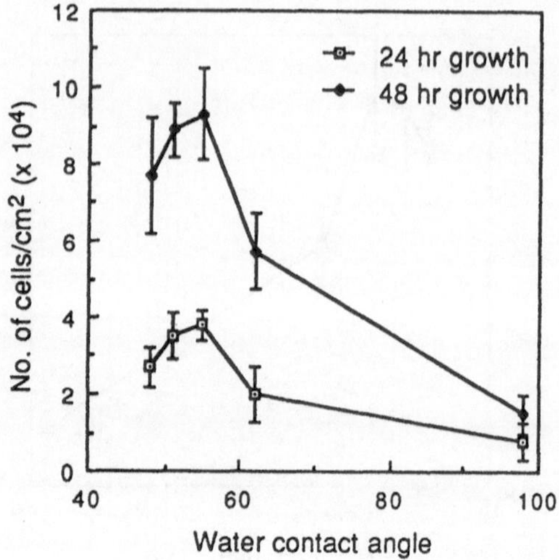

Fig. 8. CHO cell growth on corona-treated PE surface (Number of seeded cells, 4 x $10^4/cm^2$). n = 3.

PROTEIN ADSORPTION AND DESORPTION

Human albumin (crystallized powder, Sigma, USA), the major constituent of blood plasma, as a model protein was used to investigate the effect of surface wettability on protein adsorption and desorption. The corona-treated PE surfaces were immersed in 1mg/ml albumin solution (prepared with PBS at pH 7.3-7.4) for 1 hr and washed with PBS three times, followed by washing with purified water twice and vacuum drying. The albumin-adsorbed surfaces were analyzed by ESCA. The nitrogen signal from the surface was used as an indicator of the protein adsorption since we could see that little nitrogen is incorporated onto the surface by corona treatment in air (16). The nitrogen content of human albumin was about 14-15 atomic % as determined by ESCA. It is mainly derived from peptide bonds in the structure of the albumin. From Fig. 9, we can see that the nitrogen peaks (binding energy of ~399 eV) from the survey scan spectra of ESCA decreased gradually along the wettability gradient. This means that the albumin adsorption increased gradually with the increasing hydrophobicity of the surface. This is probably due to the increased hydrophobic interactions of the protein molecules with the hydrophobic sections of the gradient surface. The nitrogen content determined by ESCA was 9.3 $\pm$ 0.7 % on the section 1 (contact angle, about 98°) of the gradient surface and 5.0 $\pm$ 0.5 % on the section 5 (contact angle, about 48°) for more than three samples. The nitrogen content of 9.3 % on the section 1 is nearly the value of the monolayer coverage of albumin (26). As the albumin-adsorbed gradient surfaces were analyzed by FTIR-ATR (Fig. 10), we could see that the amide II (1500-1600 cm^{-1}) and amide I (1600-1680 cm^{-1}) absorption bands from the protein adsorbed on the surface also decrease gradually along the wettability gradient.

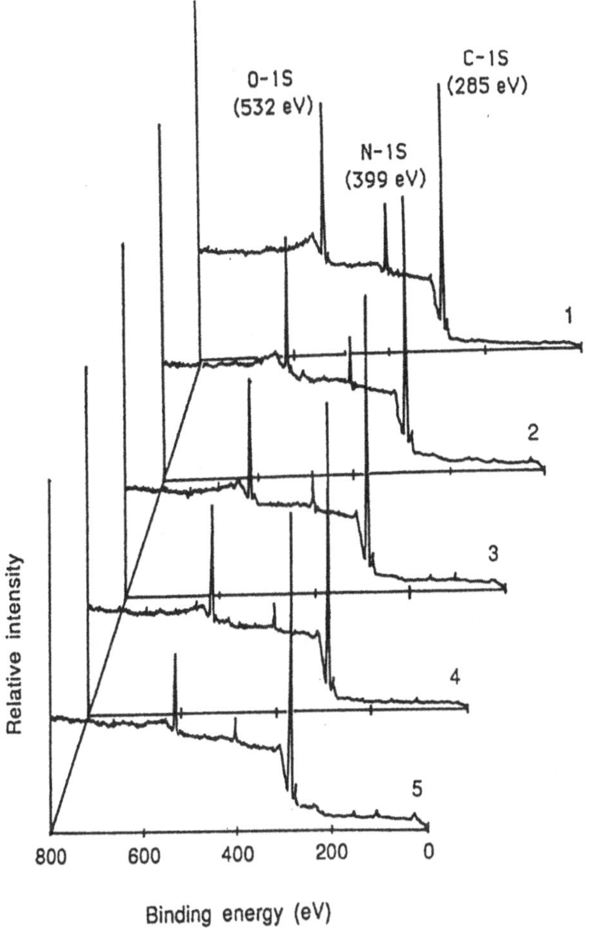

Fig. 9. ESCA survey scan spectra of corona-treated PE surface along the wettability gradient after albumin adsorption (1 hr adsorption in 1 mg/ml albumin solution). Numbers labeled on the spectra (1 to 5) represent the sections from the untreated hydrophobic end of the gradient surface.

For the desorption study, some albumin-adsorbed gradient surfaces were exposed to a nonionic polymeric surfactant, Tetronic 1504 (BASF-Wyandotte, USA), which was an effective protein removal agent (27,28). The amount of albumin remaining on the surfaces were analyzed by ESCA. The exposure of the albumin-adsorbed gradient surface to 1 mg/ml Tetronic 1504 solution for 30 min resulted in partial displacement of the protein (Fig. 11). This displacement was much greater on the hydrophobic sections of the gradient surface than the hydrophilic ones. The results of increasing protein adsorption and desorption toward the hydrophobic sections of the wettability gradient surfaces in this study agree with those of others' works (1,2,5,9,15), even though the functional groups introduced on the substrates are different.

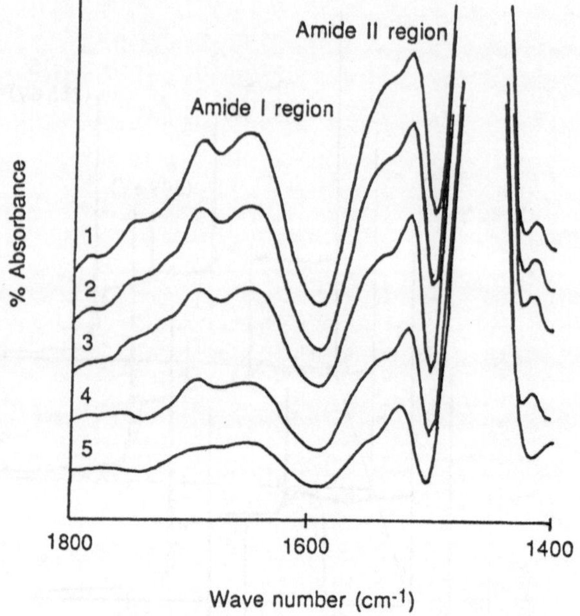

Fig. 10. FTIR-ATR spectra of corona-treated PE surface along the wettability gradient after albumin adsorption (1 hr adsorption in 1 mg/ml albumin solution). Numbers labeled on the spectra (1 to 5) represent the sections from the untreated hydrophobic end of the gradient surface.

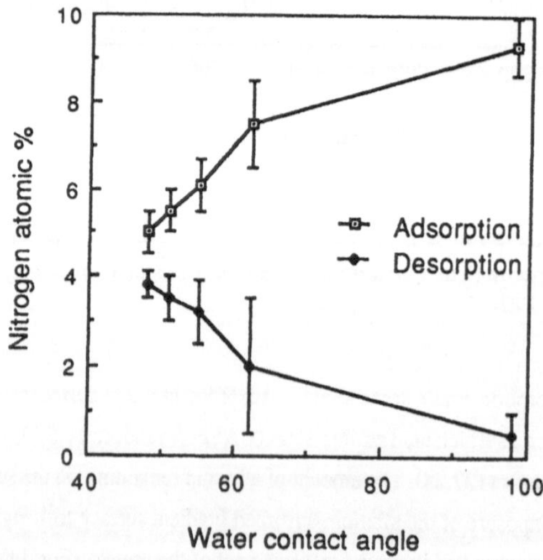

Fig. 11. Adsorption and desorption behavior of albumin on corona-treated PE surface. Nitrogen atomic % represents the amount of the protein adsorbed (for adsorption) and remaining (for desorption) on the surface (Adsorption; 1 hr adsorption in 1 mg/ml albumin solution: Desorption; 30 min desorption in 1 mg/ml Tetronic 1504 solution after 1 hr adsorption in 1 mg/ml albumin solution). n = 3.

CONCLUSIONS

The interactions of different types of cells or proteins with various solid substrates depend mainly on the surface characteristics such as wettability (hydrophilicity/hydrophobicity or surface free energy), chemistry, charge, roughness, and rigidity. We focused on the surface wettability in this study. We prepared wettability gradient surfaces by treating polymer sheets with corona whose power is changed gradually along the sample length. The major advantage to use wettability gradient surfaces as tools to study the interactions of biological species with solid surfaces is that several wettabilities are effectively compared at the same time and on the same surface. We studied the behaviors of the adhesion and proliferation of CHO cells and the adsorption and desorption of albumin on the wettability gradient surfaces prepared. The results clearly demonstrated that surface wettability is an important factor for cell adhesion/proliferation and protein adsorption/desorption. The wettability gradients prepared on various polymer surfaces by this corona treatment method can be used to systematically investigate the interactions of the different types of biological or other species in terms of the surface wettability of polymeric materials.

REFERENCES

1. Elwing H, Welin S, ASkendal A, Nilsson U, Lundstrom I (1987) J. Colloid Interface Sci. 119: 203
2. Elwing H, Askendal A, Lundstrom I (1987) Prog. Colloid Polym Sci. 74: 103
3. Elwing H, Askendal A, Lundstrom I (1987) J. Biomed. Mater. Res. 21: 1023
4. Elwing H, Nilsson B, Svensson KE, Askendal A, Nilsson UR, Lundstrom I (1988) J. Colloid Interface Sci. 125: 139
5. Elwing H, Askendal A, Lundstrom I (1989) J. Colloid Interface Sci. 128: 296
6. Klintstrom SW, Wikstrom M, Askendal A, Elwing H, Lundstrom I, Karlsson JO, Renvert S (1990) Colloids Surf. 44: 51
7. Elwing H, Golander CG (1990) Adv. Colloid Interface Sci. 32: 317
8. Hlady V, Golander C, Andrade JD (1988) Colloids Surf. 33: 185
9. Golander CG, Lin YS, Hlady V, Andrade JD (1990) Colloids Surf. 49: 289
10. Hlady V (1991) Appl. Spectrosc. 45: 246
11. Lee HB, Andrade JD (1988) Trans. 3rd World Biomaterials Congr. p.43
12. Lee HB (1989) In: Saegusa T, Higashimura T, Abe A (eds) Frontiers of macromolecular science. Blackwell, Oxford, p.579
13. Lee JH, Park JW, Lee HB (1990) Polymer (Korea) 14: 646
14. Pitt WG (1989) J. Colloid Interface Sci. 133: 223
15. Golander CG, Pitt WG (1990) Biomaterials 11: 32
16. Lee JH, Kim HG, Khang GS, Lee HB, Jhon MS (1992) J. Colloid Interface Sci. 151: 563
17. Lee JH, Kim HW, Pak PK, Lee HB (1993) J. Polymer Sci, Polymer Chem. submitted

18. Owens DK (1975) J. Appl. Polymer Sci. 19: 265
19. Iwata H, Kishida A, Suzuki M, Hata Y, Ikada Y (1988) J. Polymer Sci., Polymer Chem. 26: 3309
20. Steinhauser H, Ellinghorst G (1984) Angew. Makromol. Chem. 120: 177
21. Thompson LH (1979) In: Jakoby WB, Pastan IH (eds) Methods in enzymology: Cell culture. Academic Press, New York, p.308
22. van Wachem PV, Beugeling T, Feijen J, Bantjes A, Detmers JP, van Aken WG (1985) Biomaterials 6: 403
23. van Wachem PV, Hogt AH, Beugeling T, Feijen J, Bantjes A, Detmers JP, van Aken WG (1987) Biomaterials 8: 323
24. Lee JH, Park JW, Lee HB (1991) Biomaterials 12: 443
25. Kang IK (1987) Ph. D. Thesis, Kyoto University, Japan, Chs. 6-7
26. Lee JH (1988) Ph. D. Thesis, University of Utah, USA, Ch. 6
27. Lee JH, Kopecek J, Andrade JD (1989) J. Biomed. Mater. Res. 23: 351
28. Lee JH, Kopeckova P, Kopecek J, Andrade JD (1990) Biomaterials 11: 455

Rheological Properties of Blends: Facts and Challenges

P.J. Carreau, M. Bousmina and A. Ajji
Centre de Recherche Appliquée sur les Polymères, CRASP
Ecole Polytechnique de Montréal, Québec, Canada, H3C 3A7

INTRODUCTION

Three forces will continue to provide incentive for the industry to produce new polymer blends: the ability to tailor product properties to a particular performance cost balance; fast entry into new markets; and it is easier to develop products for small or niche markets than by synthesizing new polymers. The utilization of polymer blends in commerce represents some 16% of all plastics and the sector is growing at a rate of approximately 10% per year. We expect that polymer blends will grow at a much faster pace in the next decade. The most important class of blends from the commercial viewpoint are the so-called immiscible blends, where the presence of a distinct minor phase may be observed. The minor phase in an immiscible polymer blend is deformable and a wide range of sizes and shapes can thus be obtained for this dispersed phase during processing [1]. Both the size and the shape of the minor phase can influence a multiplicity of physical properties. At present the processing/ morphology/ property relationships in immiscible and partly miscible blends remain poorly understood and a given processing strategy may result in high value added blends. Rheological methods are powerful tools to elucidate such relationships. The morphology of blends is strongly dependent on the rheological properties of the components and rheological properties are needed to understand the processing of blends and changes occurring during processing. Conversely, the blends' morphology affects the rheological properties.

Scope

The objective of this work is to present a critical review of the recent work on the rheological properties of polymer blends. In particular, we will examine in detail how deformability of the droplets of the minor phase and miscibility of the components affect the rheological properties of blends using experimental data, in light of recent theories.

The rheological behaviour of polymer blends is generally very complex. In the case of immiscible blends, the rheological properties depend in a complex manner on composition and properties of the components, morphology, interactions between phases, interfacial tension, and the strain history during processing [1 to 8]. Under high deformation encountered in processing, which

K.P. Ghiggino (Ed.)
Progress in Pacific Polymer Science 3

is usually a complex transient flow combining shear and extensional strains, the morphology of blends changes and measurements done using a sophisticated rheometer under controlled flow kinematics (simple shear or purely elongational flow) may be a very incomplete description of the blends' rheology. Measurements in the so-called non linear viscoelastic domain, although very complex, are essential for understanding and predicting morphology changes during processing. On the other hand, measurements in the linear viscoelastic domain, such as small amplitude oscillatory shear flow, can yield very useful informations on the blends' characteristics. For small enough strain, the structure of multiphase polymeric systems is expected not to change during flow and linear viscoelastic functions such as the storage modulus G' and the loss modulus G'' are enough to describe the equilibrium properties of the blends.

Many difficulties are encountered in measuring the rheology of polymer blends, especially at high temperatures. For immiscible blends, the morphology may not be stable due to coalescence of neighbouring droplets. The polymer components may also show variations over long time periods due to re-entanglement of polymeric chains [9]. These difficulties will lead to rheological data which are highly difficult to analyze. The shear properties of molten polymers are determined either using a capillary rheometer or a rotational rheometer with a cone-and-plate or parallel plates (concentric disks) geometry. For multiphase systems, the gap or the smallest dimension encountered in the flow field must be much smaller than the size of the particles or droplets, d_p. The usually accepted criterion is that $d_p/h << 1$, where h is the gap of the measuring device. With concentric disks, large gaps (1 mm and more) can be used and it is possible to make measurements on suspensions of rather large particles. Also with the concentric disks geometry, the positioning and the control of the gap with temperature is much less critical than with the cone-and-plate for which a typical gap is of the order of 80 μm. However, one must recall that the flow field in concentric disks is not homogeneous, that is the shear rate varies linearly with the radial position. For small amplitude oscillatory flow, this should be of no consequence since the linear viscoelastic properties are independent of the rate of deformation. Nevertheless, caution should exerted to ensure that the measurements are carried out in the linear range, that is the response to an imposed sinusoidal strain or stress is independent of the strain amplitude, and that for all the range of frequencies investigated. Also the output signal should be of the same quality as the input signal. Quite often the response signals for multiphase systems (for example highly concentrated suspensions) are distorted and the measurements are no longer in the linear viscoelastic domain.

Even in small amplitude oscillatory shear flow, the rheological behaviour of polymer blends is unusual and quite complex. The rheology of molten blends is generally characterized by enhanced elastic properties (larger storage modulus) and longer relaxation times at low frequencies, compared to the properties of the matrix. This has been observed for most multiphase systems, and it has been discussed by many authors. For example, Taylor [10], Froelich and Sack [11], Oldroyd [12], and Choi and Schowalter [13] have attributed the enhancement of elasticity to the deformability of the suspended droplets under a strain in the flow field. They theoretically described the mechanism through a force balance between the hydrodynamic forces that tend to deform the droplets and the interfacial forces that tend to maintain their equilibrium spherical shape. It is only in the equations proposed by Oldroyd and by Choi and Schowalter that the elasticity of the suspensions is explicitly related to the interfacial tension. Their models are restricted to dilute or semi-dilute suspensions or emulsions made of two Newtonian fluids. The dispersed droplets are assumed to be monodisperse in size. Recently, Palierne [2] proposed an extension of the Oldroyd model for concentrated suspensions, including viscoelasticity of the components, and polydispersity of the droplets' size. For monodisperse suspensions in size, the Palierne model is equivalent to the Oldroyd model.

The Palierne model has been shown to predict adequately the linear viscoelastic properties of several blends. However, Brahimi et al. [14] have reported a good agreement between the model predictions and the experimental data for high density polyethylene/high impact polystyrene (HDPE/HIPS) blends, but they have shown that the model fails when the blends are modified by a copolymer. Discrepancies between the model predictions and the data have also been reported by Kim and Denn [15] for blends with liquid crystalline polymers. Bousmina and Muller [3] and Bousmina and Carreau [16] have demonstrated that the Palierne model gives good predictions of the linear properties of blends consisting of two viscoelastic components, but the model fails in the case of concentrated suspensions of undeformable particles. This failure is attributed to particle-particle interactions, which are neglected in the model. The Palierne model can also take into account the effect of a third (compatibilizing agent) component at the interface of the two major components. This could make the model quite powerful for discussing miscibility and immiscibility in polymer blends. However, the parameters describing the deformability of the interface (interphase) are not easily measurable.

POLYMER BLENDS STUDIED

In this work, we examine the rheological properties of four different systems.

i) Immiscible blends made up of two viscoelastic components. The specific example will be blends of polystyrene/ and polymethylmethacrylate (PS/PMMA). The PMMA is a commercial polymer (Altulite 2773) of molecular weight M_w = 133,000 kg/kmol and M_w/M_n = 1.9. The polystyrene is Gedex 1541 GA100 of M_w = 23,000 kg/kmol and $M_w/M_n = 3.6$. The average molecular weight M_w and the polymolecularity index (M_w/M_n) of the polymers were determined by gel permeation chromatography (GPC) in tetrahydrofuran (THF).

ii) Immiscible blends made up of an elastic phase suspended in a viscoelastic matrix, poly(butylacrylate-co-styrene)-PMMA/PMMA blends. The rubber particles have a core/shell type structure. The soft core consists of a crosslinked copolymer of styrene and butylacrylate. The hard shell is a grafted PMMA, which ensures a good adhesion between the particles and the matrix. The PMMA is the same as in i).

iii) Poly(butylacrylate-co-styrene)-PMMA/PS blends, for which the rubber particles with the grafted PMMA hard shell is immiscible with the polystyrene matrix. The PS is the same as in i).

iv) Polystyrene/polyvinyl methyl ether (PS/PVME) blends, for which the components can be miscible or immiscible depending on the temperature. For these blends a monodisperse PS of M_w = 50,000 kg/kmol and a PVME of M_w = 46,000 kg/kmol of polymolecularity of 2.2 were used.

The rheological measurements were carried out using either a Rheometrics system-4 or a RMS rheometer with parallel arrangements and gaps varying from 0.5 mm to 1.5 mm depending on the plate radius used. Preliminary strain sweep experiments were performed in order to determine the range of the linear viscoelastic domain. For the high frequency measurements, the use of a strain equal to 0.1 allowed us to be in the linear domain. Nitrogen was used as the heating fluid and no appreciable degradation nor changes due to coalescence could be detected for experiments carried on for periods up to 2 h in cases.

EMULSION MODEL

For the sake of simplicity we retain here the simpler form of the emulsion model proposed by Palierne [2]. For a dispersion of viscoelastic incompressible inclusions in a viscoelastic impressible matrix, the model expresses the complex shear modulus $G^*(\omega) = G'(\omega) + iG''(\omega)$ of the blend as a function of the complex shear moduli of the individual phases (G_m^*: *matrix*; G_i^*: *inclusions*) and the ratio of interfacial tension and droplets size (α/R).

$$G^*(\omega) = G_m^*(\omega)\,\frac{1+3\Sigma_i \phi_i H_i(\omega)}{1-2\Sigma_i \phi_i H_i(\omega)}$$

where:

$$H_i(\omega) = \frac{4(\alpha/R_i)\,[2G_m^*(\omega) + 5G_i^*(\omega)] + [G_i^*(\omega) - G_m^*(\omega)]\,[16G_m^*(\omega) + 19G_i^*(\omega)]}{40(\alpha/R_i)\,[G_m^*(\omega) + G_i^*(\omega)] + [2G_i^*(\omega) + 3G_m^*(\omega)]\,[16G_m^*(\omega) + 19G_i^*(\omega)]}$$

The summation is carried out over the distribution of the droplets size (ϕ_i is the volume fraction of droplets of radius R_i).

For known viscoelastic properties of the polymeric components, interfacial tension and particles' radius, the dynamic moduli G' and G'' of a two-phase polymer blend in the melt is easily calculated from the above equation.

RESULTS

i) PS/PMMA Blends

Figure 1 compares experimental data for G' and G'' with the predictions of the emulsion model for a 90/10 (mass %) PMMA/PS at 200°C. For this blend, the viscosity of the minor component is considerably smaller than that of the major component, $\eta_{0(PS)}/\eta_{0(PMMA)} = 0.035$. The morphology was characterized by transmission electronic microscopy (TEM) and the average particle radius was estimated by digital image analysis to be around 0.22 μm. The interfacial tension has been taken as 1.5 mN/m [4]. It should be mentioned that the morphology of the blend was found to be the same before and after the rheological experiments carried out at 200°C for about 2 h.

It is clear from *Figure 1* that the emulsion model describes well the dynamic moduli of the blend over the whole frequency range and predicts the increase of elasticity. The low-frequency plateau in G' is a characteristic of a long-time relaxation process of about 100 s, which is much higher than the terminal relaxation times of the phases (2 s for PMMA and 0.5 s for PS). Notice that the loss modulus for the blend is identical to that of the matrix, even at low frequencies.

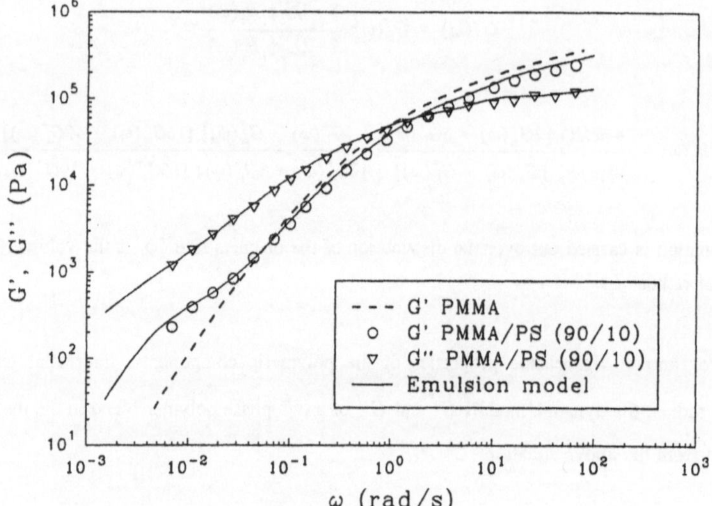

Figure 1 Comparison between experimental data and Palierne's model predictions for the 90/10 PMMA/PS blend at 200°C.

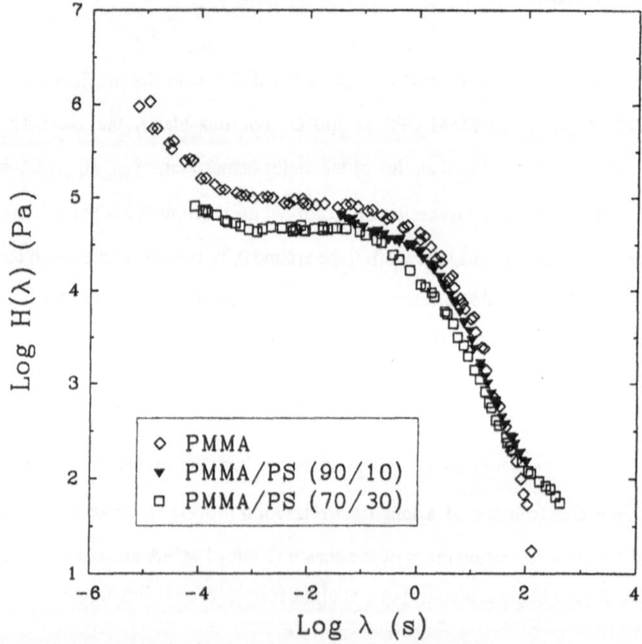

Figure 2 Relaxation spectra for PMMA and its blends with PS at 200°C.

As is seen in *Figure 2*, the relaxation spectrum of the blend is shifted to longer times compared with that of the matrix. This corresponds to the shift the lower frequencies of the terminal zone for the storage modulus G' of *Figure 1*. The model quantitatively describes this phenomenon and ascribes it to the deformability and relaxation of the suspended PS droplets which are less viscous than the PMMA matrix.

The Cole-Cole plot of the blend (*Figure 3*) shows two frequency regions corresponding to two different relaxation mechanisms; at high frequencies, the relaxation phenomenon is essentially due to the phases' relaxation, whereas at low frequencies the relaxation mainly stems from the deformability of the suspended droplets.

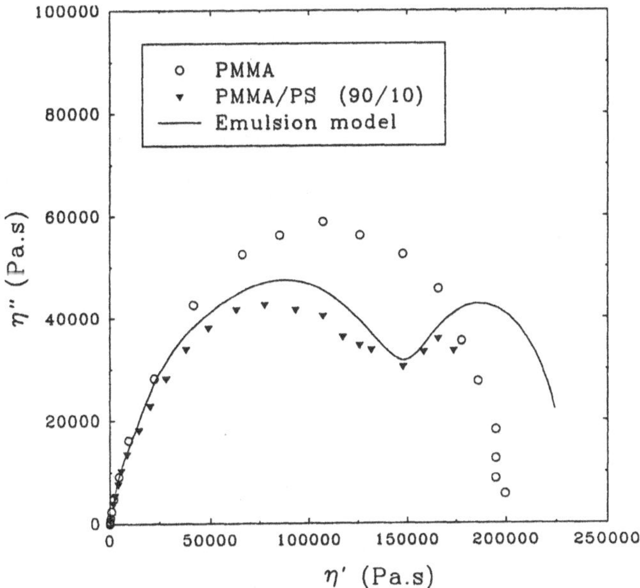

Figure 3 Cole-Cole plots for PMMA and the 90/10 PMMA/PS blend at 200°C.

To confirm that the low frequency elasticity of blends is related to the deformability of the minor phase particles, we present in *Figure 4* the linear viscoelastic properties of the inverse composition, 10/90 PMMA/PS blend. In this case, the dispersed PMMA inclusions are much more viscous than the PS matrix $((\eta_{0(PMMA)}/\eta_{0(PS)} = 30)$. As expected and predicted by the model, the secondary plateau does not appear confirming that the deformability of the suspended droplets are responsible for the increase in blend elasticity in the low frequency range, when the dispersed phase is less viscous than the matrix.

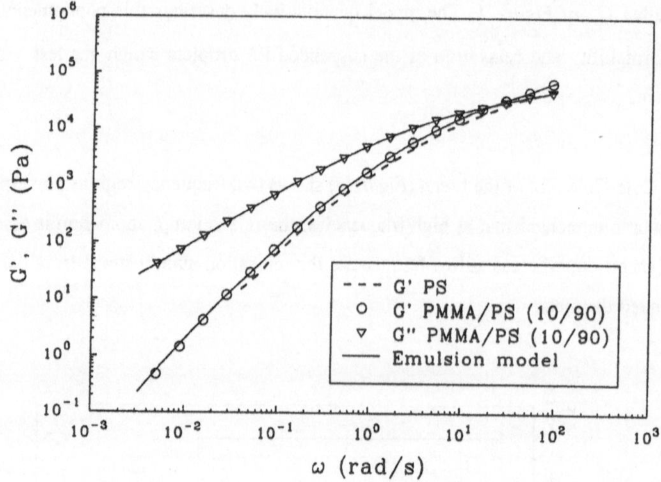

Figure 4 Comparison between experimental data and Palierne's model predictions for the 10/90 PMMA/PS blend at 200°C.

ii) PMMA/Rubber Blends

One question of interest at this stage is how would the linear viscoelastic properties be affected if the PS dispersed phase is replaced by an elastic component. The same PMMA used before has been blended with rubber particles over a wide range of PMMA/Rubber composition (95/05, 90/10, 85/15, 80/20, 70/30, and 50/50. As mentioned before the rubber particles consist of a core-shell poly(butylacrylate-co-styrene)-PMMA type.

The storage and loss moduli of both (unblended) components are presented in *Figure 5*. It is important to notice that the values for the rubber are almost independent of the frequency and about equal to 10^5 Pa in the terminal zone of PMMA. The dynamic moduli, G', G'', of the rubber are much larger than those of the PMMA, which is the matrix for the blends. That is the rubber particles are much more difficult to deform and no secondary plateau for the storage modulus for the blends is to be expected in the low frequency range. Nevertheless, *Figure 6* shows an obvious secondary plateau at low frequencies for volume fraction of rubber particles larger than 15%. The morphology reveals (Figure 7) that for volume fraction larger than 15%, the particles form a three dimensional network structure (percolation). This is not taken in to account in the Palierne model, which assumes isolated particles interacting only through hydrodynamic forces.

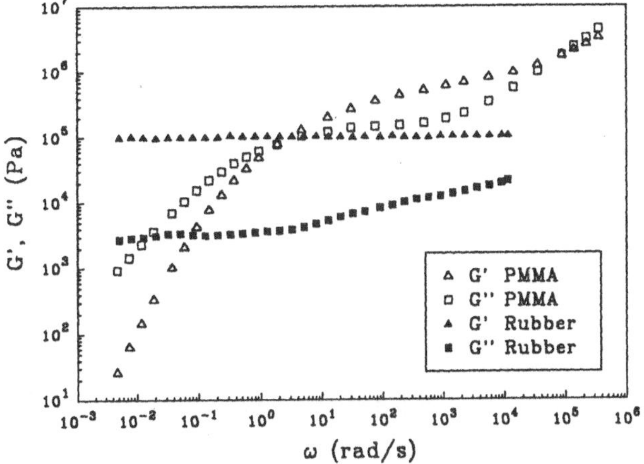

Figure 5 Storage and loss moduli of rubber and PMMA (unblended) components at 200°C.

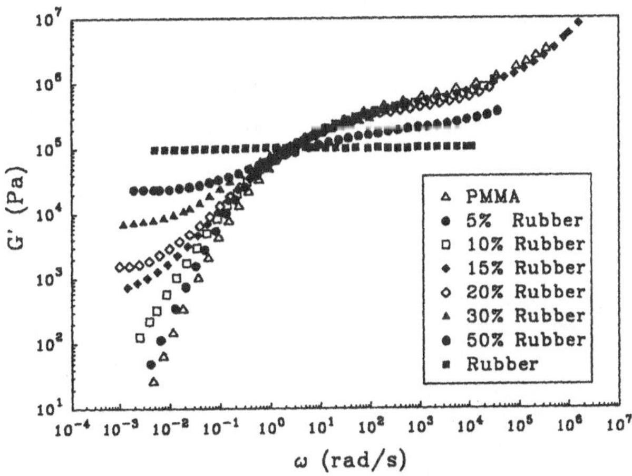

Figure 6 Effect of rubber particles content on the storage modulus of PMMA/Rubber blends.

The comparison between the model predictions and the data for the PMMA/Rubber blends for compositions ranging from 5 to 50 % in rubber. is presented in *Figure 7.* Also shown are the micrographs for the corresponding morphologies. As expected the model is in quantitative agreement only for blends containing a small fraction of rubber. For high rubber content (larger than 15 %), the model cannot describe the secondary plateau experimentally observed at low frequencies. Above this critical concentration, one may observe from the micrographs that the particles start forming a network structure. This is clearly seen for the larger rubber concentrations. The low frequency plateau may be described by a percolation approach [17], using an equation of the type [18]:

$$G' \sim |\phi - \phi_c|^{\alpha}$$

The percolation concept of de Gennes [17] for suspended monodisperse spheres predicts a critical volume fraction approximately equal to 16 %, above which a three-dimensional network starts to take place. This is consistent with our results presented in *Figure 7*.

iii) PS/Rubber Blends

When the rubber particles are blended with PS which is incompatible with the PMMA grafted shell, poor adhesion between the inclusions and the matrix is expected. The dynamic moduli of PS/Rubber blends for the compositions 95/05, 90/10, and 80/20 are reported in *Figure 8*. The most striking observation is that a secondary plateau appears at low frequencies for blends containing 10 % and more of rubber particles. The micrograph showing the morphology of the blend containing 10 % rubber is also presented in the figure. From this experimental evidence (based on morphological and rheological analyses), percolation appears at a threshold concentration of about 10 %, which is lower than that predicted by the percolation theory.

As illustrated by these different examples, the linear viscoelastic behaviour of blends made up of immiscible components is not simple at all. The emulsion model gives good predictions only in absence of complications such as particle-particle interactions, or partial miscibility, which are not taken into account in the model.

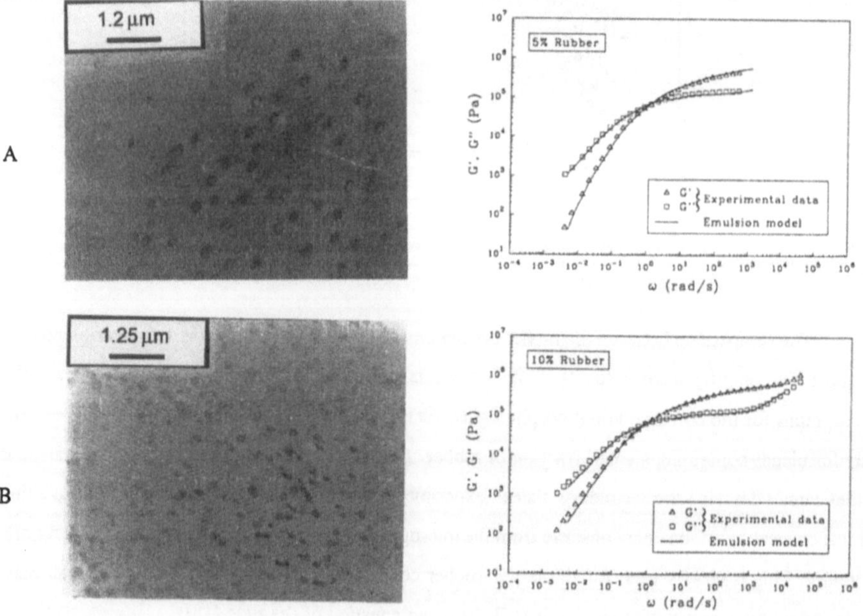

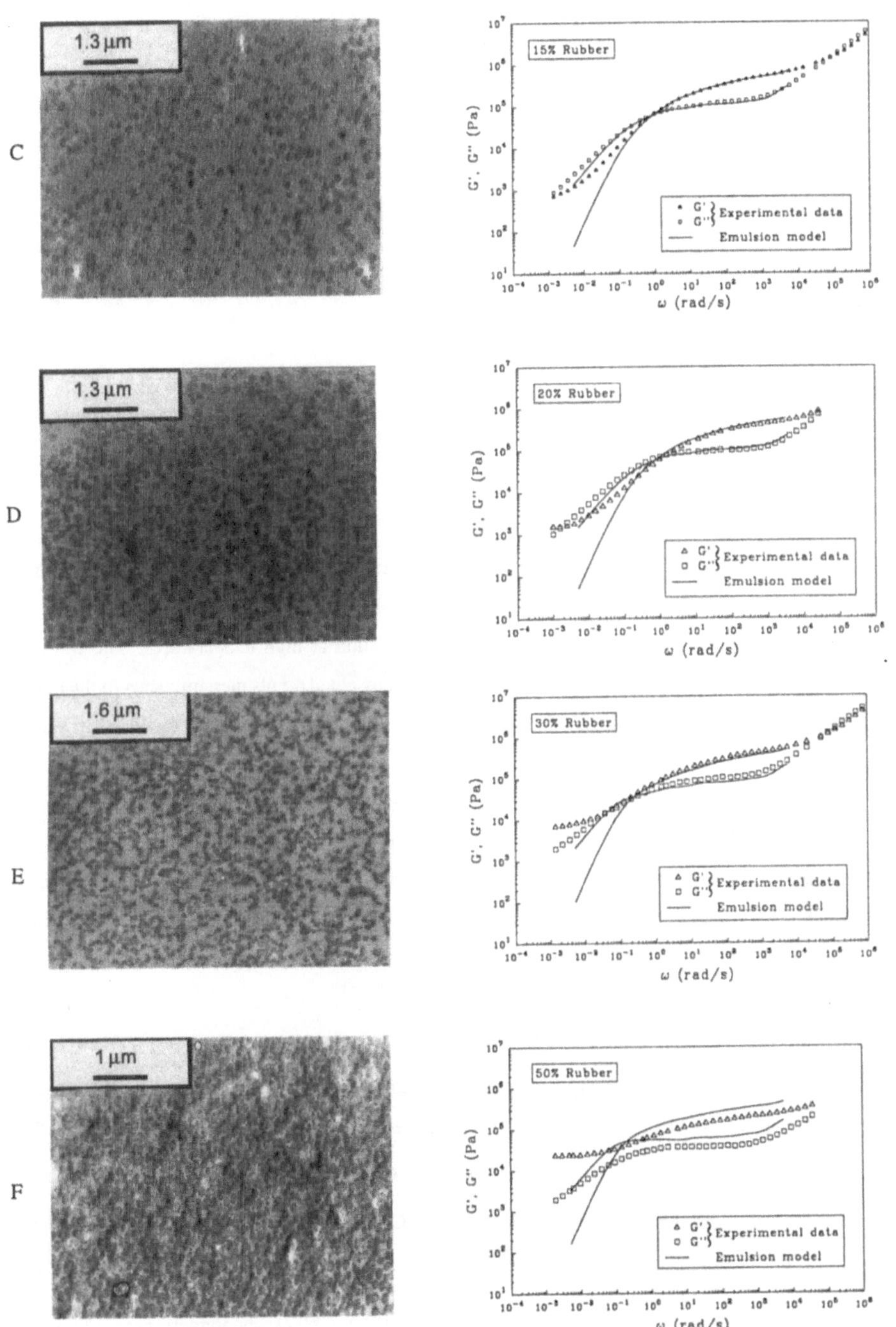

Figure 7 Micrographs and linear viscoelastic properties of PMMA/Rubber blends for various values of rubber content; emulsion model's predictions.

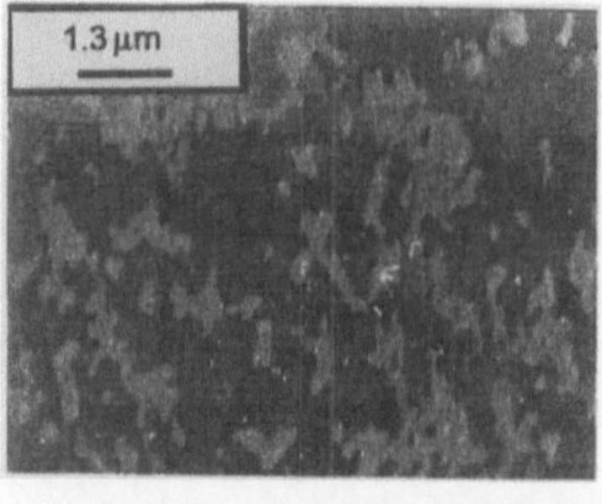

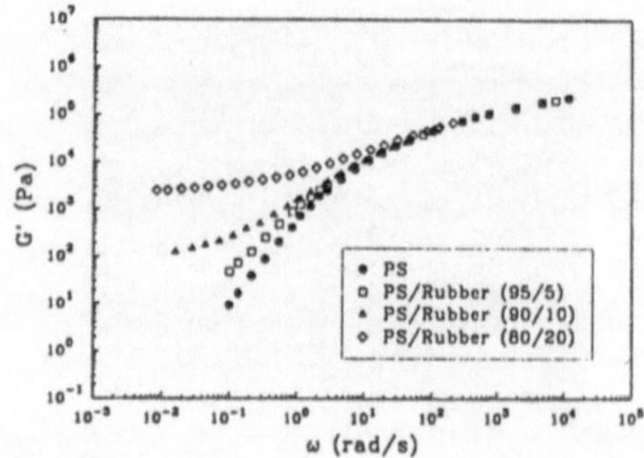

Figure 8 Storage modulus of PS/Rubber blends for various values of rubber content; micrograph for the 90/10 PS/Rubber blend.

iv) PS/PVME Blends

Another interesting phenomenon is phase separation, which for some blends depends on temperature. We discuss a typical example here. The polystyrene and the polyvinyl methyl ether are miscible at low temperatures and exhibit a phase separation at high temperatures. The transition temperature, T_s, is about 140°C, obtained under no strain by cloud points measurements in the heating mode. *Figures 9* reports the dynamic viscosity of a 60/40 PS/PVME as a function of angular frequency for different temperatures. For temperatures lower than T_s , the dynamic viscosity curve is typical of that for homogeneous molten polymers, with a low frequency plateau followed by a power-law behaviour at high frequencies. For temperatures close or higher than the transition temperature, the low frequency plateau disappears and the viscosity increases rapidly with decreasing frequency. This is a clear indication of phase separation with considerable particle-particle interactions. This behaviour cannot be predicted by the Palierne emulsion model, but possibly by a percolation theory.

The time-temperature superposition principle has been applied to the loss and storage moduli. For the homogeneous blend (one phase at temperature equal to 115°C), the superposition method works very well. Typical low frequency behaviours of G' and G'' are shown by the lines in *Figure 10*. For temperatures close to T_S (125, 135 and 140°C), a shoulder develops in the low frequency region for the storage modulus and becomes more important as the temperature is closer to T_S. This behaviour is similar to that observed by Bates et al. [19] for block copolymers near T_S in the homogeneous region (disordered zone). In fact, these temperatures are well below T_S as determined by cloud points measurements, both in the cooling and heating modes. However, for the loss modulus G'', lesser effects are observed in the frequency range covered by our experiments.

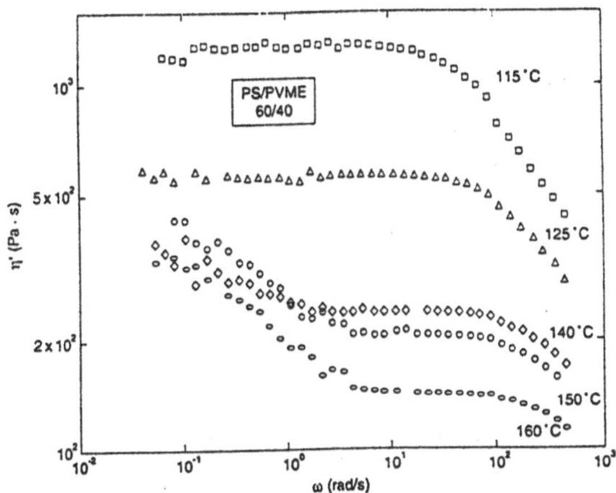

Figure 9 Dynamic viscosity for a 60/40 PS/PVME blend at various temperatures.

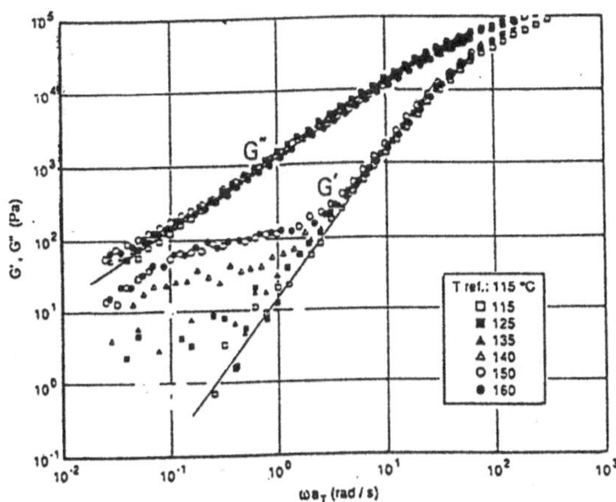

Figure 10 Reduced dynamic moduli for the 60/40 PS/PVME blend showing phase separation.

For temperature above T_S, i.e. 150 and 160°C, that is in the two-phase region, the low-frequency storage modulus presents a plateau and the superposition no longer holds. In fact, the results obtained are typical of those of an immiscible blend, similar to the results presented above for the PS/PMMA blends. In contrast, the G'' values superpose with those obtained at lower temperatures, below T_S, even in the terminal zone.

CONCLUDING REMARKS

The results reported in this work on some selected polymer blends demonstrate that the melt rheology of multiphase systems is very complex, even in the linear viscoelastic domain. The main features exhibited by molten immiscible polymer blends are an increase of elasticity at low frequencies and longer relaxation times compared with that of the matrix. The linear viscoelastic properties of blends are satisfactorily described by the Palierne emulsion model and the enhancement of elasticity is ascribed to the deformability of the minor phase's droplets.

Experimental data for 10/90 and 90/10 PS/PMMA blends clearty demonstrate that the increase in blend's elasticity and the longer relaxation times observed in the terminal zone are due to the deformability of the suspended droplets. On the other hand, results obtained for PMMA/Rubber and PS/Rubber blends illustrate limitations of the Palierne model. For these blends, the model does not even qualitatively predict the secondary plateau arising at low frequencies for high rubber contents. The model does not account for particle-particle interactions. For volume fraction of rubber larger than 15%, the particles form a network-type structure, For rubber particles concentration of 15% and larger the elasticity of the network structure is satisfactorily described by the percolation theory. For PS/Rubber blends, a network is observed at a particles concentration of 10%. This is not predicted by the percolation theory.

The same enhancement of elasticity is observed for PS/PVME blends at temperatures above the critical temperature of phase separation. For temperatures lower than 140°C, the components are totally miscible and the linear viscoelastic behaviour of these blends is typical of an homogeneous polymer melt.

For high deformation flows, the behaviour of immiscible blends is even more complex due to morphology changes during flow (coalescence, break-up, orientation and deformation of droplets). For partially miscible blends, we have observed, in many situations, significant decreases of the steady shear viscosity, even for low particles contents. This cannot be described by emulsion models or by other valid theories. This could possibly be explained by thermodynamic considerations and rationalized in light of the free energy of mixing. This will be reported elsewhere.

REFERENCES

1 Gonzalez-Numez R, Favis BD, Carreau PJ, Lavallée C (1994) Polym. Eng. Sci., in press
2 Palierne JF (1990) Rheol. Acta, 29:204
3 Bousmina M, Muller R (1993) J. Rheol. 37(4): 663; and (1992) In: Moldenaers P, Veunings R (eds) Proc. XIth Inter. Congress on Rheology. Elsevier, Amsterdam
4 Wu S, (1982) Polymer Interface and Adhesion, Dekker, New York
5 Ajji A, Choplin L, (1991) Macromolecules 24:5221
9 Ajji A, Schreiber HP, Rudin A, Teh JW, (1985) J. Appl. Polym. Sci. 30:731
10 Taylor GI, (1934) Proc. Roy. Soc. A146:501
11 Froelich H, Sack R, (1946) Proc. Roy. Soc. A185:145
12 Oldroyd JG, (1953) Proc. Roy. Soc. A218:122, (1955) proc. Roy. Soc. A232:567
13 Choi SJ, Schowalter WR, (1975) Physics of Fluids 18(4):410
14 Brahimi B, Ait-Kadi A, Ajji A, Jerome R, (1991) J. Rheol. 35:1069
15 Kim WN, Denn M, (1992) J. Rheol. 36(8):1477
16 Bousmina M, Carreau PJ, (1993) In: Utracki LA (ed) Proc. Polyblends Symposium, Boucherville, Canada
17 De Gennes PG, (1979) Scaling concepts in polymer physics, Cornell University Press
18 Bousmina M, (1992) Ph.D. Thesis, ULP, Strasbourg, France
19 Bates FS, (1984) Macromolecules 17:2607

Hydroboration Polymerization

Yoshiki Chujo

Division of Polymer Chemistry, Graduate School of Engineering, Kyoto University, Yoshida, Sakyo-ku, Kyoto 606-01, Japan

ABSTRACT: This paper describes "Hydroboration Polymerization" as a novel methodology for the preparation of organoboron polymers. A polyaddition between diene and thexylborane produced a polymer consisting of C-B bonds in the main chain. The resulting organoboron polymers can be regarded as a polymer homologue of trialkylborane and can be expected as a novel type of reactive polymers. On the other hand, hydroboration polymerization of diene with monobromoborane produced poly(organoboron halide)s as a polymeric Lewis acid. Furthermore, hydroboration polymerization of dicyano compounds such as isophthalonitrile with *t*-butylborane produced an air-stable boron-containing polymer (polycyclodiborazane) having B-N four-membered rings via dimerization of iminoborane species.

INTRODUCTION

Hydroboration reaction produces alkylborane compounds almost quantitatively, which are versatile and useful reagents in organic synthesis. The chemistry of hydroboration has been established by Brown and his group [1]. Recently, Chung and his coworkers have explored new field, in which organoboron-branched polymers were prepared from boron-containing monomers [2]. However, the direct use of hydroboration as a unit reaction in polymer synthesis has been very limited so far. We describe here a novel methodology for the preparation of various types of organoboron polymers by means of hydroboration [3].

HYDROBORATION POLYMERIZATION OF DIENES WITH THEXYLBORANE

In the present hydroboration polymerization, thexylborane was employed as a monoalkylborane (bifunctional hydroborane) component because thexylborane is known to be stable enough after distillation. As a diene monomer, relatively longer chain dienes were used to avoid the

K.P. Ghiggino (Ed.)
Progress in Pacific Polymer Science 3

competitive cyclization reaction. The general scheme of this polymerization is shown in Scheme 1, which proceeds in THF at 0°C under nitrogen without catalyst [4]. For example, the resulting organoboron polymer by the reaction between thexylborane and 1,7-octadiene was isolated as a colorless gum after coagulation and was soluble in common organic solvents such as chloroform, THF and benzene. The structure of the polymer was supported by its IR and ^{1}H-, ^{13}C-, and ^{11}B-NMR. The molecular weight of the polymer obtained increased as approached the feed ratio to unity, which is taken as a normal behavior of polyaddition reaction.

Scheme 1

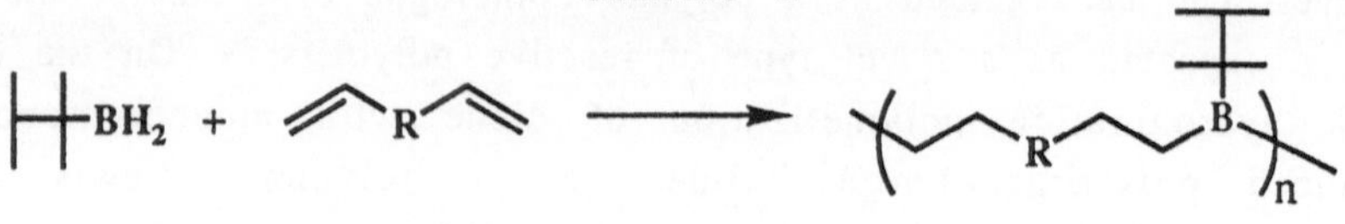

BH_2 + R → R B n

As summarized in Table 1, this hydroboration polymerization can be applied to various combinations of dienes such as 1,9-decadiene, *p*-divinylbenzene, *p*-diallylbenzene, bis(allyl ether)s of ethylene glycol, tetramethylene glycol, triethylene glycol, hydroquinone and bisphenol-A with thexylborane to produce the corresponding organoboron polymers.

Table 1. Hydroboration Polymerization between Thexylborane and Various Dienes.[a]

Run	Diene	$\overline{M}_n$ [b]	$\overline{M}_w$ [b]
1		19,000	27,700
2		18,400	27,400
3		19,000	29,200
4	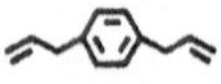	9,400	16,900
5	 O O	1,200	2,600
6	 O O	1,900	4,500
7	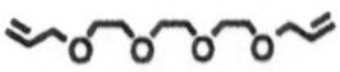 O O O O	1,900	3,200
8	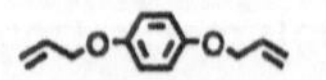 O O	5,100	11,200
9	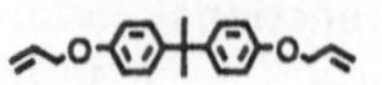 O O	7,600	15,400

a) Polymerizations were carried out by adding small excess of thexylborane to the 1M THF solution of diene at 0°C.

b) GPC (dry THF), polystyrene standard.

These organoboron polymers were stable enough against protic solvents such as water and alcohol under nitrogen. However, under air, these polymers were unstable to be decomposed as usual for organoboron compounds. This decomposition was monitored by GPC after air bubbling in THF solution of the polymer at room temperature. After 3 min of air-bubbling, the peak became broader and moved to the low molecular weight region. At 30 min, the molecular weight decreased to almost several hundreds. However, it should be noted here that the obtained organoboron polymer was a little more stable toward air in comparison with conventional trialkylboranes.

Organoboron copolymers were prepared by polyaddition of a mixture of dienes with thexylborane [5]. When a mixture of dienes such as 1,2-diallyloxyethane and *p*-diallylbenzene was polymerized with thexylborane, the peaks in GPC using both UV and RI detectors were shifted to higher molecular weight regions with the increase of the amount of thexylborane. The molecular weight distribution of the copolymer obtained by this method indicated clear difference from that of a mixture of two homopolymers. This result supports the formation of copolymers. In a similar manner, other copolymers were prepared using various combinations of 1,7-octadiene, 1,2-diallyloxyethane, *p*-divinylbenzene, and *p*-diallylbenzene.

REACTIONS OF ORGANOBORON POLYMERS

The organoboron polymers prepared by hydroboration polymerization have a new structure consisting of C-B bonds in the main chain. Generally, these organoboron polymers can be regarded as a polymer homologue of trialkylborane, which is known to be a very versatile compound in organic synthesis [6]. In other words, the obtained polymers by hydroboration polymerization can be expected as a novel type of reactive polymers.

Scheme 2 demonstrates the versatile reactions of organoboron polymers prepared by hydroboration polymerization. For example, the polymer prepared from thexylborane and 1,7-octadiene was reacted with carbon monoxide at 120°C followed by treatment with NaOH and H_2O_2 to produce a polyalcohol [7]. This conversion includes the migrations of polymer chain and thexyl group from boron atom to carbon as shown in Scheme 3. When this reaction was carried out under milder condition, polyketone segments were included in the polymer structure due to the incomplete migration of thexyl group. These conversions may offer a new synthetic method for poly(alcohol)s from the corresponding dienes.

Scheme 2

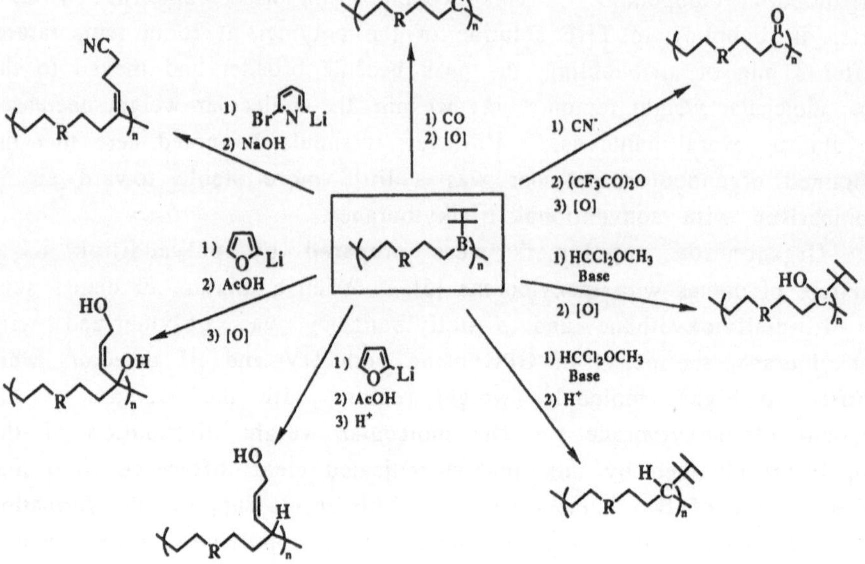

Scheme 3

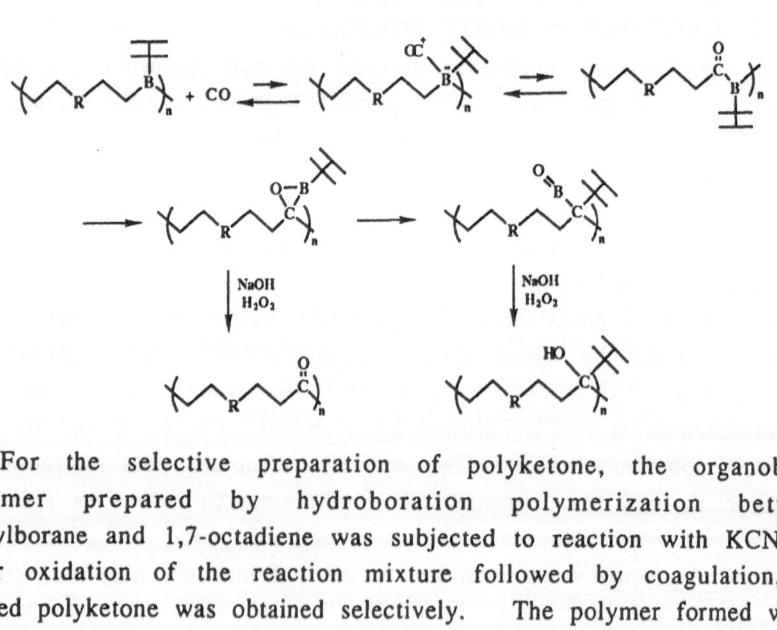

For the selective preparation of polyketone, the organoboron polymer prepared by hydroboration polymerization between thexylborane and 1,7-octadiene was subjected to reaction with KCN [8]. After oxidation of the reaction mixture followed by coagulation, the desired polyketone was obtained selectively. The polymer formed was a white solid and stable under air. This stability indicates no C-B bonds in the main chain of the resulting polymer, that is, complete conversion. This reaction also includes the migration of main chains of organoboron polymer from boron to carbon.

As an alternative method for the preparation of polyalcohol from organoboron polymer, dichloromethyl methyl ether (DCME) can be used [9]. Various organoboron polymers prepared by hydroboration polymerization of dienes with thexylborane were reacted with DCME in the presence of lithium triethylmethoxide at 0°C in THF. After oxidative treatment with NaOH and H_2O_2, the corresponding poly(alcohol)s were obtained. The structures of the resulting poly(alcohol)s were same as those prepared by the reaction with carbon monoxide described above. Because the reaction with carbon monoxide requires relatively severe conditions (e.g., 120°C, 30kg/cm^2) [7], the reaction with DCME provides more facile way for the conversion of organoboron polymers into poly(alcohol)s under mild conditions.

We also explored novel reactions of organoboron polymers via ring-opening of pyridine [10] or furan moieties [11]. Organoboron polymer prepared by hydroboration polymerization between 1,7-octadiene and thexylborane was reacted with 6-bromopyridine-2-lithium followed by treatment with NaOH to produce cyano group-containing polymer. On the other hand, the reaction of organoboron polymer with furan-2-lithium was followed by treatment with acetic acid and then with NaOH and H_2O_2 to form the polymer having primary and tertiary alcohols. These conversions include the migrations of polymeric main chains and ring-opening reactions of pyridine and furan, respectively.

In general, all the obtained polymers having functional groups described here are hardly prepared by means of conventional polymer synthetic methods. Consequently, these conversions are taken to demonstrate the useful and versatile characteristics of organoboron polymers prepared by hydroboration polymerization as polymeric precursors to give functional polymers.

HYDROBORATION POLYMERIZATION OF DIYNES

Acetylenes are known to be reactive toward hydroboration reaction as well as dienes. Accordingly, hydroboration polymerization with monoalkylborane can also be applied to diynes. Generally, the reactivity of a terminal acetylene group toward hydroboration is quite different from that of an internal acetylene group. That is, the terminal acetylenes preferentially give dihydroboration products via a further hydroboration of the initially formed vinylborane species. On the other hand, the internal acetylenes give monohydroboration products regardless of the bulkiness of borane reacted. Thus, terminal diyne and internal diyne are expected to show the different polymerization behavior due to the steric effect on the second (further) hydroboration reaction.

As a typical example of hydroboration polymerization of internal diyne, the reaction of 3,9-dodecadiyne and thexylborane was carried out in THF at 0°C under nitrogen [12]. The polymerization behavior was quite similar to hydroboration polymerization of dienes. The organoboron polymer obtained was colorless wax and soluble in common organic solvents such as benzene, THF or chloroform. IR spectrum of the polymer showed the absorption band assignable to C=C bond, which was also supported by ^{1}H- and ^{11}B-NMR spectra. Concerning the structure of this organoboron polymer, the oxidative cleavage using aqueous NaOH and H_2O_2 was examined, from which a mixture of diketones was obtained. This result means low selectivity of the addition of hydroborane to acetylene moiety. 3,8-Undecadiyne and 3,10-tridecadiyne were also used in hydroboration polymerization with thexylborane to produce the corresponding organoboron polymers.

In the case of terminal diyne such as 1,7-octadiyne, the molecular weight of the obtained polymer increased with increasing the feed ratio of hydroborane to acetylene moiety even over unity [12]. This result shows the hydroboration reaction of olefin of the resulting polymer with thexylborane occurred. In fact, when the ratio of thexylborane to acetylene unit was higher than 1.7, gelation was observed.

Hydroboration of terminal diyne gives organoboron polymer having branched structure due to the further hydroboration reaction toward the initially formed vinylborane structures. That is, terminal diynes are taken to have a potential as a multifunctional monomer. When the polymerization between 1,7-octadiene and thexylborane was carried out in the presence of a small amount of 1,7-octadiyne, the molecular weights of organoboron polymers were found to be increased when the ratio of diyne/diene was increased [5].

The organoboron polymers prepared from diynes, especially from internal diynes, consist mainly of divinylborane units. Thus, different reactivity and stability originating from this structure can be expected in comparison with the organoboron polymers prepared from dienes.

HYDROBORATION POLYMERIZATION OF DIENES WITH MONO-BROMOBORANE

Recently, we explored haloboration polymerization between boron tribromide and terminal diynes to give poly(organoboron halide)s as a polymeric Lewis acid, which has scarcely ever been known [13]. These poly(organoboron halide)s were also prepared by means of hydroboration polymerization between dienes and monobromoborane as shown in Scheme 4 [14]. As a typical example, 1,7-octadiene was added to a 1.0 M dichloromethane solution of monobromoborane-dimethyl sulfide complex with stirring at 0°C under nitrogen. The molecular weight of the polymer obtained was increased when the feed ratio of monobromoborane to diene approached or slightly exceeded unity.

Scheme 4

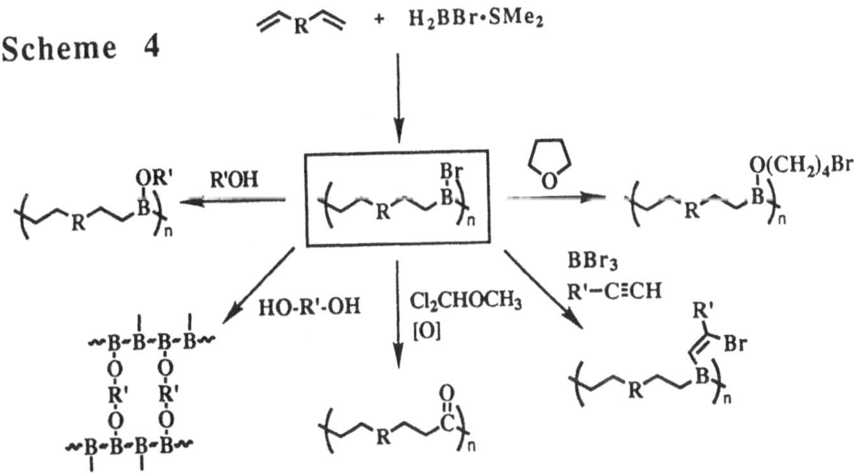

$H_2BBr{\cdot}SMe_2$; R'OH ; $O(CH_2)_4Br$; HO-R'-OH ; Cl_2CHOCH_3 [O] ; BBr_3 R'−C≡CH ; Br ; R' ; OR' ; O

The produced poly(organoboron halide) was very reactive toward moisture under air. Thus, the polymer was isolated after coagulation with dry methanol under nitrogen, by which the B-Br bonds were replaced by the B-OMe bonds. The isolated yield of the polymer was thus calculated after washing with *n*-pentane and drying in vacuo. After treatment with methanol, no change of the molecular weight was observed. This result indicates that no side-reactions such as chain scission took place during the transformation reaction. The structure of the B-OMe substituted polymer was confirmed by ^1H- and ^{11}B-NMR spectra. The results of hydroboration polymerization of various dienes with monobromoborane-dimethyl sulfide complex are summarized in

Table 2. In all cases, the corresponding poly(organoboron halide)s were obtained effectively.

Table 2. Hydroboration Polymerization between Monobromoborane-Dimethyl sulfide and Various Dienes.

Run	Diene	Yield (%)[a]	$\overline{M}_n$[b]	$\overline{M}_w$[b]
1		75	6,600	10,300
2		78	23,500	31,500
3		88	16,400	23,500
4		86	8,800	14,800
5	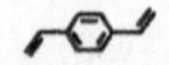	76	5,600	9,800
6	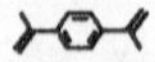	90	11,400	15,300
7	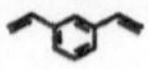	85	4,100	7,200
8	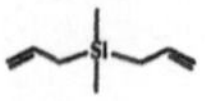	85	5,400	7,900
9	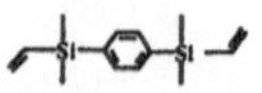	86	2,100	3,100

a) After treatment with MeOH.
b) Determined by GPC ($CHCl_3$, PSt standards).

The poly(organoboron halide) was reacted with water or diol compounds such as ethylene glycol to produce the corresponding gel via intermolecular substitution reactions. The B-Br group in the polymer chain has a potential to be subjected to haloboration reaction toward terminal acetylene. Thus, haloboration reaction with phenylacetylene or 1-hexyne was carried out by adding an equimolar amount of boron tribromide, which removed dimethyl sulfide from the polymer to generate in situ the free B-Br group. As a result, the alkenyl-branched organoboron polymer was obtained. The poly(organoboron halide) was also reacted with tetrahydrofuran via ether cleavage reaction to give 4-bromobutoxy-substituted organoboron polymer. These reactions are summarized in Scheme 4, which demonstrates the characteristic property of the polymer as a polymeric Lewis acid.

HYDROBORATION POLYMERIZATION OF DICYANO COMPOUNDS

The organoboron polymers prepared by hydroboration polymerization of dienes or diynes are relatively unstable toward air as mentioned before. On the other hand, the preparation of stable boron-containing polymers has been an attractive target as polymeric materials, for which a few examples such as borazine, phosphinoborane, and decaborane polymers have been reported [15].

It is known that hydroboration reaction of cyano groups gives iminoborane species which dimerize to form B-N four-membered ring (cyclodiborazane) in the case of appropriate borane used [16]. When this reaction is used for the bifunctional monomers, formation of polymeric materials consisting of cyclodiborazane units can be expected. Concerning the preparation of poly(cyclodiborazane)s, we have recently reported two new methods, i.e., allylboration polymerization of dicyano compounds with triallylborane [17], and polycondensation between bis(silylimine)s and chlorodialkylboranes [18].

Scheme 5

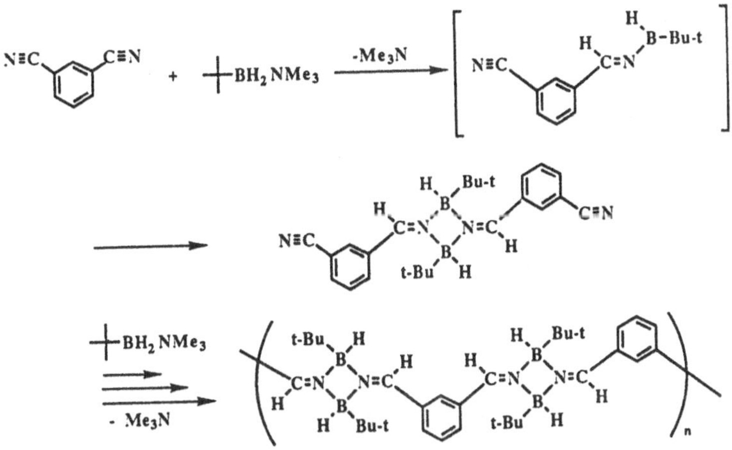

N≡C–C₆H₄–C≡N + t-BuBH_2NMe_3 —($-Me_3N$)→ [N≡C–C₆H₄–CH=N–B(H)–Bu-t]

→ dimer: N≡C–C₆H₄–CH=N / B(H)(Bu-t) four-membered ring / N=CH–C₆H₄–C≡N

t-BuBH_2NMe_3, $-Me_3N$ → polymer $(\ldots)_n$ of cyclodiborazane rings linked by m-phenylene units

As a typical example of hydroboration polymerization, isophthalonitrile was reacted with *t*-butylborane-trimethylamine in diglyme at 100°C under nitrogen to produce a boron-containing polymer as shown in Scheme 5 [19]. The obtained white solid polymer was soluble in various organic solvents such as THF, chloroform and benzene. This polymerization includes the dimerization of iminoborane to form the four-membered ring as a key essential step. In this hydroboration polymerization, terephthalonitrile and 1,5-di(4,4'-cyanophenoxy)pentane also gave the corresponding polymers. Aliphatic dicyano compounds, however, resulted in the formation of oligomers with very low molecular weights. These results are summarized in Table 3. The structures of these new boron-containing polymers were supported by spectroscopic data in comparison with those for model compound, which was prepared from benzonitrile and *t*-butylborane. The obtained boron-containing polymer has enough stability to bear at handling under air and moisture. Thermogravimetric analysis (TGA) of this boron-containing polymer shows that the weight loss started at 140°C and completed at 700°C to form a black solid. This finding suggests that the resulting poly(cyclodiborazane) prepared by hydroboration polymerization of dicyano compound can be used as a polymeric precursor for the preparation of boron-containing inorganic materials.

Table 3. Synthesis of Boron-Containing Polymers from Various Dicyano Compounds.[a)]

Run	Dicyano Compound	$\frac{[\text{t-BuBH}_2\ \text{NMe}_3]}{[\text{C≡N}]}$	$\overline{M}_n$ [b)]	$\overline{M}_w$ [b)]	Yield (%)
1	NC-C_6H_4-CN (meta)	1.55	7,590	27,200	96[c)]
2	NC-C_6H_4-CN (para)	1.94	3,280	12,000	84[c)]
3	NC-C_6H_4-$O(CH_2)_5O$-C_6H_4-CN	1.73	6,440	21,400	93[c)]
4	$NC(CH_2)_4CN$	0.89	810	1,380	d)
5		1.65	300	400	d)
6	$NC(CH_2)_6CN$	0.93	780	990	d)
7	$NC(CH_2)_8CN$	1.03	750	1,710	d)

a) Reactions were carried out at 100°C for 1h in diglyme.
b) GPC, polystyrene standard.
c) Isolated yield after precipitation into $EtOH/H_2O$ (v/v=1/1).
d) Not isolated.

Similarly to the above polymerization with *t*-butylborane, thexylborane also produced poly(cyclodiborazane)s by the reaction with aliphatic dicyano compounds, although the molecular weights of the polymers obtained were relatively low [20]. On the other hand, when polymerization was examined with less hindered monoalkylboranes (i.e., *n*-$BuBH_2{\cdot}NMe_3$ or *i*-$PrBH_2{\cdot}NMe_3$), crosslinking reaction took place due to the further hydroboration of iminoboranes [21].

In hydroboration polymerization with dicyano compounds, dialkylboranes can also be used as borane monomers (Scheme 6). In this case, the substituents at boron atoms in poly(cyclodiborazane)s should be two alkyl groups, while alkyl and hydrogen groups in the case of monoalkylborane monomer used. When polymerization between di(*n*-butyl)borane and adiponitrile was carried out in bulk under nitrogen, the reaction was completed within several hours, and the corresponding poly(cyclodiborazane) was obtained in high yield after reprecipitation into $EtOH/H_2O$. The structure of the polymer was confirmed by spectroscopic analyses such as ^1H-, ^{11}B-NMR and IR. Other dicyano monomers having different methylene chains also produced the corresponding poly(cyclodiborazane)s. Table 4 summarizes the results of hydroboration polymerization between di(*n*-butyl)borane and various dicyano compounds. In comparison with the case of monoalkylboranes, the hydroboration polymerization of dialkylboranes produced the high molecular weights polymers even from aliphatic dicyano compounds.

Scheme 6

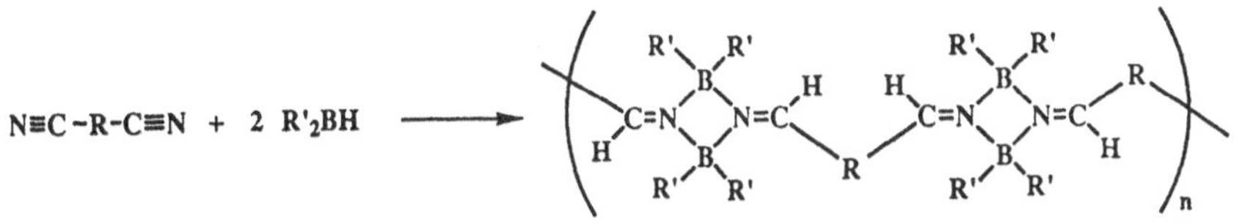

Table 4. Hydroboration Polymerization of Various Dicyano Compounds with Di(*n*-butyl)borane.[a]

Run	Dicyano Compounds	Yield[b] (%)	M_n[c]	M_w[c]
1[d]	$NC(CH_2)_2CN$	88	16,800	49,900
2	$NC(CH_2)_3CN$	90	10,100	23,500
3	$NC(CH_2)_4CN$	78	17,100	51,700
4	$NC(CH_2)_5CN$	77	25,000	69,900
5	$NC(CH_2)_6CN$	76	3,080	5,830
6	$NC(CH_2)_8CN$	88	5,930	9,070
7	m-NC-C_6H_4-CN	25	3,540	14,700

a) Polymerizations were carried out in bulk at r.t.
b) Isolated yields after precipitation into $EtOH/H_2O$ (1/1).
c) GPC (THF), polystyrene standard.
d) Polymerization was carried out at 70°C for 30min., then at r.t. for 2h.

On the other hand, dialkylthexylboranes are known to generate B-H species by the elimination of 2,3-dimethyl-2-butene upon heating. Thus, the in-situ generated dialkylboranes are expected to produce the same polymers by the reaction with dicyano compounds as those from isolated dialkylboranes described above. Both of aromatic and aliphatic dicyano compounds produced the corresponding boron-containing polymers. This method provides a useful and simple procedure for the synthesis of poly(cyclodiborazane)s, especially when dialkylborane monomer is hardly isolated.

References

1 Brown HC (1962) *Hydroboration*, Benjamin WA, Inc., New York
2 For example, Chung TC, Rhubright D (1991) *Macromolecules*, 24:970
3 Chujo Y, Tomita I, Saegusa T (1993) *Makromol. Chem., Macromol. Symp.*, 70/71:47
4 Chujo Y, Tomita I, Hashiguchi Y, Tanigawa H, Ihara E, Saegusa T (1991) *Macromolecules*, 24:345
5 Chujo Y, Tomita I, Saegusa T (1992) *Polym. Bull.*, 27:375
6 Pelter A, Smith K, Brown HC (1988) *Borane Reagents*, Academic Press, London
7 Chujo Y, Tomita I, Hashiguchi Y, Saegusa T (1991) *Macromolecules*, 24:3010
8 Chujo Y, Tomita I, Hashiguchi Y, Saegusa T (1991) *Polym. Bull.*, 25:1
9 Chujo Y, Tomita I, Hashiguchi Y, Saegusa T (1991) *Polym. Bull.*, 26:165
10 Chujo Y, Morimoto M, Tomita I (1993) *Polym. J.*, 25:891
11 Chujo Y, Morimoto M, Tomita I (1992) *Polym. Bull.*, 29:617
12 Chujo Y, Tomita I, Hashiguchi Y, Saegusa T (1992) *Macromolecules*, 25:33
13 Chujo Y, Tomita I, Saegusa T (1990) *Macromolecules*, 23:687
14 Chujo Y, Takizawa N, Sakurai T (1994) *J. Chem. Soc., Chem. Commun.*, in press
15 Teach WC, Green J (1965) *Encyclopedia of Polym. Sci. and Tech.*, 2:581
16 Hawthorne MF (1962) *Tetrahedron*, 17:112
17 Chujo Y, Tomita I, Saegusa T (1992) *Macromolecules*, 25:3005
18 Chujo Y, Tomita I, Asano T, Saegusa T (1994) *Polym. J.*, 26:87
19 Chujo Y, Tomita I, Murata N, Mauermann H, Saegusa T (1992) *Macromolecules*, 25:27
20 Chujo Y, Tomita I, Saegusa T (1993) *Polym. Bull.*, 31: in press
21 Chujo Y, Tomita I, Saegusa T (1993) *Polym. Bull.*, 31: in press

Stereospecific Polymerization of Styrene

M. Kuramoto

Polymer Research Laboratory, Idemitsu Petrochemical Co.,Ltd.
Anesaki-kaigan, Ichihara, Chiba, 299-01 Japan

ABSTRACT: Homogeneous titanium compound and methylaluminoxane (MAO) system is an effective catalyst for syndiospecific polymerization of styrene. The molecular weight distribution (MWD) is narrow (MWD≒2). A wide MWD can be obtained by using two kinds of titanium compound and MAO catalyst. Heterogeneous titanium compound containing halogen makes a mixture of isotactic and syndiotactic components. The ratio is dependent on the molar ratio of Al to Ti. The result of ESR measurement suggests that Ti(3+) species are important as a highly active site for producing syndiotactic polystyrene (SPS).

INTRODUCTION

The control of stereoregularity is very important technology in the field of polyolefins or polyvinylmonomers. Isotactic polymers are well known by the discovery of Ziegler-Natta catalyst. However, syndiotactic polymers, especially highly syndiotactic polymers, are rare. Furthermore, it is a very rare case that highly isotactic polymer and highly syndiotactic polymer can be obtained from the same monomer. We have succeeded in the synthesis of highly syndiotactic polystyrene in 1985 in Central Research Laboratories of Idemitsu Kosan Co.Ltd. We have already reported some data of our work.[1)-6)] As a result of our research, now we can get three types of polystyrene, namely atactic, isotactic and syndiotactic polymers, from styrene monomer by catalyst technologies. The ^{13}C-NMR spectra of these polystyrenes are given in Figure 1.

SPS is entirely different from conventional polystyrene (GPPS) in structure, physical properties, and synthetic method. The material is a new crystalline thermoplastic. Its melting point (Tm) is 270 ℃, the crystallization rate is very fast in comparison with isotactic polystyrene (IPS). And SPS have some superior properties such as heat resistance, chemical resistance, water/steam resistance and so on. A brief summary of these polystyrenes is given in Figure 2.

RESULTS AND DISCUSSION

It is well known that atactic polystyrene (APS) is obtained with the catalyst such as radical initiater and IPS is obtained with Ziegler- Natta catalyst. SPS can be produced by a transition metal compound and MAO catalyst. [2) 6)] As a transition metal, titanium(Ti) is excellent for the catalytic activity. We studied the stereospecific polymerization of styrene with homogeneous Ti and heterogeneous Ti compound and MAO catalyst.

Heterogeneous titanium catalyst:

Heterogeneous titanium compounds such as $TiCl_3$, titanium supported on magnesium compound (Ti-Mg) were used as a catalyst component in the titanium compounds and MAO catalyst. When these catalysts were used as a catalyst, polymers having a mixture of highly isotactic(mmmm) and highly syndiotactic(rrrr) components were mainly obtained. The examples are shown in Figure 3. The ratio

K.P. Ghiggino (Ed.)
Progress in Pacific Polymer Science 3

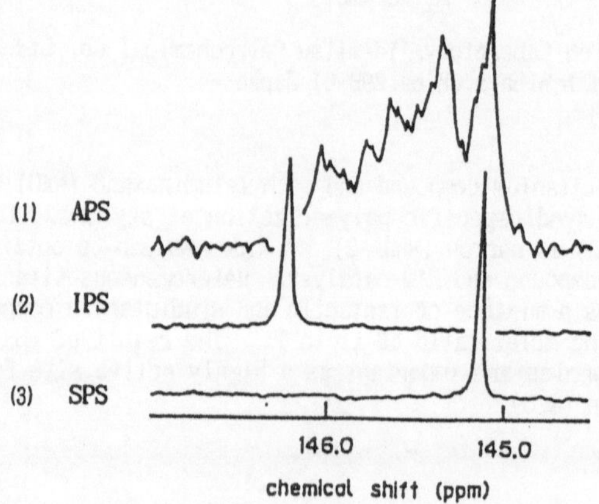

Fig.1 Aromatic C1-Carbon spectra of Polystyrenes in 1,2,4-trichlorobenzene at 130°C

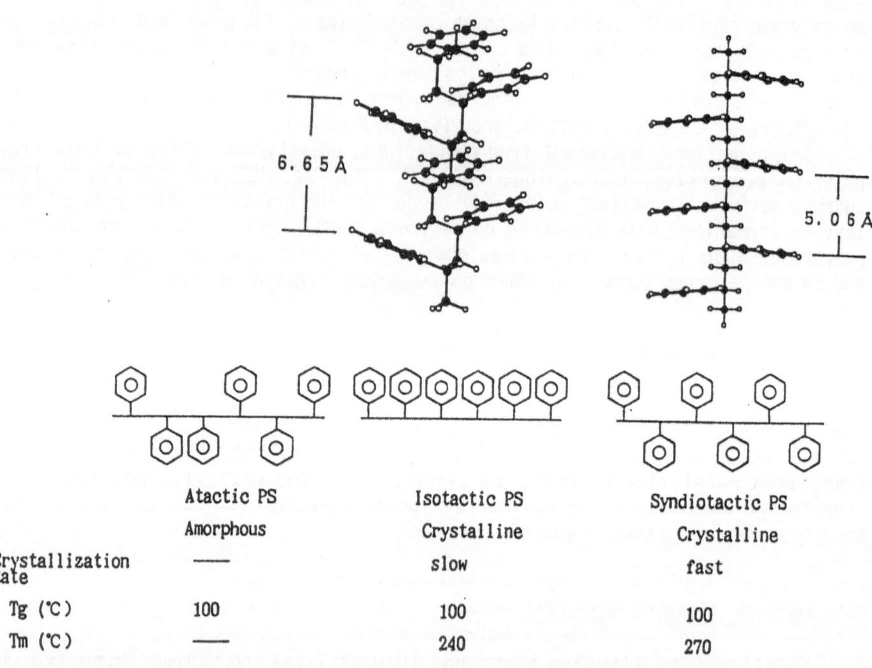

Fig.2 Brief summary of different polystyrenes

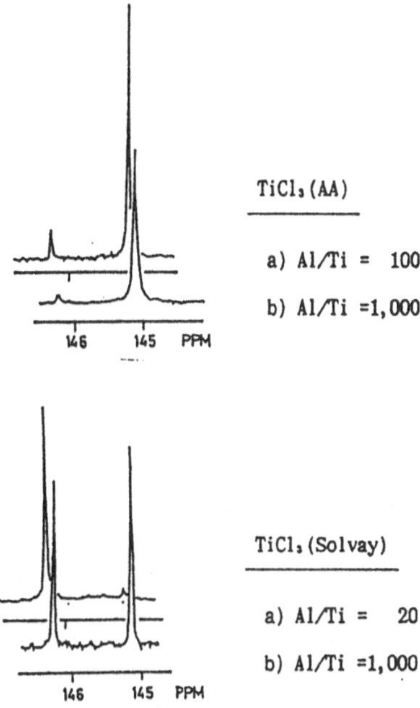

Fig. 3 Aromatic C1-Carbon spectra of Polystyrenes in 1,2,4-trichlorobenzene at 130°C

Table 1 Polymerization of styrene using various Ti compounds with MAO

Catalyst	(mmol)	Al/Ti	Conversion (wt%)	Stereospecificity
$TiCl_3$(AA)	1.0	100	8.2	Iso.PS + Synd.PS
	0.2	1,000	2.0	Iso.PS + Synd.PS
$TiCl_3$(Solvay)	1.0	20	1.9	Iso.PS + Synd.PS
	0.2	1,000	0.9	Iso.PS + Synd.PS
$Mg(OEt)_2/EB/TiCl_4$	2.0	50	2.9	Iso.PS(84) + Synd.PS(16)
	0.02	500	1.1	Iso.PS(12) + Synd.PS(88)
	0.02	1,000	1.4	Iso.PS(10) + Synd.PS(90)
$TiCl_4$	40	10	7.2	Iso.PS
	5	40	0.4	Iso.PS + Synd.PS
	0.2	500	0.7	Synd.PS
$Ti(OEt)_4$	2	10	0.3	(Atact.PS)
	2	50	2.5	Synd.PS
	0.2	500	0.9	Synd.PS

Polymn. conds. St=50ml, toluene=100ml, 50°C, 2hr.

is dependent on the molar ratio of Al (MAO) to Ti. The results are summarized in Table 1. To clarify the reason, some experimentations were done by using the titanium supported on magnesium compound as a titanium component. The catalyst system was prepared in the following manner. Thus, 2.0g of magnesium diethoxide; $Mg(OEt)_2$ were dispersed in 100ml of n-heptane followed by the addition of 0.53g of ethylbenzoate;EB at room temperature and then dropwise addition of 34g of titanium tetrachloride and the reaction mixture was heated for 3 hours under reflux. After completion of the reaction, the supernatant was discarded and the solid matter was repeatedly washed with n-heptane by decantation to give a solid material. The solid contained 40mg of titanium per g of the solid. The solid was mixed with MAO in touene and then was stirred for 30 minutes at room temperature. After then, the solid part and the solution part were separated by filtration from the reactant, respectively. And then, polymerization of styrene were done with the insoluble (solid) part and the soluble (solution) part, respectively. The ^{13}C-NMR of these polymers are given in Figure 4. The result suggests that IPS and SPS are made by heterogeneous Ti and heterogeneous Ti catalysts, respectively. From this experiment, it was found that a part of titanium on the carrior was migraded in solution. Secondly, the effect of the molar ratio of Al to Ti on the stereospecificity was investigated. The result is given in Table 2. The overall activity increased with increasing the molar ratio of MAO to Ti. Each activity was calculated on the basis of the ^{13}C-NMR data. Though the activity of isotactic polystyrene decreased, that of syndiotactic polystyrene increased with increasing the molar ratio MAO to Ti. Furthermore, the titanium concentration of the solid and the solution were investigated. The results are given in Figure 5. The titanium concentration in solution increased with the amount of MAO to titanium.

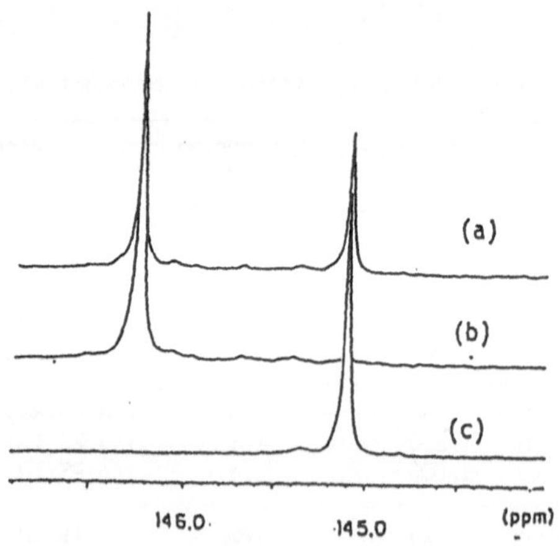

Fig.4 ^{13}C-NMR spectra of polystyrenes obtained with Ti supported on Mg compound and MAO catalyst

(a) polymer obtained with catalyst without filtration
(b) polymer obtained with insoluble part in toluene
(c) polymer obtained with soluble part

It is suggested that a part of titanium atoms on the carrier is solublized by MAO treatment in aromatic solvent. As a result, the amounts of homogeneous active site for syndiospecific polymerization increased. When the titanium compound without halogen such as titanium tetraethoxide;$Ti(OEt)_4$ was used as a catalyst, atactic or syndiotactic polystyrene were obtained and isotactic polystyrene was not obtained even if the molar ratio of Al to Ti is low level.

Table 2 Polymerization of styrene by using heterogeneous titanium [$Mg(OEt)_2$/EB/$TiCl_4$] and MAO

Ti/mmol	MAO/mmol	Al/Ti	Yield/g	MIP/wt% *	[Iso.]/[Synd.]
0.2	10	50	1.33	72	84 / 16
0.02	10	500	0.49	60	12 / 88
0.02	20	1,000	0.62	52	10 / 90

Al/Ti	Activity / Kg/gTi		
	[Iso.] + [Synd.]	[Iso.]	[Synd.]
50	4.79	4.02	0.77
500	14.70	1.76	12.94
1,000	16.12	1.61	14.51

*MIP;methyl ethyl ketone insoluble part by Soxhlet extraction method
Polymn.conds. toluene 100ml, St 50ml, 50℃, 2hrs

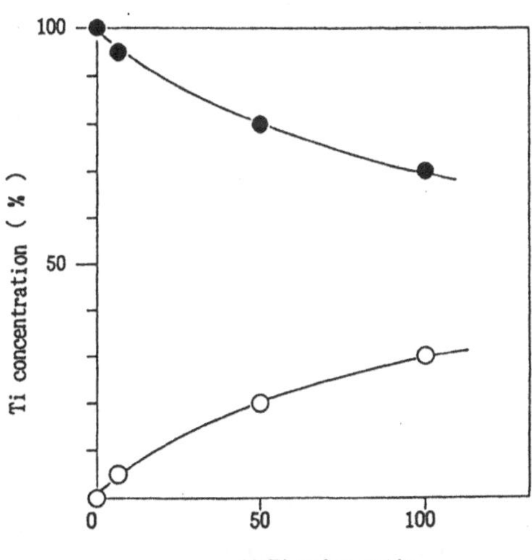

Fig.5 Effect of Al /Ti molar ratio on Ti concentration in the solution (○) and on the solid (●)

Homogeneous titanium catalyst:

Examples of the polymerization with various titanium compounds and MAO catalysts are given in Table 3. All of these Ti compound and MAO catalysts gives SPS. The molecular weight distribution (MWD) of these polymers are narrow, nearly equal two. This indicates a homogeneous polymerization. Catalitic activities of half-metallocene such as cyclopentadienyl titanium trichloride ($CpTiCl_3$), pentamethylcyclopentadienyl titanium trichloride ($Cp^* TiCl_3$) are greater than that of bent-metallocene such as Cp_2TiCl_2, $Cp^*{}_2TiCl_2$. Bent-metallocene has two bulky ligands. Therefore, the insertion of styrene having bulky group (Phenyl group) is disturbed.

Molecular weight of polymer is controlled by means of not only polymerization temperature but also Ti-ligand species. Influence of σ-ligand on variation of molecular weight is small in comparison with that of π- ligands. It suggests that the properties of the active sites arised from the reaction a titanium compound having only σ-ligand with MAO are similar to each other.

Table 3 Results of styrene polymerization with Ti compound and MAO catalyst

	Ti compounds	Yield/g	MIP/wt%	Mw	Mn	Mw/Mn
TiX_4	$TiCl_4$	1.42	93.5	313,000	151,000	2.08
$TiLX_3$	$Ti(OEt)Cl_3$	3.42	95.7	391,000	191,000	2.05
	$Ti(O\phi)Cl_3$	1.95	94.9	328,000	164,000	2.00
	$Ti(Cp)Cl_3$	12.03	97.6	120,000	55,000	2.18
	$Ti(Cp^*)Cl_3$	8.84	98.5	559,000	282,000	1.98
TiL_2X_2	$Ti(OEt)_2Cl_2$	4.25	95.1	291,000	136,000	2.14
	$Ti(Cp)_2Cl_2$	2.68	37.4	89,000	41,000	2.20
	$Ti(Cp^*)_2Cl_2$	0.85	30.1	331,000	149,000	2.23
	$Ti(acac)_2Cl_2$	1.56	91.1	275,000	128,000	2.16
$Ti(OR)_4$	$Ti(OEt)_4$	5.08	98.0	400,000	190,000	2.10

Polymn.conds. Ti;0.02mmol, MAO; 10mmol, 50°C, 2hrs
Solv.; toluene 100ml, St;100ml

Cotrol of Molecular Weight Distribution:

Styrene polymerization were done with two kinds of half-metallocene; $LTiCl_3$(L=Cp,Cp*) and MAO catalyst. The GPC curves of polystyrenes obtained by the catalyst are shown in Figure 6. The results of styrene polymerization with $CpTiCl_3$,Cp^*TiCl_3/MAO catalyst is given in Table 4. And the shape of GPC curves is dependent on the $CpTiCl_3$/ Cp^*TiCl_3 molar ratio.The results indicates that the polymerization proceeds on each homogeneous active site.

Table 4 Results of styrene polymerization with $CpTiCl_3$/ Cp^*TiCl_3- MAO catalyst

$CpTiCl_3$ /mmol	Cp^*TiCl_3 /mmol	Cp / Cp*	Yield /g	MIP /wt%
0.01	0	100 / 0	15.7	98.6
0.075	0.0025	75 / 25	14.6	98.2
0.025	0.0075	25 / 75	9.5	98.4
0	0.01	0 / 100	4.5	94.9

Cp / Cp*	Mw	Mn	Mw/Mn
100 / 0	136,200	63,300	2.15
75 / 25	569,400	85,200	6.68
25 / 75	1,240,000	227,000	5.51
0 / 100	2,046,000	856,800	2.39

Polymn.conds. MAO; 4mmol, 50°C, 1hr
Solv.; toluene 100ml, St;100ml

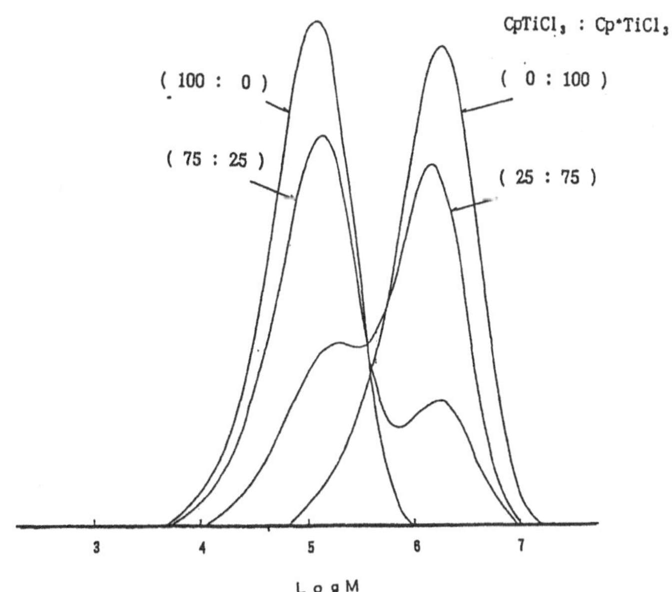

Fig.6 GPC curves of polystyrenes obtained with $CpTiCl_3$/Cp^*TiCl_3 and MAO catalyst

The effect of MAO to titanium molar ratio on catalitic activity in the case of $LTiCl_3$(L=Cp,Cp*) and MAO catalyst system was investigated. The catalitic activity increases with the amount of MAO. The amount of Ti(3+) species measured by ESR increases with the amountof MAO, too. This suggests that MAO acts as reducting agent from Ti (4+) to Ti (3+) and Ti (3+) is a very highly active species for producing SPS. MAO acts as a weak reductant. Therefore, much amount of MAO is needed to incease the activity. The addition of triisobutylaluminum (TIBA) to homogeneous $CpTiCl_3$-MAO catalyst was investigated. The results of styrene polymerization are given in Table 5 and in Figure 7. As a result, it was found that the addition of TIBA increases the activity. The amount of Ti (3+) measured by ESR increased by TIBA addition. However, excess addition of TIBA decreased the activity and the amount of Ti (3+) decreased, too. Excess addition of TIBA causes overreduction of Ti valency. Though Ti (3+) species results from the addition of TIBA to $CpTiCl_3$, the resulting Ti (3+) species is inactive site for syndiospecific polymerization. This suggests the participation of MAO such as MAO coodination to Ti in the reaction of the active site formation for producing SPS. The reactin of active site formation for syndiospecific polymerization is summerized in Scheme 1.

Table 5 Results of styrene polymerization with $CpTiCl_3$ and MAO (TIBA)

Catalyst /mmol			Al/Ti (molar ratio)	Yield /g
$CpTiCl_3$	MAO	TIBA		
0.0125	3.75		300	1.2
0.0125	5.0		400	1.7
0.0125	7.5		600	3.0
0.0125	20		1,600	4.0
0.0125	30		2,400	6.6
0.0125	5.0		400	1.7
0.0125	5.0	1.0	480	2.8
0.0125	5.0	2.5	600	4.5
0.0125	3.75	3.75	600	3.2

Polymn.conds. Solv.; toluene 100ml, St;150mmol, 50℃, 1hr

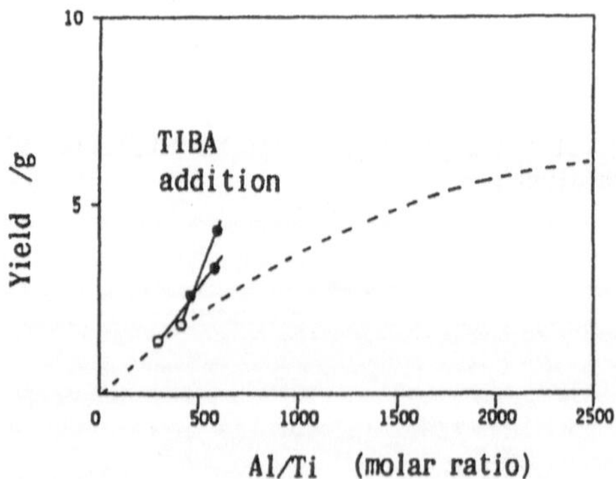

Fig.7 The effect of MAO / Ti molar ratio on catalitic activity

Scheme 1

Reaction of active site formation

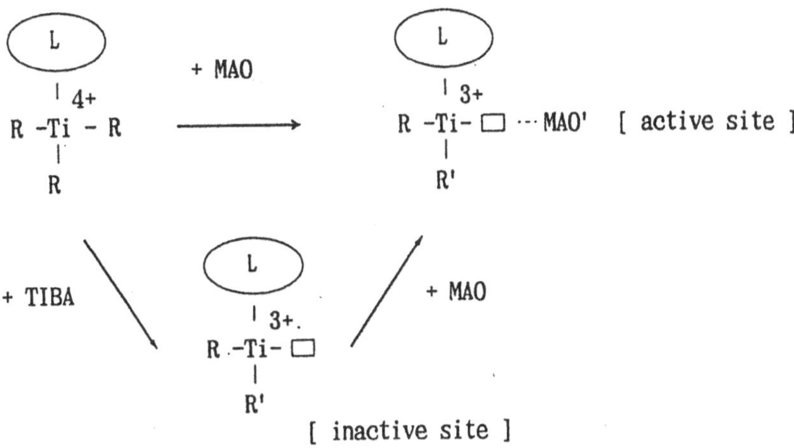

Concerning the mode of addition of the growing chain end to the double bond of the monomer, it has been reported that the mode for syndiospecific polymerization of styrene is cis-opening.[7,8] And styrene insertion has been reported as secondary (2-1) insertion.[9,10,11] On the other hand, styrene insertion has been reported as primary (1-2) insertion in the case of isospecific polymerization with Zieglar-Natta catalyst such as $TiCl_3$-R_3Al system.[12,13] When the polymerization of propylene with syndiospecific catalyst for styrene such as $CpTiCl_3$ or $Ti(OEt)_4$/MAO system, the steroregularity of the resulting polymer was atactic. The electron donation from π-electron of phenyl of the growing chain end and / or that of phenyl of monomer to titanium metal is considered on the syndiospecific polymerization of styrene.[14]

By using these syndiospecific catalyst, syndiotaciic polymers can be obtained from styrene derivatives.[2] And it is possible to modify SPS ,e.g.,Epimerization[15,16],Chrolomethylation,[17] Sulfonation.[18] SPS in itself not only have some superior properties but also its modification is possible.

REFERENCES

1 N. Ishihara, T. Seimiya, M. Kuramoto, M. Uoi (1986) Macromolecules 19:2464
2 N. Ishihara, M. Kuramoto, M. Uoi (1988) Macromolecules 21:3356
3 M. Kuramoto, N. Ishihara, M. Uoi (1987) Polymer Preprint Japan 36:209
4 N. Ishihara, M. Kuramoto, M. Uoi (1987) Polymer Preprint Japan 36:210
5 M. Kuramoto, N. Tomotsu, H. Maezawa, M. Takeuchi (1992) 4th SPSJ-IPC Preprint 28
6 A. Grassi, C. Pellecchia, P. Longo, A. Zambelli (1987) Gazz Chim Ital 117:249
7 N. Mitani, N. Ishihara, T. Seimiya, T. Ijitsu, T. Takyu (1988) Polymer Preprint Japan 37 :1152
8 P. Longo, A. Grassi, A. Proto, P. Ammendola (1988) Macromolecules 21:24
9 A. Zambelli, P. Longo, C. Pellecchia, A. Grassi (1987) Macromolecules 20:2035
10 N. Ishihara (1990) D Phill Thesis Oxford
11 N. Mitani, N. Ishihara (1992) Polymer Preprint Japan 41:1800
12 A. Benaboura, A. Deffieux, P. Sigwalt (1987) Macromol Chem 188:21
13 P. Ammendola, T. Tancredi, A. Zambelli (1986) Macromolecules 19:307
14 A. Zambelli, C. Pellecchia, L. Oliva, P. Longo, A. Grassi (1991) 192:223
15 N. Mitani, T. Seimiya, T. Ijitsu, T. Takyu (1987) Polymer Preprint Japan 36:906
16 S. J. Tsai, E. R. Santee, H. J. Harwood (1990) Polymer Preprint A.C.S. 31:472
17 Eur Pat 318833
18 E. B. Orler, D. J. Yontz, R. B. Moore (1993) Macromolecules 26:5157

Polymer Design for Minimally Adhesive Surfaces

K. J. Wynne[1], T. Ho[2], R. A. Nissan[3], X. Chen[4,5], and J. A. Gardella, Jr.[4]

[1]Chemistry Division, Office of Naval Research, Arlington, VA 22217-5660 and Materials Chemistry Branch, Naval Research Laboratory, Washington, D.C. 20375-5320
[2]Department of Chemistry, George Mason University, Fairfax, Virginia 22030-4444
[3]Chemistry Division, Naval Air Weapons Center, China Lake, CA 93555
[4]Department of Chemistry, State University of New York at Buffalo, Buffalo, NY 14214

Abstract: In connection with our interest in the development of minimally adhesive surfaces to discourage the settlement of marine organisms, we have investigated polyurethanes and polyureas containing polydimethylsiloxane (PDMS) segments. A two-step polymerization method was used to prepare dimethylsiloxane-urea-urethane copolymers with 1,4-benzenedimethanol as the chain extender. Thermal and mechanical properties of copolymers with chain extenders were found to be superior to those without chain extender, due to the additional hydrogen bonding interactions for the former. Surface composition was determined by angle-dependent electron spectroscopy for chemical analysis (ESCA). Effects of segmental length and annealing on the surface composition were investigated. One of the siloxane containing copolymers was used as the minor component (1.6, 2.5 and 6.0 wt %) in a series of blends with a poly(ether-urethane), which preserved the mechanical properties of the poly(ether-urethane) as well as the surface properties of the poly(siloxane-urea-urethane).

INTRODUCTION

Our interest in the formation of minimally adhesive polymer surfaces has led us to an investigation of poly(urethane-ureas) containing unusual diols. Our goal is to discern the compositional and morphological features which create a surface minimally attractive to the settlement of marine organisms, an area pioneered by Griffith, et al.(1)

In addition to low surface energy, we postulate that a surface phase with a low T_g is desirable for minimizing mechanical locking of a prospective adherent to the surface. With the dual criteria of low surface energy and a low T_g surface phase in mind, we have prepared and characterized

[5]Current address: Moore Research Center, Grand Island, NY 14072

K.P. Ghiggino (Ed.)
Progress in Pacific Polymer Science 3

a series of polydimethylsiloxane containing poly(urethane-ureas) and have examined the effects of chain extenders on properties.

The synthesis of siloxane urea urethane copolymers through step-growth polymerization has been explored previously,(2 - 5) and work performed before 1988 has been reviewed.(6) We have used a "hard segment first" two-step polymerization procedure related to that of Harrell(7) to achieve hard segment compositional control. The synthesis of these polymers(8) and initial studies of surface characterization(9) have been reported.

The pioneering work of Clark demonstrated the power of angle-dependent electron spectroscopy for chemical analysis (ESCA) in polymer surface analysis.(10) Subsequently, surface composition and morphology of block copolymers containing poly(dimethylsiloxane) (PDMS) segments have been studied using angle-dependent ESCA and other techniques. Polyurethane-PDMS,(11 - 13) Nylon-6-PDMS,(14) poly(α-methylstyrene)-PDMS,(15) polycarbonate-PDMS,(16,17) polysulfone-PDMS(18) and polystyrene-PDMS(19,20) of various block architectures and overall compositions have been investigated. Enrichment of PDMS segments in surface region was detected in each case due to the lower surface energy of PDMS component in these block copolymers.

The present work builds on the observation that surface composition of multicomponent polymers depends not only on structure, but also sample history. Thus, surface composition of solution cast films of block copolymers can be controlled by using selective solvents(14,21) or by annealing.(14,15,17,22)

The water repellency of silicone is usually attributed to the extreme flexibility of the siloxane linkage, which facilitates the optimum orientations of the hydrophobic methyl groups.(23) By anchoring PDMS onto a neutral or hydrophilic substrate, hydrophobicity can be imposed upon that substrate.(6) We investigated the feasibility of blending the siloxane copolymer with a poly(ether-urethane) prepared also by the "hard-segment first" scheme. Blending a base polymer and a siloxane containing copolymer to improve the surface properties of the base polymer has been investigated by various researchers. Many such systems have been reported in the literature and summarized in a review by Yilgör and McGrath.(6) The base polymers investigated include polystyrenes, polycarbonates, commercial polyurethanes, polymethacrylates, polysulfones, poly(vinyl chloride)s, poly(ethylene terephthalate)s, and nylons. Our materials differ from previous efforts in two aspects: the siloxane containing copolymers are segmented copolymers, which have different morphologies than the commonly used block copolymers; and the hard segments of the poly(ether-urethane) are based on IPDI and BDM, which have a less tendency to crystallize than the hard segments of commercial polyurethanes often based on 4,4'-methylene bis-(phenyl isocyanate) (MDI) and butanediol (BD). The design of our materials offers an opportunity to exploit the crystallinity of the soft segments of the base polymer, which is usually suppressed due to the high crystallinity of the hard segments. Our characterization of bulk and surface properties are reported below.

EXPERIMENTAL

Materials: Isophorone diisocyanate (IPDI, **1**, 5-isocyanato-1-(isocyanatomethyl)-1,3,3-trimethylcyclohexane, a mixture of isomers), and 1,4-benzenedimethanol (BDM), **2**, were purchased from Aldrich. 3-Aminopropyl endcapped dimethylsiloxane oligomers, $H_2N(CH_2)_3(Si(CH_3)_2)O)_nSi(CH_3)_2(CH_2)_3NH_2$, with different average molecular weights were kindly provided by Dr. İ. Yilgör of Goldschmidt Chemical Corporation, Hopewell, VA. The nominal molecular weights of Tegomer A-Si 2120, **4a**, Tegomer A-Si 2320, **4b**, and Tegomer A-Si 2920, **4c**, were 1000 (n = 11), 2400 (n = 30) and 10,000 (n = 133), respectively. In addition, aminopropyl end-capped PDMS oligomer Hüls PS 513 (viscosity 2000 cs, MW ≅ 27,000, n = 363) was employed. Polytetrahydrofuran (PTMO), MW ≅ 2 000, was purchased from Aldrich Chemical Co. All chemicals were used as received.

Polymerization: The two-step polymerization is shown in Scheme 1. Synthetic details have been described previously.(8,24) Polymers without chain extender were prepared in THF solution using only the second step of Scheme 1.

$$\underset{\mathbf{1}}{(x+1)\ \text{IPDI}} + \underset{\mathbf{2}}{(x)\ \text{BDM}} \xrightarrow{105-118^\circ C} \underset{\mathbf{3}}{\text{IPDI-(BDM-IPDI)}_x} \quad (1)$$

$$\text{IPDI-(BDM-IPDI)}_x + \underset{\mathbf{4}}{\text{PDMS oligomer}} \xrightarrow{\text{THF}, 50^\circ C} \text{PDMS-urea-urethane copolymer} \quad (2a)$$

$$\text{IPDI-(BDM-IPDI)}_x + \text{PTMO} \xrightarrow{\text{THF}, 50^\circ C, \text{DBTDL}} \text{PTMO-urethane copolymer} \quad (2b)$$

Scheme 1

The blends were prepared in tetrahydrofuran (THF) solution (typically 6 g polymer / 90 ml of THF). The solutions were prepared by vigorous stirring with a magnetic stirrer at room temperature for more than two hours; the resulting mixtures were slightly turbid due to the poor solubility of siloxane in THF. Films were formed by pouring the solutions onto a Teflon sheet. Solvent was allowed to evaporate in the hood for one day. The films were then dried in a vacuum oven at 60° C for at least one more day. The films were translucent indicating phase separation between the two components. Films of the polyetherurethane (from a solution of same concentration in THF as above) were also prepared as the control samples.

Characterization: The new polymers were characterized by infrared and NMR spectroscopy, viscometry, and mechanical properties as previously described.(8) Molecular weights were determined by gel permeation chromatography, using a Hewlett-Packard Series 1050 pump, two Altex μ-Spherogel columns (size 10^3 and 10^4Å, respectively) connected in series, and THF as the solvent. Polymer in the effluent was detected with a Wyatt/Optilab 903 interferometric refractometer, and average molecular weights were determined relative to polystyrene standards. As previous workers have noted, polymers containing high weight fractions of poly(dimethyl siloxane) segments have low specific refractive increments (*dn/dc*) in THF solution.(25) High concentrations (about 2-3 wt%) were therefore used to enhance signals. Chromatograms were smoothed by the method of Fourier transform convolution.(26)

Advancing contact angle measurements were performed with an NRL A-100 Contact Angle Goniometer (commercially available through Ramé-Hart, Inc). Surface and interfacial tensions of the liquids were determined with a Cahn Dynamic Contact Angle Analyzer (DCA 312). Distilled water was prepared in-house, methylene iodide was purchase from Aldrich Chemical Co. and used as-received.

Film preparation for ESCA analysis involved solvent pretreatment to remove adventitious low molecular weight materials, exposure to high vacuum for solvent removal, and film casting from THF utilizing aluminum weighing pans.(9) Solvent was removed from the films (thickness ~50 μm) by a 2-day vacuum treatment at ambient temperature. Film annealing was accomplished by heating in a vacuum oven at 120°C for 15 minutes.

Angle-dependent ESCA experiments were performed on a Perkin-Elmer Physical Electronic Model 5300 ESCA with a hemispherical analyzer and a single channel detector.(9) For quantifying ESCA signals in carbon 1s, nitrogen 1s, silicon 2p, and oxygen 1s regions spectra were recorded at high resolution conditions. ESCA peak areas were measured by a Perkin-Elmer 7500 computer with the PHI ESCA version 2.0 software. An average of three independent runs was taken for all ESCA measurements. The ratio of N/C, instead of Si/C, is chosen to calculate the PDMS surface concentrations because of the larger variation in nitrogen concentration at different sampling depths. This results in higher sensitivity in PDMS wt % calculation. Concentrations were also calculated from Si/C ratios for a selected set of samples. The resultant concentration data fell within error limits as equivalent to that calculated from N/C ratios.

Photoelectron intensities detected by ESCA are convoluted signals, i.e. all atoms within the path of the probing X-ray contribute to the signal but the contribution of each decreases exponentially with the distance from the free surface.(10) The convoluted nature of the signal distorts depth profiles for samples with compositional gradients. To recover the depth profiles for such samples, a deconvolution procedure was applied to the ESCA data.(9)

RESULTS AND DISCUSSION

Hard Segment Characterization

The preparation of copolymers of aminopropyl endcapped dimethylsiloxane oligomers and isophorone diisocyanate with 1,4-benzenedimethanol as the chain extender outlined in Scheme 1 is similar to the method used by Pascault and co-workers to prepare alkoxy-silane terminated macromers.(27) This "hard segment first" method allowed the separate isolation and characterization of the hard block. Thus, GPC analysis for the intermediate IPDI-BDM-IPDI-BDM-IPDI (desired "3:2" composition), also revealed peaks for IPDI-BDM-IPDI, and BDM-IPDI-BDM. The intensity ratio of the curves indicate that more than 85% of the molecules in the mixture were of the target "3:2" composition.

Tegomer Characterization

Assignment and determination of relative intensities of the NMR peaks due to the dimethylsilyl and the propyl amine end group protons was carried out to confirm GPC determined macromonomer molecular weight.(8) In addition to expected ^{1}H nmr signals, additional peaks were observed in 1, 2.4 and 10K Tegomer samples from Goldschmidt Chemical Co. A careful investigation of Tegomer A-Si 2120 was carried out to determine the identity of any additional materials present. All results, including chemical shifts and coupling patterns, indicate that the "impurity" is a 2-amino-1-methylethyl end group.(8)

Integrated ^{1}H spectra for $SiCH(C\underline{H}_3)CH_2NH_2$ vs. $Si\underline{CH_2}CH_2CH_2NH_2$ revealed the relative amounts of these two end groups. In the respective Tegomers, the proportion of $SiCH_2CH_2CH_2NH_2$ is: 72%, Tegomer A-Si 2120; 73%, Tegomer A-Si 2320; 76%, Tegomer A-Si 2920. The presence of $SiCH(CH_3)CH_2NH_2$ end groups could not be detected in the aminopropyl end-capped PDMS oligomer Hüls PS 513. Molecular weights determined by NMR and GPC were found to be in excellent agreement.(8)

Polymer Synthesis and Characterization

Diamines or diols can be used as chain extenders in copolymerization reactions, but urea linkages formed from diamines often lead to polymer insolubility.(28) Poor solubility is avoided by using diols, but this requires a two stage preparation (Scheme 1) because of the disparity in reactivity of alcohols and amines with isocyanates. To form urea and urethane linkages separately, the reaction between isophorone diisocyanate and benzene dimethanol was carried out first in bulk at 115 °C without catalyst (Equation 1). Secondly, reaction between aminopropyl endcapped dimethylsiloxane oligomers and diisocyanate intermediates was effected in THF at room temperature (Equation 2). Completion of the polymerization reaction was established by NMR and infrared spectroscopy.(8)

Since the capability to form films is an important requirement for coating materials, we have focused on those compositions that form flexible films (Table 1). Composition designations are as follows: in the formula PDMS2.4K-IP-B0, the segmental average molecular weight of the siloxane oligomer is given first; the two letters following the first hyphen identify IPDI as "IP"; the letter after the second hyphen represents BDM, the chain extender, "B"; and the next number shows the molar ratio of chain extender to siloxane oligomer (0, i.e. no chain extender).

Mechanical properties: Stress-strain tests were performed at room temperature at a strain rate of 1.67 min^{-1} on specimens cast from THF solutions (Figure 1, Table 2). The average molecular weight of the siloxane segments has a dominant effect on the rigidity of the copolymers. Young's moduli for copolymers based on oligomer **4a** (MW≈1000) are above 40 MPa, those for copolymers based on **4b** (MW≈2400) are in the range of 1 to 12 MPa, and the moduli for copolymers based on **4c** (MW≈10,000)and Hüls PS 513 (MW≈27,000) are below 0.4 MPa.

Table 1 Molecular Weights of the Segmented Copolymers

No.	Sample ID	PDMS MW[a]	Reaction Stoichiometry PDMS: IPDI: BDM	Copolymer MW[b] M_w	M_n	M_w/M_n
1	PDMS1K-IP-B0	1,000	1 : 1 : 0	16,700	12,000	1.4
2	PDMS1K-IP-B0.5	1,000	2 : 3 : 1	17,600	11,800	1.5
3	PDMS2.4K-IP-B0	2,400	1 : 1 : 0	65,000	30,700	2.1
4	PDMS2.4K-IP-B1	2,400	1 : 2 : 1	76,500	27,700	2.8
5	PDMS2.4K-IP-B2	2,400	1 : 3 : 2	42,700	19,100	2.2
6	PDMS10K-IP-B0	10,000	1 : 1 : 0	109,000	54,500	2.0
7	PDMS10K-IP-B2	10,000	1 : 3 : 2	81,200	29,200	2.8
8	PDMS27K-IP-B2	27,000	1 : 3 : 2	198,000	149,000	1.3

(a). Amino terminated siloxane oligomers with molecular weight provided by the supplier.
(b). Polymer molecular weight determined by GPC.

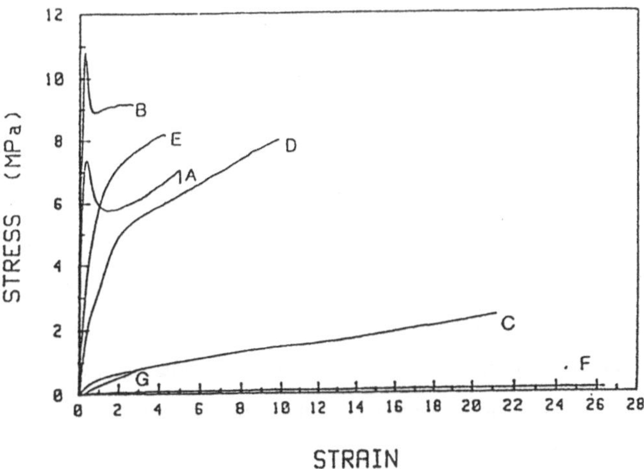

Figure 1. Stress-strain curves for the PDMS/urea/urethane co-polymers. A: PDMS1K-IP-B0; B: PDMS1K-IP-B0.5; C: PDMS2.4K-IP-B0; D: PDMS2.4K-IP-B1; E: PDMS2.4K-IP-B2; F: PDMS10K-IP-B0; and G: PDMS10K-IP-B2. (Reproduced with permission from ref. 8. Copyright 1993 ACS.)

Table 2. Mechanical Properties of PDMS/Urea/Urethane Copolymers.

Designation	Young's Modulus (MPa)	Strain at Break	Area under Curve (MPa)
PDMS1K-IP-B0	42.68	5.0	31.0
PDMS1K-IP-B0.5	70.43	2.7	24.1
PDMS2.4K-IP-B0	1.02	21.2	30.5
PDMS2.4K-IP-B1	6.94	9.8	57.8
PDMS2.4K-IP-B2	12.38	4.2	28.0
PDMS10K-IP-B0	0.03	26.8	2.82
PDMS10K-IP-B2	0.38	2.8	1.05
PDMS27K-IP-B2	0.18	6.26	2.08

The incorporation of BDM as the chain extender increases the Young's modulus significantly. The addition of two moles of BDM per mole of siloxane oligomer in copolymers based on **4b** and **4c** results in a tenfold increase in the modulus. On the other hand, the strain at break of the polymers decreases with the incorporation of BDM. For copolymers based on **4a**, the strain at break drops from 5.0 (500 %) to 2.7 with the incorporation of one half mole of BDM per mole of siloxane oligomer. In the series of copolymers based on **4b**, the strain at break decreases from over 20 to 4.2 with an increase in the content of the chain extender from none to two moles of BDM per mole of **4b**. Similarly, incorporation of BDM into copolymers based on **4c** at the ratio of two moles of BDM per mole of **4c** leads to a decrease in strain at break from 26.8 to 2.8. Toughness of a material is proportional to the area under the stress-strain curve. By this measurement, polymers with average siloxane segmental molecular weights at 1K and 2.4K are tougher, by about one order of magnitude, than those with siloxane segmental molecular weights at 10K and 27K, regardless of the composition of the hard segment.

It has been recognized that many of the unique properties of the polyether or ester based polyurethanes are due to their phase-separated structure: the hard segment (urethane or urea) rich domains provide the "solid" attributes, while the soft segment (polyether or ester) rich domains account for the elastomeric behavior.(29) Previous work has shown that mechanical properties of PDMS urea (or urethane) copolymers also follow that pattern.(2,3,5) It has been shown, based on dynamic mechanical data, that phase separated morphologies exist in the current copolymers.(8) In such systems, an increase in the average molecular weight of the soft segments, which increases the average distance between two hard segment domains, will cause a decrease in the rigidity of the material, while an increase in chain extender content, which increases the proportion of the hard segment domain, will enhance the rigidity. The observed mechanical properties are in agreement with this reasoning. Based on this model the observed variation in toughness can also be explained. Materials based on **4a** are tough because of the proportionately greater contribution from the hard segment, which also leads to high modulus. The toughness of materials based on **4b** is mainly due to the ductility of the materials facilitated by the longer soft segment. In contrast, the weakness of materials based on **4c** is probably due to failure of the long soft segment to transmit the external force to be born by the hard segment.

Surface Composition of PDMS-PU Films by ESCA

Structural effects: Previous studies have shown that surfaces of polymers containing PDMS segments display surface enrichment of the PDMS component.(11-20) PDMS surface phase separation is attributed to the lower surface energy of the PDMS segments at the polymer air interface, as compared with a surface presenting polar hard segments.(30) Our results for PDMS-PU segmented copolymer films cast from THF in air show a similar enrichment. ESCA results for the PDMS-PU segmented copolymers are angle dependent indicating a gradient in concentration of PDMS over the surface region. Through deconvolution of the ESCA data insights into effects of chain structure and processing effects have been obtained.

Some trends are apparent without extensive data analysis. Thus, Figure 2 shows surface compositions derived from direct measurement of N/C ESCA peaks as a function of take-off angle. The N/C ratio goes to zero when the surface is pure PDMS. The data in Figure 2 for as-cast films of PDMS2.4K-IP-B2, PDMS10K-IP-B2 and PDMS27K-IP-B2 shows that the copolymer with the highest average molecular weight for the soft segments has the highest PDMS surface concentration. This might be expected, as the PDMS weight fraction goes from 0.708 to 0.966 for this series of copolymers.

Figure 3 illustrates the change in PDMS surface concentration with increase in the length of hard segment. In this group of PDMS-PU samples, the average molecular weights for the PDMS

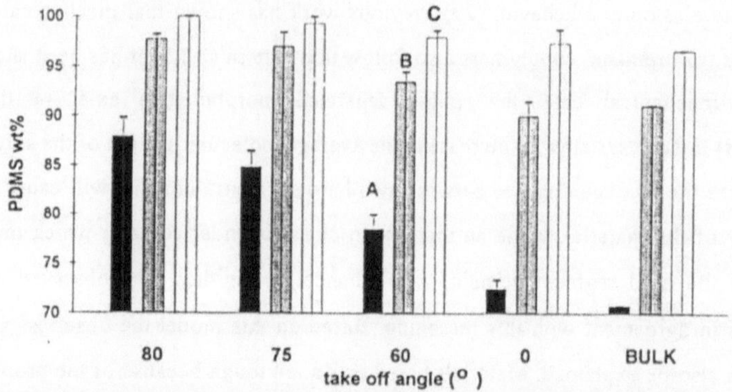

Figure 2. PDMS concentration in the surface region measured by ESCA for three copolymers. The distance between the horizontal bar and the top of the column represents one standard deviation. Data series: A, PDMS2.4K-IP-B2; B, PDMS10K-IP-B2; and C, PDMS27K-IP-B2.

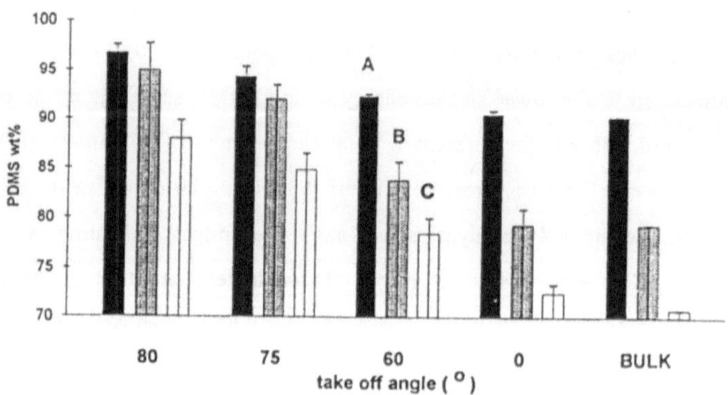

Figure 3. PDMS concentration in the surface region measured by ESCA for three copolymers. The distance between the horizontal bar and the top of the column represents one standard deviation. Data series: A, PDMS2.4K-IP-B0; B, PDMS2.4K-IP-B1; and C, PDMS2.4K-IP-B2.

segments are the same (2.4K), but the average lengths for the hard segment are different. The surface concentration of PDMS decreases as the length of the hard segment increases; the weight fraction of PDMS is in the 0.708 to 0.903 range for this series. Tezuka and co-workers have used ESCA and contact angle measurement to study the surface morphology of a series of polyetherurethane-PDMS segmented copolymers,(11) and their results are in good agreement with ours.

A deconvolution of the ESCA data has been carried out to recover depth profiles for the hard segment in each copolymer.(31) The profiles reveal some features which are not obvious in the ESCA data derived from direct measurement of C/N peak intensities (Figures 2 and 3 discussed above). The deconvolution method takes into account the differing C atom densities in the PDMS (25.7M) and hard segment (ca. 54M) phases and utilizes an optimal fitting of the data.(9) The results of this analysis are seen qualitatively in Figure 4, which shows concentration depth profiles for hard segment in PDMS2.4K-IP-B2. The solid squares represent ESCA data derived from direct measurement of C/N peak intensities; curves A and B result from a continuous and a discrete model, respectively.

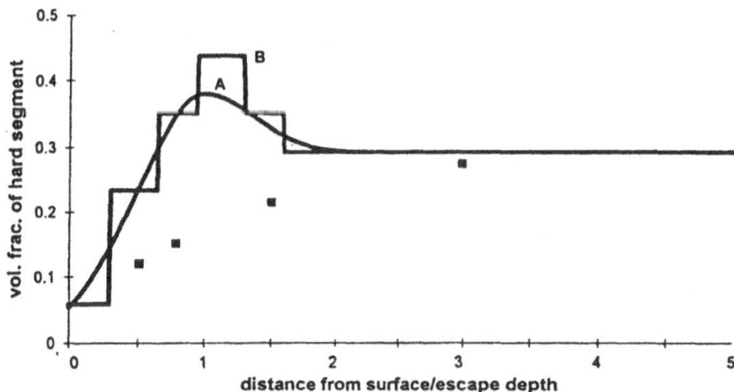

Figure 4. Concentration depth profiles for the hard segment in as-cast films of copolymer PDMS2.4K-IP-B2. Profiles: A, constructed based on a continuous model; B, constructed based on a discrete model; and (■) data measured by ESCA.

Both the discrete and continuous models indicate a ***maximum*** in the hard segment profile in contrast to the monotonously increasing profile suggested by N/C peak-ratio data. The peak ratio ESCA data also give a lower hard segment content in the surface region than the models. These differences emphasize the fact that for polymers with partially or completely phase separated surface regions, the convoluted nature of the photoelectron intensity signals prevents a direct measurement of concentration depth profiles. Profiles produced by both models, when inserted into the convolution equations, lead to photoelectron intensity ratios in very good agreement with experimental data.

The usefulness of the continuous model may be seen in an analysis of surface profiles for samples with similar *bulk* PDMS concentrations. For such samples, the length ratios of PDMS segment to hard segment are similar. Copolymers PDMS2.4K-IP-B0 and PDMS10K-IP-B2 have similar bulk concentrations of PDMS (90.3% and 91.1%). Results based on ESCA peak-ratio data for these two copolymers are compared in Figure 5. Peak ratio data do not allow differentiation in surface composition within error limits although a slight trend is observed. The continuous deconvolution model, on the other hand, resolves differences in these two concentration depth profiles, as illustrated in Figure 6.

In addition to entropically driven phase separation for longer soft segments, phase separation is driven by the enthalpic effect of stronger hydrogen bonding of larger hard segments. These two effects may act in a synergistic fashion, but we believe the enthalpic effect is of greater importance.

Annealing effects: The surface compositions of PDMS-PU copolymer samples changed after a short annealing. The extent of this change depends on the length of the soft segment. For samples with the shortest PDMS segments (MW_{soft} = 1000), annealing does not have any detectable effect on the surface concentration of PDMS. For samples with PDMS segments of moderate length (MW_{soft} = 2.4K), an increase in PDMS surface concentration was observed in the sample with the longest hard segments (Figure 7). If the PDMS segments are long (MW_{soft} = 10K), significant changes in the PDMS surface concentrations are observed regardless the composition of the hard segments. For example, the concentration of PDMS in the as-cast films of PDMS10K-IP-B0 and PDMS10K-IP-B2 measured at the lowest take off angle are 98.9 and 97.7 %, respectively, after annealing, the corresponding values are 99.9 and 99.4 %. Figure 7 reveals increases in the thickness of PDMS and hard-block enriched phases after annealing. The thickness of the topmost PDMS layer with less than 5% hard segment content increases from zero to 0.2 λ. (The escape depth λ is approximately 3.4 nm). By a comparison of peak widths, the thickness of the layer enriched with hard segments, which is responsible for the maximum in the profile, doubles from ~λ to ~2λ. These observations again illustrate the positive effects of annealing in enhancing the surface concentration of PDMS.

The dominance of PDMS segments in the surface region is even more pronounced in copolymer PDMS27K-IP-B2. No hard segments were detected in measurements taken at the lowest take-off

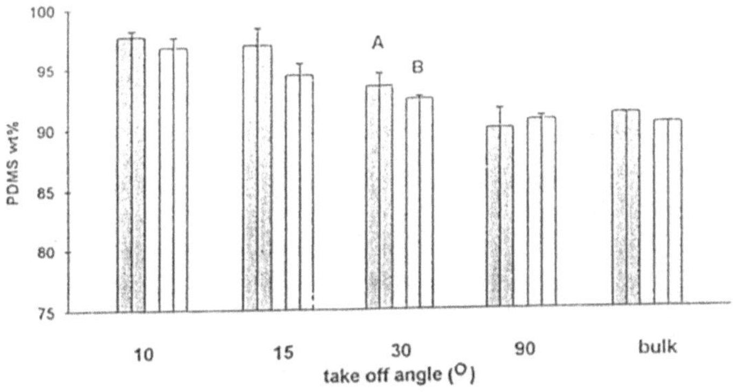

Figure 5. PDMS concentration in the surface region measured by ESCA for two copolymers of similar bulk PDMS concentration. The distance between the horizontal bar and the top of the column represents one standard deviation. Data series A, PDMS10K-IP-B2; and B, PDMS2.4K-IP-B0.

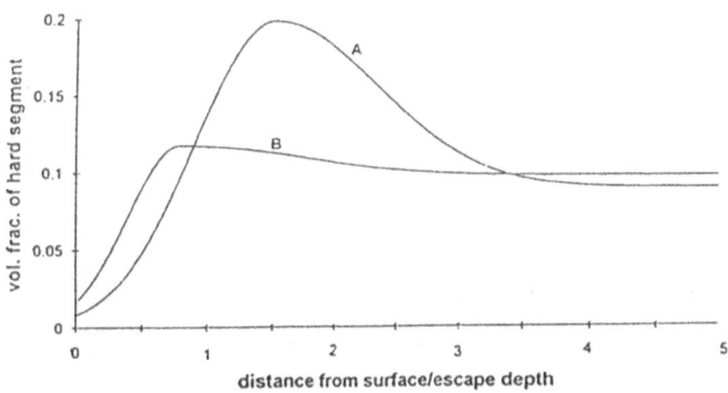

Figure 6. Hard segment concentration in the surface region for two copolymers of similar bulk PDMS concentration. The profiles are constructed based on a continuous model. Data series: A, PDMS10K-IP-B2; and B, PDMS2.4K-IP-B0.

angle, and the topmost surface regions of both as-cast and annealed films contain only PDMS component.

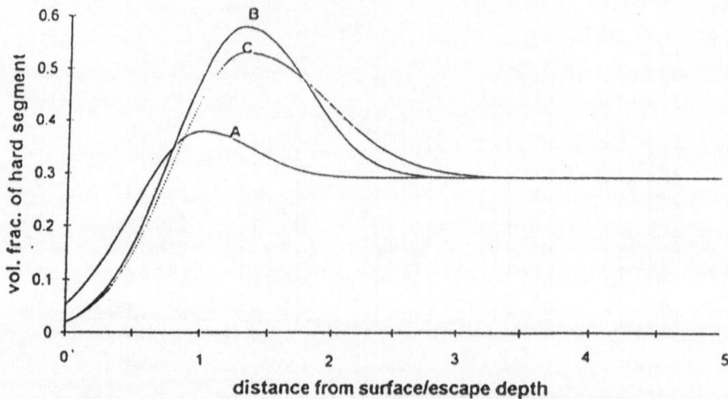

Figure 7. Concentration depth profiles of the hard segments in the surface region for the copolymer PDMS2.4K-IP-B2 under different thermal treatments. The profiles are constructed based on a continuous model. Curves: A, as-cast; B, annealed at 120°C for 15 minutes; and C, for two hours.

Effects of Structure and Molecular Weight on the Distribution of the Hard Segment

As mentioned above, the difference in surface energies is the underlying driving force for the enrichment of PDMS segments at the surface. In addition, long soft segments amplify entropically driven phase separation, and increased hydrogen bonding in hard segments augments the enthalpic driving force for phase separation. The interplay of these factors under different situations is discussed below. We define $v_N(x)$ as:

$$v_N(x) = \{v(x)/[1-v(x)]\} \; / \; \{v(\infty)/[1-v(\infty)]\} \qquad (3)$$

Equation 3 represents the normalization of the volume fraction of the hard segment at the surface region, $v(x)$, with respect to the bulk value, $v(\infty)$. That is, $v_N(x)$ is the ratio of the volume fraction of hard/soft segments at the surface divided by the same ratio for the bulk. The quantity $v_N(x)$ can be visualized in the following way. Suppose a unit volume of soft segments with the entrapped hard segments moves from the bulk (location ∞) to the surface (location x), and during the movement the hard segments gradually drop out this volume due to the progress of phase separation. Then $v_N(x)$ represents the fraction of initial hard segments remaining at the end of the movement. Using this parameter, materials can be compared based on their "effectiveness" in separating the soft and the hard segments.

In Figure 8, v_N depth profiles in the region less than one λ (ca. 3.4 nm) from the surface for copolymers PDMS10K-IP-B0 (A, C) and PDMS10K-IP-B2 (B, D) are shown. The two solid lines, representing concentration profiles in the as-cast films, indicate that the incorporation of chain extender promotes surface phase separation at ambient temperature presumably due to the increased interactions among the hard segments. This advantage is lost in annealing as both after-annealing profiles (dashed lines) are similar and shift to lower hard segment concentrations. Since the soft segments are the same and relatively large compared to the hard segments, the occurrence of the similar profiles in the surface region after short annealing is driven by enthalpy. Thus, the rate of disadvantaged phase separation for PDMS10K-IP-B0 catches up to that of PDMS10K-IP-B2, as an increase in temperature enhances the higher-energy-barrier phase separation process for PDMS10K-IP-B0.

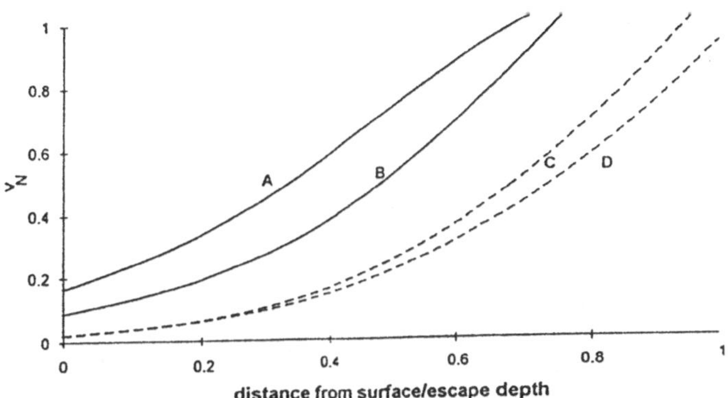

Figure 8. v_N as a function of depth in the topmost surface region for two copolymers. Curves: A, PDMS10K-IP-B0 as-cast, B, PDMS10K-IP-B2 as-cast; C, PDMS10K-IP-B0 annealed at 120°C for 15 minutes; and D, PDMS10K-IP-B2 annealed at 120°C for 15 minutes.

In Figure 9, v_N profiles for copolymers PDMS2.4K-IP-B2, PDMS10K-IP-B2, and PDMS27K-IP-B2 are plotted. Cast at ambient temperature, the hard segment concentration in the surface region decreases with increase in the average molecular weight of the soft segments. After annealing, the curves shift to lower hard segment concentrations while the relative positions to one another are maintained. The inverse correlation between surface hard segment concentration and bulk soft segment molecular weight as well as the insensitivity of the shape of the curve to temperature suggests entropic origins for the observed differences.

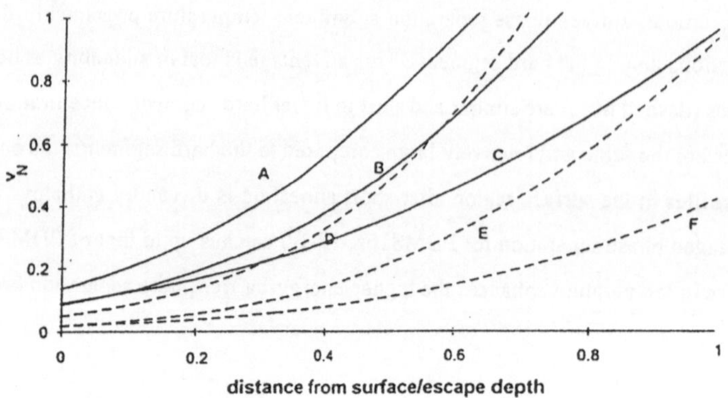

Figure 9. v_N as a function of depth in the topmost surface region for three copolymers. Curves: A, PDMS2.4K-IP-B2 as-cast; B, PDMS10K-IP-B2 as-cast; C, PDMS27K-IP-B2 as-cast; D, PDMS2.4K-IP-B2 annealed at 120°C for 15 minutes; E, PDMS10K-IP-B2 annealed at 120°C for 15 minutes; and F, PDMS27K-IP-B2 annealed at 120°C for 15 minutes.

Properties of Blends

Three blends of copolymers PDMS27K-IP-B2 and PTMO2K-IP-B2 were prepared. The compositions were 1.6, 2.5, and 6.0 % of PDMS27K-IP-B2 by weight.

Dynamic mechanical properties of the blends: Storage modulus as a function of blend composition is shown in Figure 10. The general features of all traces are the same reflecting the dominance of the major component PTMO2K-IP-B2 in the mechanical properties. There is no discernible change in the curves at -120°C, the glass transition temperature of the minor component (PDMS). The gradual decline between -80 and 0 °C reflects the effects of the glass transition of the PTMO segments moderated by the existence of ordered regions of both the hard segments and the PTMO segments. The melting transition of the PDMS segments at about -46°C caused no measurable changes. The steeper decline between 0 and 30 °C indicates the melting transition of the PTMO segments. The effects of the melting of the hard segments were indicated by the incipient steep decline at about 70°C. The moduli reached the lower limit of the instrument, ~1 MPa, at that temperature.

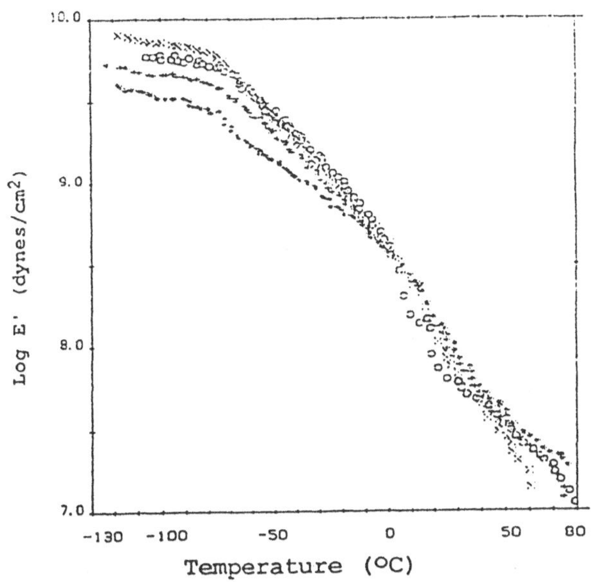

Figure 10. Storage moduli as functions of temperature for the blends. Traces: x , PDMS27K-IP-B2; o , blend of PDMS27K-IP-B2 and 1.6 wt% PDMS27K-IP-B2; * , 2.5 wt%; and + , 6.0 wt%.

Surface properties: Contact angle measurements indicated enrichment of the siloxane containing copolymer at the surface (Table 3). Contact angles of water and methylene iodide on copolymer PTMO2K-IP-B2 are values expected for a polyetherurethane, while contact angles of the same

Table 3 Contact angles and surface tensions of the blends

	PTMO2K-IP-B2	PTMO2K-IP-B2/ *1.6%* PDMS27K-IP-B2	PTMO2K-IP-B2/ *2.5%* PDMS27K-IP-B2	PTMO2K-IP-B2/*6%* PDMS27K-IP-B2
θ_{H2O}(degree)	79.9	100.5	100.9	102.3
θ_{CH2I2}(degree)	40.9	80.6	80.5	83.1
γ^h(dyne/cm)	12.3	8.8	8.5	8.4
γ^d(dyne/cm)	23.2	10.8	10.9	10.2
γ_s(dyne/cm)	35.5	19.6	19.4	18.6

Contact angle measured by goniometry.
Surface tension calculated by harmonic-mean method.

liquids on the blends are typical for siloxanes. Contact angles on copolymer PDMS27K-IP-B2 were not measured due to the extreme softness of the film. The methylene iodide drops tend to distort the topography of the film.(32) Contact angles on a copolymer of similar structure but shorter soft segments, PDMS10K-IP-B2, were measured. The results were θ_{H2O}=103°, and θ_{CH2I2}=70°.

The surface tensions of the copolymer and blends were calculated from the cosine of the contact angles of the two probing liquids using the harmonic-mean method.(30)

$$1+\cos\theta=4(\gamma_s^d\gamma_l^d/(\gamma_s^d+\gamma_l^d)+\gamma_s^p\gamma_l^p/(\gamma_s^p+\gamma_l^p))/\gamma_l \qquad (4)$$

$$\gamma_l=\gamma_l^d+\gamma_l^h \qquad (5)$$

$$\gamma_s=\gamma_s^d+\gamma_s^h \qquad (6)$$

In equations 4-6, γ_l is the surface tension of the liquid, γ_l^d denotes the component due to dispersion forces, and γ_l^h the component due to hydrogen bonding and dipole-dipole interactions. γ_s is the surface tension of the polymer, γ_s^d its dispersion force component, and γ_s^h the hydrogen bonding and dipole-dipole interaction component.

Surface tensions of water and methylene iodide measured in our laboratory were 71.8 and 42.8 dyne/cm, respectively. The interfacial tension between these two liquids was 33.2 dyne/cm. The values for surface tension of methylene iodide and the interfacial tension differ from frequently cited literature values, viz. 50.8 and 41.6 dyne/cm, respectively.(33) The difficulties in determining the surface tension of methylene iodide consistently has been alluded to in the literature.(33,34) On the other hand, the DCA instrument we used to determine the surface tension of the liquids is quite reliable. For example, the measured surface tension for toluene, 28.2 dyne/cm, and the interfacial tension between toluene and water, 34.6 dyne/cm, are in good agreement with literature values, 28.5 and 36.1 dyne/cm.(30). With this demonstrated accuracy, the values obtained in our laboratory were used in the calculations described by equations 4-6.

Literature values for γ_l^d and γ_l^h for water are 22.1 and 50.7 dyne/cm.(33) In our calculation, we split the measured surface tension of water based on the ratio γ_l^d / γ_l^h = 22.1 / 50.7. The two components for the surface tension of methylene iodide were obtained by solving the harmonic-mean equation of interfacial tension, Eq. 7. Thus, for water, γ_l^d is 21.8 dyne/cm, and γ_l^h 50.0 dyne/cm, while for methylene iodide, γ_l^d is 34.7 dyne/cm, and γ_l^h 8.1 dyne/cm.

$$\gamma_{12}=\gamma_1+\gamma_2-(4\gamma_1^d\gamma_2^d)/(\gamma_1^d+\gamma_2^d)-(4\gamma_1^p\gamma_2^p)/(\gamma_1^p+\gamma_2^p) \tag{7}$$

Surface tension of the copolymer PTMO2K-IP-B2 and blends were calculated using equations 4-6. The resulting values are listed in Table 3. The inclusion of copolymer PDMS27K-IP-B2 at a level as low as 1.6% by weight reduced the surface tension of the blend from a level typical for polar materials (~35 dyne/cm) to a level common to siloxane homopolymers (~20 dyne/cm). A detailed study on the siloxane concentration depth profile using angle dependent ESCA is in progress.

CONCLUSIONS

A hard-segment first two-step polymerization was used to prepare PDMS-urea-urethane copolymers with 1,4-benzenedimethanol as the chain extender. The chain extended copolymers were found to be superior to non-chain extended copolymers in thermal and mechanical properties due to the reinforcement effect of additional hydrogen bonding interactions between urea and urethane linkages. Details of thermal and mechanical behavior were consistent with those expected of a phase separated morphology.

ESCA analysis of films cast from THF surface show that surface concentration of PDMS increases with PDMS segment length. Annealing as-cast samples increases the PDMS surface concentration. ESCA data were deconvoluted to give depth profiles which typically revealed a hard segment maximum beneath the PDMS enriched surface layer. Annealing increases the thickness of both regions. Both the length of the soft segments and the interactions among the hard segments affect the phase separation in the surface region. Long soft segments amplify entropically driven phase separation, while increased hydrogen bonding in hard segments augments the enthalpic driving force for phase separation.

Blends of PDMS27K-IP-B2 and a poly(ether-urethane) with the former as the minor component (1.6, 2.5, and 6.0 wt%) were prepared. The identical hard segments provide a driving force for compatibility between the two components and the disparity in the soft segment surface tensions ensures a surface enriched in siloxanes. Inclusion of poly(siloxane-urea-urethane) even at 1.6%

by weight reduces the surface tension of the blend from a level typical for polar materials (~ 35 dyne/cm) to a level common to siloxane homopolymers (~20 dyne/cm). The poly(ether-urethane) controls the mechanical properties.

Acknowledgment: This research was supported in part by the Office of Naval Research. We are grateful to Dr. İ. Yilgör of Goldschmidt Chemical Corporation for a gift of the aminopropyl terminated polydimethylsiloxanes.

References:

1 Brady RF, Griffith JR, Love KS, Field DE (1987) J Coatings Tech 59:113
2 Tyagi D, Yilgör İ, McGrath JE, Wilkes G L (1984) Polymer 25:1807
3 Yu XH, Nagarajan MR, Grasel TG, Gibson PE, Cooper SL (1985) J Polym Sci Polym Phys Ed 23:2319
4 Oishi Y, Kakimoto M, Imai Y (1987) J Polym Sci, Part A: Polym Chem 25:2185
5 Chen L, Yu XH (1991) Functional Polymer 4: 19
6 Yilgör İ, McGrath JE (1988) Adv Polym Sci 86:1
7 Harrell LL Jr (1969) Macromolecules 2: 607
8 Ho T, Wynne KJ, Nissan RA (1993) Macromolecules 26:7029
9 Chen X, Gardella JA Jr, Ho T, Wynne, KJ in press
10 Clark DT (1977) Advances in Polymer Sciences 24:126
11 Tezkuka Y, Kazuma H, Imai K (1991) J Chem Soc Faraday Trans 87:147
12 Shibayama M, Suetsugu M, Sakurai S, Yamamoto T, Nomura S (1991) Macromolecules 24:6254
13 Benrashid R, Nelson GL, Linn JH, Hanley KH, Wade WR (1993) J Appl Poly Sci 49:523
14 Chen X, Gardella JA Jr (1992) Polym Prepr Am Chem Soc Div Polym Chem 33(2):312
15 Chen X, Gardella JA Jr, Kumler PL in press
16 Mittlefehldt ER, Gardella JA Jr (1989) Appl Spectrosc 43:1172
17 Chen X, Lee HF, Gardella J A Jr (1993) Macromolecules 26:4601
18 Pertsin AJ, Gorelova MM, Levin V Yu, Makarova LI (1992) J Appl Poly Sci 45:1195
19 Chen X, Gardella JA Jr, Kumler PL (1992) Mcromolecules 25:6621
20 Chen X, Gardella JA Jr, Kumler PL (1992) Macromolecules 25:6631
21 Thomas, H. R.; O'Malley, J. J. *Macromolecules*, 1979, *12*, 323.
22 Coulon G, Russell TR, Deline VR, Green PF (1989) Macromolecules 22:2581
23 Bass RL, Porter MR (1963) In: Moilliet JL (ed) Water Proofing and Water Repellency, Elsevier, Amsterdam London New York

24 Ho T, Wynne KJ in press
25 Veith CA, Cohen RE (1991) Makromol Chem Macromol Symp 42/43: 241
26 We used programs included in Mathematica to perform this operation; Wolfram S (1991) Mathematica: a system for doing mathematics by computer, 2nd ed, Addison-Wesley Publishing Company, Inc, Redwood City, California
27 Surivet F, Lam TM, Pascault JP (1991) J Polym Sci Polym Chem Ed 29:1977
28 (a) Kajiyama M, Kakimoto M, Imai Y (1990) Macromolecules 23:1244 (b) The authors unpublished results
29 Cooper SL, Tobolsky AV (1966) J Appl Polym Sci 10:1837
30 Wu S (1982) Polymer Interface and Adhesion, Marcel Dekker, New York
31 Previous efforts in de-convoluting ESCA data include (a)Iwasaki H, Nishitani R, Nakamura S (1978) Jpn J Appl Phys 17:1519 (b) Pijolat M, Hollinger G (1981) Surf Sci 105:114 (c) Nefedov VI, Baschenko OA (1988) J Electron Spectrosc Relat Phenom 47:1 (d) Tyler BJ, Castner DG, Ratner BD (1989) Surf Interface Anal 14:443 (e) Jisl R (1990) Surf Interface Anal 15:719 (f) Holloway PH, Bussing TD (1992) Surf Interface Anal 18: 251
32 Phillips MC, Riddiford AC (1972) J Colloid Interface Sci 41: 77
33 Owens DK, Wendt RC (1969) J Appl Polym Sci 13:174
34 Kobayashi H, Owen MJ (1990) Macromolecules 23:4929

Some Recent Developments in the Aminoxyl Radical Trapping Technique

S.E. Bottle, W.K. Busfield, I.D. Grice, K. Heiland, I.D. Jenkins*, W. Meutermans, M. Monteiro

Faculty of Science and Technology,
Griffith University, Nathan, Brisbane,
Queensland, 4111, Australia

Abstract: The radical trapping technique employing the stable aminoxyl (nitroxide) **1** as a radical scavenger, has been used to study (a) the initiation stage in the mechanism of formation of alternating copolymers (b) the reaction of diphenylphosphinoyl and dimethyl phosphoryl radicals with monomers (c) the unusual polymerisation characteristics of maleates and fumarates.

Introduction: The initiation process in free radical polymerisation plays an important role in determining polymer properties such as stability, molecular weight distribution and composition.[1] In particular, initiator-derived end groups can have profound effects on the stability of the polymer towards thermal or photochemical degradation. The early stages of free radical polymerisation involve reaction between an initiating radical (often an oxygen-centred radical) and a monomer, to produce a carbon-centred radical, which then adds to a second molecule of monomer to produce a 'second generation' carbon-centred radical which then adds to a third monomer molecule and so on. Through the pioneering studies of Rizzardo and Solomon[2] of the CSIRO, it has been possible to identify the structures of the first-formed ('first generation') carbon-centred radicals. These studies have revealed a rich variety of first-generation radicals and, by implication, a range of possible end groups in the resulting polymers. The work of the CSIRO group has recently been reviewed.[1]

About ten years ago, Rizzardo and Solomon initiated a collaboration between the CSIRO and Griffith University. Employing the radical trapping technique, the Griffith University group have carried out a number of studies complementary to those of the CSIRO. For example, we have investigated the mechanism of initiation in the radical polymerisation of certain 'difunctional' monomers such as allyl methacrylate[3], allyl acrylate and diallyl ether[4]. We have also carried out investigations with less common monomers and monomers that exhibit unusual behaviour such as N-vinylcarbazole and N-vinylpyrrolidone[5], acetylenes[6], diethyl maleate and diethyl fumarate[7], and we have carried out a comparison of initiation mechanisms for polymerisations initiated by primary, secondary and tertiary alkoxyl radicals.[8] Some of our more fundamental studies have included the stereochemistry of addition of *tert*-butoxyl[9] and benzoyloxyl[10] radicals to alkenes, the structure of the aminoxyl **1** and an alkoxy amine product **2**[11], the unusual stereomutation process undergone by alkoxy amines **2**[12], and the reaction of organophosphorus radicals with alkenes in the presence of the aminoxyl **1**.[13] The

K.P. Ghiggino (Ed.)
Progress in Pacific Polymer Science 3

use of **1** as a probe to study the mechanism of formation of alternating copolymers has also been investigated.[14]

The radical trapping technique relies on the almost diffusion-controlled[15,16] trapping of carbon-centred radicals by stable aminoxyls (nitroxides) such as **1** (1, 1, 3, 3-tetramethyl-2,3-dihydro-IH-isoindol-2-yloxyl) to form stable alkoxyamine products **2** (Scheme 1).

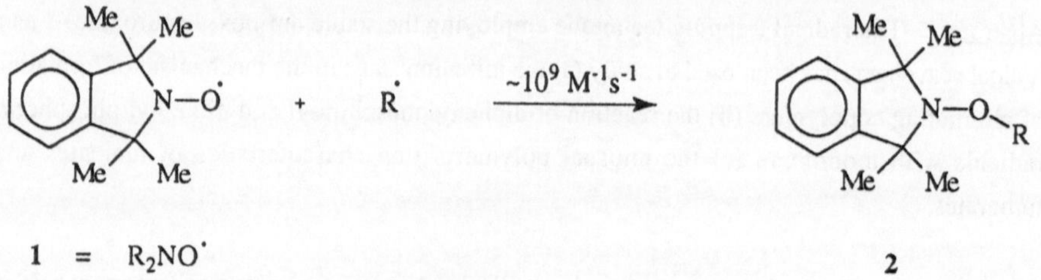

Me Me N—O˙ Me Me + R˙ $\xrightarrow{\sim 10^9\ M^{-1}s^{-1}}$ Me Me N—O—R Me Me

1 = $R_2NO^\bullet$ **2**

Scheme 1

Compound **1** is a readily prepared,[17] stable, yellow crystalline solid with a flat structure.[11] It is readily soluble in most monomers and is one of the most potent chain terminators known. Thus, if a monomer such as styrene containing **1** and the initiator di-*tert*-butyl peroxalate (a source of *tert*-butoxyl radicals) is heated, no polymer is formed. The major products isolated are unreacted styrene and **4** (Scheme 2).

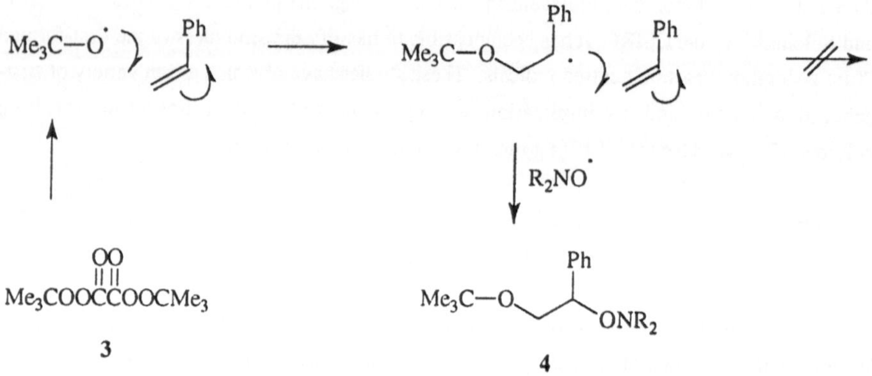

$Me_3C—O^\bullet$ + Ph(styrene) → $Me_3C—O—CH_2—CH^\bullet(Ph)$ + Ph(styrene) ⇸

↓ $R_2NO^\bullet$

$Me_3COOC(O)C(O)OOCMe_3$ **3** ; $Me_3C—O—CH_2—CH(Ph)—ONR_2$ **4**

Scheme 2

Two minor products are also formed, **5** & **6**, both resulting from β-fragmentation of *tert*-butoxyl radicals to give methyl radicals (Scheme 3).[18] The products can all be isolated by reverse-phase HPLC and their structures determined by ^1H & ^{13}C NMR.

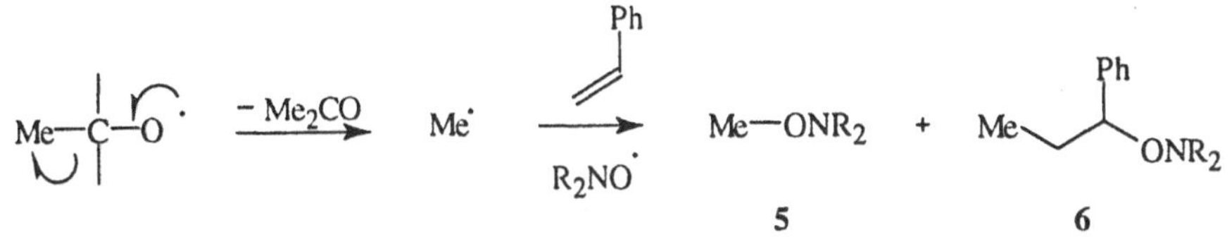

$$\mathrm{Me{-}C(Me_2){-}O^{\bullet}} \xrightarrow{-\,\mathrm{Me_2CO}} \mathrm{Me^{\bullet}} \xrightarrow[\mathrm{R_2NO^{\bullet}}]{\mathrm{CH_2{=}CHPh}} \underset{\mathbf{5}}{\mathrm{Me{-}ONR_2}} + \underset{\mathbf{6}}{\mathrm{MeCH_2CH(Ph)ONR_2}}$$

Scheme 3

Copolymerisation Studies: Alternating copolymers are formed when a monomer unit M_1 adds to a monomer unit M_2 in a srictly regular manner to form extended -$M_1M_2M_1M_2$- sequences, thus instilling a much higher degree of molecular order than exists in random copolymers. This feature can strongly influence the macroscopic properties.[19]

A high degree of alternation in copolymers is achieved when the two monomers involved consist of an electron acceptor and an electron donor. In these cases, there has been considerable speculation as to whether the alternating copolymer is a result of the formation of a charge-transfer complex. The idea that alternation is related to the ability of the two monomers to form a charge-transfer complex was first suggested by Bartlett and Nozaki[20], nearly fifty years ago. However, the so-called 'complex mechanism', in which the charge-transfer complex itself acts as a new kind of monomer capable of reacting with the growing polymer chain, appears to be a more recent invention that has received considerable criticism. A mechanism involving separate addition of donor and acceptor molecules ('free monomer' mechanism) appears to be much more likely.[21]

We thought that the radical trapping technique might provide a simple means of distinguishing between the 'complex' mechanism and the 'free monomer' mechanism. For example, if the charge-transfer complex is much more reactive towards radical addition than either of the separate monomers, then it might be expected that a dimeric species of the type In-M_1M_2-ONR_2 (where In is the initiating radical, M_1 and M_2 are the two monomers that undergo copolymerisation, and R_2NO is the aminoxyl radical trap) should be the major product of the reaction.

Treatment of a mixture of styrene and acrylonitrile (1:1 Molar ratio) with di-*tert*-butyl peroxalate (2.4×10^{-5}M) and the aminoxyl **1** (0.04M) at 60°C for 70 mins. gave the same products as those expected from the two monomers reacting separately. The relative yields of products were **4**(79.1%), **7**(15.9%), **8**(3.0%), **5**(1.5%) and **6**(0.5%). Other products accounted for less than 0.1%.

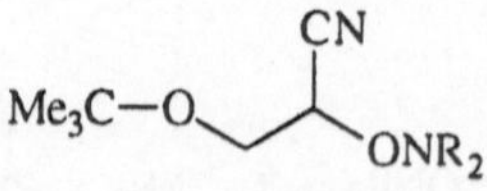

Although the rate of trapping by **1** is very fast, we considered that it might be possible to detect 'dimeric' products if the concentration of the aminoxyl was kept very low. In order to maintain a low and constant concentration of **1**, a syringe pump was employed. A solution of **1** in the two monomers was delivered to the reaction vessel at such a rate that the aminoxyl concentration was maintained at approximately 5 x 10^{-3} M during the course of the reaction (at 60°C). Under these conditions, small amounts of dimeric products **9** and **10** were observed. Both **9**(~2%) and **10**(<1%) were formed as approximately 1:1 diastereomeric mixtures. The question then arose as to whether **9** was formed via a charge-transfer complex [path (a)] or by the free monomer mechanism [path (b)] as outlined in Scheme 4 (B = *t*-butoxyl, S = styrene, A = acrylonitrile, T = aminoxyl 1).

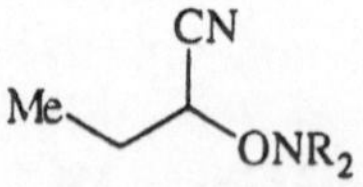

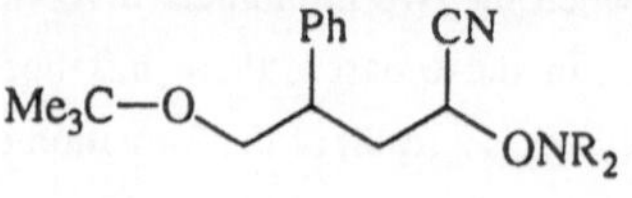

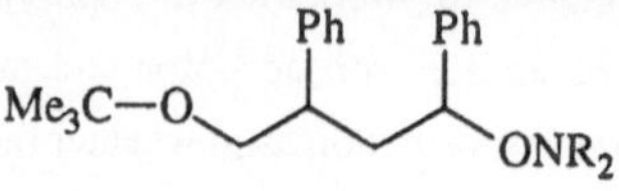

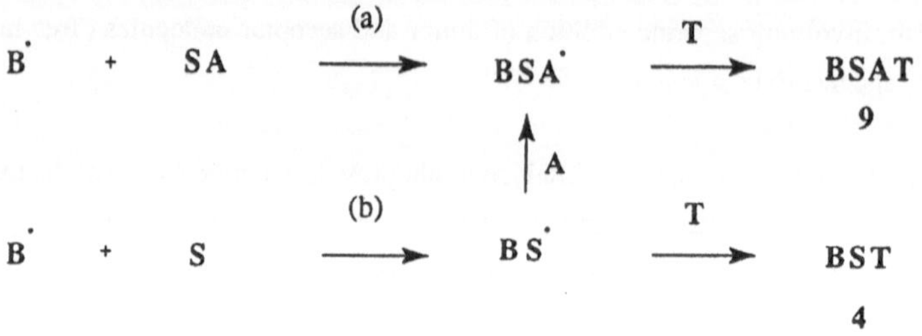

$$B^{\cdot} + SA \xrightarrow{(a)} BSA^{\cdot} \xrightarrow{T} \underset{\mathbf{9}}{BSAT}$$

$$BS^{\cdot} \xrightarrow{A} BSA^{\cdot}$$

$$B^{\cdot} + S \xrightarrow{(b)} BS^{\cdot} \xrightarrow{T} \underset{\mathbf{4}}{BST}$$

Scheme 4

It is clear from Scheme 4 that if **9** (BSAT) is formed exclusively via path (a), then the ratio of BSAT to BST should be independent of the concentration of T (aminoxyl). Conversely, if path (b) were exclusive, the product ratio BSAT : BST will be dependent on [T] as the reaction of BS· with A and with T will be competitive. The effect of a six-fold change in [T] is shown in Fig. 1. Clearly, there is a strong dependence on aminoxyl concentration showing that a charge-transfer complex between styrene and acrylonitrile plays little or no part in the formation of the dimeric product **9**.

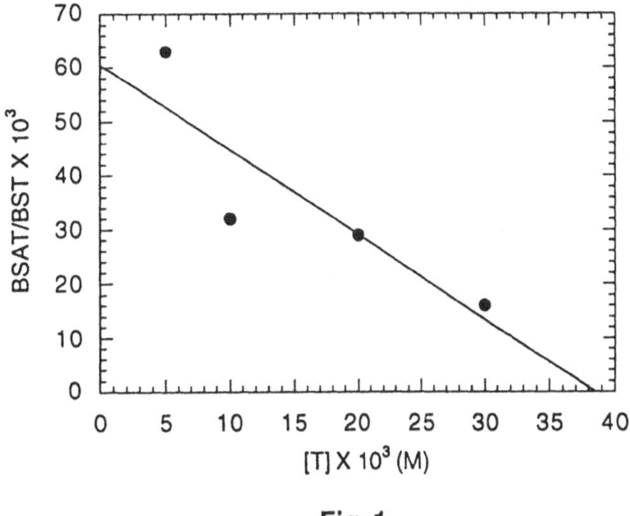

Fig 1

Analogous results were obtained with vinyl acetate/acrylonitrile[14] and with ethyl vinyl ether/acrylonitrile, however with these systems, the relative yields of dimeric products were much higher (20 - 30%).

$Me_3C{-}O{-}CH_2{-}CH(ONR_2){-}OAc$ **11** $Me_3C{-}O{-}CH_2{-}CH(ONR_2){-}CN$ **12**

$Me{-}ONR_2$ **5** $Me{-}CH_2{-}CH(ONR_2){-}CN$ **13** $CH_2{=}CH{-}O{-}C(=O){-}CH_2{-}ONR_2$ **14**

In the vinyl acetate/acrylonitrile system, we have also examined the effect of added Lewis acids as these are known to enhance alternation in alternating copolymerisation.[19] In the presence of zinc chloride, no dimeric products were observed. Nor were the normal addition products **11** and **12** observed. The major products formed were **5** (55%), **13** (20%) and **14** (25%). Similar results were obtained with ferric chloride. Thus with 4% $FeCl_3$, the ratio of **5** : **13** : **14** was 84 : 10 : 6; with 60% $FeCl_3$, the ratio changed to 25 : 3 : 72 respectively. Apparently Lewis acids catalyse the β-fragmentation of *tert*-butoxyl radicals (or of di-*tert*-butyl peroxalate) and/or slow the rate of addition of the (electrophilic) *tert*-butoxyl radicals to both vinyl acetate and acrylonitrile. The product **14** is thought to arise via hydrogen-abstraction from vinyl acetate by the aminoxyl **1** complexed to the Lewis acid. This product was also formed in the absence of di-*tert*-butyl peroxalate.

Studies with Organophosphorus Radicals: The success of the radical trapping technique has been due in part to the very efficient and selective scavenging of carbon-centred radicals in the presence of oxygen-centred radicals used as initiators. It was not known if the technique could also be used if the initiators were phosphorus-centred radicals. Indeed, there has been one report[22] of the trapping of the diphenylphosphinoyl radical **16** by the aminoxyl TEMPO, suggesting that the radical trapping technique might not be useful for the study of the addition of phosphorus-centred radicals to alkenes. Part of our interest in phosphorus-centred radicals stems from the claim that acylphosphine oxides are effective photoinitiators for the free radical polymerisation of olefins and might be useful in the production of surface coatings.[23]

Diphenylphosphinoyl radicals **16** were generated by hydrogen-abstraction from diphenylphosphine oxide **15** by *tert*-butoxyl radicals. Thus, heating a mixture of di-*tert*-butyl peroxalate (5×10^{-3} M), the aminoxyl **1** (1.2×10^{-2} M) and diphenylphosphine oxide (1×10^{-2} M) in acrylonitrile at 60°C for 70 mins. gave a mixture of the addition product **17** and the phosphinate ester **18** in the ratio of 1 : 0.28 (Scheme 5). Clearly, the addition of diphenylphosphinoyl radicals to acrylonitrile is competitive with aminoxyl trapping, and as expected, increasing the aminoxyl concentration (to 5×10^{-2} M) resulted in an increase in the ratio of **17** : **18** (to 1.0 : 0.78).

Analogous results were obtained when diphenylphosphinoyl radicals were generated by photolysis of the acylphosphine oxide **19**. In addition to the products **17** and **18**, the ester **20** was also formed as expected[22] (Scheme 6).

That diphenylphosphinoyl radicals exhibit nucleophilic behaviour was demonstrated by employing vinyl acetate instead of acrylonitrile. Thus, treatment of diphenylphosphine oxide (1×10^{-2} M) with di-*tert*-butyl peroxalate (5×10^{-3}M) and the aminoxyl **1** (5×10^{-2} M) in vinyl acetate at 60°C for 70 mins. gave a mixture of the addition products **21** and **22** as well as the phosphinate ester **18** in a ratio of 1.02 : 0.33 : 1.0 respectively. The much lower ratio of addition product **22** to phosphorus radical trapping product **18**, compared with the corresponding ratio of **17** : **18** in the case of acrylonitrile, shows that diphenylphosphinoyl radicals are nucleophilic in nature.

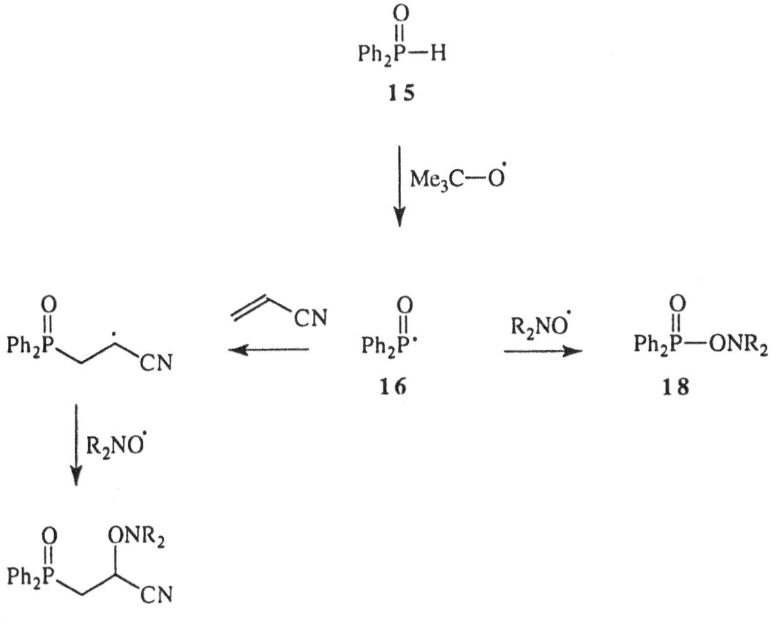

Scheme 5

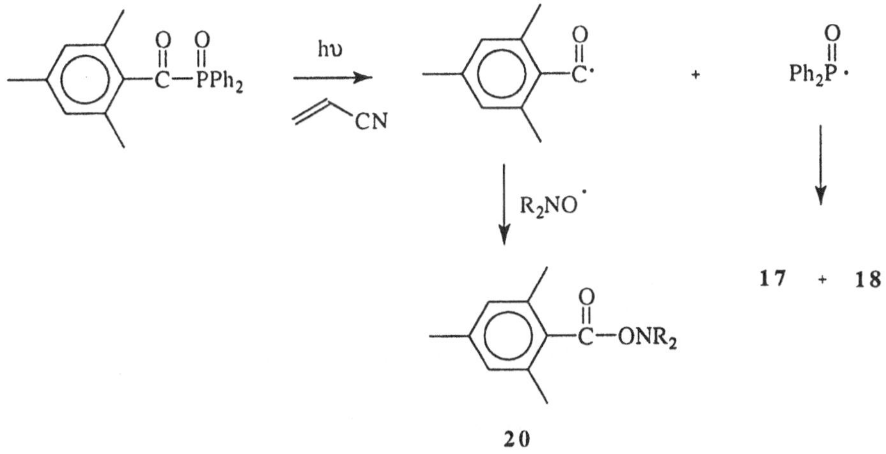

Scheme 6

The relative reactivity of diphenylphosphinoyl radicals towards acrylonitrile and vinyl acetate was also determined in a competition experiment using a mixture of the two monomers. Thus,

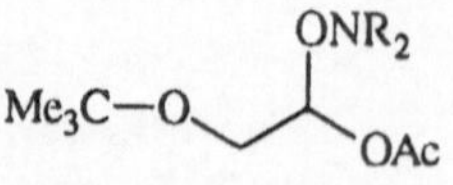

21 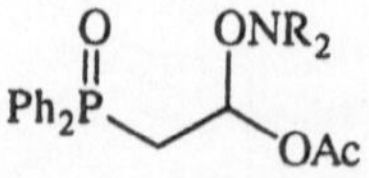 **22**

when a 2 : 1 molar mixture of vinyl acetate and acrylonitrile was employed and in the presence of a five-fold excess of aminoxyl **1** (relative to diphenylphosphine oxide), the products **17** and **22** were formed in the ratio of 6 : 1, indicating that acrylonitrile is 12 times more reactive than vinyl acetate towards addition of diphenylphosphinoyl radicals.

Under similar reaction conditions, dimethoxyphosphoryl radicals were found to be much less nucleophilic than diphenylphosphinoyl radicals. Thus, a 2 :1 molar mixture of vinyl acetate and acrylonitrile gave approximately equal amounts of the two addition products **23** and **24** indicating that acrylonitrile is only twice as reactive as vinyl acetate toward addition of dimethoxyphosphoryl radicals. Surprisingly, the ratio of *tert*-butoxyl radical addition

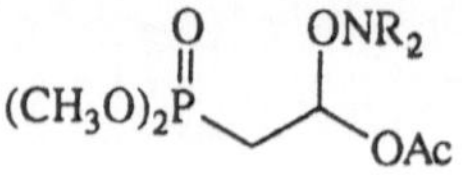

23 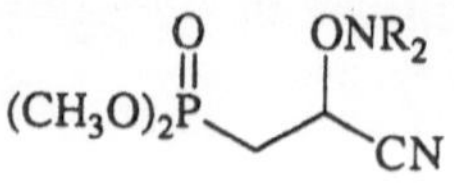 **24**

products (**25** and **26**) was also approximately equal, and yet tert-butoxyl radicals are generally considered to be electrophilic.

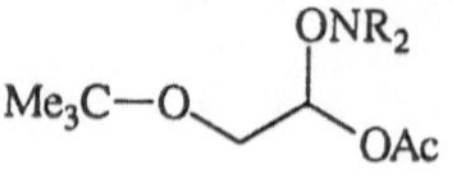

25 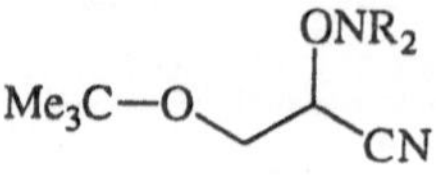 **26**

We suggest that the explanation for this apparent anomaly is that the transition state for attack by *tert*-butoxyl radicals is more advanced than that for attack by less sterically hindered radicals and that as a result, the product radical stability plays a more significant role than it usually does in determining radical reactivity.[24] A recent paper by the CSIRO group[25] provides some support for this. Thus, while the *tert*-butoxyl radical was found to behave as an electrophilic species towards a series of substituted styrenes, *para*-substituents generally resulted in higher reactivities than those expected on the basis of the results for meta-substituents, suggesting a resonance contribution to the rate of addition.

We have also examined the addition of phosphorus-centred radicals to acetylenes. Thus, treatment of dimethylphosphite (2 mmol) with di-*tert*-butyl peroxalate (1 mmol) and the aminoxyl **1** (2.1 mmol) in phenylacetylene (100 mmol) at 60°C for 70 mins. gave the expected addition product **27** (85%) as well as **28** (15%). The structure of **27** was confirmed by

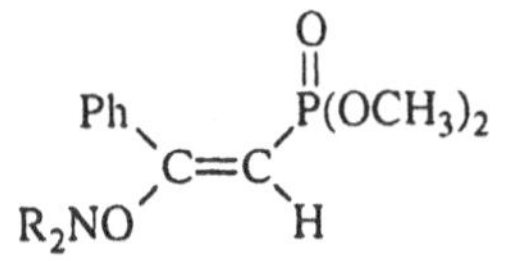

$$\underset{R_2NO}{\overset{Ph}{}}\!\!>C=C<\!\!\underset{H}{\overset{P(O)(OCH_3)_2}{}}$$

27

$$(CH_3O)_2P(=O)-ONR_2$$

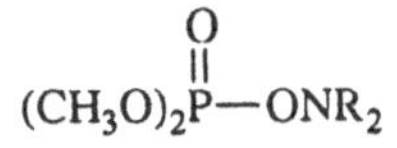

28

single-crystal X-ray structural analysis. Analogous results were obtained with diphenylphosphinoyl radicals.

<u>Studies with Fumarates and Maleates:</u> Diethyl fumarate can be polymerised under free radical conditions to give a high molecular weight polymer. Strangely, the isomeric compound diethyl maleate doesn't polymerise under similar conditions,[26] although it can be copolymerised with styrene.[21] Steric hindrance has been suggested as the reason, however the closely related (and relatively unhindered) maleic anhydride is also rather unreactive and undergoes polymerisation only under extreme conditions (to give a low molecular weight polymer).[21] We thought that the radical trapping technique might help to elucidate the lack of reactivity of maleic acid esters towards polymerisation.

Treatment of diethyl fumarate with di-*tert*-butyl peroxalate and the aminoxyl **1** (0.03 M) at 60°C for 70 mins. gave a mixture of the expected addition and abstraction products **29** (9%), **30** (56%), and **31** (9%), as well as the methyl radical-derived products **5** (7%) and **32** (<1%) and the 'dimeric' product **33** (13%) (Scheme 7). The ratio of H-abstraction to addition of *tert*-butoxyl radicals was approximately 9 : 1.

Under similar conditions, diethyl maleate gave analogous products (Scheme 8), however the ratio of products was quite different : **29** (1%), **34** (64%), **35** (12%), **5** (18%) and **36** (1%) (the methyl addition product **32** was not detected). In this case, the ratio of H-abstraction to addition of *tert*-butoxyl radicals was approximately 77 : 1. As the H-abstraction rate would be expected to be very similar for both the maleate and the fumarate, it can be seen that diethyl maleate is approximately an order of magnitude less reactive than diethyl fumarate towards addition of *tert*-butoxyl radicals. A similar result was obtained by comparison of the *tert*-butoxyl addition product **29** with the products (**5** and **32**) derived from ß-fragmentation of the *tert*-butoxyl radical (a radical clock[27]).

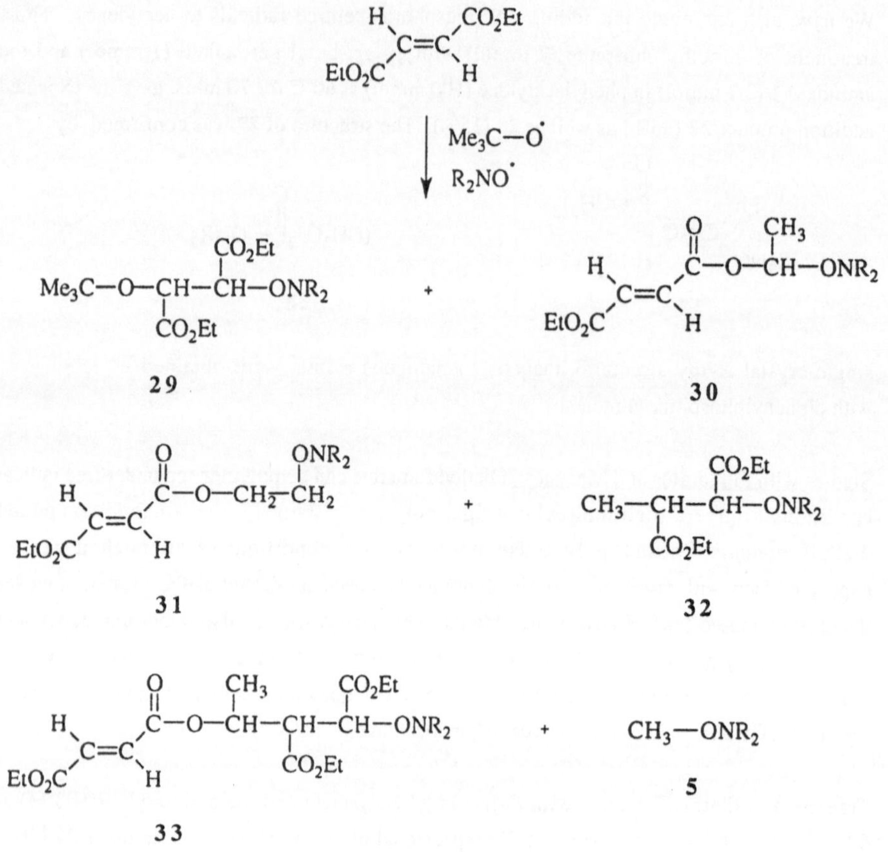

$H(CO_2Et)C{=}C(H)CO_2Et$ (diethyl fumarate, EtO_2C–CH=CH–CO_2Et)

$\xrightarrow{Me_3C{-}O^{\bullet},\ R_2NO^{\bullet}}$

$Me_3C{-}O{-}CH(CO_2Et){-}CH(CO_2Et){-}ONR_2$ (**29**) + $EtO_2C{-}CH{=}CH{-}C(=O){-}O{-}CH(CH_3){-}ONR_2$ (**30**)

$EtO_2C{-}CH{=}CH{-}C(=O){-}O{-}CH_2{-}CH_2{-}ONR_2$ (**31**) + $CH_3{-}CH(CO_2Et){-}CH(CO_2Et){-}ONR_2$ (**32**)

$EtO_2C{-}CH{=}CH{-}C(=O){-}O{-}CH(CH_3){-}CH(CO_2Et){-}CH(CO_2Et){-}ONR_2$ (**33**) + $CH_3{-}ONR_2$ (**5**)

Scheme 7

We suggest that the much lower reactivity of diethyl maleate towards radical addition is due to polar factors. Consider the addition of an electrophilic radical (R) to diethyl maleate. In the formation of the transition state, there is an increase in the positive charge on the carbon-carbon double bond, resulting in an increase in dipolar repulsion between the two ester moieties (Fig. 2). This will raise the energy of the transition state and therefore decrease the rate of the reaction. Similar arguments apply to the attack of nucleophilic radicals, where the increase in the negative charge on the β-carbonyl group results in an increase in the dipolar repulsion (Fig. 3).

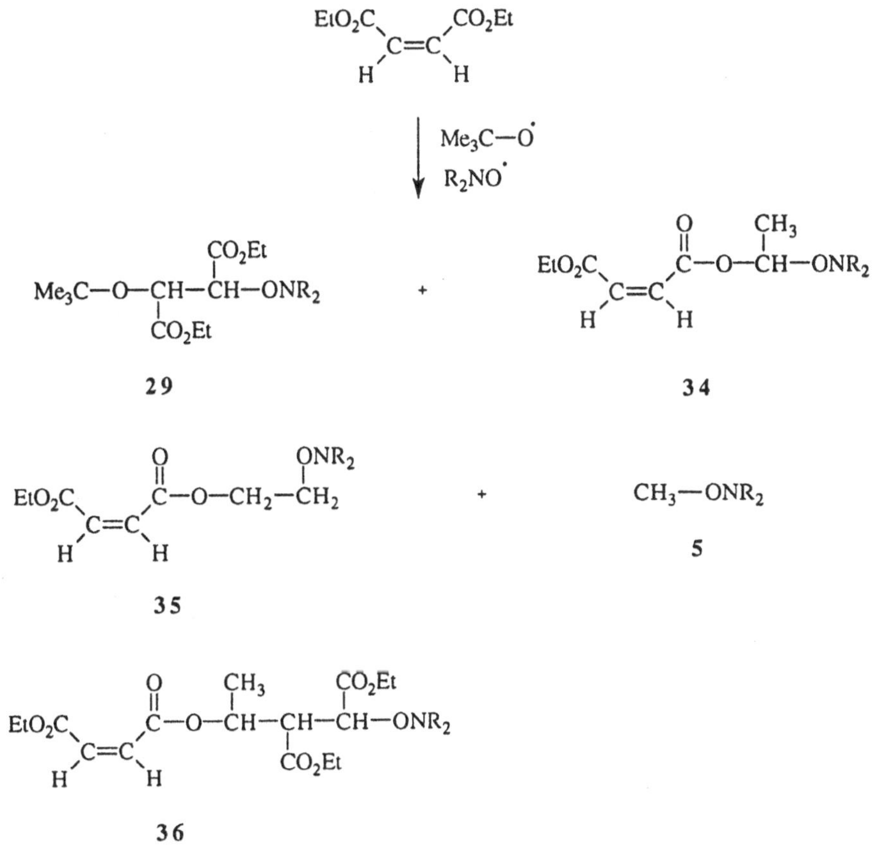

$EtO_2C(H)C{=}C(H)CO_2Et$

$\xrightarrow{Me_3C{-}O^{\bullet},\ R_2NO^{\bullet}}$

$Me_3C{-}O{-}CH(CO_2Et){-}CH(CO_2Et){-}ONR_2$ + $EtO_2C(H)C{=}C(H){-}C(=O){-}O{-}CH(CH_3){-}ONR_2$

29 34

$EtO_2C(H)C{=}C(H){-}C(=O){-}O{-}CH_2{-}CH_2{-}ONR_2$ + $CH_3{-}ONR_2$

35 5

$EtO_2C(H)C{=}C(H){-}C(=O){-}O{-}CH(CH_3){-}CH(CO_2Et){-}CH(CO_2Et){-}ONR_2$

36

Scheme 8

R δ−

δ+

H C C H

EtO—C C—OEt

O O

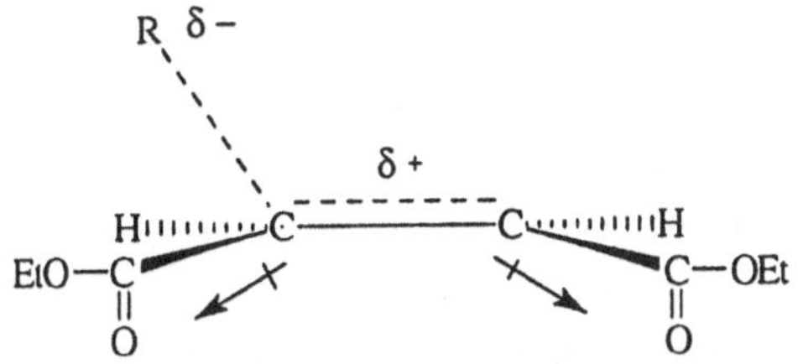

Fig 2

As diethyl maleate and diethyl fumarate are both strong electron acceptors, it is clear that attack by a nucleophilic radical should be much more favourable than attack by an electrophilic radical. This was reflected in the formation of significant amounts of 'dimeric' addition products **33** and **36** from the fumarate and maleate respectively. Both **33** and **36** (each formed as a set of four diastereomers) became the major products formed (> 55%) at very low aminoxyl concentrations (4.6 x 10^{-3}M). We suggest that these products are significant because of an enhanced nucleophilicity of the secondary H-abstraction radicals. This enhanced nucleophilicity results from donation of a lone pair on the adjacent oxygen atom into the semi-occupied orbital on the carbon atom, i.e. a radical "α-effect".[28] The formation of dimers **33** and **36** is consistent with the known[21,29] ability of both diethyl fumarate and diethyl maleate to undergo copolymerisation with electron-donor monomers that can form nucleophilic propagating radicals.

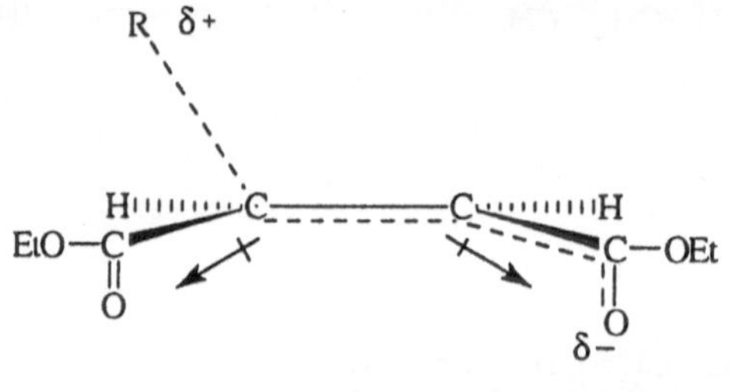

Fig 3

In contrast, no 'real' dimeric products were observed, even at very low aminoxyl concentrations. This shows that the addition of radicals with an adjacent carbonyl group (i.e. those formed by addition of radicals to diethyl fumarate or diethyl maleate) must be a very unfavourable process, as expected for the attack of an electrophilic radical on a strong electron acceptor. Presumably this is the reason for the slowest propagation rate ever reported for diethyl fumarate (0.015 $Lmol^{-1}s^{-1}$).[30]

<u>Conclusions</u>: The radical trapping technique has been applied to the study of alternating copolymerisation and the results do not support a mechanism involving the formation of charge-transfer complexes between the respective monomers. A mechanism involving kinetically-controlled addition of the separate monomers is supported.

The radical trapping technique can be used to study the initiation step in polymerisation where the initiating species is a phosphorus-centred radical. Addition of phosphorus-centred radicals to alkenes was found to be competitive with direct trapping by the aminoxyl.

The radical trapping technique has demonstrated that diethyl maleate is an order of magnitude less reactive than diethyl fumarate towards radical addition and that using *tert*-butoxyl radicals as initiators results in H-abstraction from the ester moiety as the major reaction pathway. It is suggested that the inability of diethyl maleate to undergo homopolymerisation is due to polar factors, in particular, an unfavourable dipolar repulsion in the transition state for free radical addition.

Acknowledgements: We thank Griffith University, the ARC and the CSIRO for financial assistance.

References:

1 Moad G, Solomon D (1990) Aust J Chem 43:215
2 Rizzardo E, Solomon D (1979) Polym Bull 1:529
3 Busfield WK, Jenkins ID, Thang SH, Rizzardo E, Solomon DH (1985) Aust J Chem 38:689
4 Busfield WK, Jenkins ID, Thang SH, Rizzardo E, Solomon DH (1988) J Chem Soc Perkin Trans 1:485
5 Bottle, S, Busfield WK, Jenkins ID, Thang SH, Rizzardo E, Solomon D (1989) Eur Polym J (1989) 25:671
6 Bottle S, Busfield WK, Jenkins ID, Skelton BW, White AH, Rizzardo E, Solomon D (1991) J Chem Soc Perkin Trans 2:1001
7 Busfield WK, Jenkins ID, Heiland K. (1993) Eur Polym J accepted for publication
8 Busfield WK, Jenkins ID, Thang SH, Rizzardo E, Solomon D (1993) Eur Polym J 29:397
9 Busfield WK, Jenkins ID, Thang SH, Rizzardo E, Solomon D (1985) Tet Lett 26:5081; Busfield WK, Grice ID,Jenkins ID, Thang SH (1991) Aust J Chem 44:1407
10 Bottle SE, Busfield WK, Jenkins ID, (1992) J Chem Soc Perkin Trans 2:2145
11 Busfield WK, Engelhardt LM, Healy PC, Jenkins ID, Thang SH, White AH, (1986) Aust J Chem 39:357
12 Busfield WK, Jenkins ID, Thang SH, Moad G, Rizzardo E, Solomon DH (1985) J Chem Soc Chem Commun 1249
13 Busfield WK, Grice ID, Jenkins ID in preparation
14 Busfield WK, Grice ID, Jenkins ID (1992) Polymer International 27:119
15 Beckwith ALJ, Bowry VW, Moad G, (1988) J Org Chem 53:1632
16 Beckwith ALJ, Bowry VW, Ingold KU (1992) J Am Chem Soc 114:4983; Bowry VW, Ingold KU ibid 114:4992
17 Griffith PG, Moad G, Rizzardo E, Solomon DH (1983) Aust J Chem 36:397
18 Moad G, Rizzardo E, Solomon D (1982) Macromolecules 15:909
19 Cowie JMG (1985) Alternating Copolymers.Plenum,New York and London
20 Bartlett PD, Nozarki K (1946) J Am Chem Soc 68:1495
21 Ratzsch M, Vogl O (1991) Prog Polym Sci 16:279
22 Baxter JE, Davidson RS, Hageman HJ, Overeem T (1987) Makromol Chem Rap Commun 8:311
23 Baxter JE, Davidson RS, Hageman HJ (1988) Eur Polym J 24:419
24 Tedder JM, Walton JC (1980) Tetrahedron 36:701
25 Jones MJ, Moad G, Rizzardo E, Solomon DH (1989) J Org Chem 54:1607
26 Otsu T (1987) Makromol Chem Macromol Symp 10/11:235
27 Griller D, Ingold KU (1980) Accounts Chem Res 13:317
28 March J (1992) Advanced Organic Chemistry, 4th ed.Wiley,New York
29 Mayo FR, Walling C (1950) Chem Rev 46:191
30 Otsu T, Yamada B, Ishikawa T (1991) Macromolecules 24:415

The Surface Excess Structure in 1,4-Polyisoprene-Polyacetylene Copolymer Solutions

J.W. White*, J.C.H. Kerr‡, J. Penfold†, P.M. Saville*, R.K. Thomas§

* Research School of Chemistry, The Australian National University, Canberra, 0200, Australia.

‡ European Vinyls Corporation, ICI Chemicals and Polymers, The Heath, Runcorn, Cheshire, UK

† The Rutherford Appleton Laboratory, ISIS Facility, Chilton, Didcot, Oxon, OX11 0QX, UK

§ Physical Chemistry Laboratory, South Parks Road, Oxford, OX1 3QZ, UK.

Abstract: Copolymers of 1,4-polyisoprene-polyacetylene with up to 25% polyacetylene are soluble in organic solvents and, like native polyacetylene, have interesting electrical and non linear optical properties. Because the molecules are amphiphilic techniques analogous to Langmuir-Blodgett methods may prove useful for self assembly and so here we study the surface excess of toluene solutions by the specular reflection of neutrons. A surface layer structure with excess PA at the air/solvent interface is characterised using the isotopic replacement method in neutron scattering.

INTRODUCTION

By using standard block copolymerisation methods [1] and a low valent metal catalyst [2,3] polyacetylene-polyisoprene and related copolymers of defined molecular weight in each moiety may be produced [4]. The solubility of these polymers in many organic solvents opens up a number of interesting chemical, electrochemical [5] and electrical applications. Particularly interesting are the amphiphilic properties of the molecules which in some respects resemble those of soaps, the insoluble acetylenic fraction being analogous to the insoluble fatty acid chain. The effect of this amphiphilic character on the morphology of any deposited or adsorbed layer should be very important in device applications. The recent measurement of the surface excess by specular reflection of neutrons from diblock 3,4-polyisoprene (MW 8000): deuteropolyacetylene (MW 2000) in toluene-d_8 [6] made possible a preliminary identification of the nature of the surface activity associated with the amphiphilic structure.

In the present paper specular neutron reflection has been used to probe the structure of the surface excess in more detail, using 1,4-polyisoprene-polyacetylene (PIPPA) solutions in toluene, freshly prepared, and after a further 20 days of ageing. The 1,4- isomer was used to eliminate any effects of solvent change between preparation and measurement, the 3,4-system having been prepared in tetrahydrofuran but the surface studied in toluene. Furthermore, having eliminated ageing and solvent effects it was possible to apply stricter constraints to the interpretation of the reflectivity profiles from different isotopic species, which revealed more detail of the surface structure. Finally the concentration dependence of the surface excess was studied in the range 0.1 to 4% by weight of polymer.

K.P. Ghiggino (Ed.)
Progress in Pacific Polymer Science 3

EXPERIMENTAL DETAILS

The synthesis of PIPPA was based [4] on the anionic to Ziegler-Natta route. As ageing effects were found to be important the provenance of the polymers is given here. The living polyisoprenyl carbanion was prepared by anionic polymerisation using n-butyl lithium as the initiator with all monomers and solvents used purified by recommended procedures [7]. The anionic polymerisation was carried out in toluene and the polyisoprene molecular weight was controlled by varying the ratio of isoprene monomer to initiator with a target molecular weight of ~8000.

Anhydrous $CoCl_2$ dissolved in dry toluene was used as the "Ziegler-Natta" catalyst. The resultant blue solution was cannulared into the magnetically stirred mixture in the reaction vessel, which had previously been cooled to -80°C to minimise catalyst degradation. The solution turned from the initial pale yellow of the polyisoprenyl solution to a dark green over an ageing period of ~9 hrs. Following this ageing period, measured amounts of acetylene either deuterated or protonated, were admitted into the rapidly stirred reaction vessel, maintained at -80°C, from a glass bulb of known pressure and volume (note that acetylene may explode spontaneously at pressures above 2 atm). The deuterated acetylene was prepared by dripping D_2O onto calcium carbide chips while the protonated form was commercially available and contained phosphine and acetone as stabilisers. Both samples of acetylene were purified by passage through dry ice/ethanol cold traps, concentrated H_2SO_4 bubblers, a KOH drying column, and an activated alumina drying column. Upon addition of the acetylene, a deep, bright red colour at the surface indicated the formation of the polyacetylene block. The appropriate stoichiometric amount of acetylene (to obtain 25% by weight PA in the copolymer) was taken up by the reaction within 24 hrs. The reaction vessel was sealed and left at -80°C for one week to ensure completion of the reaction which was finally terminated using dry methanol saturated with HCl gas. At no stage of the synthesis or purification was the product exposed to the atmosphere.

An important piece of information needed in the analysis of the reflectivity data is the ratio of PA to PiPPA in the copolymer. For copolymer produced by a titanium catalyst [8] the maximum wt% PA in PiPPA where the copolymer remained soluble was found to be about 25%. It was also noted by GPC [8] that at least 25% of the reaction mixture was PiP homopolymer due to premature termination or low conversion rates and the distribution in MW of the PA is broad [8]. FTIR and NMR produced useful information about the cis- to trans- ratio of PA and the ratio of 3,4- to 1,4-PiP, respectively, however, neither technique was effective in determining the PiP to PA ratio. In the first case IR bands were observed for both moieties but there were large uncertainties in their extinction coefficients and hence stoichiometries totally inconsistent with the quantities of material used in the reaction. In the second case, no NMR peaks were observed for the PA block at any concentration of the solution. From the uptake of acetylene and the minimal amount of insoluble PA, the molecular weight of this block is on average 2500 or 33% by weight. Overall, given all the error associated with the estimation of the PiP:PA ratio it was decided to settle on the 25% previously found as the limit of solubility [8]. The difference between this value and the value of 33%, from the volume of gas added, should not greatly affect the analysis of the reflectivity.

The neutron reflectivity measurements were done on the reflectometer CRISP at the neutron spallation source ISIS, at the Rutherford-Appleton Laboratory, UK. The instrument has been described fully elsewhere [9]. Neutrons were reflected from the surface of the liquid sample at an incident angle of 1.5°. The incident neutron beam contained neutrons of wavelengths from 0.5 to 6.5 Å which were analysed by time of flight. The reflectivity was determined by forming the ratio of the reflected neutron intensity at each wavelength to the number incident on the sample. The absolute intensity was calibrated by fitting a calculated profile to the observed reflectivity profile of toluene-d_8. This is possible because the scattering length density of toluene is accurately known.

The beam dimensions were 4 mm high and 40 mm wide and the illuminated area of the sample 160 x 40 mm^2. The sample was contained in a Teflon trough 5 mm deep with a liquid surface area of 200 x 80 mm^2. The trough was cleaned by soaking in heptane followed by toluene and toluene-d_8. It was enclosed in an air tight container with quartz windows using indium seals. Solutions of 1,4-polyisoprene-polyacetylene-h (PiPPAH), MW 8000:2000, and 1,4-polyisoprene-polyacetylene-d (PiPPAD), MW 8000:2000 of concentrations, (w/w) 0.12%, 1.2%, 2.5% and 4.0% in toluene-d_8 and toluene-h_8 were studied at ambient temperature (25° C).

Specular Neutron Reflection

The variation in the reflectivity with momentum transfer normal to the interface between two media depends on the variation of η, the neutron refractive index, normal to the interface. η is related directly to composition by

$$\eta = 1 - (\lambda^2 / 2\pi)\rho_s$$

where ρ_s is the scattering length density given by

$$\rho_s = \sum n_i b_i$$

where n_i is the number density of the ith nucleus and b_i its scattering length. The reflectivity profile may be deconvoluted exactly to the scattering length density-thickness profile normal to the interface, using the optical matrix method [10]. A particular feature of the neutron scattering technique is isotopic substitution which changes the scattering length density of the system with little or no change to the chemical nature. By adequate variation in the contrast of the system, ie variation of the scattering length density of substrate and polymer, it becomes possible to distinguish between the various models which fit the data. A single model must fit the reflectivity profiles for all contrast variations [11]. In the present work, H/D substitution in the PA group and toluene was used to provide a variety of contrasts. Measurements with the copolymer in toluene-h_8 were not very informative because of the contrast between PiPPAH and toluene-h_8 is low and the incoherent scattering background from the toluene is high. The densities and scattering length densities of the various isotopicly labelled compounds used are given in Table 1.

Table 1. Densities and scattering length densities of PiPPA and toluene isotopes.

	$\bar{\rho}$/g/cm^{-3}	ρ_s/10^{-6} Å^{-2}
Polyisoprenene (PiP)	0.913	0.27
Polyacetylene-d (PAD)	1.24 ± 0.1	7.1 ± 0.5
Polyacetylene-h (PAH)	1.16 ± 0.1	1.5 ± 0.1
Toluene-d8	0.87	5.7
Toluene-h8	0.94	0.9

RESULTS

Ageing Effects

Previous measurements [6] on 3,4-PIPPA/toluene solutions were taken 4 months apart. There were differences in the profiles, especially on changing the isotopic composition, which were attributed to sample ageing effects. To set some lower limit to ageing in the present series of experiments, 1,4-PiPPAD solutions from the same preparative batch were measured after 7 days and again after 27 days, having been stored under argon at 20°C. Figure 1 shows the reflectivity of a 2.5% solution of 1,4-PiPPAD in toluene-d8 at the beginning and end of the ageing period. No significant difference within the accuracy of the measurement can be detected over the whole range of κ (κis the scattering vector defined as $\kappa = 4\pi/\lambda \sin\theta$ where θ is half the deviation between the incident and reflected beams).

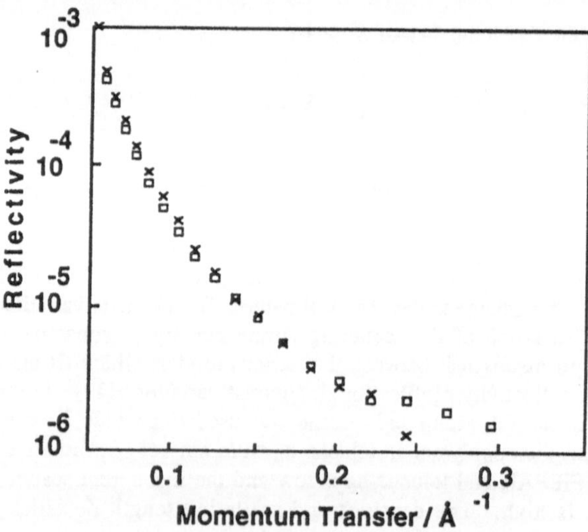

Figure 1. Neutron reflectivity profiles from 2.5% polyisoprene-polyacetylene solutions in toluene-d8 after ageing at 20°C under Argon for 7 days (x) and 27days (boxes).

That ageing does occur was indicated by the gradual deepening in colour of the solutions as monitored by the changes in the visible absorption spectrum. A possible cause of this is cis-trans isomerisation in the polyacetylene component but the reflectivity data show qualitatively that from the point of view of the surface structure this ageing has no detectable effect at the concentrations studied here.

Reflectivity Profiles and their Interpretation

Figure 2 shows the reflectivity profiles of a 2.5% PiPPAH and PiPPAD in toluene-d_8 (markers) and the fitted profiles of the best model discussed below, (represented by the lines). The reflectivity profile for a 2.5% solution of PiPPAD on toluene-h_8 was not very intense and was similar to the profile for pure toluene-h_8.

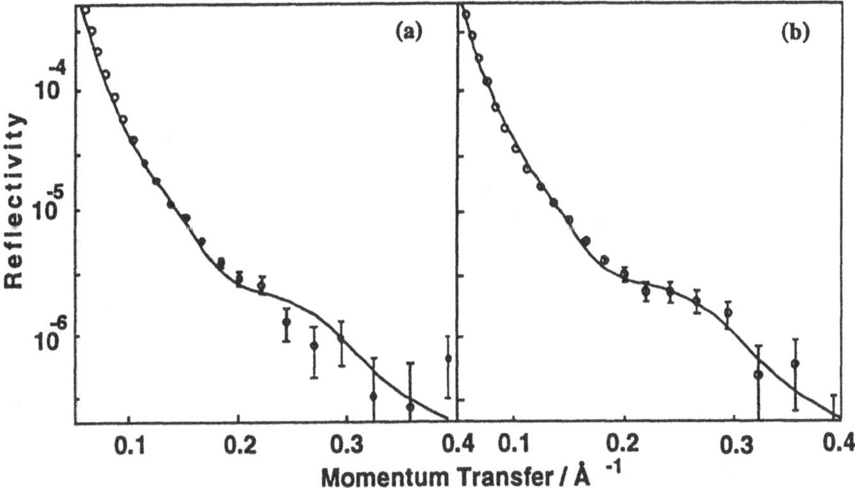

Figure 2. Neutron reflectivity from (a) 2.5% PiPPAH in toluene-d_8 and(b) from 2.5% PiPPAD in toluene-d_8 at 25°C.

The dependence of the reflectivity on the polymer concentration in the range 0.1 to 4 % by weight of 1,4-PiPPAD in toluene-d_8 at 298 K is shown in Figure 3. The same data ratioed to the reflectivity of pure toluene-d_8 shows a peaked structure and a steady increase in reflectivity in the region $0.1 < \kappa < 0.3$ Å^{-1} was observed as the concentration increased.

Reflectivity profiles of deuterated and protonated 3,4-PiPPA [6] were fitted independently to simple models by least squares methods. The simplest model was of a single homogeneous polymer layer of thickness (t) 30 Å and scattering length density (ρ_s) 4.8×10^{-6} Å^{-2}, pure toluene-d_8 having a ρ_s of 5.7×10^{-6} Å^{-2}. This did not produce an extremely good fit to the observed reflectivities and so the simplest model consistent with the data, a three layer model with a thin (about 8.5 Å) PAD layer at the surface, was adopted. This model will be discussed in detail later. No attempt was made to fit a single model to both polymer isotopes because it was thought that the protonated sample had aged and therefore would not necessarily have the same surface structure.

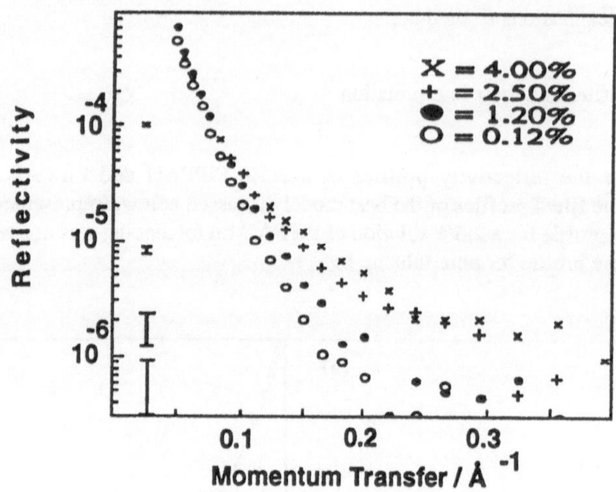

Figure 3. Dependence of the neutron reflectivity from PiPPAD solutions in toluene-d$_8$ at 25°C on the copolymer concentration.

As for the 3,4-isomer, the model of a single blended surface layer containing a homogeneous mixture of PiP, PA and toluene does not fit the reflectivity profile of the 2.5 % solution of PiPPAD. No model consisting of such a blended layer of whatever composition could give the reflectivity profile of the type shown in Figure 2. It was therefore deduced that there is some separation of PiP and PA between the surface and solvent.

In this partially segregated model it is assumed that PiPPAD and PiPPAH have identical surface structures, that is, there are no isotope effects on the surface structure. Then, given the known scattering length densities of the single components PiP (ρ_I), PAH (ρ_{Ah}), PAD (ρ_{Ad}) and toluene-d$_8$ (ρ_{dT}) (see Table 1), the scattering length density of a layer at the surface will be given by

$$\rho_d = x\rho_{Ad} + y\rho_I + (1\text{-}x\text{-}y)\rho_{dT} \qquad (1)$$

$$\rho_h = x\rho_{Ah} + y\rho_I + (1\text{-}x\text{-}y)\rho_{dT} \qquad (2)$$

where x and y are the respective volume fractions of PA and PiP in a given layer. The equations refer to PiPPAD and PiPPAH respectively in toluene-d$_8$. Equations (1) and (2) will hold for any layer in a multilayer structure but with different values of x and y. The values of x and y for the different layers are not independent because the total amounts of PA and PiP in the surface region must be in the stated ratio 1 to 4 of their component molecular weights. Thus, for a surface region consisting of two layers we start with four unknowns, x_1, x_2, y_1, and y_2 but stoichiometry reduces this to three. Fitting a two layer structure to the reflectivity data for PiPPAD alone gives values of the scattering length density of each of the layers, but this is not sufficient to determine the three unknowns which determine the composition of the two layers. However, a similar fit to PiPPAH gives two extra parameters, more than sufficient to determine x_2, y_1, and y_2 independently.

Previous experience [6] of the reflectivity data for the related 3,4-PiPPAD system suggests that the surface structure must have at least two features, a layer adjacent to the vapour containing a excess of PA and a rather thicker layer on the solution side of the interface containing a excess of PiP. We now make the simplifying assumption that the structure consists of only these two layers at the surface and that there is polyacetylene only in the layer on the vapour side. The model is a physically reasonable one given the amphiphilic nature of the copolymer. Taking the PA layer to be layer 1 and using equation (1) then

$$\rho_1 = x_1\rho_A + y_1\rho_I + (1-x_1-y_1)\rho_T \tag{3}$$

and

$$\rho_2 = y_2\rho_I + (1-y_2)\rho_T \tag{4}$$

where y_2 is the volume fraction of PiP in layer 2. Because of the stoichiometry of the polymer x_1, y_1 and y_2 it is easy to show that y_2 is given by

$$y_2 = \left[\frac{x_1\bar{\rho}_A}{R\bar{\rho}_I} - y_1\right]\tau \tag{5}$$

where t is the ratio t_1/t_2 and t_1 and t_2 are the thicknesses of the two layers, R = 0.25 is the molecular weight ratio of PA to PiP, and $\bar{\rho}_A$ and $\bar{\rho}_I$ are the densities of the bulk polymer components. With this simplified model the reflectivity of one isotopic species is determined unambiguously by the parameters obtained from the reflectivity of the other. Substituting the values of ρ_A, ρ_I, and ρ_T from Table 1 into equations (3) and (4) and eliminating y_2, using (5), the following values of the scattering lengths of the two layers of each combination of polymer and solvent are:

PiPPAD/toluene-d_8: $\rho_{1d} = 5.7 + 1.4x_1 - 5.4y_1$

$$\rho_{2d} = 5.7 - 5.4\tau(5.4x_1 - y_1) \tag{6}$$

PiPPAH/toluene-d_8: $\rho_{1h} = 5.7 - 4.2x_1 - 5\ 4y_1$

$$\rho_{2h} = 5.7 - 5.4\tau(5.1x_1 - y_1) \tag{7}$$

PiPPAD/toluene-h_8: $\rho'_{1d} = 0.9 + 6.2x_1 - 0.6y_1$

$$\rho'_{2d} = 0.9 - 0.6\tau(5.4x_1 - y_1) \tag{8}$$

These equations are now used in conjunction with the reflectivity data to make self consistent deductions about the surface structure.

In testing this approach the model for the interfacial region for the 3,4-PiPPAD and 3,4-PiPPAH in toluene-d_8 [6] is taken first of all. This model was fitted independently to each profile and no account was taken of the stoichiometry. All the PA was assumed to be in the first layer. From the figures given in the final table of reference 6 a mean value for x_1 from PiPPAD and PiPPAH in toluene-d_8 of about 0.7 was obtained. By using the scattering length densities of subsequent layers to calculate the total amount of PiP a value for the ratio R close to 1 is obtained, instead of the stoichiometrically reasonable 0.25. Thus the structure proposed in reference 6 cannot be quantitatively correct, though the essential feature of the thin layer containing PA is qualitatively correct.

Fitting the profiles of all three 1,4-PIPPA/toluene contrasts, taking into account the stoichiometry, suggests that the surface excess of polymer is small because, for all three, the reflectivities are close to the values for the pure solvent at low κ. This is particularly the case for PiPPAD in toluene-h_8 where a large surface excess would increase the reflectivity by an amount much larger than observed. On the other hand, for PiPPAD in toluene-d_8, it is necessary to include one layer of higher scattering length density than the solvent to fit the data, and so there must be a sufficient excess of PA to bring this about. Since the scattering length density of PiP is much lower than the solvent there can only be at most a small amount of PiP in the PA containing layer, i.e. y_1 is small. Fits to the data also suggest that t_1 is less than t_2, as expected from the stoichiometric ratio and relative solubility of the two polymers in toluene. Figure 4 uses equations (6) to (8) to calculate the variation of the scattering length densities with x_1 at small values of y_1 and values of $\tau < 1$. Figure 4(a) shows the variation of ρ_1 with x_1 for the three different isotopic compositions for values of y_1 of 0 and 0.05. Figure 4(b) shows the corresponding variation in ρ_2 at $y_1 = 0$ and at three values of τ, 0.2, 0.5, and 1.

There are two key features to note. First, ρ_1 varies strongly with x_1 and quite differently for the three isotopes; thus the comparison of the reflectivities of PiPPAD and PiPPAH can be used to fix x_1 reasonably accurately. Secondly, ρ_{2d} decreases very rapidly with increasing x_1 and quite differently at different values of τ. The shape of the reflectivity will be very sensitive to this variation and therefore, once x_1 has been fixed, τ can be determined from the shape of the reflectivity profiles of either isotopic species in toluene-d_8. Bearing these two features in mind, the data was fitted by varying the three parameters x_1, y_1, and τ, and using the plots shown in Figure 4 to obtain stoichiometrically consistent sets of scattering length densities. This procedure is demonstrated for the 2.5% by weight solution.

For $y_1 = 0$ and a value of τ of 0.5 the effects of varying x_1 (volume fraction of PA in layer 1) on the reflectivities of PiPPAD and in toluene-d_8 are as shown in Figure 5(a) and (b) respectively. For both isotopes $x_1 = 0.07$ does not fit the data well but there is less to choose between $x_1 = 0.03$ or 0.05, the latter value producing the shoulder at $\kappa = 0.25$ Å^{-1} rather better. For the constant value of $t_1 = 8$ Å used for Figure 5(a) and (b), a change in x_1 corresponds to a change in the surface excess. For constant surface excess the product t_1x_1 is constant. The result of varying x_1 at constant surface excess is shown for PiPPAD in toluene-d_8 in Figure 5(c). Now $x_1 = 0.05$ clearly gives the best fit to the data. These fits establish the approximate value for the surface excess.

The effect of including some PiP in layer 1, i.e. a non-zero value of y_1 is to reduce ρ_1 and increase ρ_2. This reduction in contrast reduces the intensity of the shoulder at $\kappa = 0.25$ Å^{-1}. For the shoulder to be produced in the calculated profile y_1 must be close to zero.

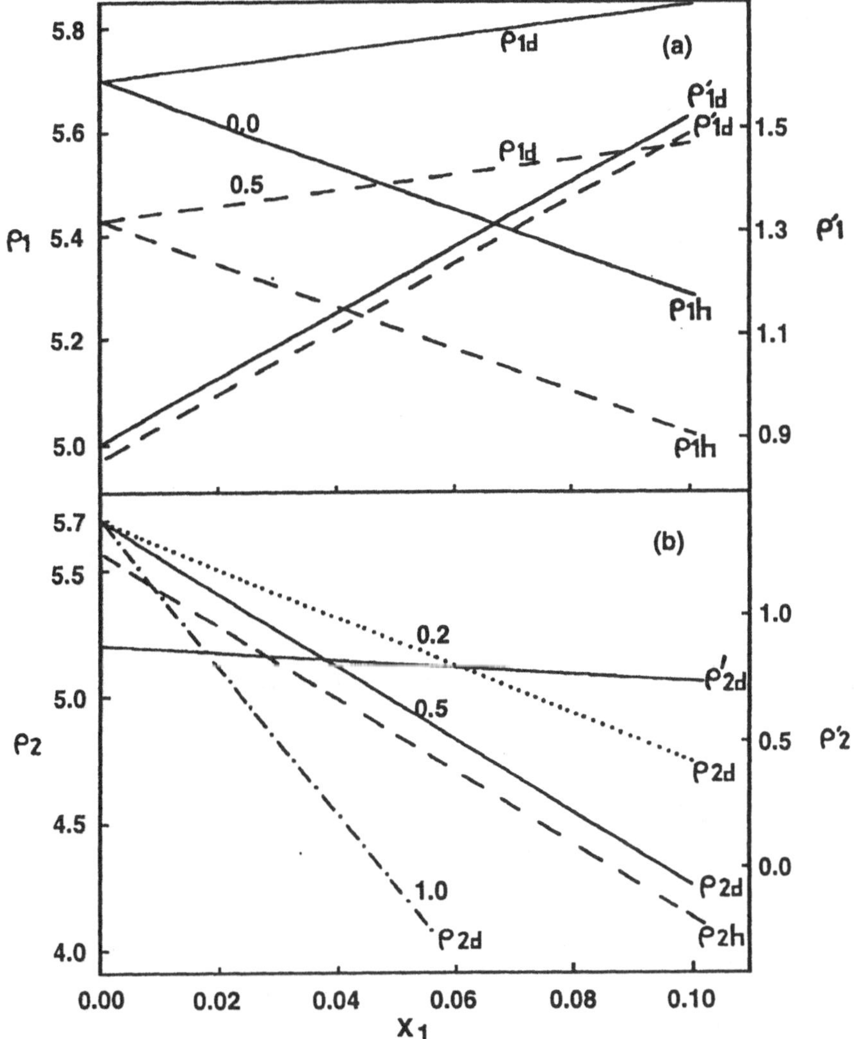

Figure 4. Interdependence of the layer scattering length densities on the content of polyacetylene in the top layer. See text for a complete description.

In the next iteration of the model, t_1 and t_2 were varied either by changing the ratio τ while keeping $(t_1 + t_2)$ constant, or by keeping τ constant and varying the overall thickness. Both variations are given in Figure 6. Figure 6(a) shows the calculated profiles for = 0.33 and 0.7 at the fixed values of x_1 (0.05) and y_1 (0) determined above, and for ($t_1 + t_2 = 24$ Å). The low value of τ suppresses the shoulder at $\kappa = 0.25$ Å^{-1} while the high value shifts it to too low a value of κ. Although a decrease of τ at constant x and $(t_1 + t_2)$ corresponds to a decrease in the surface excess, doing the same calculation at constant surface excess only has a small effect on the calculated profile of Figure 6(a). t must therefore be close to 0.5. Taking this value of τ and the earlier fixed values of x_1 and y_1, $(t_1 + t_2)$ was varied. Figure 6(b) shows calculations for $(t_1 + t_2) = 18$ and 30 Å, neither of which fits the data at all well. Further refinement shows that $(t_1 + t_2)$ is about 24 Å and therefore the thickness of the bilayer is 24 Å, see Figure 5c.

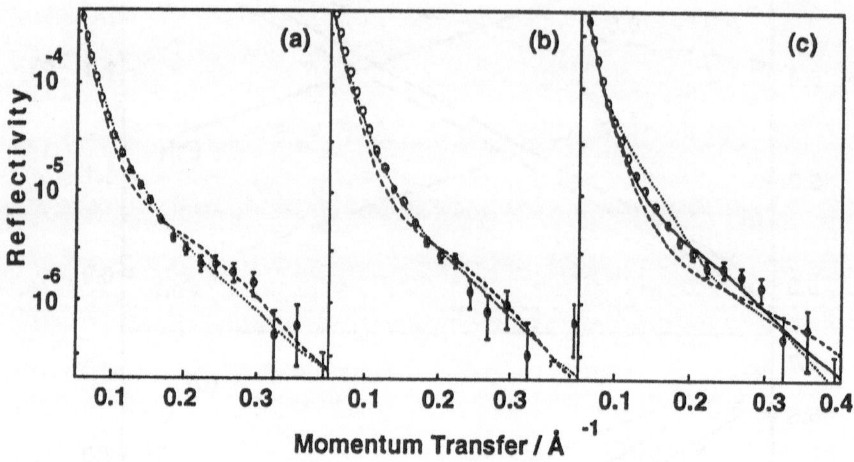

Figure 5. Self consistent variation of the parameters in fits to the reflectivity data for 2.5% PiPPA solutions. (a) PiPPAD/toluene-d_8, (b) PiPPAH/toluene-d_8 for $\tau = 0.5$ and $x = 0.03$, (dotted line), $x = 0.05$ (dashed line) with $t_1 = 8$Å, $t_2 = 16$Å. (c) PiPPAD/toluene-d_8 best fits for variation of x, the mole fraction of PAD in the top layer at constant surface excess x_1t_1, see Table 2 for parameters of each fit.

Table 2. Parameters used in fitting the neutron reflectivity data in Figure 5c.

	ρ_1 / Å^{-2}	t_1 / Å	ρ_2 / Å^{-2}	t_2 / Å
x=0.03 (dotted line)	5.74	13.5	5.26	27
x=0.05 (solid line)	5.77	8	4.97	16
x=0.07(dashed line)	5.80	6	4.67	12

The best fitting set of parameters obtained so far is for a model with PA in the upper layer. Reversal of the layers does not fit the data at all. Thus PA is adsorbed preferentially on the vapour side of the interface as was concluded earlier [6]. This qualitative aspect can be refined further if some PA were allowed to be in layer 2, carrying with it the stoichiometric amount of PiP, ρ_2 would be lower than for any of the calculations shown in Figures 5 and 6, thus lowering the reflectivity unacceptably. This reinforces the assumption that all of the PA is in layer 1.

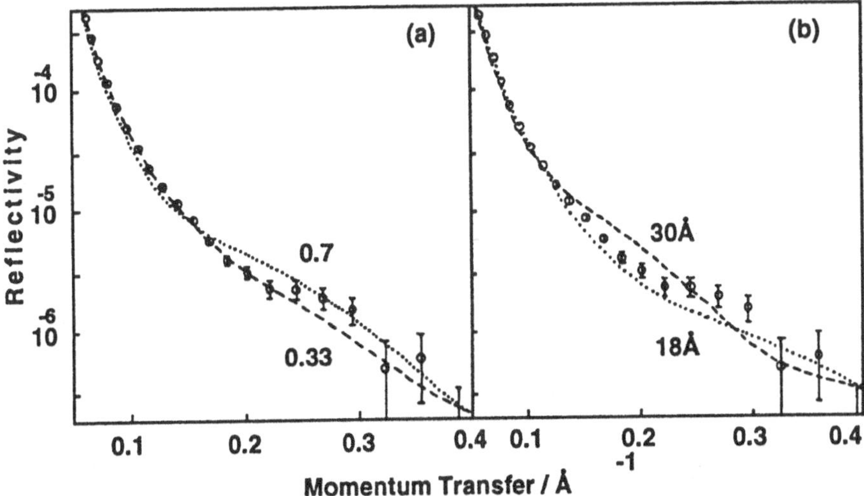

Figure 6. Profile fits to the neutron reflectivity from 2.5% PiPPAD in toluene-d8 at a fixed value of the polyacetylene content ($x_1 = 0.05$) (a) for ratios of $\tau = 0.33$ (dashed line) and $\tau = 0.70$ (dotted line) and (b) at constant $\tau = 0.5$ and for t_1+t_2 of 18Å (dotted line) and 30Å (dashed line).

Multilayering at the Interface

The initial assumption, made for simplicity, was that only one bilayer is present. The model can be refined by supposing that there are further bilayers each successively more weakly adsorbed than the first. The best fits to the data when a second bilayer is included at half the excess of the first are shown in Figure 2. The addition of a second bilayer improves the fit to the reflectivity significantly in that the shoulder becomes more pronounced, but the fits to the data for the 2.5% solution (Figures 5 and 6) are adequate and the improvement in Figure 2 is not such as to establish unambiguously that a second bilayer is present. At higher concentrations, however, the presence of a second bilayer is essential to explain the data.

The best fit parameters and their uncertainties for 2.5% PiPPAD and PiPPAH in toluene-d8 are $x_1 = 0.06 \pm 0.02$, $x_2 = 0.03 \pm 0.02$, $y_1 = y_2 < 0.03$, $t_1 = 8 \pm 3$ Å and $t_2 = 16 \pm 4$ Å. The errors in these parameters are strongly coupled and depend more on the model than on the fits to the data. There is particular uncertainty in the parameters of the second bilayer. For completeness the fitted parameters for the two isotopes of PIPPA in toluene-d8 are given in Table 3.

This model overestimates the reflectivity from the solution of PiPPAD in toluene-h8, but not by more than can be accounted for by the uncertainties of making an accurate background subtraction. The values of x_1 and x_2 can be used to calculate the surface excess of polymer and the area occupied per molecule. The molar volume of a 2000 MW polyacetylene unit is 2700 $Å^3$ and, for $t_1 = 8$ Å and $x_1 = 0.06$, the area per molecule in layer 1 is 5600 $Å^2$, giving a mean lateral distance between molecules of about 85 Å. The area per molecule is doubled in the second bilayer.

Table 3. Fitted scattering length densities and layer thicknesses for 2.5% solutions of PiPPAD and PiPPAH in toluene-d_8 after 7 days ageing.

	Layer	ρ / 10^{-6} Å^{-2}	Thickness / Å
PiPPAD/Toluene-d_8	1	5.8	8
	2	5.0	15
	3	5.7	9
	4	5.2	14
PiPPAH/Toluene-d_8	1	5.4	8
	2	5.0	15
	3	5.5	9
	4	5.2	14

DISCUSSION

Surface activity in 1,4-polyisoprene-polyacetylene, AB, block copolymer solutions was to be expected from the amphiphilic properties of such a diblock system with one moiety so insoluble because of the strong polyacetylene-polyacetylene attractive interactions. The present experiments allow access, for the first time, to some of the thermodynamic parameters of these interactions and give a structural model for the surface excess above and below the critical micelle concentration. This has been identified as about 10^{-3} moles/L for the related polymer 1,4-polyisoprene-polyacetylene (MW 8000:520) in toluene at 20°C using the drop weight method to determine surface tension. From ca. 10^{-4} molar to 10^{-3} molar the surface tension drops by about 3.5% to a constant value of ca. 28.4 dyne cm^{-1} at concentrations above 10^{-3} molar (ca. 1% w/w). Referring to Figure 3 we see that it is above ca 1% that a broad peak develops in the solution/solvent reflectivity ratio for $0.15 < \kappa$ / Å$^{-1} < 0.25$. The area per polyacetylene group (MW 2000) of about 5600 Å^2 show that even in the top most layer there is by no means specific adsorption of the polyacetylene fragment - the distances between groups being quite consistent with the known radii of gyration of 8000 molecular weight 1,4-polyisoprene homopolymers in toluene at the measuring temperature [6,13] (ca. 80 Å). It is not unreasonable therefore to suppose relatively well packed, solvent swollen polyisoprene coils in the second layer.

REFERENCES

1 Stowell JA, Amass AA, Beevers MS, Farren TR (1989) Polymer 30:195
2 Masuda T, Higashimura T (1984) Acc Chem Res 17:51
3 Edwards JH, Feast WJ (1980) Polymer 21:595
4 Armes SP, Vincent B, White J W (1986) J Chem Soc Chem Comm 1525
5 Hamnett A, Kerr J, White JW (1992) J Chem Soc Faraday Trans 89:277
6 Dai L, White JW, Kerr J, Thomas RK, Penfold J, Aldissi M (1989) Syn. Metals 28:D69
7 Fetters LJ (1966) J Res Natl Bur Standards 70A:5
8 Aldissi M, Bishop AR (1986) Synthetic Metals 14:13
9 Penfold J, Williams WG (1987) J Phys E 20:1411
10 Born M and Wolf E (1975) Principles of Optics (5th Edition) pp 51-72 Pergamon, Oxford
11 Lee EM Thomas RK Penfold J and Ward RC (1989) J Phys Chem 93:381
12 Crowley TL, Lee EM, Simister EA, Thomas RK, Penfold J, Rennie AR (1990) Colloids and Surfaces 52:85
13 Dai L and White JW (1993) J Polymer Sci B 31:3

Helix-Sense-Selective Polymerization of Methacrylate and Conformational Characteristics of the Polymers

Yoshio Okamoto and Tamaki Nakano

Department of Applied Chemistry, Faculty of Engineering, Nagoya University Furo-cho, Chikusa-ku, Nagoya 464-01 JAPAN

INTRODUCTION

We have been studying the helix-sense-selective polymerization of bulky methacrylates such as triphenylmethyl methacrylate (TrMA),[1] diphenyl-2-pyridylmethyl methacrylate (D2PyMA),[2] phenyl-2-pyridyl-*o*- and *m*-tolylmethyl methacrylates (PPyoTMA and PPymTMA),[3,4] and (*S*)-(-)-diphenyl(1-methylpyrrolidin-2-yl)methyl methacrylate (DMPMA)[5] with chiral anionic initiators. In these polymerizations, optically active polymers with one-handed helical structure have been obtained and some of the polymers exhibited characteristic conformational transition in solution.[6] In the present paper, we describe helix-sense-selective polymerization of 1-phenyldibenzosuberyl methacrylate (PDBSMA), diphenyl-3-pyridylmethyl methacrylate (D3PyMA) and phenyl[bis(2-pyridyl)]methyl methacrylate (PB2PyMA) by means of chiral anionic initiators and conformational transition of the obtained polymers. Asymmetric radical polymerization of PDBSMA is also reported because this monomer can also afford a helical, highly isotactic polymer by radical mechanism.

EXPERIMENTAL

The three monomers were prepared from methacryloyl chloride and corresponding alcohol. 1-Phenyldibenzosuberyl alcohol for PDBSMA was prepared by reaction of phenylmagnesiumbromide with dibenzosuberone. Diphenyl-3-pyridylmethanol for D3PyMA was synthesized from 3-lithiopyridine and benzophenone, and phenyl[bis(2-

K.P. Ghiggino (Ed.)
Progress in Pacific Polymer Science 3

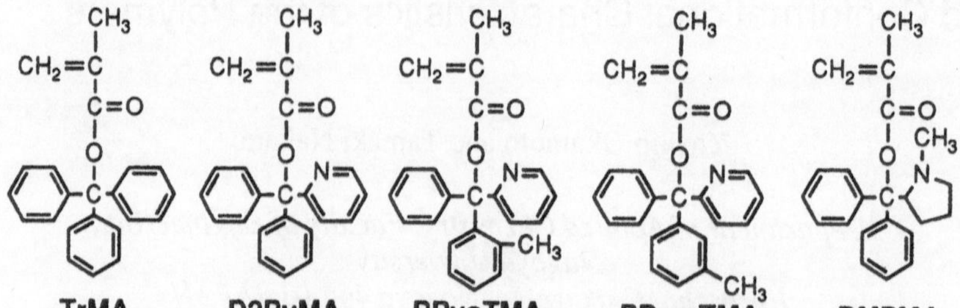

TrMA D2PyMA PPyoTMA PPymTMA DMPMA

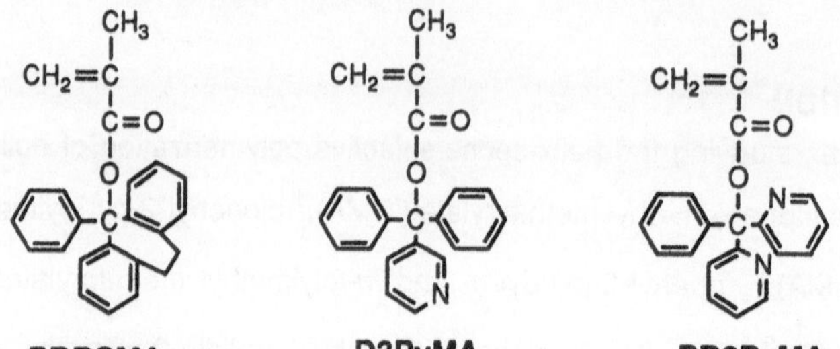

PDBSMA D3PyMA PB2PyMA

pyridyl)]methanol for PB2PyMA from 2-lithiopyridine and 2-benzoylpyridine. The asymmetric anionic polymerization was carried out with the complexes of *N,N'*-diphenylethylenediamine monolithium amide (DPEDA-Li) or fluorenyllithium (FlLi) with chiral ligands, (-)-sparteine (Sp), (+)- or (-)-2,3-dimethoxy-1,4-bis(dimethylamino)butane ((+)- or (-)-DDB), and (*S*)-(+)-1-(2-pyrrolidinylmethyl)pyrrolidine (PMP) in toluene at -78°C. Radical polymerization was performed in toluene with α,α'-azobisisobutyronitrile (AIBN) at 60°C or $(i\text{-PrOCOO})_2$ at 40°C.

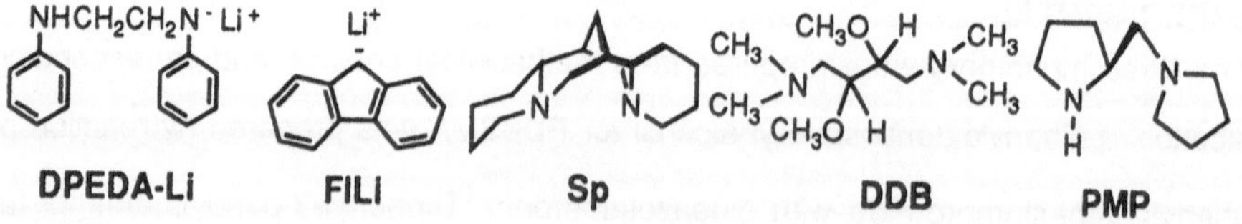

DPEDA-Li FlLi Sp DDB PMP

RESULTS AND DISCUSSION

Rate of ester solvolysis for the three monomers: The one-handed helical poly(TrMA) shows a high chiral recognition ability,[7] but it has a defect that the ester linkage is readily solvolized by methanol, which is an effective HPLC solvent. Because this problem could be also the case for the polymers of PDBSMA, D3PyMA, and PB2PyMA, solvolysis rate for the three monomers was estimated by monitoring the reaction shown in **Scheme 1** by means of ^1H NMR spectroscopy prior to the polymerization. **Table 1** summarizes the first-order rate constant of the methanolysis reaction and half-life period of the monomers in CD_3OD-$CDCl_3$ (1/1 v/v) at 35°C. The rate constants for PDBSMA,[8] D3PyMA,[9] and PB2PyMA[10] are smaller than that for TrMA;[11] the novel monomers are more durable against methanolysis than TrMA. The durability of PDBMSA can be attributed to 1-phenyldibenzosuberyl group which prevents a stable planer structure of an intermediate tertiary cation species in the solvolysis reaction and that of D3PyMA and PB2PyMA to electron-withdrawing effect of pyridyl group which destabilizes the tertiary cation. The remarkably small k value for PB2PyMA is explained as follows. The solvolysis of the esters must be catalyzed by acid because the reaction is obviously accelerated by the addition of a small amount of hydrochloric acid. Basic 2-pyridyl group can catch the acid proton to depress the sol-

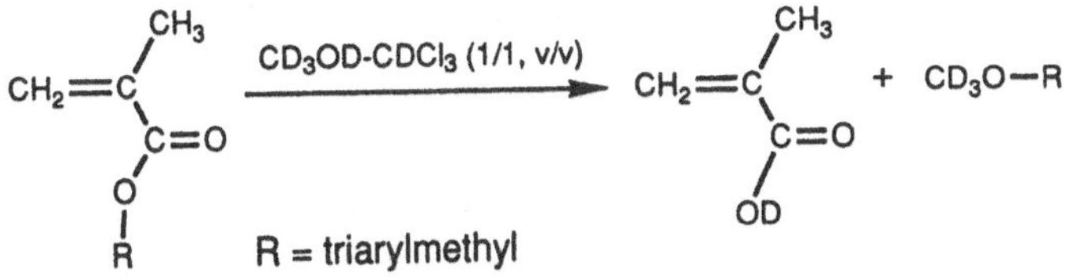

$$CH_2{=}C(CH_3){-}C({=}O){-}O{-}R \xrightarrow{CD_3OD\text{-}CDCl_3\ (1/1,\ v/v)} CH_2{=}C(CH_3){-}C({=}O){-}OD \ + \ CD_3O{-}R$$

R = triarylmethyl

Scheme 1. Methanolysis of bulky methacrylates with triarylmethyl ester group.

Table 1. Methanolysis rate and half-life period of methacrylates in CD_3OD-$CDCl_3$ (1/1, v/v) at 35°C[a]

Methacrylate	k^b (hr^{-1})	Half-life period (hr)
PDBSMA	4.66×10^{-1}	1.48
D3PyMA	2.91×10^{-2}	24
PB2PyMA	1.24×10^{-5}	55900 (6.38 years)
TrMA[c]	2.86	0.24

[a] Conc., 0.39-0.41 mol/ l.
[b] Pseudo first-order rate constant.
[c] Data from ref. 11.

volysis. The effect may be much more pronounced by coopreative effect of two adjacent 2-pyridyl groups of PB2PyMA.

Polymerization of PDBSMA: Asymmetric anionic polymerization of PDBSMA with the complexes of DPEDA-Li with the three chiral ligands afforded optically active polymers. A small amount of oligomers contained in the products was successfully removed by fractionation with benzene-hexane (1/1, v/v). The benzene-hexane-insoluble polymers showed large positive rotation and were nearly 100% isotactic (**Table 2**).[8] These specific rotation values ($[\alpha]_D$) are about 100° larger than that of purely one-handed helical poly(TrMA), suggesting that these poly(PDBSMA)s probably have 100% one-handed helical structure. Furthermore, the polymers obtained by (+)-DDB and PMP initiator systems had narrow molecular weight distribution. The optically active poly(PDBMSA) showed chiral recognition ability to some racemic compounds.

Radical polymerization of PDBSMA with (i-PrOCOO)$_2$ at 40°C also yielded an exclusively isotactic polymer (mm 98%).[12] Although TrMA also gives an isotactic polymer by radical polymerization, the value of isotacticity is much smaller (mm 63%).[13] The iso-

Table 2. Asymmetric anionic polymerization of PDBSMA with DPEDA-Li complexes in toluene at -78°C[a]

Run	Chiral ligand	Time (hr)	Yield[b] (%)	Benzene-hexane (1/1)-insoluble Part					
				Yield (%)	$[\alpha]^{25}_{365}$ [c]	$[\alpha]^{25}_{D}$ [c]	DP[d,e]	Mw/Mn[e]	Tacticity (%)[f] mm
1	Sp	24	100	71	+1670°	+442°	80	1.30	>99
2	(+)-DDB	3	100	86	+1778°	+477°	48	1.07	>99
3	PMP	3	100	95	+1755°	+468°	43	1.10	>99

[a]Conditions: PDBSMA 1.0 g, toluene 20 ml, [PDBSMA]/[Li]=20.
[b]Methanol-insoluble part.
[c]Measured in chloroform (conc. 1.0 g/dl).
[d]DP: degree of polymerization.
[e]Determined by GPC of poly(MMA) derived from original polymer.
[f]Determined by ^{1}H NMR analysis of poly(MMA) derived from original polymer.

tactic specificity of the radical polymerizations of PDBSMA and TrMA is contrast to the syndiotactic specificity of polymerization of conventional methacrylates and this has been ascribed to a helical structure of growing polymer radical. The higher isotactic specificity of PDBSMA polymerization may come from a higher rigidity of the helix. The 1-phenyldibenzosuberyl group must be conformationally less flexible than the triphenylmethyl group.

On the basis of above results, we performed the helix-sense-selective polymerization of PDBSMA by radical mechanism.[14] When the radical polymerization of PDBSMA was carried out with $(i\text{-PrOCOO})_2$ in (+)- and (-)-menthol containing a small amount of toluene at 50°C , optically active (-)- and (+)-poly(PDBSMA)s ($[\alpha]_{365}$ ca. 200°) with high isotacticity (mm >97%), respectively, were obtained. CD spectral pattern of these polymers was similar to that of the anionically obtained (+)-poly(PDBSMA). Therefore, one of right- and left-handed helical structures must be produced preferentially over the other, though optical yield is about 12%. This is the first example of helix-sense-selective radical polymerization to our knowledge.

Radical copolymerization of PDBSMA with a small amount of an optically active monomer was also investigated.[14] The copolymerization was carried out with optically active (-)-PPyoTMA (98.7 % e.e.) at its ratio in feed monomer mixture of 5-20%. The polymerization products were fractionated into tetrahydrofuran (THF)-soluble and insoluble parts, after being washed with a mixture of benzene and hexane (1/1 v/v) to remove oligomers. The THF-insoluble part was highly isotactic (mm 95~97%) and contained only small amount of PPyoTMA monomeric unit in the polymer chain. In contrast to this, the THF-soluble polymer was less isotactic (mm 82~87%) and contained larger amount of PPyoTMA monomeric units (13~22%). These results suggest that at least two kinds of propagating species exist in the copolymerization system, and one of them preferentially adds PDBSMA rather than PPyoTMA to form the THF-insoluble polymer with high isotacticity. The optical activity of the THF-soluble part depended sensitively on the content of PPyoTMA monomeric unit in the copolymer and was different from that expected from the content. The optical activity may be ascribed to helical conformation of the copolymer with preferential screw-sense induced by the chirality of PPyoTMA unit incorporated into the copolymer chain.

Polymerization of D3PyMA: The results of asymmetric anionic polymerization of D3PyMA are summarized in **Table 3.**[8] The polymers obtained with the Sp- and DDB complexes of DPEDA-Li showed much smaller optical activity than one-handed helical poly(TrMA)[1] and poly(D2PyMA)[2b,c] (runs 1 and 2 in **Table 3**); these polymers are considered to be mixtures of (+)- and (-)-polymers. The polymerization with the PMP complex gave better results (runs 3 and 4 in **Table 3**). In the polymerization at [D3PyMA]/[Li] ratio of 20, a highly isotactic polymer with large positive optical activity and narrow

Table 3. Asymmetric anionic polymerization of D3PyMA with DPEDA-Li complexes in toluene at -78 °C for 24 hr [a]

Run	Chiral ligand	[D3PyMA]/[Li]	Time (hr)	Yield[b] (%)	$[\alpha]_{365}^{25}$ [c] (deg)	DP[d,e]	M_w/M_n [e]	Tacticity(%)[f] mm	mr	rr
1	Sp	20	163	64	+73	57	1.56	89	8	3
2	(+)-DDB	20	24	70	+280	70	1.23	97	2	1
3	PMP	20	24	95	+882	46	1.06	~100	~0	~0
4	PMP	50	126	58	+505	78	1.45	98	1	1

[a]Conditions: D3PyMA 1.0g, toluene 20ml, [Ligand]/[Li] = 1.2.

[b]Methanol-insoluble part.

[c]Measured in $CHCl_3$ immediately after being dissolved.

[d,e,f]See captions d, e, and f of **Table 2.**

molecular weight distribution was obtained. However, in the polymerization with the same chiral ligand at a higher [D3PyMA]/[Li] ratio, the optical activity and isotacitiy of the obtained polymer were lower. This may be due to the ligation to the polar monomer to a Li cation of growing end as will be described for PB2PyMA polymerization. The obtained poly(D3PyMA)s were not conformationally stable, and the optical activity of the polymer with $[\alpha]_{365}$ +882° (run 3 in **Table 3**) slowly changed in chloroform at 60 °C to $[\alpha]_{365}$ -260°. During the mutarotation, no obvious change was observed in the ^1H NMR spectral pattern of the polymer. Therefore, this mutarotation may be based on helix-helix transition of the polymer chain which has been found for optically active poly(D2PyMA)s.[15] The change in sign of rotation suggests that a transition occurred from a less stable helix predominantly formed in the polymerization reaction at low temperature into a more stable one at 60°C.

The polymerization system of D3PyMA at low temperatures showed characteristic changes in optical rotation. **Figure 1** shows the change in optical activity of the polymerization system of D3PyMA with PMP-DPEDA-Li as an initiator at -78°C and the

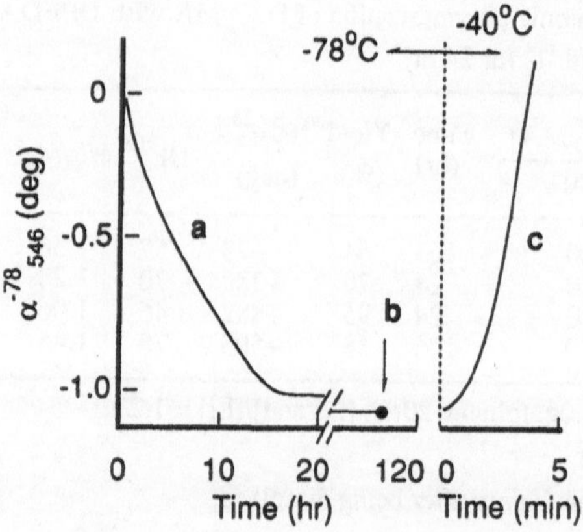

Figure 1. Change in optical activity of polymerization system of D3PyMA with PMP-DPEDA-Li (cell length 1 cm, D3PyMA 0.15 g, toluene 3 ml, [D3PyMA]/[Li] = 20): change during polymerization reaction at -78°C (curve **a**), termination reaction (point **b**), and change at -40°C after termination reaction (curve **c**).

change at -40°C after termination reaction by addition of a small amount of methanol at -78°C. The optical activity of polymerization system at -78°C slowly increased in negative direction after initiation reaction and reached a large negative value ($[\alpha]^{-78}_{546}$ -221°). This seems to correspond to the formation of a helical polymer anion with preferred screw sense. However, on warming the system to -40°C after termination reaction, the optical rotation quickly moved in positive direction. This observation and the large positive specific rotation of the isolated polymer suggest that some conformational transition occurred in isolation process. Because the mutarotation at -40°C is much faster than that of the isolated polymer at 60°C described above, change in propeller structure of side group without inversion of helix may be responsible for it.

The radical polymerization of D3PyMA at 40°C gave an isotactic polymer with mm 79%.

Polymerization of PB2PyMA[10]

Sp- and DDB-complexes were not effective in obtaining a one-handed helical polymer also in the anionic polymerization of PB2PyMA; the polymers obtained by using Sp and DDB as chiral ligands showed only small optical activity (runs 1 and 2 in **Table 4**). In contrast to these results, the results of radical polymerization were similar to those of TrMA;[12] the obtained polymers were rich in isotacticity (mm 59% in polymerization with AIBN in toluene at 60°C). These results suggest that the existence of two polar 2-pyridyl groups in PB2PyMA monomer causes difficulty in the stereocontrol of the anionic polymerization. The chiral ligands may be unable to exist around the growing end by coordinating to lithium cation due to the existence of the polar monomer.

Fortunately, PMP was found to be an effective chiral ligand for the helix-sense-selective polymerization of PB2PyMA. The highest optical rotation ($[\alpha]_{365}$ +1355°) was ob-

Table 4. Asymmetric anionic polymerization of PB2PyMA in toluene at -78°C for 24 hr[a]

Run	Initiator	[PB2PyMA]/[Li]	Yield[b] (%)	$[\alpha]^{25}_{365}$ [c] (deg)	DP[d,e]	Mw/Mn[e]	Tacticity (%)[f] mm	mr	rr
1	DDB-DPEDA-Li	20	34	+7	57	1.95	59	35	6
2	Sp-DPEDA-Li	20	37	-24	35	2.07	65	31	4
3	PMP-DPEDA-Li	20	94	+1355	73	1.58	97	2	1
4	PMP-DPEDA-Li	30	90	+960	97	2.22	86	12	2
5	PMP-DPEDA-Li	50	42	+152	52	2.31	71	24	5
6[g]	PMP-DPEDA-Li	15	90	+904	54	1.47	95	4	1
7[g]	PMP-DPEDA-Li	30	93	+1302	81	2.26	94	5	1
8	PMP-FlLi	20	91	+820	70	1.93	87	11	2
9	PMP-FlLi	40	48	+490	81	2.11	76	17	7

[a]Conditions: PB2PyMA 1.0 g, toluene 20 ml, [ligand]/[Li] = 1.2.
[b]Methanol-insoluble part.
[c]Measured in $CHCl_3$-TFE (9/1, v/v).
[d,e,f]See captions d, e, and f of **Table 2**.
[g]Polymerization was initiated at -98°C and maintained at the same temperature for 6 hr. Then the reaction system was warmed to -78°C and maintained for 24 hr.

served for the polymer obtained with the complex with DPEDA-Li at [PB2PyMA]/[Li] ratio of 20 at -78°C (run 3 in **Table 4**). The value of this rotation is comparable to that of poly(TrMA) having almost pure one-handed helicity. This polymer also showed a large circular dichroism absorption: [θ] ~+20000° at 260 nm in $CHCl_3$-2,2,2-trifluoroethanol (9/1, v/v).

The optical activity and isotacticity of the polymer depended on the type of organolithium compound, the ratios of monomer to initiator, and the temperature for initial stage of polymerization. The DPEDA-Li complex afforded the polymer with higher optical activity and isotacticity than FlLi complex under the same reaction condition (runs 3-5, and runs 8 and 9 in **Table 4**). For both polymerizations with DPEDA-Li and FlLi complexes at -78°C, higher specific rotation, yield, and isotacticity were attained at lower [PB2PyMA]/[Li] ratio. The existence of higher amount of the polar monomer per the amount of growing end at higher [PB2PyMA]/[Li] ratio appears to make the control of the polymerization more difficult. By conducting the initial stage of polymerization at -98°C, the better results were obtained: a polymer of higher optical activity and isotacticity was obtained at [PB2PyMA]/[Li] ratio of 30 (runs 4 and 7). The initiation reaction and initial stage propagation (formation of oligomer anions) at lower temperature may produce a larger amount of the oligomer anion which has a stereostructure suitable for producing a one-handed helix.[1d,e]

The optically active poly(PB2PyMA)s were also conformationally unstable and showed mutarotation (**Figure 2**). Optical rotation of the polymer decreased with time and the rate of the change depended on DP of the polymer. It is noteworthy that the stereomutation of poly(PB2PyMA) appears to proceed more rapidly than that of poly(D2PyMA). This suggests that the conformation of poly(PB2PyMA) is less stable

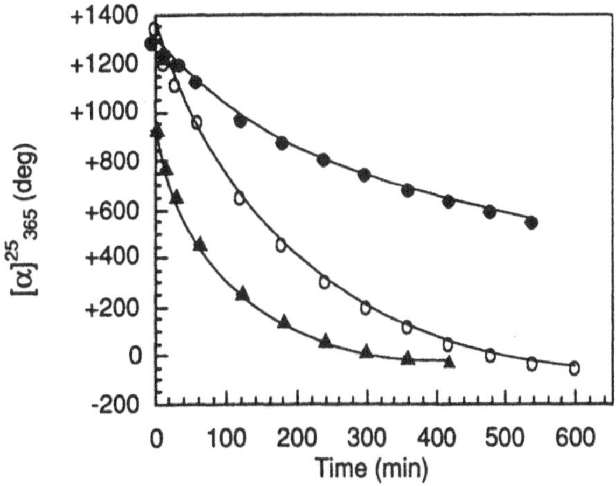

Figure 2. Change in optical activity of poly(PB2PyMA)s prepared with PMP-DPEDA-Li complex in $CHCl_3$-2,2,2-trifluoroethanol (9/1):● DP=81 (run 7 in **Table 4**), ▲ DP=73 (run 3 in **Table 4**), ○ DP=54 (run 6 in **Table 4**).

than that of poly(D2PyMA). The stability of helix is considered to be closely related with flexibility of the triarymethyl side group. A phenyl[bis(2-pyridyl)]methyl group seems more flexible than a diphenyl-2-pyridylmethyl group because a 2-pyridyl group is slightly smaller than a phenyl group. This may be the reason of lower conformational stability of poly(PB2PyMA).

The optically active poly(PB2PyMA) showed chiral recognition ability when used as a stationary phase for HPLC; however, the ability was lower than that of poly(TrMA).[7] The high chiral recognition ability of poly(TrMA) is related to the conformation of side group (propeller structure). The existence two polar 2-pyridyl groups and the flexibility of the phenyl[bis(2-pyridyl)]methyl group may be responsible to the lower recognition ability of poly(PB2PyMA).

REFERENCES

1 (a) Okamoto Y, Suzuki K, Ohta K, Hatada K, Yuki H (1979) J Am Chem Soc 101:4763

(b) Okamoto Y, Suzuki K, Yuki H (1980) J Polym Sci, Polym Chem Ed 18:3043

(c) Okamoto Y, Shohi H, Yuki H (1983) J Polym Sci, Polym Lett Ed 21:601

(d) Okamoto Y, Yashima E, Nakano T, Hatada K (1987) Chem Lett 1987: 759

(e) Nakano T, Okamoto Y, Hatada K (1992) J Am Chem Soc 114:1318

2 (a) Okamoto Y, Ishikura M, Hatada K, Yuki H (1983) Polym J 15: 851

(b) Okamoto Y, Mohri H, Hatada K (1988) Chem Lett 1988:1879

(c) Okamoto Y, Mohri H, Nakano T, Hatada K (1991)Chirality 3:277

3 (a) Okamoto Y, Yashima E, Hatada K (1987) J Polym Sc., Part C, Polym Lett 25:297

(b) Yashima E, Okamoto Y, Hatada K (1987) Polym J 19: 897

(c) Yashima E, Okamoto Y, Hatada K (1988) Macromolecules 21:854

4 (a) Okamoto Y, Nakano T, Asakura T, Mohri H, Hatada K (1991) J Polym Sci, Part A, Polym Chem 29:287

(b) Okamoto Y, Nakano T, Asakura T, Mohri H, Hatada K (1990) Polym Prepr 30(2):437

5 Okamoto Y, Nakano T, Ono E, Hatada K (1991) Chem Lett 1991:525

6 See for a review: Okamoto Y, Nakano T Chem Rev in press.

7 (a) Okamoto Y, Hatada K (1986) J Liq Chromatogr 9:369

(b) Okamoto Y, Hatada K (1988) In: Zief M, Crane LJ (ed) Chromatographic Chiral Separations. Dekker, New York

8 Nakano T, Mori M, Okamoto Y (1992) Polym Prepr Jpn (English Edition) 41:E880

9 Taniguchi K, Nakano T, Okamoto Y (1993) Polym Prepr Jpn (English Edition) 42:E106

10 C. Ren C, Chen C, Xi F, Nakano T, Okamoto Y J Polym Sci, Part A, Polym Chem in press.

11 Okamoto Y, Yashima E, Ishikura M, Hatada K (1987) Polym J 19:1183

12 Nakano T, Mori M, Okamoto Y (1993) Macromolecules 26: 867

13 Yuki H, Hatada K, Niinomi T, Kikuchi Y (1970) Polym J 1:36

14 Shikisai Y, Nakano T, Okamoto Y (1993) Polym Prepr Jpn (English Edition) 42:E48

15 Okamoto Y, Mohri H, Nakano T, Hatada K (1989) J Am Chem Soc 111:5952

Spectroscopic Studies of Conducting Polymers on Oxide Catalysts

G.J. Millar, G.F. McCann, G.A. Bowmaker and R.P. Cooney*

Department of Chemistry, University of Auckland,
Private Bag 92019, Auckland, New Zealand.

Abstract: A variety of spectroscopic techniques have been utilised to study the polymerisation of monomers such as acetylene or pyrrole within zeolite channels. The nature of the cations located in the zeolite framework was demonstrated to be a significant factor in the design of a successful catalyst. In particular cesium ions enhanced the degree of acetylene polymerisation relative to the situation with sodium, and copper ions were active for the polymerisation of pyrrole. Other factors such as the presence of Lewis acid sites and the coverage of adsorption sites by acetylene were also determined to be important in the mechanism for acetylene polymerisation. Resonance Raman studies indicated that polypyrrole formed within a zeolite channel was partially charged *i.e.* in the conducting state. Furthermore, this charge was associated with quinoid structures in the polymer chains. X-ray photoelectron spectroscopy (XPS) and photoacoustic infrared spectroscopy (PAIRS) data also provided evidence for the presence of a charged polymer segment. UV/VIS analysis allowed a detailed band description of the conducting polymer to be assimilated, and the level of doping to be estimated. ESR investigations further demonstrated formation of polaron charge carriers at the expense of Cu^{2+} .

INTRODUCTION

In recent years a vast amount of research has been devoted to the study of electrically conducting organic polymers(*1*,*2*). Applications for these materials include batteries, solar cells, electromagnetic shielding, sensors and "smart" windows, in addition to a myriad of other potential uses. Synthesis of conducting polymers can be achieved electrochemically (*3*,*4*) or catalytically (*5*). The basic concept in each case is to produce a conjugated polymer system which has been "doped" by exposure to an appropriate oxidising or

K.P. Ghiggino (Ed.)
Progress in Pacific Polymer Science 3

reducing agent(*6*). Polyacetylene in particular shows extremely high electrical conductivity in the "doped" state, with conductivities approaching that of copper in the some instances (*1*). However, its use as a conducting polymer has been limited due to its instability in air, therefore systems such as polypyrrole, polythiophene and polyaniline have received considerable attention as a consequence of their enhanced resistance to air and moisture (*7-9*). Electrochemical generation of for example polypyrrole films, results in formation of a disordered structure which is detrimental to attainment of optimium conductivity. One possible solution is to synthesise conducting polymers within the confines of a zeolite channel (*10-14*). An important aspect of this preparation procedure is that it allows the preparation of "molecular wires"with potential future application in shielding films and electrical circuits of the smallest conceivable size(*15*).

The polymerisation of various monomers within mordenite and ZSM-5 channels has been investigated by a variety of spectroscopic techniques, including resonance Raman, Photoacoustic Infrared (PAIRS) and X-ray photoelectron spectroscopy (XPS)(*16-19*). This paper summarises previous data and reports recent results from ESR and UV/VIS studies. Collation of this research allows the factors affecting the polymerisation reaction within zeolite channels to be identified.

EXPERIMENTAL

Cesium-exchanged mordenite zeolite (CsM) was prepared by exchange of sodium mordenite (Toyo Soda Manufacturing Co., TSZ 640, silicon to aluminium ratio 7.4, surface area 400 m^2g^{-1}) with 0.5 M aqueous solution of cesium chloride (BDH AnalaR Grade) at 295 K. Likewise the copper-exchanged form of mordenite was obtained by refluxing at 363 K, a 0.3 M aqueous solution of copper (II) nitrate (BDH AnalaR Grade) with sodium mordenite. ZSM-5 zeolite was supplied by Conteka Research and Development Centre, Sweden (silicon to aluminium ratio 33, surface area 397 m^2g^{-1}), and subsequently ion-exchanged by contact with an aqueous solution of copper (II) nitrate at 295 K. Pyrrole (Aldrich, 99%) was further purified by a series of freeze-pump-thaw cycles. Acetylene (NZIG, industrial grade) was purified by passage through a three-stage system which removed acetone contaminant (*20*).

Raman spectra were acquired using Jobin-Yvon U1000 and modified Cary 81 spectrometers coupled with a cooled GaAs photomultiplier, and samples excited with either a Spectra Physics Model 2016 argon ion laser or krypton ion laser. XPS measurements were performed using a Kratos XSAM800 spectrometer, and curve-fitting of the XPS data was executed assuming that the peaks were 80 % Gaussian and 20 % Lorentzian. Charging effects were taken into account by referencing peak positions to the strongest peak in the C1s region which was assigned a binding energy of 284.4 eV (*21*). FT-IR spectra were recorded on a Digilab FTS-60 spectrometer coupled to an MTEC PAC 200 photoacoustic detector at 4 cm^{-1} resolution. ESR spectra were recorded at room temperature with a Varian E-4 EPR spectrometer operating at 5 mW power, and UV/VIS spectra were acquired by a Shimadzu UV-2101PC spectrometer.

A typical experiment involved placing *ca.* 200 mg of zeolite powder into a reaction cell which was designed to connect to a conventional vacuum line, whereupon *in situ* pretreatment and reaction could occur. Zeolite samples were initially calcined in a flow of oxygen at 773-873 K for a period of up to 17 h, in order to remove any residual organic impurities. Following prolonged evacuation the appropriate monomer was contacted with the treated catalyst at ambient temperature, and the relevant spectra subsequently recorded.

RESULTS AND DISCUSSION

The mechanism for the polymerisation of acetylene is inherently different from that of aromatic monomers such as pyrrole or thiophene. Whereas the polymerisation of pyrrole or thiophene involves a redox reaction,(*11,14*) the corresponding reaction of acetylene is probably initiated by acidic properties of the catalyst.(22) In the case of polyacetylene evidence has been obtained to suggest that the nature of the cations in the zeolite lattice is also important.(*18*) Fig. 1 shows a series of Raman spectra which illustrate the influence of various cations upon the extent of polymerisation, demonstrate the effect of elevating the acetylene pressure and indicate a role for Lewis acid sites in the reaction mechanism. Exposure of acetylene (0.1 MPa) to sodium-mordenite (NaM) at 295 K gave the spectrum displayed in Fig. 1(a). Bands at 398 and 468 cm^{-1} are ascribed to lattice modes of the mordenite structure(*23*), whereas the peak at *ca.* 1958 cm^{-1} can be attributed to the ν_2 vibration of adsorbed monomeric acetylene bound in a "side-on" manner to cation sites (*16,23*). Relatively small maxima at 1112 and 1502 cm^{-1} are characteristic of *trans*-polyacetylene (*5,18,24,25*). Exchange of cesium for the sodium ions in mordenite was found to be beneficial for the formation of polyacetylene, as can be seen in Fig. 1 (b). In addition to the noted intensification of bands typical of *trans*-polyacetylene at 1112 and

1488 cm^{-1}, peaks developed at 1011 and 1297 cm^{-1} which could be ascribed to ν_7 and ν_2 modes of *trans*-polyacetylene, respectively(*18*). The promoting effect of cesium upon the polymerisation of acetylene is suggested to be due to two main factors. Firstly, the probability of finding two adsorbed acetylene monomers within sufficient proximity for reaction to occur is significantly greater with cesium, as supported by the data of Schlenker and co-workers (*26,27*) which shows a comparitively even distribution of cesium ions in the main mordenite channel and a more disordered arrangement in the case of sodium. Secondly, the separation between acetylene monomers on adjacent sodium sites in mordenite is greater than the corresponding situation with cesium, which again leads to a diminished chance for polymerisation.

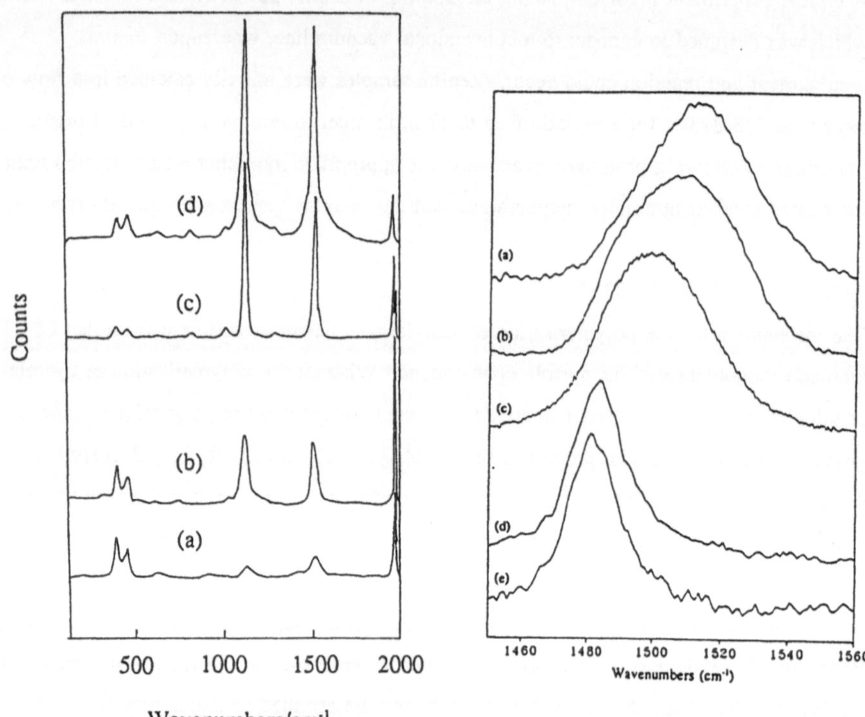

Fig.1 Raman spectra resulting from exposure of acetylene (0.1 MPa) at 295 K for 24 h to (a) NaM, (b) CsM. (c) 2 MPa acetylene on CsM and (d) 0.1 MPa acetylene on CCl_4 pretreated CsM.

Fig. 2 Variations in the ν_1 polyacetylene band on CsM (a) 457.9, (b) 488.0, (c) 514.5, (d) 647.1 and (e) 676.4 nm.

Justification for the hypothesis that reaction depends upon the coverage of monomeric acetylene was given by experiments involving the use of elevated acetylene pressure (2 MPa) [Fig. 1 (c)]. The concentration of adsorbed monomeric species was almost doubled by the application of high-pressure conditions, and concomitantly the quantity of *trans*-polyacetylene increased accordingly. However, the fact that substantial amounts of adsorbed monomeric acetylene were still observed suggested that another condition had to be satisfied for activation of the polymerisation reaction. As Dutta and Puri(22) have indicated that Lewis acid sites facilitate formation of *trans*-polyacetylene, the acidity of cesium mordenite catalyst was modified by pretreatment in CCl_4 vapour at 573 K for 1 h. In accord with the increased number of Lewis acid sites present,(*16*) the polymerisation reaction was significantly promoted [Fig. 1 (d)].

Variation of the Raman excitation wavelength used caused distinct changes in both the shape and position of bands due to *trans*-polyacetylene [Fig. 2]. For example the ν_1 mode was almost symmetrical in appearance, and is relatively broad (FWHM, *ca.* 45 cm^{-1}) when excited with shorter wavelength laser lines [Fig. 2 (a), (b) and (c)]. However, when excited by the red-lines of a krypton laser the band narrowed significantly (FWHM, 17 cm^{-1}). In addition the frequencies of the ν_1 and ν_3 modes for *trans*-polyacetylene shifted to lower frequencies as the excitation frequency was concomitantly decreased. For example, the ν_1 peak moved from 1510 to 1480 cm^{-1} as the excitation wavelength varied from 457.9 to 676.4 nm, respectively. This behaviour is interpreted as evidence that the *trans*-polyacetylene produced is composed of polymer segments of different conjugation lengths, a theory also proposed by several other authors (*5,22*). The segments of longer conjugation length (lower ν_1 and ν_3 frequencies) will have a π-π^* transition in resonance with longer excitation wavelengths, and those of shorter chain length with short excitation wavelengths. An estimate of the conjugation lengths involved was obtained by using the empirical relationships given by Harada *et al.*(*28*) and Baruya *et al.*(*29*). It was calculated that the *trans*-polyacetylene formed in this system consisted of chains containing a distribution of conjugation lengths ranging from 6 to over 30 double bonds.

Pyrrole Polymerisation

In contrast to the situation with acetylene, a successful catalyst for the polymerisation of pyrrole usually contains reducible cations. Bein and Enzel(*11*) have shown that copper exchanged zeolites were active for the polymerisation of pyrrole. In order to elucidate the mechanism for this process a comprehensive study of pyrrole interaction with both copper exchanged ZSM-5 and mordenite zeolites was performed.

The Raman spectrum of pyrrole in contact with CuZSM-5 at 295 K is shown in Fig. 3. Bands apparent at 1614, 1559 (sh), 1414, 1358, 1231, 1089, 1055, 983 and 938 cm^{-1} can all be ascribed to modes of polypyrrole (*8,17,30*). From the results of Umapathy and Hester(*31*) it can be concluded that the maxima at 1614, 1414, 1231, 1055 and 938 cm^{-1} provide evidence for an aromatic polypyrrole structure associated with neutral polymer segments. Furthermore, bands at 1559 (sh), 1089 and 983 cm^{-1} can be assigned to vibrations of a quinoidal type structure (*31*). The detection of quinoidal segments suggests that the polypyrrole is partially charged since theoretical studies predict that if a charge carrier such as a polaron or bipolaron is created, the aromatic polymer structure should relax locally around the charge to give a quinoidal arrangement (*32*).

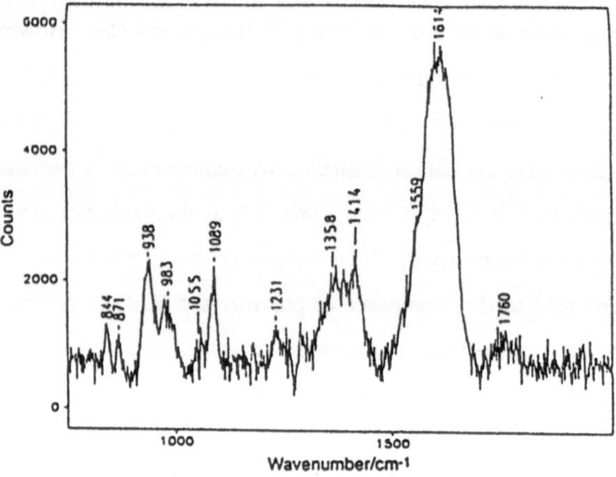

Fig. 3 Raman spectrum of polypyrrole within the channels of CuZSM-5.

Support for the hypothesis that the polymer produced within a zeolite lattice is partially charged is provided by the XPS profile displayed in Fig. 4, which displays the N1s profile obtained after exposure of cesium-mordenite to pyrrole at 295 K. The major peak at 400.5 eV can be assigned to a neutral nitrogen species in an aromatic polypyrrole chain (*19,33*), whereas the high-energy 402.5 eV peak is typical of a positively charged nitrogen species as might be associated with formation of a charge carrier. Finally the peak at 398.7 eV may be attributed to either aza-type nitrogen species which are formed by deprotonation of nitrogen atoms in the polypyrrole structure, or to monomeric pyrrole species (*33*).

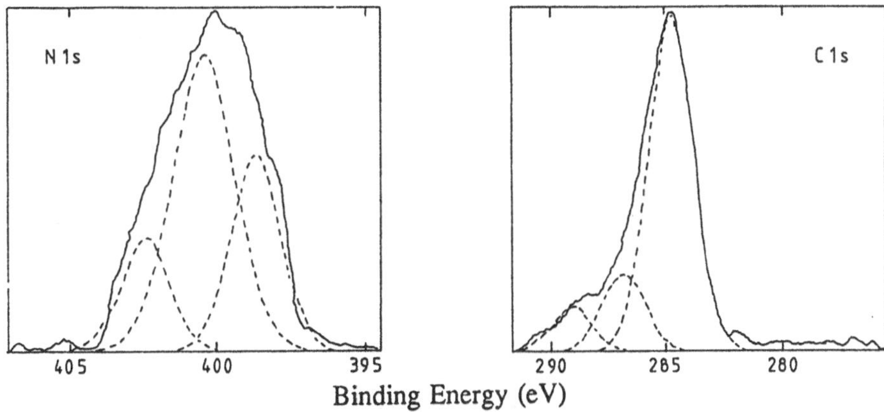

Fig. 4 Sub-band analysis of the XPS profile obtained after exposure of copper-exchanged mordenite to pyrrole vapour at 295 K.

The PAIRS spectrum depicted in Fig. 5 also substantiated these assignments as peaks were detected at 1572 and 1360 cm^{-1} characteristic of polypyrrole (*11*). Notably no monomer species were discerned, even though XPS and Raman data indicated their existence. However, this behaviour can be rationalised in terms of the ability of the appropriate modes on the charged polymer chain (note that a large dipole moment is created by the presence of the charge) to couple with the incident infrared radiation to produce a large enhancement in signal.

UV/VIS spectra of polypyrrole in copper-exchanged zeolite were obtained by mixing the powdered sample with glycerol in an analogous manner to that outlined by Bein and Enzel (*11*). Exposure of pyrrole vapour to two copper exchanged mordenite samples of differing copper concentration at 295 K gave the UV/VIS spectra shown in Fig. 6. Absorptions at 1.6, 1.9, 2.7, 2.9 and 3.6 eV were evident in the case of the sample consisting of a relatively large number of copper species [Fig. 6(a)]. Whereas, only absorptions at 1.6, 2.9 and 3.6 eV were determined for the catalyst containing a low concentration of copper ions [Fig. 6 (b)]. Bredas and Street (*34*) reported that the bandgap

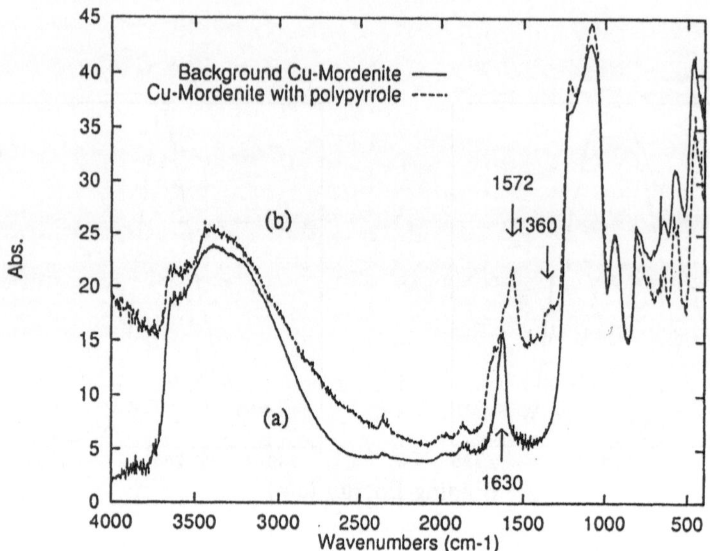

Fig. 5 PAIRS spectrum of pyrrole in contact with copper-exchanged mordenite (a) zeolite alone (b) zeolite and pyrrole

for neutral polypyrrole was 3.2 eV, and this value was predicted theoretically and verified experimentally to increase to 3.6 eV upon formation of the conducting state. Consequently the absorption at 3.6 eV determined in this study can be ascribed to a bandgap transition for the conducting form of polypyrrole. Bredas and Street(*34*) suggested that absorptions at 0.7, 1.4 and 2.1 eV present for polypyrrole doped at low levels, were characteristic of polaron species. Furthermore, as the doping levels increased, features at 1.0 and 2.7 eV developed which were taken to be indicative of bipolaron formation. Batz *et al.*(*35*) used HREELS to study doped polypyrrole films, and they noted that a low doping levels polarons caused loss features to be established at 2.0 and 2.5 eV. This structure was attributed to the fact that isolated polarons interacted with each other to produce a band with an appreciable width.

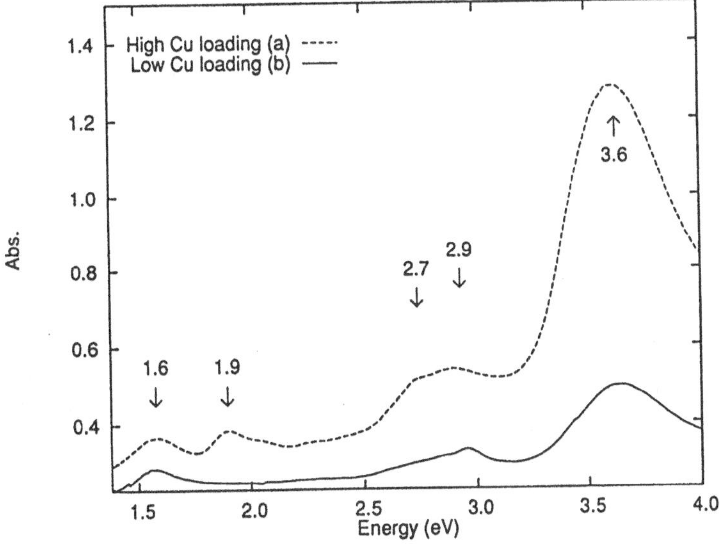

Fig. 6 UV/VIS spectrum of copper-exchanged mordenites exposed to pyrrole at 295 K.

Batz *et al.*(*35*) reported that the Fermi level initially shifted by 0.5 eV towards the valence band at very low doping concentration and then became pinned in the gap over a wide range of doping levels. This behaviour cannot be reconciled with bipolaron formation as the presence of these species would be expected to move the valence band maximum closer to the Fermi level. Hence Batz *et al.*(*35*) proposed that at high doping levels a metallic state was created and that the observed Fermi edge structure was caused by repulsive interactions between polaron species. In the present study absorptions were apparent at 1.6 and 2.9 eV for the catalyst containing only a relatively small number of copper species [Fig. 6 (b)]. These features can be assigned to transitions characteristic of isolated polaron species which are found at doping concentrations of < 1 % [Fig. 7 (a)]. It is noted that the valence band maximum is predicted to be *ca.* 1.2-1.4 eV below the Fermi level, by use of the displayed model, which is in excellent agreement with the value of 1.2 eV obtained using UPS by Batz *et al.*(*35*).

Increasing the concentration of copper ions in the zeolite lattice would be expected to enhance the doping of the polypyrrole chains. Thus additional absorptions which appeared at 1.9 and 2.7 eV are taken to be typical for transitions between polaron bands associated with extensively doped polypyrrole. The 2.7 eV absorption is ascribed to a transition

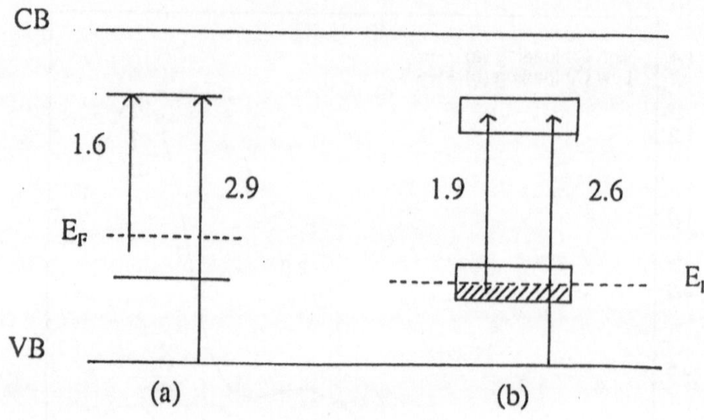

Fig. 7 Band diagram of levels in polypyrrole formed within copper-exchanged mordenite (a) "low" copper sample (b) "high" copper sample

from the valence band to the upper polaron band, and the 1.9 eV feature to a transition between the lower and upper polaron bands [Fig. 7 (b)]. This data is consistent with the observation of Batz *et al.*(*35*) that a shift in the Fermi level of *ca.* 0.5 eV occurred upon creation of polaron bands. Furthermore, the lowering of the upper polaron band by 0.2 eV was in accordance with theoretical calculations (*34*) which predict a shift in the upper polaron band of 0.2-0.3 eV closer to the valence band upon increasing the doping from low to moderate concentration.

It is relevant to re-examine data acquired using XPS, of the polypyrrole formed within the channels of copper mordenite. The C1s region [Fig. 4 (b)] contained a profile which could be deconvoluted into three peaks at 284.7, 286.9 and 289.1 eV. The main peak at 284.7 eV is ascribed to the α and β carbons of polypyrrole(*36*), and the feature at 289.1 eV to a "shake-up" peak due to the interband π-π^* transition (*19*). Finally the peak at 286.9 eV has been previously assigned to a "shake-up" structure caused by simultaneous electron excitation and transition from the valence band to a polaron level in the band gap (*19*). Since the "shake-up" process corresponds to a transition between an initially neutral state and a final ionized species the value obtained for the band gap using XPS would not initially correspond to that resulting from UV/VIS. This statement is verified by the fact that the "shake up" due to the interband transition was of 4.2-4.4 eV higher binding energy relative to the main C1s polypyrrole signal [Fig. 4 (b)], whereas UV/VIS measurements indicated a value of 3.6 eV for the band gap. Interestingly, correction of the "shake-up" value to 3.6 eV (a shift of 0.6-0.8 eV) would also change the value for the "shake up" due to the polaron band (normally at 2.2-2.4 eV) to *ca.* 1.6-1.8 eV which is in good agreement with UV/VIS data [Fig. 6].

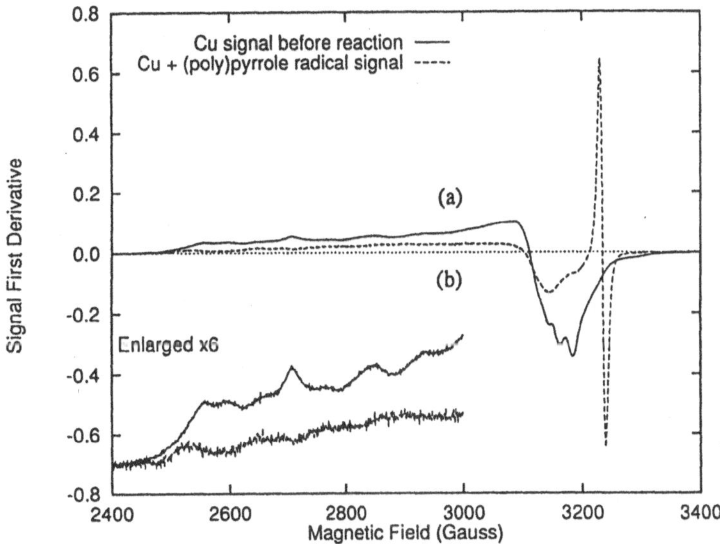

Fig. 8 ESR spectra of copper-exchanged mordenite (a) zeolite alone and (b) after exposure to pyrrole vapour at 295 K.

The formation of radicals during the polymerisation of pyrrole on copper mordenite was followed by ESR spectroscopy. Two mordenites of different Si/Al ratios were used to prepare host materials with identical structures, but containing different amounts of Cu^{2+} ions, identified here as "high Cu loading" and "low Cu loading" materials.

Cu^{2+} has a single unpaired electron of spin 1/2 and, in the randomly oriented zeolite powder, gave an anisotropic ESR signal with two distinct regions : the applied magnetic field is either perpendicular or parallel to the symmetry axis of the ion (*37*). Both the resonance components were split into quartets due to hyperfine coupling. The signals arising from the perpendicular interactions were weak but had a large hyperfine splitting constant and thus were well resolved. In contrast the signals from parallel interactions were inherently stronger but were not resolved due to a smaller hyperfine coupling constant. These features are illustrated in Fig. 8 (a) which shows the Cu signal measured before reaction with pyrrole monomer. The perpendicular resonances have been enlarged for clarity. These spectra indicated that there were two types of copper site in mordenite, the resonances overlap to give the complex pattern observed. Two copper sites have also been observed by other workers (*38-40*) and modelling of the spectra in Fig. 8 (a) was in accordance with parameters for one of the two sites. This is site (I) in the centre of the small channels in mordenite. The second site is likely to be site (IV) in the main channel, however there is no general agreement about the ESR spectrum arising from this site.

When pyrrole was introduced as a vapour at ambient temperature into the evacuated precalcined samples a radical signal was observed, as shown in Fig. 8 (b), which increased in intensity as the copper signal decreased. This behaviour suggested that the Cu^{2+} was directly associated with the formation of the conducting polymer and that the ESR radical signal is probably due to conducting polaron species. The peak area of the ESR signals are directly proportional to the number of spins for each species, therefore by using these figures the number of Cu^{2+} ions reacting to give each polymer radical could be calculated. For pyrrole on the low Cu loading sample *ca.* 100 Cu^{2+} ions were lost to form one radical (70 % of copper reacted) and for the high copper loading sample the figure drops to *ca.* 37 Cu^{2+} ions (50 % of copper reacted). These values indicate a doping level of approximately 1 % and 3 %, respectively for the two materials, which are in excellent agreement with the analysis of the UV/VIS spectra displayed in Fig. 6, and illustrated in Fig. 7.

CONCLUSIONS

The formation of conducting polymers within the confines of zeolite channels has been shown to depend on several factors, including the nature of the cation, reaction pressure and the presence of acid sites. From a detailed study of pyrrole polymerisation it is possible to postulate the following reaction mechanism. Interaction of pyrrole with a Cu^{2+} site initially gave a radical monomer species, H^+ (which presumably became associated with AlO_4^- sites in the framework) and a Cu^+ site. Further creation of radical polymers leads to a polymerisation reaction in which aromatic polymer chains were formed. As the polymer chain lengthens the energy levels for the π system are lowered and thus oxidation of the polymer becomes more facile. Therefore removal of an electron from the π system results in the formation of a polaron species associated with a quinoid type structural defect.

ACKNOWLEDGEMENTS

One of us (GJM) would like to thank the New Zealand Vice Chancellors Committee for a postoctoral fellowship. We would also like to thank the University of Auckland Research Commitee and the New Zealand Lottery Science Grants Board for financial contribution to this project.

REFERENCES

1. Kanatzidis MG, *Chem. Eng. News*, **68**, 36 (1990).

2. Skotheim TA (Ed.), *Handbook of Conducting Polymers*, vol. 1, Marcel Dekker, New York (1989).

3. Ofer D, Crooks RM and Wrighton MS, *J. Am. Chem. Soc.*, **112**, 7869 (1990).

4. Wang JH, *Surf. Int. Anal.*, **15**, 635 (1990).

5. Rives-Arnau V and Sheppard N, *J. Chem. Soc. Faraday Trans. 1*, **76**, 394 (1980).

6. Roncali J, *Chem. Rev.*, **92**, 711 (1992).

7. Pfluger P, Krounbi M, Street GB and Weiser G, *J. Chem. Phys.*, **78**, 3212 (1983).

8. Cheung KM, Smith BJE, Batchelder DN and Bloor D, *Synth. Met.*, **21**, 249 (1987).

9. Patil AO, Heeger AJ and Wudl F., *Chem. Rev.*, **88**, 183 (1988).

10. Tsai P, Cooney RP, Heaviside H and Hendra PJ, *Chem. Phys. Lett.*, **59**, 510 (1978).

11. Bein T and Enzel P, *Angew. Chem. Int. Ed. Engl.*, **28**, 1692 (1989).

12. Larsen G, Haller GL and Marquez M, *J. Phys. Chem.*, **96**, 4145 (1992).

13. Enzel P and Bein T, *J. Phys. Chem.*, **93**, 6270 (1989).

14. Enzel P and Bein T, *J. Chem. Soc. Chem. Commun.*, 1326 (1989).

15. Carter FL (Ed.), *Molecular Electronic Devices II*, Marcel Dekker, New York (1987).

16. Millar GJ, Lewis AR, Bowmaker GA and Cooney RP, *J. Mater. Chem.*, **3**, 867 (1993).

17. Blanking JR, Millar GJ, Bowmaker GA and Cooney RP, *J. Raman Spectrosc.*, **24**, 523 (1993).

18. Lewis AR, Millar GJ, Cooney RP and Bowmaker GA, *Chem. Mater.*, in press.

19. Millar GJ, McCann GF, Hobbis CM, Bowmaker GA and Cooney RP, *J. Chem. Soc. Faraday Trans.*, submitted for publication.

20. Tam NT, Cooney RP and Curthoys G, *J. Chem. Soc. Faraday Trans. 1*, **72**, 2577 (1976).

21. Barr TL and Lishka MA, *J. Am. Chem. Soc.*, **108**, 3178 (1986).

22. Dutta PK and Puri M, *J. Catal.*, **111**, 453 (1988).

23. Pechar F and Rykl D, *Zeolites*, **3**, 329 (1983).

24. Lefrant S, Litchmann LS, Temkin H and Fitchen DB, *Solid State Commun.*, **29**, 191 (1979).

25. Takeuchi H, Furukawa Y, Harada I and Shirakawa H, *J. Chem. Phys.*, **84**, 2882 (1986).

26. Schlenker JL, Pluth JJ and Smith JV, *Mater. Res. Bull.*, **14**, 751 (1979).

27. Schlenker JL, Pluth JJ and Smith JV, *Mater. Res. Bull.*, **13**, 901 (1978).

28. Harada I, Furukawa Y, Arakawa T, Takeuchi H and Shirakawa H, *Mol. Cryst. Liq. Cryst.*, **117**, 335 (1985).

29. Baruya A, Gerrard DL and Maddams WF, *Macromolecules*, **16**, 578 (1983).

30. Tian B. and Zerbi G., *J. Chem. Phys.*, **92**, 3887 (1990).

31. Umapathy S and Hester RE, *J. Mol. Struct.*, **224**, 113 (1990).

32. Bredas JL, Themans B, Fripiat JG, Andre JM and Chance RR, *Phys. Rev. B*, **29**, 6761 (1984).

33. Zeller MV and Hahn SJ, *Surf. Int. Anal.*, **11**, 327 (1988).

34. Bredas JL and Street GB, *Acc. Chem. Res.*, **18**, 309 (1985).

35. Batz P, Schmeisser D and Gopel W, *Phys. Rev. B*, 43, 9178 (1991).

36. Pfluger P and Street GB, *J. Chem. Phys.*, **80**, 544 (1980).

37. Pinnavaia TJ in *"Advanced Techniques for Clay Mineral Analysis, Developments in Sedimentology"*, vol 34, page 139, Elsevier, Amsterdam, 1982.

38. Schoonheydt RA, *Catal. Rev. Sci. Eng.*, **35**, 129 (1993).

39. Sass CE and Kevan L, *J. Phys. Chem.*, **93**, 4469 (1989).

40. De Tavernier S and Schoonheydt RA, *Zeolites*, **11**, 155 (1991).

Synthesis of Poly (p-Epoxystyrene-co-N-Phenylmaleimide-co-styrene)

B. Pulpoka[1], P.C. Ternai[1], F. Schue'[2], S. Tantayanon[1]*

[1] Department of Chemistry, Faculty of Science, Chulalongkorn University, Thailand.
[2] Laboratoire Chimie Macro-moleculaire, Universite Montpellier II, France.

Abstract: The photosensitive negative resists containing p-epoxystyrene were synthesized. The synthesis was performed starting from chloromethylation of 2-phenylethyl bromide, then dehydrobromination and copolymerization with N-phenylmaleimide and styrene to yield p-chloromethylstyrene-N-phenylmaleimide-styrene copolymer. The comparison of glass transition temperature of these copolymers and with poly(p-chloromethylstyrene-co-styrene) exhibited that the existing N-phenylmaleimide moiety in the polymer chain affected on the glass transition temperature. The more N-phenylmaleimide moiety was in the polymer chain, the higher the glass transition temperature was. The conversion of the chloromethyl group to the carboxaldehyde group and then to the epoxide group were consequently carried out. Accordingly, poly(p-epoxystyrene-co-N-phenylmalcimidc-co-styrene) could be synthesized. However, by using this synthetic route, the chloromethyl and carboxaldehyde groups would be retained partially in the copolymer.

INTRODUCTION

The modern world is becoming more and more dependent upon sophisticated methods of information acquisition, storage and retrieval, and the sheer volume of information processed dictates the methods be developed for very high density recordings and rapid manipulations of data. Solid state electronics plays a key role in this function through integrated circuitry (IC). A tiny modular silicon "chip" contains all the elements of an electronic circuit such as transistors, capacitors, etc. associated and connected to each other by metallic interconnects on the surface of the chip. Today, upward of several thousand circuits, each incorporating a dozen transistors or other devices, can be put onto a silicon chip a few millimeters square. Within a few years, up to 50,000 circuits will be integrated on a similar chip.

One of the mainstays of this technology is the evolution of ever new methods and materials for resists. In these microfabrication methods a resist is a radiation sensitive material that

* corresponding author

K.P. Ghiggino (Ed.)
Progress in Pacific Polymer Science 3

is coated on a Si/SiO_2 wafer and exposed to radiation through a patterned mask. The exposure causes differential solubility of the resist betweeen exposed and unexposed regions and on solvent development, the pattern on the mask, formed by the resist, is left on the wafer. Subsequent etching of the SiO_2 along the pattern and removal of remaining resist generates the circuit pattern on the wafer. Organic polymers have often been tailor-made to improve sensitivities, resolutions, and processabilities of these resists.

The resists used in microlithography to protect the surface of the substrate during the lithographic steps are classified into two classes depending on their behavior under irradiation. Positive resists become more soluble in the irradiated area relative to the unexposed area, whereas negative resists become less soluble in the irradiated area than in the unexposed area. Negative resists, in which crosslikings happen, are polymers bearing unsaturated double bonds (1) or other reactive groups e.g. vinyl groups, episulfur groups, and epoxide groups.

Though many classes of compounds have been examined in the past as negative e-beam resists (2), recent work has concentrated on synthesizing and evaluating polymers containing the epoxide group. In addition, It has generally been shown by many experiments that the epoxy groups as side chains of the polymer, as opposed to in the backbone, results in marked improvement in the contrast of the resist.

Resistance to deformation due to flow during high temperature processing is a property that is deficient in most currently available photoresists. This deficiency in thermal stability is primarily due to the relatively low Tg of the polymers which range between 70-120 °C depending on structure and molecular weight. The stability upto 200 °C are desired for several processes used in semiconductor manufacturing (3). In this research, the synthesis of new negative e-beam resist, poly(p-epoxystyrene-co-N-phenylmaleimide-co-styrene) was studied. It was expected that the incorporation of N-phenylemaleimide into the polymer backbone would raise its glass transition temperature which would in turn increase the thermal stability of the resist.

EXPERIMENTAL

Materials

Styrene was commercial product (Fluka AG) ; p-chloromethylstyrene was synthesized from 2-phenylethyl bromide using the method of Kondo et al (4); N-phenylmaleimide was prepared following the method described by Cava et al (5).

Polymerization Procedure

Polymerizations were performed homogeneously in benzene as solvent at 60°C in the presence of 0.3 wt% of AIBN for 7 hours with stirring rate of 75 rpm. The reaction mixture was poured into methanol. The precipitate was filtered off, redissolved in benzene and reprecipitating from methanol. This process was repeated three times. It was then dried in vacuo overnight at 60 °C.

Carboxylation

The polymer was stirred in dimethylsulfoxide with 3.00 g. of sodium bicarbonate at 155 °C for 6 hrs. The resin was then collected on a glass filter, washed with dimethylsulfoxide, hot water and 2:1 mixture of 1,4-dioxane and water then subsequently rinsed with 1,4-dioxane, acetone, ethanol, methylene chloride and benzene. The resulting product was dried under vacuum. The chloride content in the resulting polymer was determined by modified Volhard method.

Epoxidation

In an ice bath, 20 ml. of 50% sodium hydroxide was slowly added into 50 ml.of methylene chloride solution containing the polymer, 0.25 g. of benzyltriethyl ammonium chloride and 0.83 g. of trimethylsulfonium chloride*. After the complete addition, the solution mixture was brought to room temperature and continued stirring for 12 hrs. The resulting polymer was then collected and washed with water, methanol and ether. After drying at 70 °C under reduced pressure, the product was characterized and determined for the epoxide content by the modified iodometric method.

Characterization

The molecular weights and molecular weight distributions were determined by a Waters High Pressure GPC instrument (Model 600 E). DSC analysis was performed on a Dupout Instrument DSC 910s. Elemental analysis was performed on a Perkin Elmer Elemental Analyzer (Model 240). The infrared spectra were recorded on a Perkin Elmer Infrared Spectrophotometer (Model 1430).

RESULTS AND DISCUSSION

Two monomers, p-chloromethylstyrene and N-phenylmaleimide were synthesized in our laboratory with the acceptable percentage yields. Solution polymerizations of p-chloromethylstyrene, N-phenylmaleimide and styrene were then carried out with variation of the amount of N-phenylmaleimide and styrene while the moles of p-chloromethylstyrene in the feed were kept constant at 50%. At the same reaction condition, the polymer products were obtained with different composition of the incorporated monomers as revealed by the elemental analysis for percentage of nitrogen and chlorine which related to the numbers of N-phenylmaleimide and chloromethylstyrene, respectively. As the more N-phenylmaleimide was in the feed, the more of it was incorporated into the copolymer while the less chloromethylstyrene was found (Figure 1). The molecular weight of the resulting

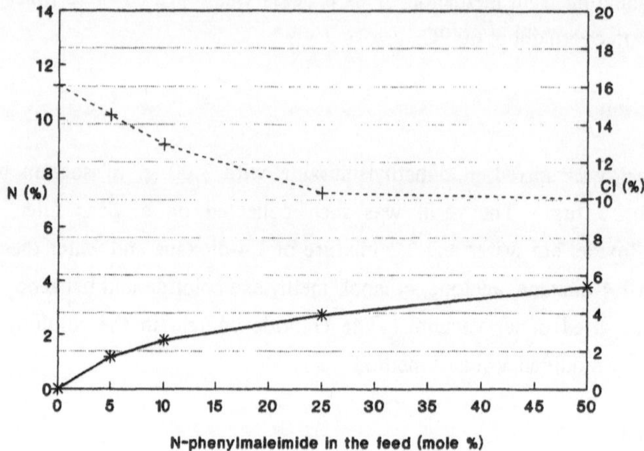

Figure 1 The incorporation of N-phenylmaleimide and p-chloromethylstyrene into the copolymer

copolymers also increased with the increasing amount of N-phenylmaleimide in the feed and their molecular weight distributions were quite large (Table1). This could be explained

Table 1 Synthesis and characterization of Poly(p-chloromethylstyrene-co-N-phenyl-maleimide-co-styrene)

Feed composition (% mole)			Mn	Mw	MWD	Tg (°C)
P-CMSt	N-PhMI	Styrene				
50	0	50	12,971	32,994	2.55	95
50	5	45	12,921	49,326	3.19	101
50	10	40	15,242	87,324	5.73	103
50	25	25	32,639	143,302	4.39	134
50	50	0	35,968	148,590	4.13	181

P-CMSt = p-chloromethylstyrene
N-PhMI = N-phenylmaleimide

that the bulkiness of N-phenylmaleimide induced the slow termination between polymer radicals similar to the observation reported by Otsu et al (6).

The conversion of the chloromethyl to the carboxaldehyde groups was performed by the treatment with dimethylsulfoxide in the presence of sodium bicarbonate. It was monitored by IR spectroscopy (Figure 2 and 3) which exhibited the aldehydic C-H

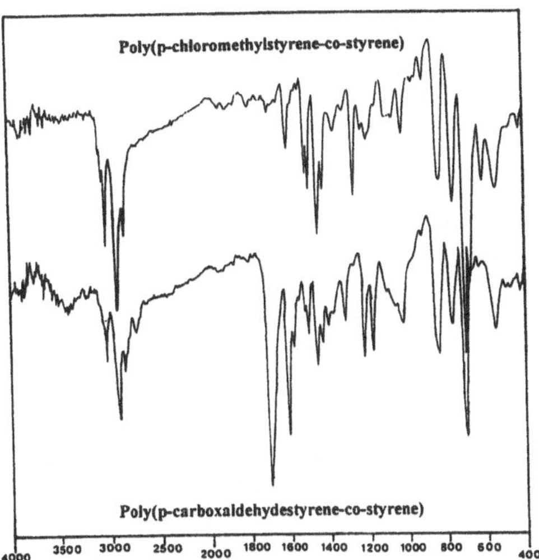

Figure 2 IR (KBr) spectra of poly(p-chloromethylstyrene-co-styrene) and poly(p-carboxaldehydestyrene-co-styrene)

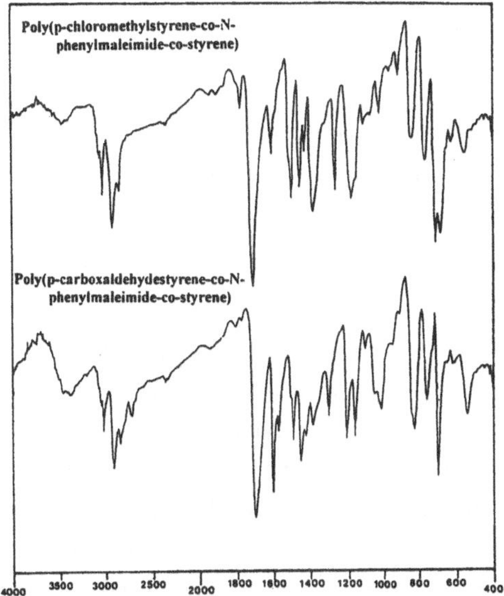

Figure 3 IR (KBr) spectra of poly(p-chloromethylstyrene-co-N-phenylmaleimide-co-styrene) and poly(p-carboxaldehydestyrene-co-N-phenylmaleimide-co-styrene)

stretching absorption band at 2,724 cm^{-1} and the increase in the intensity of carbonyl absorption band at 1,700 cm^{-1}. Concurrently, the absorption bands of C-Cl bending at 1,212 cm^{-1} and C-Cl stretching at 700 cm^{-1} disappeared.

The determination for chloride content in the polymer before and after oxidation using modified Volhard method indicated that the chloromethyl group could not be converted completely but very high percentage of conversion was achieved in each case (Table 2).

Table 2 Chloride content in the polymers before and after the conversion of the chloromethyl to carboxaldehyde groups

N-PhMI (mole %)	Cl^- content(mmole/g)		% Conversion
	Before	After	
0	4.0097	0.0874	97.81
5	3.8906	0.0642	98.35
10	3.7057	0.0736	98.06
25	3.3502	0.4526	86.49
50	3.2046	0.4317	86.53

The presence of N-phenylmaleimide in the polymer did affect on the conversion of the chloromethyl to the carboxaldehyde group since lower conversion was obtained in the case of the composition of N-phenylmaleimide in the feed was 25 and 50% more. This should be resulted from the bulkiness of N-phenylmaleimide as well.

The transformation of the carboxadehyde group to the epoxide was performed through the phase transfer reaction using trimthyl sulfoniun chloride and 50% aqueous sodium hydroxide in the presence of benzyltriethyl ammonium chloride. The formation of epoxide was confirmed by modified iodometric method which also revealed the exposide content in the polymer (Table 3). The same effect due to the bulkiness of N-phenylmaleimide moiety in the polymer caused lower conversion of the carboxaldehyde group to epoxide when the percentage of N-phenylmaleimide in the feed, in other words, its moiety in the polymer was high.

Table 3 Epoxide content in the polymers after epoxidation

N-PhMI (mole %)	Carbonyl Content (mmole/g)	Epoxide Content (mmole/g)	% Conversion
0	3.9223	3.0574	77.95
5	3.8264	3.0095	78.65
10	3.7221	2.8053	75.37
25	2.8976	1.7099	59.01
50	2.7729	1.4824	53.46

Accordingly, poly(p-epoxystyrene-co-phenylmaleimide-co-styrene) could be synthesized. The whole synthetic route was exhibited in Scheme 1.

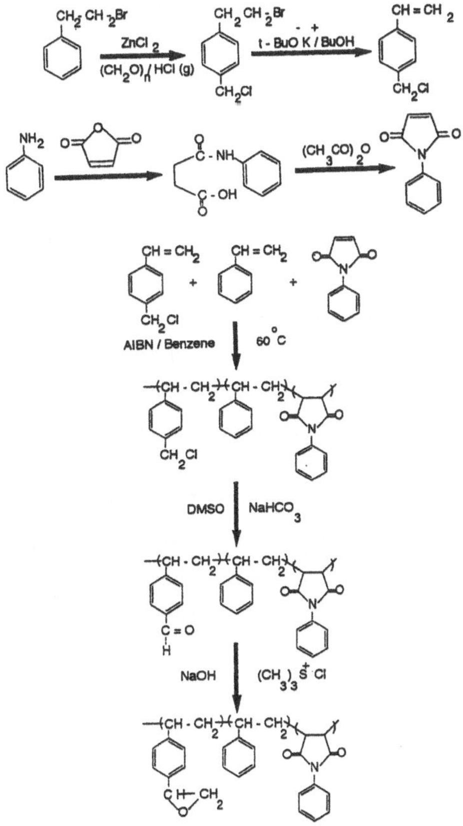

Scheme 1 The synthetic route of poly(p-epoxystyrene-co-N-phenylmaleimide-co- styrene)

As shown in Table 4, when mole % of N-phenylmaleimide in the feed was increased, the nitrogen content in the copolymer also increased and so was the glass transition temperature.

Table 4 Synthesized poly(p-epoxystyrene-co-N-phenylmaleimide-co-styrene

NPhMI* (mole %)	Nitrogen** content (%)	Functional group content (mmole/g. polymer)			Tg (^{o}C)
		$-CH_2-Cl$	-C(=O)-H	$-CH-CH_2$ (epoxide, O)	
-	-	0.0847	0.8649	3.0574	102.7
5	1.19	0.0642	0.8169	3.0095	111.8
10	1.79	0.0736	0.9168	2.8053	127.3
25	2.71	0.4526	1.1877	1.7099	148.5
50	3.77	0.4317	1.2905	1.4824	195.7

* N-phenylmaleimide in the feed
* by elemenutal analysis

Unfortunately, by using this synthetic route the chloromethyl and caboxaldelyde groups would be retained partially in the copolymer.

CONCLUSION

It was primarily expected that high percentage of epoxide could be introduced into the polymer through this new synthetic route. It was clearly shown that the conversion of the chloromethyl to carboxaldehyde and then to the epoxy groups could not occur completely. However, the incorporation of N-phenylmaleimide into poly(p-epoxystyrene-co-styrene) could raise its glass transition temperature, although this may change some lithographic properties of poly(p-epoxystyrene-co-styrene). The percentage of N-phenylmaleimide in the polymer should also be adjusted to optimize its functional characteristic as the polymeric resist system in order to be useful in the microlithographic application. This should thus be further studied and subjected to resist characterization.

ACKNOWLEDGEMENT

The authors are grateful to the National Metal and Materials Technology Center, National Science and Technology Development Agency of Thailand for fully financial support of this research work.

REFERENCES

1 Biryukova TG, Voskoboinik GA, Molodniyakov SP, Soboleva EB, Fedorov YI (1978) Vys Soed Ser R 20:565

2 Hatzakis M (1974) Polym Eng Sci 14:516

3 Turner SR, Ahn KD, Willson CG (1987) ACS Symposium Series 346: 200

4 Kondo ST, Ohtsuda K, Ogura K, Tsuda K (1979) J Macromol Sci Chem A13(6):767

5 Cava MP, Deana K, Mitchell MJ (1973) Organic Syntheses Collective Vol. 5:944

6 Otsu T, Masumoto A, Kubota T, Mori S (1990) Polym Bull 23:43

Production, Modifications and Applications of Chitosan in Vietnam

DANG VAN LUYEN
Laboratory of Biopolymers - Institute of Chemistry
National Center for Natural Science and Technology
Nghia do, Tu liem, Hanoi, Vietnam

Chitin is a cellulose-like biopolymer distributed widely in nature. Chitosan is the deacetylated product of chitin. The chemical structure of chitin and chitosan can be seen to be very similar to that of cellulose, the difference being that OH group of each glucose unit in cellulose is substituted by an acetylated amino group ($-NHCOCH_3$) in chitin and an amino group ($-NH_2$) in chitosan.

Chitin and chitosan have been in recent years of great interest to scientists and technologists because of the variety of their commercial applications. Research and development on chitin/chitosan has been conducted in our Laboratory of Biopolymers for a number of years because of the large volume of waste shrimp shells which are derived from the ever-growing frozen shrimp industry in Vietnam [1].

In this report, the production of chitin and chitosan, the preparation of their derivatives and some applications in Vietnam were described.

A. PRODUCTION OF CHITIN/CHITOSAN

After cellulose, chitin is the most abundant natural polymer. It is an essential component of the skeletal material of crustacean shells, insect cuticles as well as in cephalopods and fungi. In the seafood industry, chitin sources are the exoskeletons of shrimp, crab and lobster and also squid pens and cuttlefish bones. The important commercial sources of chitin is crab and shrimp shells.

K.P. Ghiggino (Ed.)
Progress in Pacific Polymer Science 3

The raw material for production of chitin in Vietnam, generally shrimp shells, are readily available as waste originating in much larger quantities and with less seasonal fluctuation in supply than in previous years because of the expansion and centralization of the frozen shrimp industry.

A method for the production of chitin and chitosan from shrimp waste is given in Fig.1.

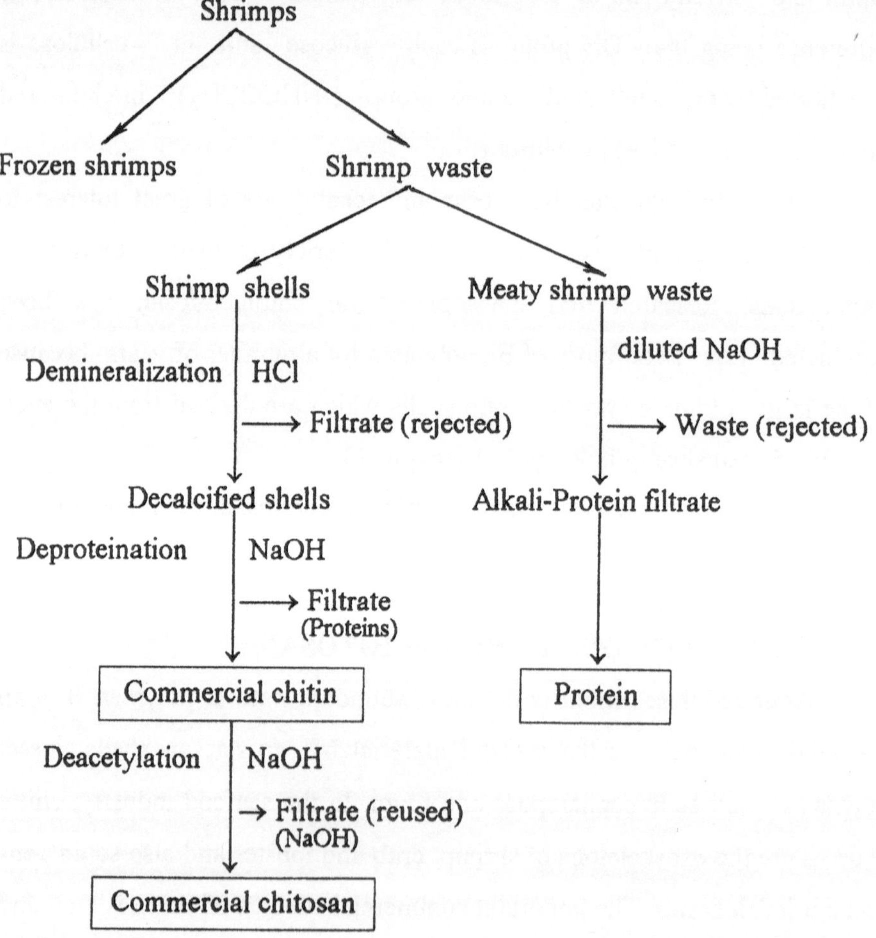

Fig. 1: Simplified flow-chart of a chitin/chitosan manufacturing process

The production of chitin involves the treatment of raw materials with dilute acid for demineralization (removal of especially $CaCO_3$) and with dilute alkali to remove the proteins. Chitin can be deacetylated using hot and concentrated alkali to form chitosan.

- Raw material

If possible, wet raw material, or if not available, dried waste one could be used. The raw material has to be of good quality.

The shrimp waste was washed to remove sand and dirt and was divided into two categories: shrimp shells and meaty shrimp waste.

- Separation of protein

The meaty shrimp waste was immersed in 0.5% aqueous sodium hydroxide and boiled for approximately 30 minutes. The alkali-protein filtrate was neutralised, initially by hydrochloric acid (HCl) and then by acetic acid (CH_3COOH) towards the end. The filtrate was then concentrated by boiling to a protein paste.

- Demineralization

The shrimp shells were immersed in 1N HCl and kept at room temperature for 12 hours. The acid-mineral solution was drained off and the residue was washed with fresh water till free of acid.

- Deproteination

The residual protein was removed from the decalcified shells by boiling with 4% NaOH solution for 3 hours. The dilute protein filtrate was drained off. The residue was washed with fresh water, collected and pressed to remove as much water as possible. The finished product i.e. commercial chitin was dried up and packaged. An overall recovery of approximately 90% of the chitin present in shrimp shells can be obtained.

• Deacetylation:

In the conventional deacetylation process of chitin, concentrated NaOH solution and high temperature have been used. The obtained chitosan cannot have a large degree of polymerization than that of the original chitin. In order to avoid the degradation of the polysaccharide, the alkali treatment should be carried out at as low a temperature as possible. We have carried out a special heat treatment prior to deacetylation for increasing the reactivity of chitin and the chitin thus treated has been converted into chitosan at temperatures always below 100°C. [2]

40% NaOH solution was added sufficiently to submerge the chitin. The water present in the chitin flakes should be taken into account to obtain 40% NaOH solution. The mixture was heated up to 80°C-85°C for approximately 2 hours. The conversion was controlled by the solubility of the material in 1% CH_3COOH. The alkali solution was stored for re-use (after adjusting to required strength). The residue was washed with clean water, pressed to remove the water and dried in a dryer or in sunlight, keeping the temperature below 60°C to give the finished product, i.e. commercial chitosan.

For a yield of 10 kg chitosan one needs 90 kg dried shrimp shells.

This modified method of chitosan production allows us to use simple equipment and to obtain chitosan of high quality and enables us to elaborate a small factory producing 10 tons of high quality chitosan per year.

The deacetylation of chitin was also carried out with strong alkali at ambient temperature without heating for several days. Chitosan was obtained by using 50% alkali concentration and ratio of 50% NaOH/chitin = 20/1. By this simple technique, different products were prepared according to the reaction time. Chitosan was produced in 5-6 days and chitosan oligomers were obtained by increasing the mixing days. The use of mixing days for producing oligomers is more economic than the use of higher alkali concentration or higher volume ratio of NaOH solution. This method will save power and the cost of equipment because of the simplicity of the handling and recovery of the alkali.

Some characteristics of chitosan are shown in Table 1.

Table 1

General characteristics of chitosan produced in pilot plant

Appearance	0.5 ÷ 8 mm dry flakes
Colour of chitosan flakes	white
Colour of 1% chitosan solution (in 1% acetic acid)	transparent, colorless
Moisture content, (of air dried product)	less than 10%
Ash content	less than 0.5%
Degree of deacetylation	85 ÷ 90%
Molecular weight	$2 \div 3 \times 10^5$

B. MODIFICATION OF CHITIN

Our special attention has been paid only on the study of two kinds of chitin derivatives, i.e., derivatives soluble in water and those soluble in commonly used organic solvents because chitin is much more cheaper than chitosan. As the matter of fact, we have concentrated our efforts only on chitin derivatives soluble in organic solvents for the reason that they are, of great fundamental research interest.

Despite its abundance in nature and numerous potential uses, chitin still remains an almost unexplored biomass resource and has limitation in practical application because of its poor solubility in commonly used solvents.

As the introduction of an alkyl chain into the chitin molecule is likely to improve the solubility in organic solvents, it will be interesting to find new derivatives for promoting its usefulness and widening its practical applications.

With a view to introducing substituents into the hydroxyl groups of chitin, the reactions with halogen carboxylic acid for preparing carboxyalkylchitin, and with carboxylic anhydrides or carboxylic chlorides for obtaining acylchitins, are promising.

6-0-carboxymethylchitin and acylderivatives of chitin have been reported [3,4]. I would like to present here the preparation and characterization of some chitin acylderivatives and chitin mixed esters in order to obtain products with good solubility in various organic solvents. [5]

The preparation of acylchitin and chitin mixed esters was performed as in [4]. Methanesulfonic acid was found to be a suitable solvent and catalyst for the homogeneous acylation reaction. A mixture of 4 parts of CH_3SO_3H and 6 parts of carboxylic anhydride for one part of chitin powder was employed. When a smaller amount of CH_3SO_3H is employed, it is difficult to obtain fully acylated chitin owing to the heterogeneity of the reaction mixture. The temperature for the reaction was kept at 0 - 5°C to prevent decomposition of the chitin by CH_3SO_3H. In the case of chitin mixed esters the products were obtained by changing the amounts of carboxylic anhydrides.

• The degree of substitution (DS) were determined by elemental analysis and were found to be very high in a range of 0.9 to 0.95 indicating that the reaction had been proceeded almost quantitatively under these conditions.

• The intrinsic viscosity of butyrylchitin was 2.5 dl/g (acetone, 20°C), higher than that of butyrylchitin in [6] 0.66 dl/g.

• The DSC-diagram of butyrylchitin shows a large endotherm in the range of 60 - 120°C due to water evaporation and does not reveal any transition to 281°C, similar to the DSC measurement of chitosan reported in [7].

• Infrared absorption spectra of various acylated chitins are shown in Fig. (2). On proceeding with the acylation the absorptions at 1750 - 1730 cm^{-1} and 1260 - 1210cm^{-1}, which are characteristic for the ester of saturated fatty acid, increased extensively but the absorption at 3600-3200cm^{-1} due to OH stretching decreased noticeably.

Furthermore, the absorptions due to the aliphatic-methylene chain also appeared at 2925, 2850cm^{-1}.

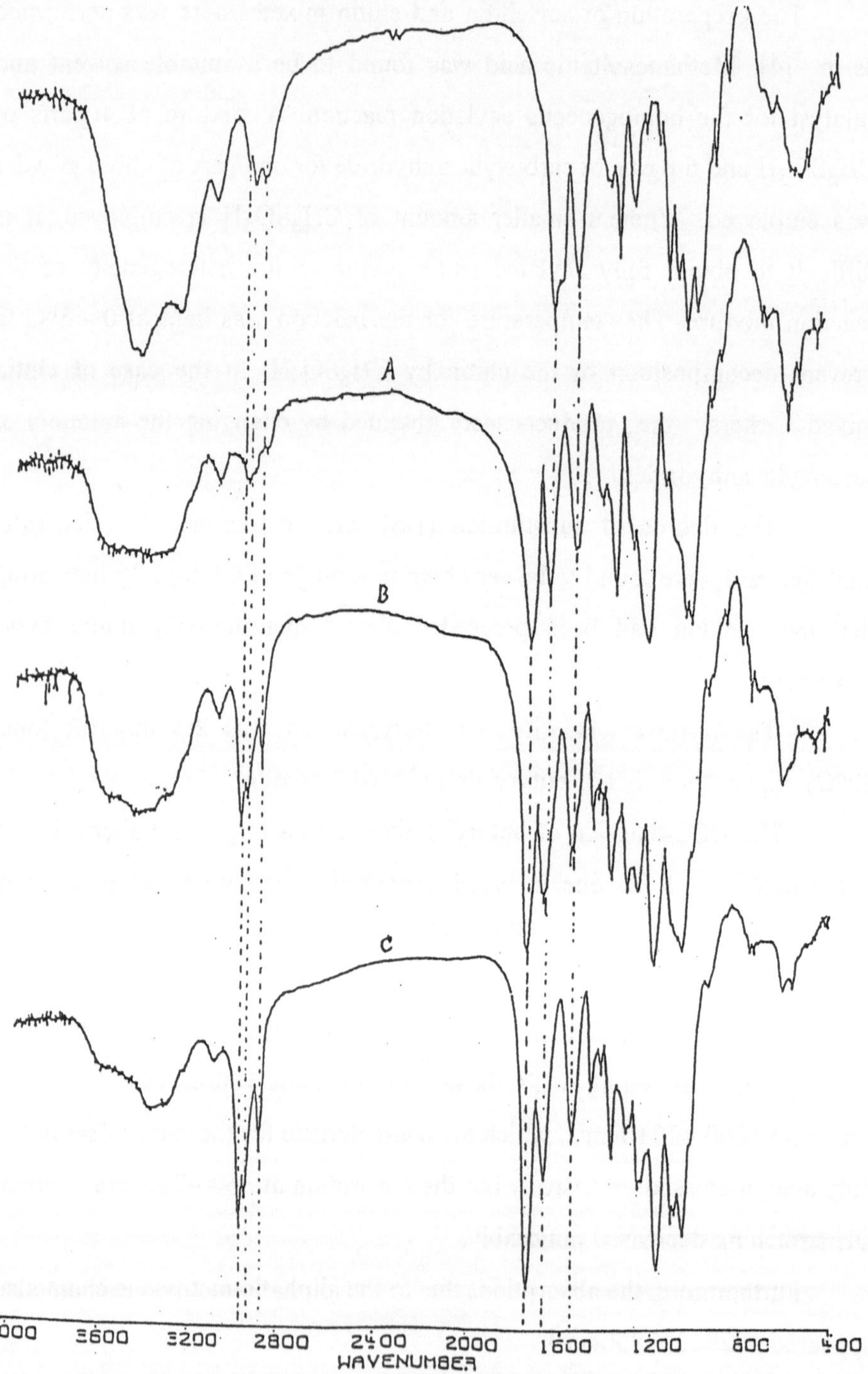

Fig 2: IR-Spectra of Chitin, Acetylchitin(A), Butyrylchitin(B), Valeroylchitin(C)

No change is observed in the spectra of chitin and acylchitin at 1670-1650cm^{-1} (Amide I) and at 1560-1520cm^{-1} (Amide II), showing that the deacetylation process does not proceed under the conditions of the acylation. Infrared absorption spectra of the chitin mixed esters are very similar, there is difference only in the absorption range of the methylene chain in various chitin mixed esters.

Acetylchitin is soluble only in acidic solvents, such as formic acid, dichloroacetic acid and methanesulfonic acid but the solubility properties of valeroylchitin and butyrylchitin are remarkable. They have very good solubility properties in commonly used organic solvents. Butyrylchitin is soluble in methanol, ethanol, dimethylformamide, formic acid, acetic acid, dioxane, acetone, tetrahydrofurane. A higher degree of substitution gave rise to better solubility, but acylchitin with DS lower than 1.5 is only swelled in high degree in acetone.

Thus, the introduction of a relatively long alkyl chain into a chitin molecule resulted in the remarkable enhancement of the solubility properties of chitin. The good solubility properties of butyrylchitin in commonly used organic solvents will enhance the possibility of its utilization.

Butyrylchitin and acetylchitin are both soluble in formic acid, whereas the mixed ester acetobutyrylchitin is not only soluble in formic acid but also in the solvent mixture of formic acid with polar solvents, such as dimethyl-formamide, methanol. The chitin mixed ester was first soluble in formic acid, then polar solvents were added to this solution. The solubility of acetobutyrylchitin is significantly better than that of acetylchitin, and thus will promote the usefulness of chitin and widen the scope of its industrial applications.

C. APPLICATIONS

We would like to describe the first applications of chitosan in Vietnam. Chitosan has been used as plant growth regulator. Diluted solution of chitosan salts are applied to crops in order to improve resistance to fungal pathogens and increase the yield.

The effect of chitosan treatment of growth and yield of rice have been recognized from the germination, the sowing to the gathering of rice as follows:

- seed germination: seeds of rice soaked in chitosan salts solution resulted in more than 95% germination, they show better germination than that of the controls.
- growth parameters: they have been increased as follows:

	Seedlings	Plants
Root length	2 - 3%	25%
Plant height	5-8%	10%
Dry weight of roots	5-10%	10%
Dry weight of plants	20-30%	20%

- Yield parameters:
  - $\frac{\text{average number of ears}}{\text{shoot}}$ (branching) increased by 15-20%
  - $\frac{\text{average number of seeds}}{\text{ear}}$ increased by 20-25%
  - dry weight of seeds increased by 1.5%.

These yield parameters lead to the average increase in yield of rice by 20%:

$$\text{Yield} = \frac{\text{shoot number}}{\text{m}^2} \times \frac{\text{ear number}}{\text{shoot}} \times \frac{\text{seed number}}{\text{ear}} \times \text{dry weight of 1000seeds} \times 10^{-5}\ \text{tons/ha}$$

Chitosan and chitinase inhibit the growth of several phytopathogens and rice seeds coated with chitosan enhanced the stimulation of the increase in chitinase activity in the seedlings. Thus chitosan seems to be a new additive to agricultural fertilizers for preventing plant from infection with pathogens, usable in agroindustry as ecologically functional material enhancing self defensive function of cells.

Chitosan has been used also as carriers for synthetic fungicides against Pyricularia oryzae and Rhizoctoria solani. A mixture of equal weight of chitosan and fungicide has the same effect on pathogens as fungicide used alone and less toxic.

In medicine, chitosan films indicated antibacterial activities against Escherichia Coli, Staphylococcus aureus and Pseudomonas aeruginosa and has been used for burn wound healing.

Chitosan is known to have a significant hypocholesterolemic activity in various experimental animals. Chitosan has been tested for reducing cholesterol, it has lipid-lowering effects similar to those of cholestyramine and has been used in clinical application.

ACKNOWLEDGMENTS

This work was supported, in part, by the Swedish Agency for Research Cooperation with developing countries (SAREC), Stockholm. For this the author was highly appreciated.

REFERENCES

1. Dang Mai Huong, Nguyen Xuan Dung, Dang Van Luyen
Tap chi Hoa hoc (Journal of Chemistry) 27, №3, 20-23 (1989)
2. Dang Van Luyen, Dang Mai Huong, Pham Thi Mai, Le Quang Tuan, Truong Dinh Tam, Nguyen Thi Nhung
Vienamese Patent № 123A, appl. 28/08/1990
3. S. Tokura, N. Nishi, A. Tsutsumi, O. Somorin
Polymer Journal 15, №6, 485-89 (1983)
4. K. Kaifu, N. Nishi, T. Komai, S. Tokura, O. Somorin
Polymer Journal 13, № 3, 241-45 (1981)
5. Dang Van Luyen, V. Rossbach, in preparation
6. J. Dutkiewicz, L. Szosland, M. Kucharska, L. Judkiewicz, R. Ciszewski
J. of Bioactive and Compatible Polymers 5, 293-304 (1990)
7. M. Pizzoli, G. Ceccorulli, M. Scandola
Carbohydrate Research 222 (1991) 205-213

A Novel Polymer Film that Controls Light Transmission

M. Honda , S. Hozumi , S. Kitayama

Sumitomo Chemical Co.,Ltd. Tsukuba Research Laboratory
6 Kitahara , Tsukuba , Ibaraki 300-32 JAPAN

ABSTRACT

This polymer film is a novel light control film which scatters only incident lights from particular angles and transmits incident lights from all other angles. This unique light control function is caused by the microstructure in the polymer film.

The optical properties described in this paper support that the principle of an angle dependent light control function can be explained mainly by diffraction due to the microstructure like a stack of various transmission volume phase gratings. The proposed mechanism of forming the unique microstructure during photopolymerization is due to the diffraction of exposure lights by the polymerization-induced microgels.

1. INTRODUCTION

This polymer film is a novel light control film (LCF) which scatters only inident lights from particular angles and transmits incident lights from all other angles as shown in Fig.1. Therefore the appearance of LCF changes from transparent to translucent ,and vice varsa , depending on an angle with which it is viewed.

Table 1 shows the light transmitting caracteristic of LCF. LCF transmits as high as 90% of total incident light. Total light transmittance is almost constant at any incident angle. LCF does not obviously show backscattering at scattering angles. However both parallel and scattering light transmittance vary dramatically with changing the incident angle. LCF scatters 74% of the transmitted light in the translucent area , making itself look like frosted glass. Since LCF has a view cut function by scattering , the optical property can be evaluated mainly by the haze percentage. It is obtained by calculation from total light transmittance and scattering light transmittance as expressed by Eq.(1). The haze percentage of scattering part is 82% ,this value is equal to frosted glass.

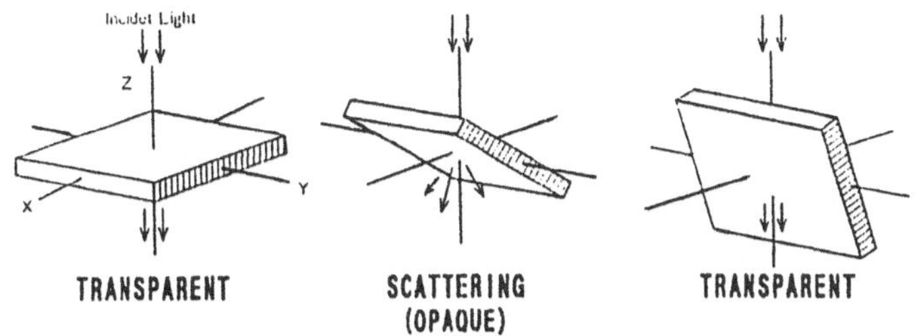

Fig.1 An example of the angle dependent property of LCF

K.P. Ghiggino (Ed.)
Progress in Pacific Polymer Science 3

Table 1 Light transmitting characteristic of LCF with 200 micron thickness

Transmittance	Transparent part	Scattering part
Total light transmittance (%)	90	90
Parallel light transmittance (%)	87	16
Scattered light transmittance (%)	3	74
Haze percentage (%)	3	82

$$\text{Haze (\%)} = \frac{\text{Scattered light transmittance}}{\text{Total light transmittance}} \times 100 \qquad \text{Eq. (1)}$$

The novel light control function of LCF is determined by the microstructure in this polymer film.

The compostion of LCF is composed of at least two photopolymerizable components having different refractive indices and reactivities. The preferred composition is ultraviolet (UV) light polymerizable acrylic monomers and/or oligomers including photoinitiators and other ingredients. The above composition is coated on a plastic film by using conventional coating methods and then exposured UV light to complete polymerization continuously.

In this paper ,the microstructure in LCF as well as the optical properties is reported and then the principle of this unique light controll function is described. Finally, the mechanism of forming the microstructure is proposed.

2. THE MICROSTRUCTURE IN LCF

LCF has the microstructure having the refractive index modulation in the film. Optical microphotographs of the cross section view as well as the plane view of LCF is revealed in Fig.2.[1] The alternating dark and bright lines like as Venetian blind represents the refractive index modulation. In this case, when LCF is rotated around Y axis , an angle dependent light control property is apparently observed. In the case of front translucent type , the microstructure is parallel to the vertical direction of the film surface.

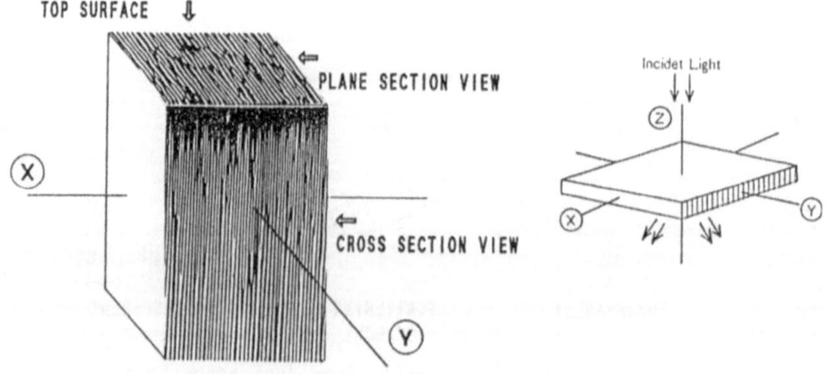

Fig.2 Optical microphotograph of LCF

As shown in Fig.3 , the direction of microstructure can be controlled by the fabrication condition. Not only the slanted microstructure but also the overlapped microstructure can be easily obtained by controlling the exposure condition. Thus the incident angles of incident lights scattered by LCF can be arbitrarily set up.

But the arrangement of the microstruture is somewhat disordered as compared with a diffraction gratings recorded by the two laser beam interference exposure technique. It is apparent from the microphotograph of cross section view that the spacing increases gradually from the top film surface.
The spacing increases from 2 micron to 3 or 4 micron with the distance from the top film surface as shown in Fig. 4.

The refractive index modulation was also measured by the interference microscopy , Interphako (Carl Zeiss Jena).[2] The refractive index modulation increases with the distance from the top surface within about 50 micron and then becomes almost constant having the maximum value of 0.04 .

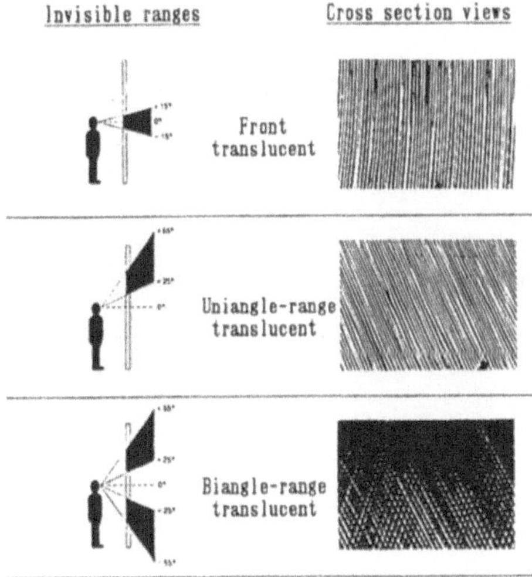

Fig.3 Three types of commercially available LCF

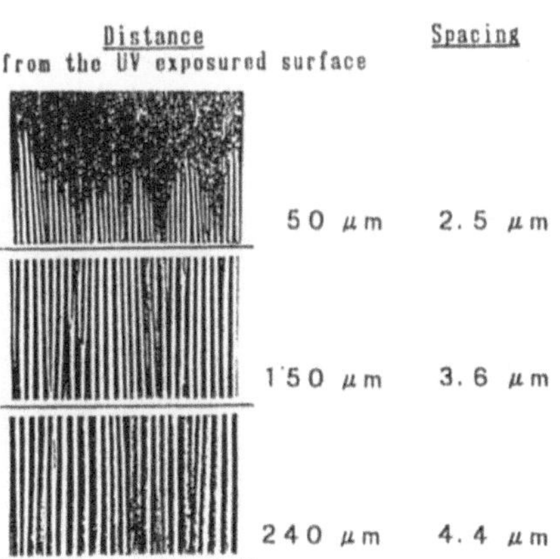

Fig. 4 Cross section views of LCF

The composition of LCF is ordinaly composed of at least two photopolymerizable components having different refractive indices and reactivities.

In order to simplify the analysis of microstructure , the composition composed of a component having " Br atom " was prepared on purpose. Fig.5 shows TEM microphotograph of the microstructure by Electron Microscope (Hitachi H-500).[2] The difference of Br content between the bright and dark region in this photograph was measured by an additional analysis equipment EDS system (Hitachi KEVEX 7000μX) . The measurement results for Br content are shown in Fig. 6. It is apparent from this figure that Br content at dark region have about twice as much as that at the bright region. This result represents that there is a large composition difference between the two regions in the microstructure and the refractive index modulation is caused by the composition modulation occurred during UV exposure.

Photopolymerizable monomer

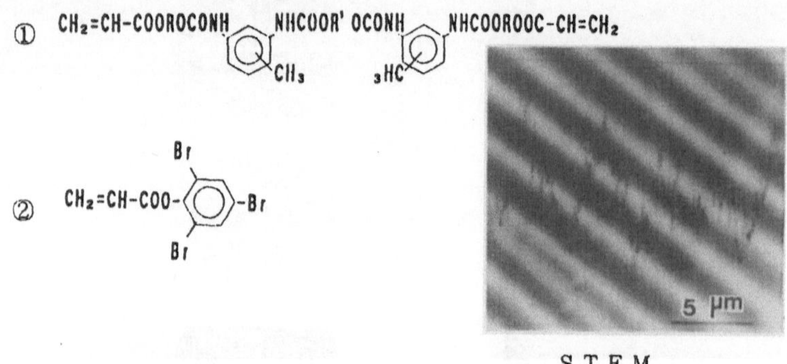

Fig. 5 T E M Photograph of the microstructure

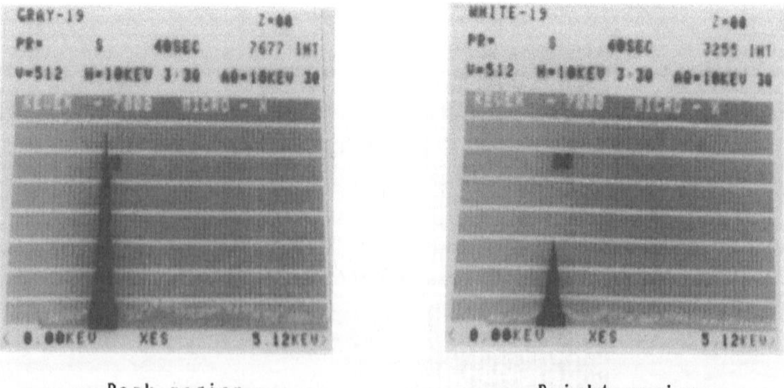

Dark region

Bright region

Fig. 6 Measurement of Br content (EDS system)

The structual parameters of LCF are listed in Table 2. The haze profile of LCF is also greatly influenced by the thickness of LCF. Not only the maximum haze but also the haze at center region of scattering angle increases with the thickness. However the incident angle showing the maximum haze is almost constant regardless of thicknes. The structual parameters such as the spacing and the thickness obtained by the analyses of microstructure support that the microstructure looks like a transmission volume phase grating recorded by the two laser beam interference exposure technique.

Table 2 Structual parameter of LCF

Film thickness	180 μm
spacing	
(Near UV exposured surface)	2 μm
(Near bottom surface)	3 − 4 μm
Refractive index modulation	0.04

3. MODEL STRUCTURE OF LCF

3.1. Principle of Light Control Function

As above mentioned, the microstructure of LCF looks like a transmission volume phase grating. Then various measurement results[3] suggest that the principle of light control function can be explained mainly by diffraction of phase grating.

Diffraction of phase grating is simply classified into two kinds of diffraction. Raman Nath diffraction and Bragg diffraction by the magnitude of Q value expressed by Eq.(2), where λ_B is Bragg wavelength, T is thickness of phase grating, n_0 is average refractive index and Λ is spacing.[4] If Q value is less than 1, the diffraction by a phase grating is classified into Raman Nath diffraction. If Q value is more than 10, it is classified into Bragg diffraction. According to Eq.(2), Q value is in direct proportion to the thickness and in inverse proportion to the square of spacing. Thus the phase grating showing Raman Nath diffraction is called a plane phase grating and the grating showing Bragg diffraction is called a volume phase grating.

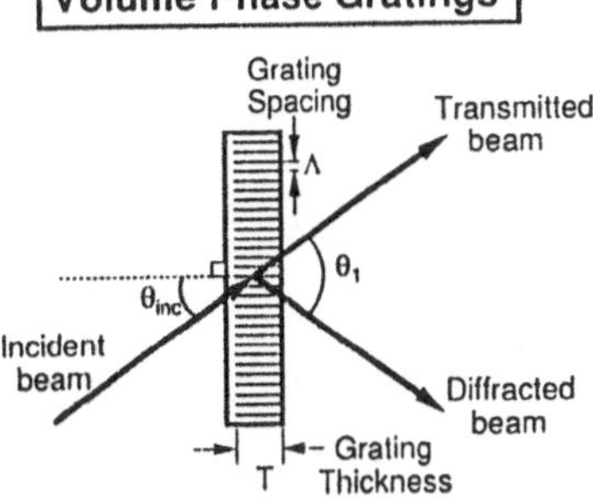

n_0 : The Average Refractive Index
n_1 : The Amplitude of the Sinusoidal Refractive index grating

$$Q = 2\pi\lambda T / n_0\Lambda^2 \qquad \text{Eq. (2)}$$

$$\left(\begin{array}{l} Q < 1 \ : \ \text{Raman-Nath diffraction} \\ Q > 10 \ : \ \text{Bragg diffraction} \end{array} \right)$$

$$\sin \theta_B = \lambda_B / 2n_0\Lambda \qquad \text{Eq. (3)}$$

$$\eta_B = \sin^2 (\pi n_1 T / \lambda_B \cos \theta_B) \qquad \text{Eq. (4)}$$

Fig. 7 Bragg diffraction by phase gratings

Raman Nath diffraction does not clearly show not only angular but also wavelength selectivity , as compared with Bragg diffraction. In addition it shows only 33.9 % diffraction efficiency as the maximum efficincy and has many order diffracted rays. Scince LCF shows apparently angular and wavelength selectivity , it is reasonable that the model structure is considered as a volume phase grating. Diffraction efficiency at Bragg angle is expressed by Eq.(4). If the amplitude of refractive index modulation , n_1 is much smaller , incident lights are not efficiently diffracted or scattered. Thus, the larger n_1 than a certain value is desired.

Bragg diffraction for a volume phase grating is defined by Eq.(3). According to Eq.(3) , the diffraction angle increases with decreasing of the spacing.

3.2. Wavelength Dependence of LCF

The wavelength dependence of LCF showing scattering at normal incidence on incident angles was measured by using VIS-NIR Recording Spectrophotometer (Shimadzu UV-3100). The measurement results on LCF with 80 micron thickness are shown in Fig.8. The incident angles showing the minimum transmittance at each wavelength are completly separated. The angles showing the minimum at each wavelength fit in well with the calculated Bragg angles by Eq. (3) that LCF with 80 micron thickness is considered to a kind of volume phase grating.

Angle distribution of scattered light through LCF was measured by using Photo Detector illustrated schematically in Fig.9.[6] If the angle of incident light at θ is equal to the Bragg diffraction angle , the scattering lights are clearly observed at 2θ. Fig.10 shows the angle distribution of scattered light by LCF with 80 micron thickness. The diffracted light are clearly observed at even normal incidence and more apparently at the slanted incident angle corresponding to the caluculated Bragg angle.

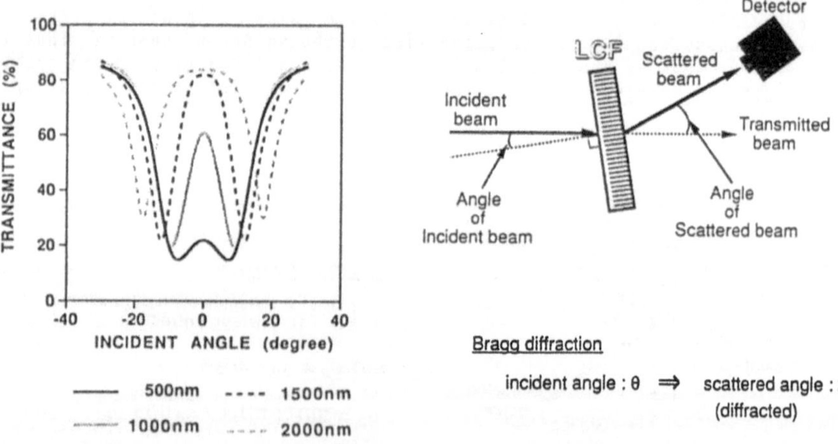

Fig. 8 Angle dependence of parallel light transmittance at each wavelength

Fig.9 Measurement method

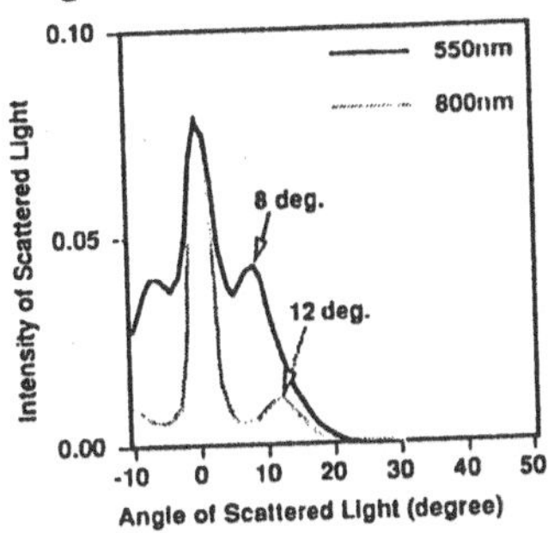

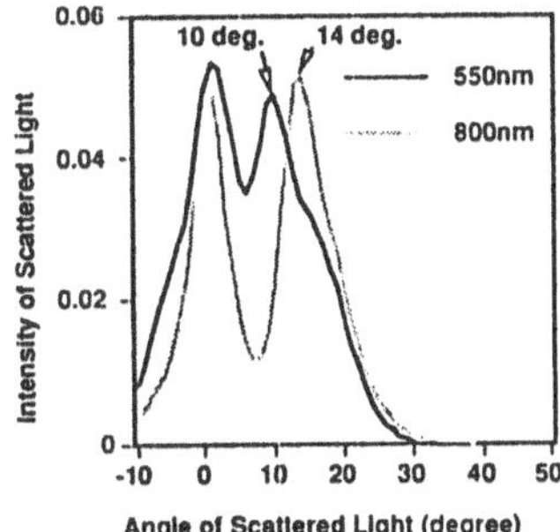

Fig.10 angle distribution of scattered light intensity
(by LCF with 80 micron thickness)

3.3. Model Structure of LCF

In the case of a thin LCF, the microstructure is analogous to a single transmission volume phase grating and optical properties are mainly explained by diffraction. On the other hand, in the case of a thick LCF, the microstructure is more complicated.

Assuming that LCF is composed of many grating components such as Fig.11, Bragg angle of each grating are calculated as Eq.(3). Consequently, the maximum diffraction angle by the model structure is determined by the grating with the minimum spacing and smaller diffraction angles than the maximum angle are due to gratings with larger spacing. As a matter of fact, each grating should not be perfectly arranged in a parallel direction. The maximum angle showing scattering is determined by the minimum spacing in the microstructure. Therfore, LCF having a view cut function can be compared to a stuck of various kinds of volume phase gratings. Since only incident lights from particular angles are diffracted to various directions within a fixed angle determined by the minimum spacing in the microstructure in LCF, an angle dependent property remains but a wavelength dependent property is eliminated due to overlapped diffraction lights. Consequently white light scattering can be observed within a certain incident angle range.[3]

Grating : $1 \quad 2 \quad \cdots\cdots \quad n$

Spacing : $d_1 < d_2 < \cdots\cdots < d_n$

Bragg Angle : $\theta_1 > \theta_2 > \cdots\cdots > \theta_n$

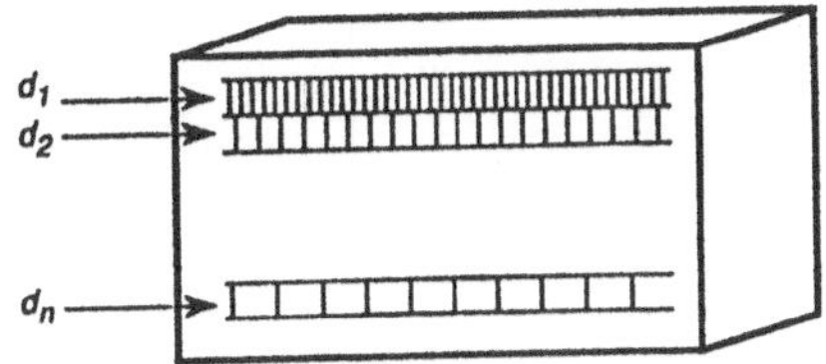

Fig.11 Model structure of LCF

4. PROPOSED MECHANISM

As mentioned in the section 2, the refractive index modulation is caused by the composition modulation occurred during UV exposure. In order to find the mechanism of composition modulation, we used the two-component model chosen to have very different rates of polymerization and refractive indices. The two-component model system used urethanediacrylate as the high-reactivity monomer and diallylphthalate as the low-reactivity monomer as shown in table 3. Fig.12 shows the photo-polymerization profile obtained the two component system and Fig.13 shows time dependence of the haze profil during exposure.

In the model system the haze profile was first observed at 10 min exposure. By this time, the acrylate monomer has mainly polymerized, but the calculated refractive index of polymer parts was nearly equal to that of remaining monomer parts. And then, as the remaining allyl monomer polymerized gradually, the refractive index modulation in LCF was more clearly observed.

These results represent that there are different intensity areas occurred in the exposure lights. The polymer will be formed preferentially from the high-reactivity monomer, produing a concentration gradient that will cause additional high-reactivity monomer to diffuse into the high-intensity areas. As the polymerization proceeds the volume fraction of polymer in the high-intensity areas increases and monomer will tend to be excluded from these areas. Since the remaining monomer mixture has been enriched in the low-reactivity monomer, the low-intensity areas will then contain an excess of low-reactivity monomer, and the high-intensity areas will contain a polymer with an excess of the high-reactivity monomer. In this way the refractive index modulation result from this modulation of chemical composition recorded by the intensity distribution in the exposure lights.

Table 3 Model composition of the two-component system

COMPOUND	
Ⓐ Acryl compound (UA) [n_D: 1.47/1.47]*	Ⓑ Allyl compound (DAP) [n_D: 1.52/1.57]*
High photopolymerizable monomer 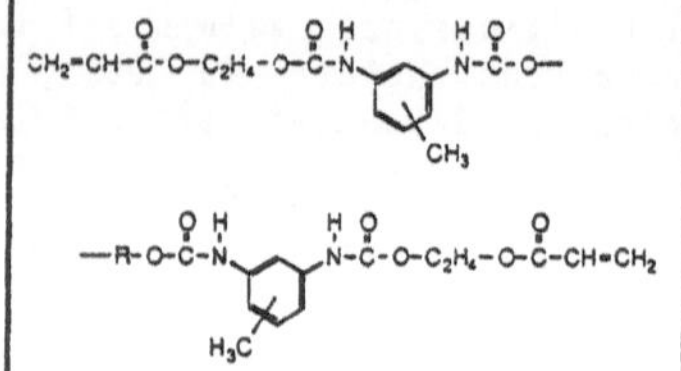	*Low photopolymerizable monomer* 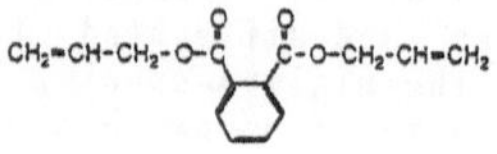
$CH_2=CH-C(O)-O-C_2H_4-O-C(O)-NH-$[$CH_3$-substituted ring]$-NH-C(O)-O-$ — $-R-O-C(O)-NH-$[H_3C-substituted ring]$-NH-C(O)-O-C_2H_4-O-C(O)-CH=CH_2$	$CH_2=CH-CH_2-O-C(O)$[benzene ring]$C(O)-O-CH_2-CH=CH_2$
	Photoinitiator
	* n_D: monomer / homopolymer

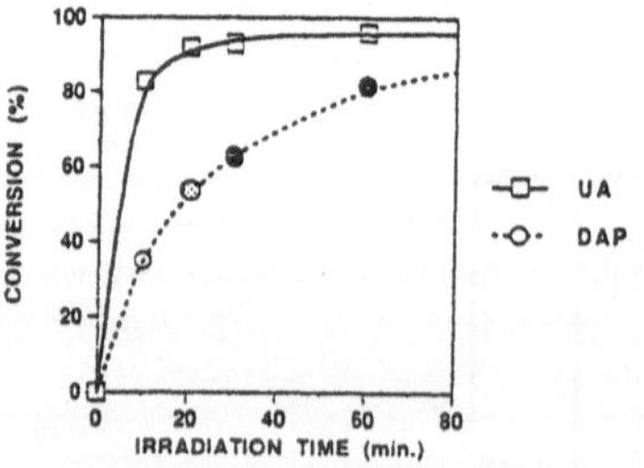

Fig.12 Photopolymerization of a polyurethanediacrylate with diallylphthalate

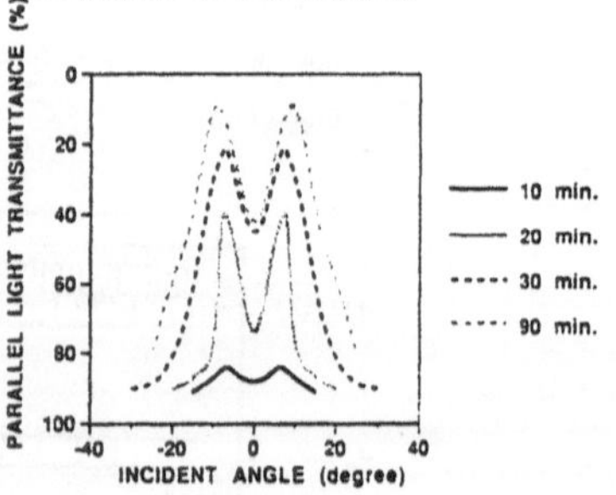

Fig.13 Haze profile change by the exposure time (UA/DAP system)

In this section , we discuss the mechanism of the intensity distribution occurred in the exposure lights. The optical microphotographs of the cross section view suggest that the refractive index modulation pattern looks like the diffraction pattern of light by an opaque sphere. In our system it is possible that the photopolymerization induce phase separation during exposure. Therefore the polymerization-induced microgels will be suited to the opaque spheres. We have not attempted to detect the microgels, but the experimental results and the computer simulations suggested that.

The diffraction pattern of the exposure UV lights by an opaque sphere can be evaluated by the Fraunhofer's equation as shown in Fig.14.[13] The interval of the shadow and illuminated regions in Fig.14 is 2 or 3 micron equal to the spacing of the microstructure in LCF. Optical microphotographs of the plane view as well as the cross section view of LCF exposed by a point light source is revealed in Fig.15. The diffraction pattern changes by exchanging the shape of light source. Fig.16 shows the caluculated diffraction pattern and the plane view of LCF exposed by a rod light source. Both observation patterns in LCF are similar to the computer results of diffraction pattern.

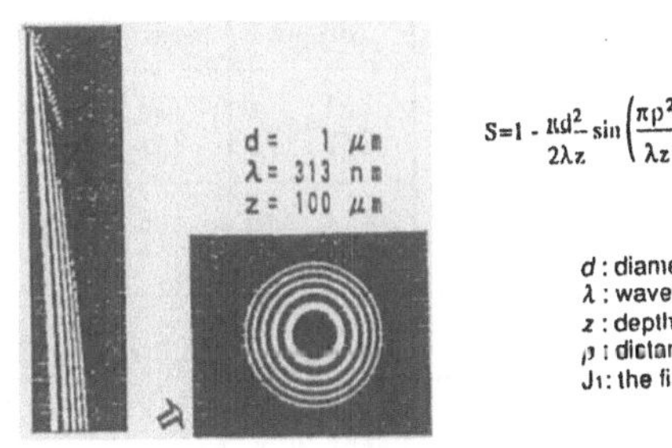

$$S = 1 - \frac{\pi d^2}{2\lambda z}\sin\left(\frac{\pi\rho^2}{\lambda z}\right)\cdot\left[\frac{2J_1\left(\frac{\pi d\rho}{\lambda z}\right)}{\frac{\pi d\rho}{\lambda z}}\right] + \left[\frac{\pi d^2}{4\lambda z}\cdot\frac{2J_1\left(\frac{\pi d\rho}{\lambda z}\right)}{\frac{\pi d\rho}{\lambda z}}\right]^2$$

d : diameter of microgel
λ : wavelength of incident beam (UV)
z : depth
ρ : distance from the center
J_1: the first order Bessel function

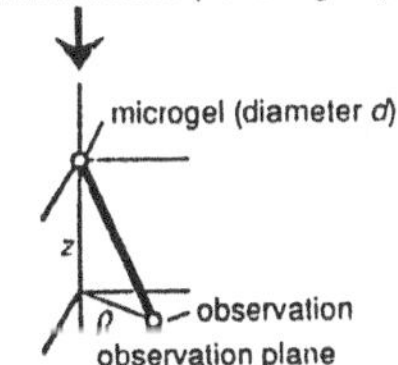

Fig.14 Intensity profile evaluated by Fraunhofer's equation of diffraction pattern (irradiated wavelength 313 nm)

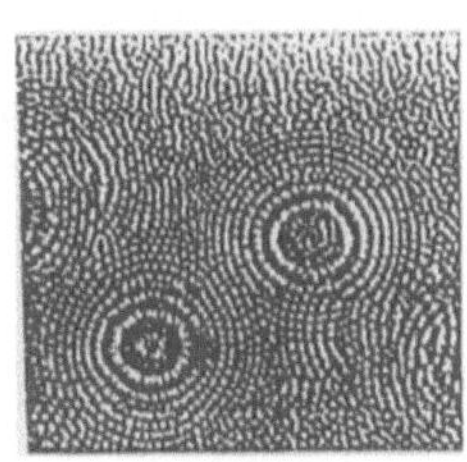

plane section view

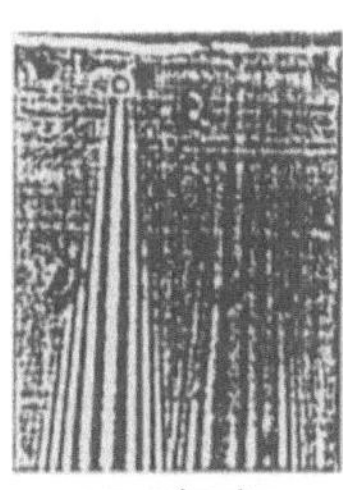

cross section view

Fig.15 Optical microphotograph of LCF exposed by a point light source

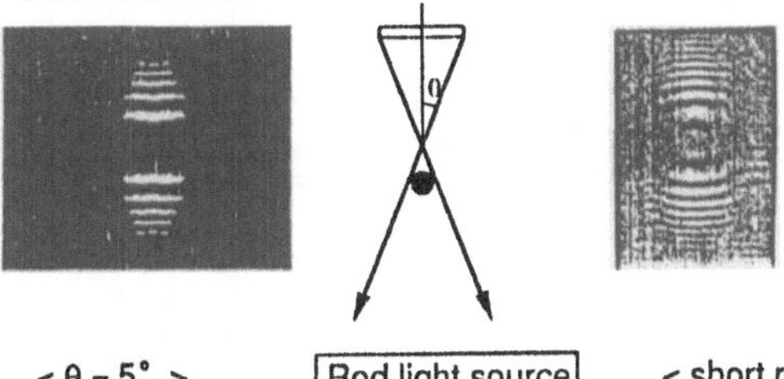

Fig.16 Analytic-experimental comparison (diffraction pattern)

The spacing of the microstructure in LCF is dependent on the wavelength of exposure UV light as shown in table 4. The observed spacing of LCF at each UV wavelength is similar to the calculated interval of diffraction patterns.

Table 4 Experimental spacing of the microstructure at each exposed wavelength

	313 nm	365 nm	405 nm
uniform region	17 μm	25 μm	31 μm
Distance from the top surface 50 μm	2.4 μm	2.9 μm	4.0 μm
Distance from the top surface 100 μm		3.3 μm	4.2 μm

Table 5 Calculated interval of diffraction pattern by an opaque sphere at each irradiated wavelength

Propagation distance		313 nm	365 nm	405 nm
50 μm	Shadow region interval Illuminated region interval	2.9 μm 2.1 μm	3.1 μm 2.2 μm	3.3 μm 2.4 μm
100 μm	Shadow region interval Illuminated region interval	4.0 μm 2.9 μm	4.4 μm 3.2 μm	4.6 μm 3.3 μm

As above mentioned ,during exposure polymerization-induced phase separation will take the form of microgel and the diffraction of exposed lights by the microgels will be occurred. In our system we find that the use of combination of monomers results in a two-way diffusion of the monomers during exposure, giving a modulation of the chemical composition of the final polymer. The resulting composition modulation of the polymer can give a large refractive index modulation. Fig.17 shows the schematic explanation of the mechanisms that we believe.

The actual situation is very complex. We have not attempted to obtain a quantitative theory, but simply to idntify the basic mechanisms and the important parameters.

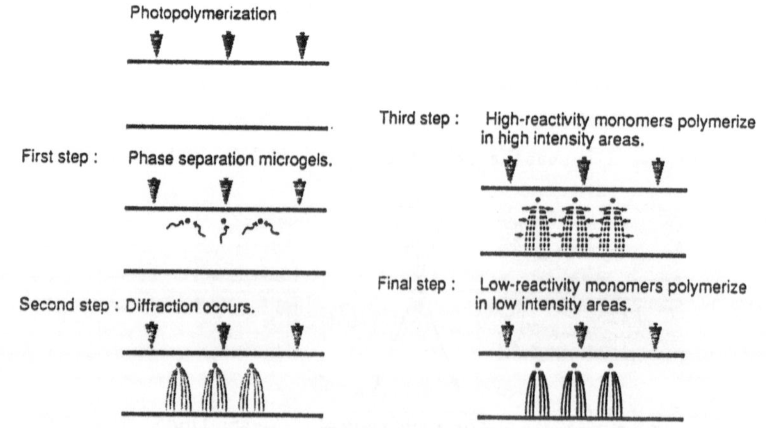

Fig.17 Schematic explanation of the proposed mechanism during UV exposure

5. APPLYCATIONS

LCF has already put on the market for various use in JAPAN, under the trade mark "LUMISTY". In LCF, the yellowing of LCF caused by ultraviolet lights is undesirable for outdoor use. However, the transparent adhesive layer which is used for LUMISTY includes a large quantity of ultraviolet absorver, so almost ultraviolet lights entering LCF are cut off.

LCF creates new visual effects in various fields of applications, Building materials for protecting privacy in buildings, Automobile parts for protecting the privacy of drivers, Industrial appliances, and Display for controlling the visible range of displays to protect information on the screens.

The maximum dimension of LCF, 100 ft.long roll by 4 ft. wide, is now commercially available.

6. ACKNOWLEDGEMENTS

The authors wish to thank to Mr.Takigawa,Mr.Maeda, and their research group Nippon Sheet Glass Co.,Ltd. for valuable disscusion and helpful suggestion in the collaborative research for LCF.

7. REFERENCES

1 Hozumi S, Kitayama S, Shiraga N, Adachi T (1989) Polymer Preprint Japan 38:3548
2 Kitayama S, Shiraga N (1991) Hyomen Kagaku Japan 12:496
3 Maeda K, Yoshida M, Takigawa A, Ishizuka S, Tsujino T, Yamamoto (1989) Polymer Preprint Japan 38:3554
4 Kogelnik H (1969) Bell Syst Tech J. 48:2909
5 Ohno T, Kawamura K, Ueda M, Ohshima T, Okita T (1989) Polymer Preprint Japan 38:3551
6 Sommargren GE, Weaver HJ (1990) Applied Optics 29:4646

The γ-Initiated Degradation of Polypropylene: Stabilization by Phenols

D.J. Carlsson,[a] S. Falicki[a], J.M. Cooke[b] and D.J. Gosciniak[b]

[a]Institute for Environmental Chemistry, National Research Council, Ottawa, Canada K1A 0R6

[b]Speciality Chemicals, ICI Americas Inc., Wilmington, Delaware 19897, USA

ABSTRACT

The effects of various phenols and some hindered amines have been compared both during the γ-irradiation process and during the post-irradiation oxidations of polypropylene films and test strips. Oxidation product formation, yellowing and embrittlement (as measured in an instrumented bend test) have been compared with product formation. A partial correlation between suppression of oxidation during the irradiation step with long term, post-irraidation oven aging at 60°C was found, but complicated by extensive destruction during irradiation of the active phenolic functionality in some additives, essential for peroxyl radical scavenging. Very long lifetimes with barely detectable yellowing were found for combinations of the amines with completely unhindered or only partially hindered phenols.

INTRODUCTION

The use of γ-irradiation to sterilize polymers for medical applications requires careful stabilization of plastics such as polyolefins to prevent severe embrittlement. Most of this embrittlement occurs as a result of post-irradiation oxidative chain scission (1-4). Many stabilizer types have been investigated to minimize this problem in highly sensitive polymers such as polypropylene (2-7).

Hindered phenols are known to be very effective at minimizing post-irradiation embrittlement, but usually at the expense of unacceptable yellow-brown discolouration (3,5). The coloration is generally believed to result from the formation of products such as stilbene quinones (8-9).

* Issued as NRCC # 35784

K.P. Ghiggino (Ed.)
Progress in Pacific Polymer Science 3

In this paper we compare the stabilization effectiveness of some less usual phenolic compounds in γ-irradiated polypropylene with that of a conventional hindered phenolic antioxidant and hindered amine stabilizers (HALS).

EXPERIMENTAL

Polypropylene (PP, Himont 6501, additive free) was compounded with the additives shown in Figure 1. HALS additives I and II are produced by Ciba-Geigy as Tinuvin 770, bis(2,2,6,6-tetramethyl-4-piperidinyl)decanedioate and Tinuvin 765, bis(1,2,2,6,6-pentamethyl-4-piperidinyl)decanedioate. HALS additives III, bis(2,2,6,6-tetramethyl-4-piperidinyl) 1,5-dioxaspiro [5.5] undecane-3,3-dicarboxylate and IV, bis(1,2,2,6,6-pentamethyl-4-piperidinyl) 1,5-dioxaspiro [5.5] undecane-3,3-dicaroxylate are produced by ICI Americas as Topanex 500 H and 516 H respectively. The phenols studied included 4,4'-[1-{4-[1-(4-hydroxyphenyl)-1-methylethyl]phenyl}ethylidene]bis[phenol] (I), 4,4'-(1-phenylethylidene)bis[phenol] (II), 4,4'-[1,4-phenylenebis(1-methylethylidene)]bis[phenol] (III), 4,4'-[1,4-phenylenebis(1-methylethylidene)]bis[2,6-dimethylphenol] (IV) (all from ICI Americas) and tetrakis-[methylene 3-(3',5'-di-tert-butyl-4' – hydroxyphenyl)-proprionate] methane (V) (Ciba Geigy Irganox 1010). Additives were used alone at 0.12 wt% or in pairs with each additive at 0.12 wt%. PP formulations were then melt pressed into film under nitrogen (60-90 μm thickness by micrometer) or injection moulded into test strips (1.25 x 12.5 x 125 mm^3). Films and test strips were γ-irradiated in air to a normal sterilization dose of 2.5 Mrad in an AECL Gammacell 220 (0.7 Mrad/h). Ambient temperature inside the cell for this standard sterilization dose was 35°C.

After irradiation, some samples were immediately analyzed, whereas others were exposed to accelerated aging condition of 60°C in a forced air oven. Films were analyzed by transmission FTIR, derivitization FTIR and esr spectroscopy and test strips by an instrumented bend test as described previously (6). Bend failure was taken as the time at 60°C at which a sample failed when bent through a 100° angle about a 1.5 mm pin in a hinge-based, 3 point bend device driven by an Instron Tensile Tester. The resolution of transmission IR was extended by the use of carefully selected derivatization reactions which allowed the identification and quantification of alcohol and specific hydroperoxide groups (after NO reaction) and ketone and carboxylic acids (after SF_4 reaction).

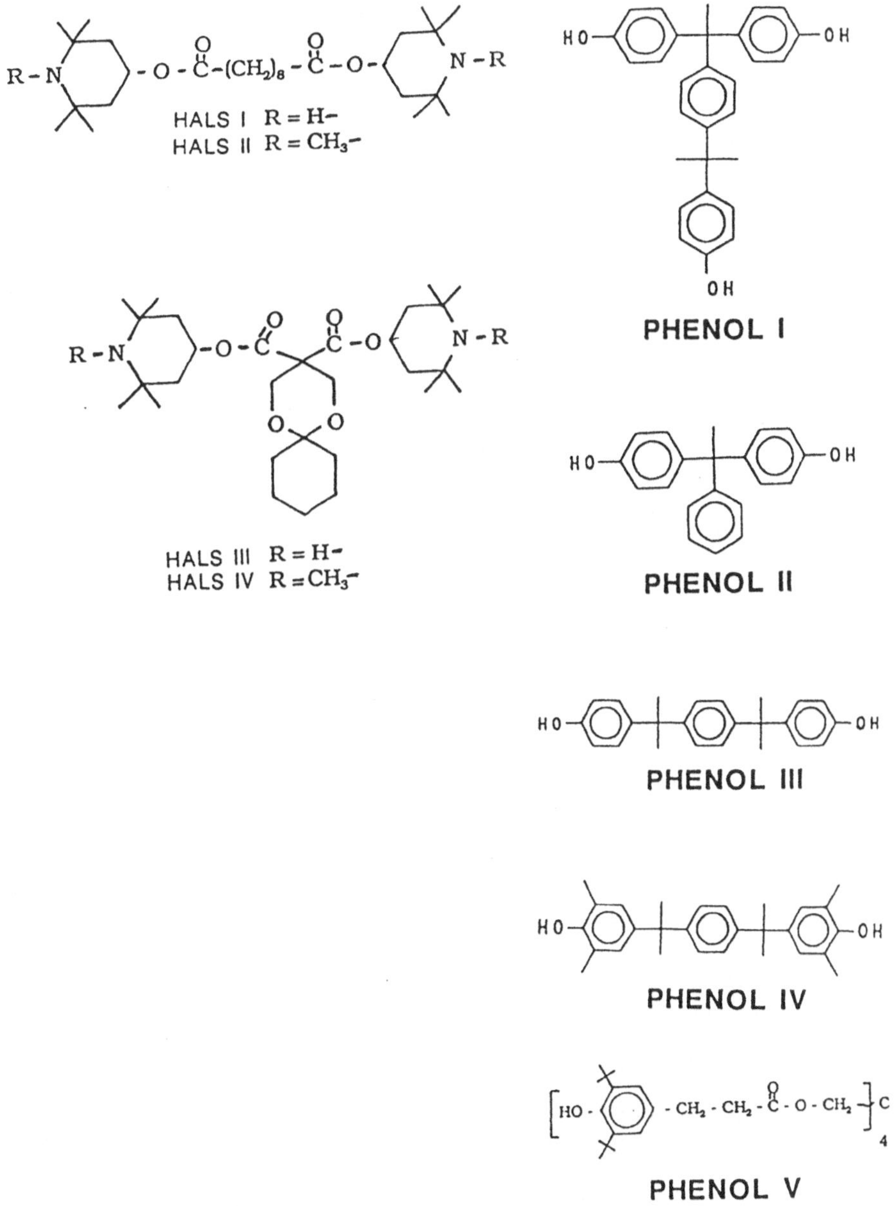

HALS I R = H-
HALS II R = CH_3-

HALS III R = H-
HALS IV R = CH_3-

PHENOL I

PHENOL II

PHENOL III

PHENOL IV

PHENOL V

Figure 1 Structures of the additives employed

Concentrations of phenols IV and V were estimated from the sharp phenol -OH FTIR absorption at 3620-3650 cm^{-1}, based upon calibration curves for each specific phenol. Concentrations of the completely unhindered phenols (I-III) could not be measured by this method as the phenolic OH of these compounds was a broad IR absorption at ~3400 cm^{-1}. This was obscured by corresponding absorption of the alcohol and hydroperoxide oxidation products from the PP.

To explore the effects of additives on the initial generation of PP peroxyl radicals, some films were irradiated at -78°C in air. At this temperature, only peroxyl radicals are detected (no carbon centred radicals) and the peroxyl radicals are indefinitely stable (10). The peroxyl yield could then be quantified by double integration of the esr spectrum at -140°C, relative to a cuprous acetylacetonate standard (10). The effects of additives on peroxyl reactions at room temperature could then be followed by decay of the esr signal.

RESULTS AND DISCUSSION

a) Immediate Effects of γ-Irradiation: None of the additives was found to significantly effect PP peroxyl radical formation during irradiation at -78°C. All samples including the stabilizer-free control showed a radical concentration variation of less than ±30% which is well within the error range of absolute radical yields by esr double integration. It would seem reasonable to expect a similar absence of effect for samples irradiated at ambient temperature. Furthermore peroxyl radical decay at room temperature was essentially unaffected by the presence or absence of the additives as we have reported previously (6).

Analyses of samples immediately after irradiation indicated both the formation of oxidation products and the partial destruction of the phenolic additives (Table 1). Values of oxidation products for the secondary and tertiary HALS additives are included for comparison. Although the conventional antioxidant, phenol V suppresses oxidation during irradiation by a factor of about 2 as compared to the additive free control, it is the partially blocked phenol IV which is most effective. However this effectiveness comes at a price as analysis of residual phenolic groups from the sharp IR absorption from the phenolic -OH showed that only ~20% of phenol IV survived the irradiation step (Table 1). In contrast 40% of phenol V was detectable after irradiation. Allen et al. have

TABLE 1

Polypropylene oxidation and embrittlement

(2.5 Mrad γ-irradiation)

Stabilizer(s)[a]	Initial Oxidation Product[b,c] mmole/kg					Post-γ Phenol Concentration (wt%)[b,c]	Bend Failure at 60°C (days)[d]
	Total OH	tert. OOH	sec. OOH	~C(OH)=O	>C=O		
———	48	32	10	3.0	9		~1.0
sec. HALS I	39	23	9	2.5	10		~220
tert. HALS II	31			2.0			~300
PHENOLS							
Phenol I	47			3.0	9		20
Phenol II	41			3.0	9		100
Phenol III	38			3.0	8		60
Phenol V	26	15	7	2.0	7	0.05[e]	395
Phenol IV	12	6	5	0.8	2.7	0.025[f]	~200

a) All initially 0.12 wt%
b) Measured immediately after irradiation
c) 60-90 µm film
d) 1.25 x 12.5 x 125 mm. test strips. Time to failure to withstand a 100° bend angle
e) From the 3649 cm^{-1} phenolic-OH absorption
f) From the 3623 cm^{-1} phenolic -OH absorption

recently shown that some loss of tert. butyl groups occurs when phenol V is irradiated with γ-rays or electron beams (9). However they found that the phenolic-OH was retained in these products.

b) Post-irradiation reactions: The formation of hydroxyl containing groups (alcohol and hydroperoxide) and carbonyl species were followed by IR at 3400 cm^{-1} and ~1720 cm^{-1} respectively. The 3400 cm^{-1} absorption is indicative of hydrogen bonded -OH species with only trace amounts of free alcohol (3600 cm^{-1}) and hydroperoxide (3550 cm^{-1}) groups detectable. For these kinetic studies, shown in Figure 2, films were stored at 60°C in separate, loosely covered glass beakers. This condition give an ~16 fold acceleration over ambient storage and prevents inter-sample infection by transfer of partly volatile additives.

From Figure 2, the unhindered phenols are quite weak stabilizers in γ-irradiated PP film. The partially hindered phenol IV is surprisingly effective but not as effective as the fully hindered phenol V. However it must be noted that phenol IV is most extensively consumed during the irradiation step (Table 1) so that it is present at about half of the concentration of phenol V at the start of the 60°C aging process. A combination of a HALS stabilizer with phenol IV does not suppress the partial destruction of the phenol during the irradiation step but does confer a very high 60°C aging stability (Figure 2).

Oxidation product accumulation may give useful information on the oxidation mechanism. However a physical test of deterioration is most important from a practical view point. Various tensile or bend test have been employed to measure failure of PP mouldings (2,5). Because PP is extensively used in syringe manufacture, a bend test appears most appropriate to simulate post-storage failure during actual use. The time period of post-irradiation storage at 60°C to produce a failure when test strips were bent through a 100° radius about a slim (1.25 mm) pin was used to compare stabilizers. This method has been described in detail previously (6). Failure times are shown in Table 1.

When discoloration was compared for the irradiated series after quite long storage at 60°C (~105 days), test strips containing phenol V were found to be less yellowed than with phenol IV. However phenols I-III gave the least yellowing of all, being barely different from the unirradiated test strips. The best combinations

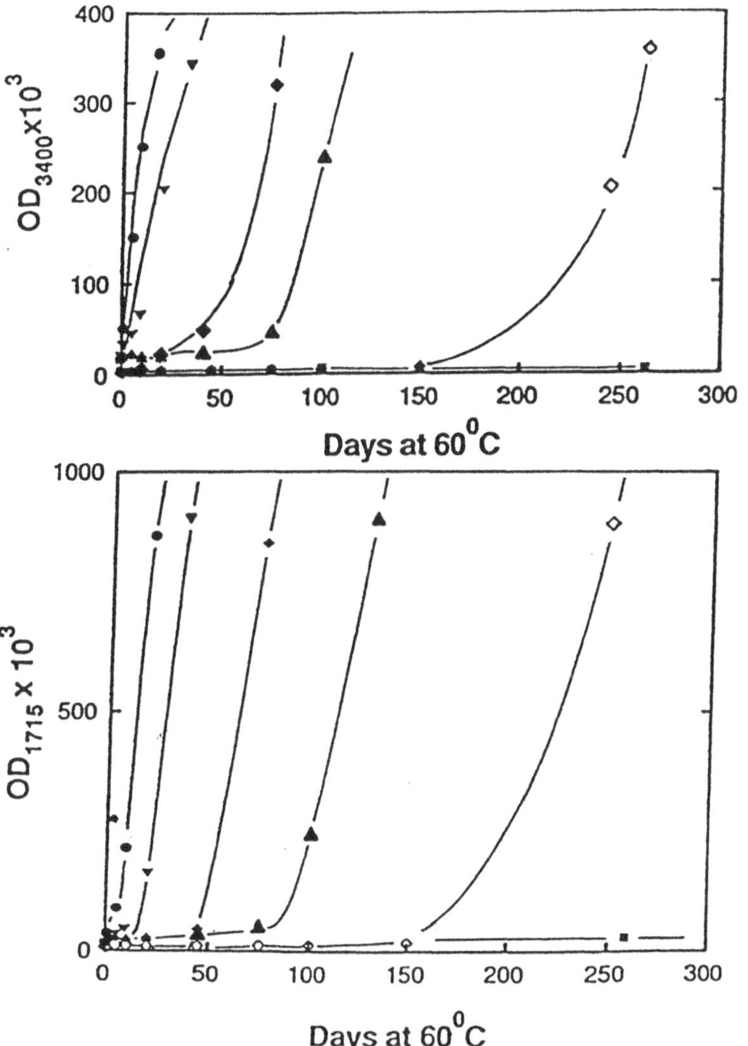

Figure 2 IR changes of PP film after γ-irradiation and storage at 60°C.
2.5 Mrad dose. Film ~60 μm thickness.

- ● Control: no additives
- ▾ Phenol I at 0.12 wt%
- ◆ Phenol III at 0.12 wt%
- ▴ Phenol II at 0.12 wt%
- ◊ Phenol IV at 0.12 wt%
- ■ Phenol V, or Phenol IV (0.12 wt%) + HALS II (0.12 wt%)

of retention of toughness with minimal yellowing came from the combination of phenol IV with the secondary HALS I or III which yellowed less than with IV alone. Similarly barely detectable yellowing occurred with phenol III and any of the HALS (I to IV). In contrast for the hindered phenol V, the presence of the HALS caused quite strong yellowing, with the secondary HALS most accentuating the effect.

Although care must be taken in comparing oxidation in thin film with that in the much thicker test strips, from Table 1 it is apparent that additive effectiveness during the irradiation stage is not of dominant importance in protecting in the post irradiation stage. Ability of the additive to survive this step and then suppress the post-irradiation process appears to be more significant. This is supported by a detailed study of the total -OH and hydroperoxide levels in irradiated film containing phenols IV and V (Figure 3). For phenol IV, a very slight drop in hydroperoxide level occurs at 60°C, albeit from a very low initial level. For phenol V a more distinct drop in -OOH level occurs (from a higher initial level) and in fact -OOH is undetectable in this sample after ~60 days at 60°C with a small corresponding increase in carbonyl species (not shown).

In a previous study of HALS additives in γ-irradiated PP, we have shown that, at the bend failure point, IR indicates that carboxylic acids are a major product in failure in contrast to the very low level in additive free or weakly stabilized samples (6). In unstabilized samples, ketonic and hydroperoxide products dominate. This same importance of carboxylic acids (measured as acid fluorides after SF_4 treatment) was found at the failure point of samples containing phenols IV and V (reaction scheme 1). The detected carboxylic acid [~ C(H)-(CH_3) - C(=O) - OH] is a backbone scission product and its formation clearly contributes to embrittlement of the test strips (1).

CONCLUSION

Although detailed product analysis immediately after irradiation gives a clear picture of the effectiveness of an additive in suppressing direct polymer oxidation, the level of unstable products such as hydroperoxide groups appears to be less important than the extent of additive deactivation. Unhindered or partially hindered phenols are more extensively consumed than conventional, fully

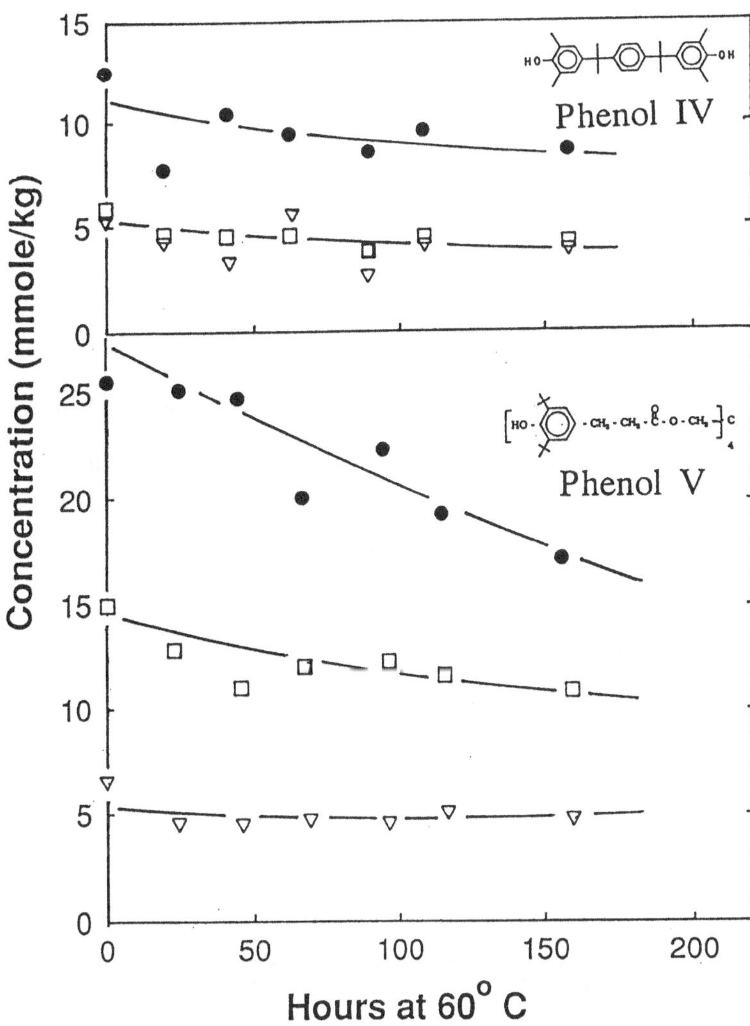

Figure 3 Changes in total -OH and hydroperoxides during oven aging

Phenol V at 0.12 wt% initially

Phenol IV at 0.12 wt % initially

● Total -OH from IR absorbance at 3400 cm^{-1}

□ Tert.-OOH from IR after derivatization with NO

▽ Sec.-OOH from IR after derivatization with NO

hindered phenols. It is the surviving phenol which protect against oxidative embrittlement during the post-irradiation oxidation. Unhindered or partially hindered phenols cause less yellowing of test strips when combined with hindered amines than does a blocked phenol. In the latter case yellowing is accentuated in the presence of the amines.

REFERENCES

1 Lacoste J, Vaillant D and Carlsson DJ (1993) J Polym Sci Polym Chem Ed 31:715-722
2 Dunn TS and Williams JL (1983) J Ind Irrad Tech 1:33-49
3 Carlsson DJ, Dobbin CJB, Jensen JPT and Wiles DM (1985) Amer Chem Soc Symp Ser 280:359-371
4 Becker RE, Carlsson DJ, Cooke JM and Chmela S (1988) J Polym Deg Stab 22:313-323
5 Klemchuk PP and Horng PL (1991) J Polym Deg Stab 34:333-346
6 Falicki S, Gosciniak DJ and Cooke JM, Cooney JD and Carlsson DJ (in press) J Polym Deg Stab
7 Falicki S, Gosciniak DJ, Cooke JM and Carlsson DJ (in press) J Polym Deg Stab
8 Pospisil J (1980) Adv Polym Sci 36:69-133
9 Allen DW, Clench MR, Growson A and Leathard DA (1993) J Polym Deg Stab 39:293-297
10 Carlsson DJ, Chmela S and Lacoste J (1990) Macromolecules 23:4934-4938

PPOOH + ~C(=O)~ + ~C(=O)-OH
(220) (50) (4)

↑ 60°C, Air, 1 day

PP —2.5 Mrad γ/Air, Unstabilized→ PPOOH + ~C(=O)~ + ~C(=O)-OH
(50) (12) (3)

PP —2.5 Mrad γ/Air, Phenol IV→ PPOOH + ~C(=O)~ + ~C(=O)-OH
(12) (3) (1)

↓ 60°C, Air, 200 days

PPOOH + ~C(=O)~ + ~C(=O)-OH
(55) (53) (13)

Reaction Scheme 1

Values in parenthesis are product concentrations in mmole/kg at the bend failure point. (i.e. times shown in days at 60°C)

Construction of Polypeptide Tertiary Structure by the Template-Assisted Synthesis

Yukio Imanishi, Shunsaku Kimura and Tatsuro Tsuchimoto

Department of Material Chemistry, Kyoto University,
Yoshida Honmachi, Sakyo-ku, Kyoto, Japan 606-01

ABSTRACT

Two helical chains of dodecapeptides were connected to a cyclic octapeptide in a parallel (F12-C8KL) or an antiparallel orientation (CH2). CD measurements revealed that two dodecapeptide chains took α-helical conformation and were non-aggregative in organic solvents, but assembled intramolecularly in a buffer solution. The formation of the supersecondary structure was also indicated by fluorescent probes introduced to the chain-end region. Since each peptide chain was designed to take an amphiphilic α-helical structure, the hydrophobic surfaces face each other in the α-helix assembly. The intramolecular assembly caused a distortion of peptide chains from α-helical conformation in a buffer solution. The distortion of CH2 was less marked than that of F12-C8KL. CH2 was bound by λ DNA, while CH1, which is a single-chain analog of CH2, and F12-C8KL, which is analoguous to CH2 without a terminal anthryl group, were not. It is considered that CH2 binds itself to DNA backbone through electrostatic interaction of two Lys residues in one chain, and that the anthryl group of the other chain is intercalated into a base pair. The mechanism of CH2/DNA interaction changed by the addition of Ca^{2+} probably due to Ca^{2+} complexation with the cyclic moiety of CH2.

INTRODUCTION

The protein tertiary structure is a typical example of supramolecular architecture. The tertiary structure is constructed on the basis of the primary structure with a delicate

K.P. Ghiggino (Ed.)
Progress in Pacific Polymer Science 3

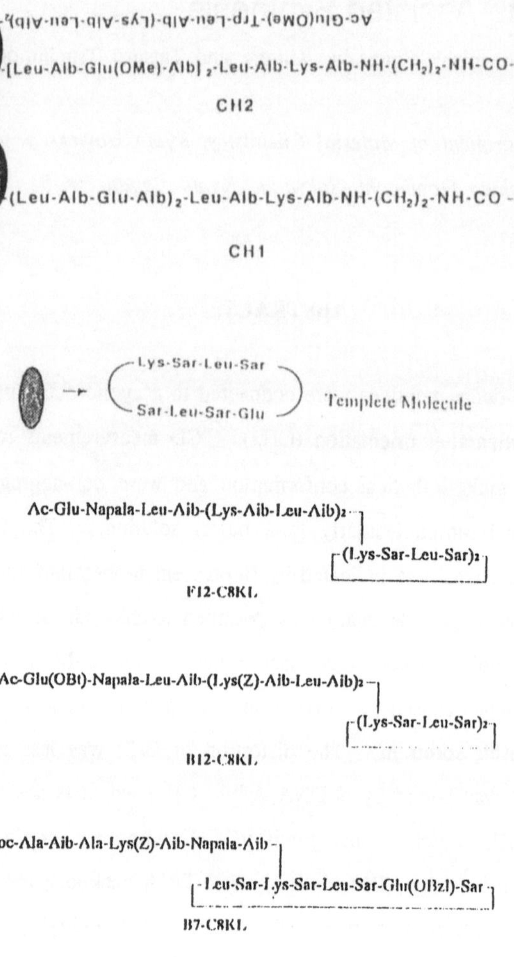

Fig. 1 Molecular structure of cyclic octapeptides having one or two helical pendant chains.

balance of enthalpy and entropy terms. However, the folding problem how a linear polypeptide chain is folded into a specific tertiary structure remains to be solved. Recently, *de novo* design of artificial proteins has attracted much attention [1,2], in which the strategy to achieve a super-secondary or a tertiary structure is ascribed to regulation of the spatial arrangement of secondary structures by designing suitable molecular interaction between the peptide segments [3]. It has also been reported that a template molecule is effective for building an ordered structure in terms of the entropy change

[4]. We are interested in using cyclic peptides for the template molecule, because allosteric proteins can be synthesized by utilizing conformational change of the cyclic peptide upon complexation with a metal ion [5]. In the present study, two basic α-helical peptides, which mimic transcriptional regulator proteins binding to DNA, were synthesized and connected to a cyclic octapeptide template.

A class of sequence-specific DNA-binding proteins form rotationally symmetric dimer [6]. A basic region that binds itself to DNA is followed by a "leucine zipper" that motives protein dimerization. The "leucine zipper" disposes the basic regions of two helical polypeptides for sequence-specific recognition of DNA. The change in orientation of the basic regions exhibits a dramatic effect on DNA binding [7].
The structures of synthetic peptides used in the present investigation are shown in Fig. 1. Two amphiphilic α-helical peptide chains containing Lys residues were connected to cyclic octapeptide. The association of two peptide chains should be promoted by the presence of cyclic-peptide template. In addition, two amphiphilic peptide chains are stabilized in water with the hydrophobic surfaces facing each other. The arrangements of hydrophobic and hydrophilic residues are shown as helical wheels in Fig. 2. The

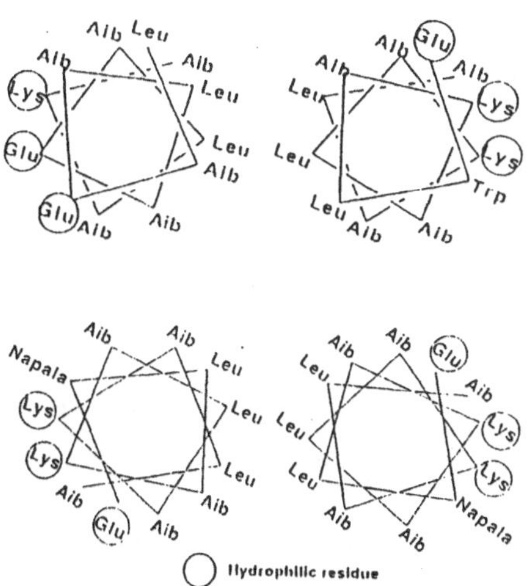

Fig. 2 Helical wheel presentation of amphiphilic helical pendant chains connected to cyclic peptides.

distance between two Lys residues in the α-helical peptide chains is 6 Å. Since adjacent phosphodiester groups are separated by 7 Å and 6 Å, respectively, in DNA and RNA, the peptides are expected to be captured by the major groove of DNA through electrostatic interaction between the Lys residue and the phosphodiester group.

RESULTS AND DISCUSSION

Conformation by CD measurement

Two chains of dodecapeptides were connected to a cyclic octapeptide in a parallel (F12-C8KL) or an antiparallel orientation (CH2). In principle, the dodecapeptides were designed to take an amphiphilic α-helical structure, which promotes association of the peptides in water.

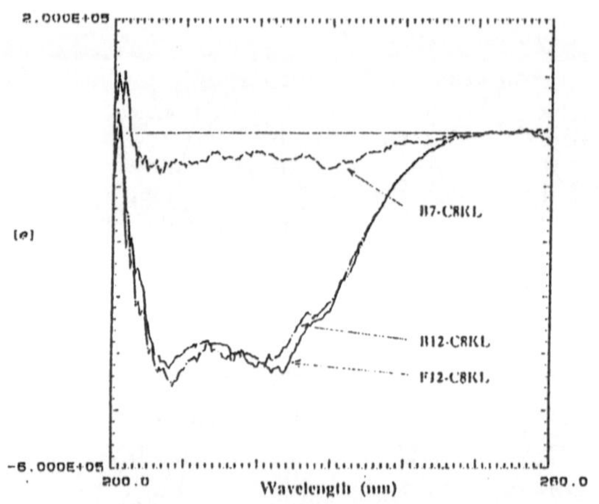

Fig. 3 CD spectra of F12-C8KL, B12-C8KL and B7-C8KL in ethanol.

F12-C8KL in ethanol showed the double-minimum-type CD spectrum, indicating the occurrence of α-helical conformation (Fig. 3). The helix content was estimated to be 59 % from the molar ellipticity at 222 nm, which is evidently higher than the pendant chain of B7-C8KL consisting of seven amino acid residues. Two pendant chains of F12-C8KL are shown free from aggregation in ethanol, because the ellipticities at 208 and 222 nm were comparable each other [8] and no significant difference between the charged F12-C8KL and the non-charged B12-C8KL is detected in the CD spectra.

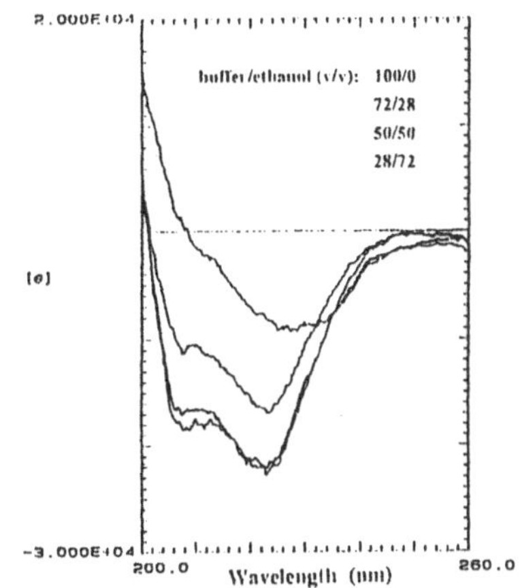

Fig. 4 CD spectra of F12-C8KL in ethanol/buffer solution mixture of various compositions. The ellipticity is represented in a mean residue value (the part of the cyclic peptide is neglected).

On the other hand, CD spectrum of F12-C8KL changed by changing solvent composition (Fig. 4). The CD spectra in the media containing a buffer solution of 28 % and 50 % are of a double-minimum pattern with stronger ellipticity at 222 nm than at 208 nm. This CD pattern is ascribed to associated helices as exemplified by rhodopsin [9]. When the content of buffer solution exceeded over 72 %, the negative signal became weaker and the minimum point shifted to longer-wavelength region. It is considered that the helical structure was seriously deformed by strong intramolecular or intermolecular association of peptide chains.

CD spectra of CH1 and CH2 in trimethyl phosphate are shown in Fig. 5, which indicate α-helical conformation without significant interaction between the helices. The helix contents of CH1 and CH2 were 97 % and 55 %, respectively, from the ellipticities at 222 nm. The reduction of the helix content in CH2 is presumably because the plane of the cyclic-peptide template is not big enough to accommodate two helices in assembly. The negative ellipticities of CH1 and CH2 decreased in a buffer solution as compared with those in trimethyl phosphate (Fig. 6). However, the extent of decreasing ellipticity

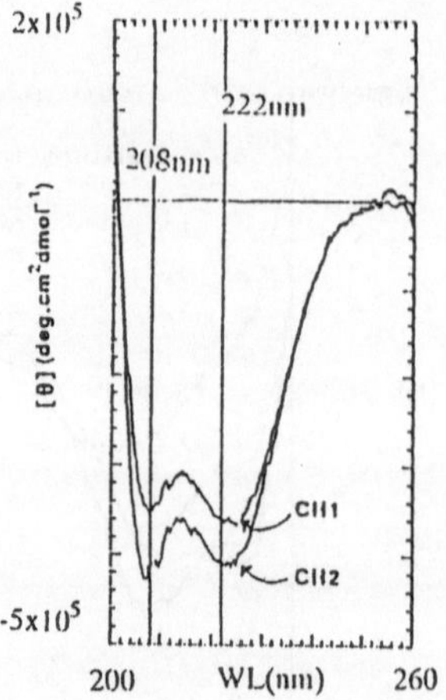

Fig. 5 CD spectra of CH1 and CH2 in trimethyl phosphate.

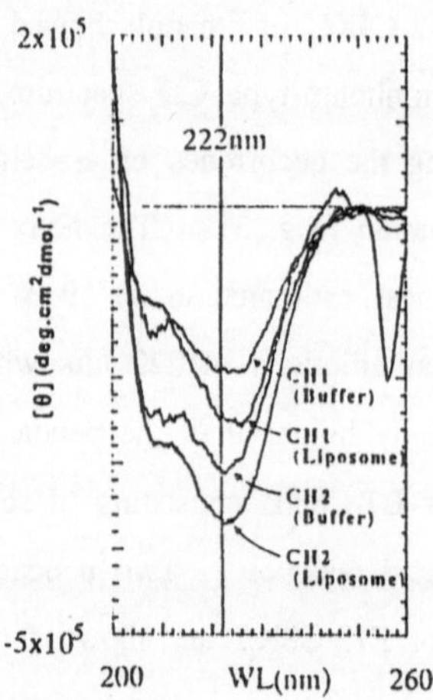

Fig. 6 CD spectra of CH1 and CH2 in a buffer solution with or without liposome.

of CH2 is less than that of CH1. The stronger negative ellipticity at 222 nm than that at 208 nm suggests that either in CH1 or in CH2 the peptide chains are associated together in a buffer solution. It is considered that intramolecular association of two peptide chains in CH2 is more resistant to the change of solvent in presering the helical structure than intermolecular association of CH1, because the reduction of the helix content of CH2 in a buffer solution is less extensive than CH1.

The intensity of the negative ellipticity of CH2 in a buffer solution was 3-fold of that of F12-C8KL, suggesting that the helical conformation of CH2 is distorted less extensively than that of F12-C8KL. The major difference in the molecular structure between CH2 and F12-C8KL is the orientation of two pendant chains. Anti-parallel packing of two helices is more stable than parallel packing due to the interaction ofhelix macrodipoles and the close packing of side chains of two helices. These points explain the stable intramolecular association of two helices of CH2.

Conformational investigation by fluorescence spectroscopy

The intramolecular association of two peptide chains of F12-C8KL and CH2 was also revealed by fluorescence spectroscopy. F12-C8KL in a buffer solution showed a very weak excimer emission of the naphthyl group, which is introduced to the N-terminal region (Fig. 7). Since the intensity ratio of the excimer and monomer emissions was independent of the peptide concentrations, intramolecular association of the two peptide chains is concluded.

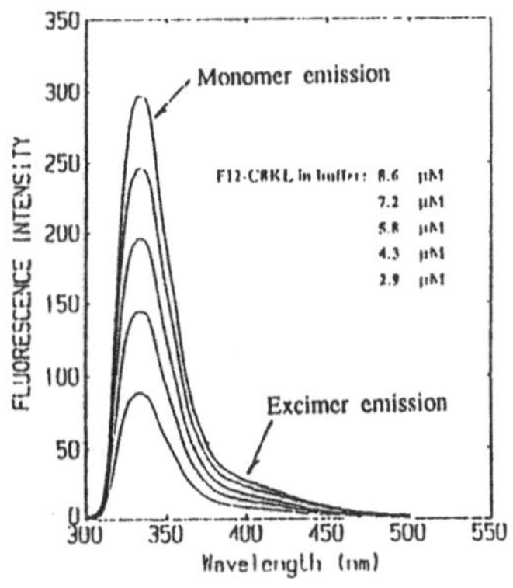

Fig. 7 Fluorescence spectra of F12-C8KL of various concentrations in a buffer solution. Excitation wavelength was 280 nm.

The association of two pendant chains of CH2 in a buffer solution was investigated in detail on the basis of the excited energy transfer from the Trp residue of one chain to the anthryl group of the other chain. The energy transfer efficiency was estimated to be 39 % from the excitation spectra, which means that the average distance between the probes is 23 Å. When the two helices take α-helical conformation and associate together as schematically shown in Fig. 8, the distance between the probes is estimated to be 22 Å. The agreement also proves that the two pendant chains of CH2 are intramolecularly aggregated in a buffer solution.

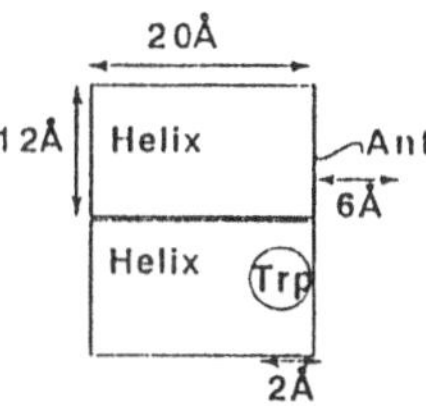

Fig. 8 Schematic presentation of the locations of fluorescent probes in CH2.

Interaction with λ DNA

Fluorescence spectrum of CH2 was measured in the presence of λ DNA of various concentrations (Fig. 9). The emission from the Trp residue of CH2 at ca. 350 nm decreased the intensity drastically by the addition of DNA, while the fluorescence

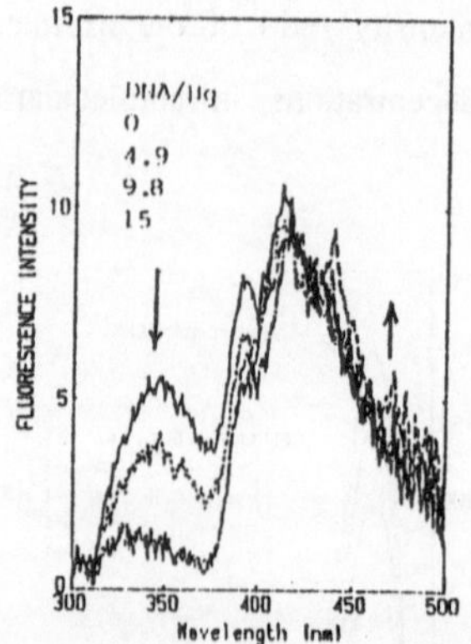

Fig. 9 Fluorescence spectra of CH2 in a buffer solution containing λ DNA of different concentrations.

intensity from the anthryl group at ca. 420 nm slightly decreased and that at ca. 440 nm slightly increased. The 420-nm emission might be ascribed to the exciplex formed between the anthryl group and nucleic acid bases. The emission from the anthryl group was not changed so much by the addition of λ DNA, indicating that the energy transfer from the Trp residue to the anthryl group was not suppressed by quenching Trp emission. This observation is not consistent with the quenching mechanism due to interaction of the indolyl group with nucleic acid bases, which would decrease the emission from the anthryl group. Therefore, only the anthryl group must have been intercalated into a base pair.

It is notable that two pendant chains of CH2 are necessary for binding to λ DNA. Fluorescence spectroscopy showed that CH1, which has the anthryl group at the chain end and a Lys residue in the middle of a pendant chain, was not bound by λ DNA (Fig. 10). One Lys residue or anthryl group alone is not enough to produce a stable complex with DNA. There are two Lys residues in the pendant helix chain of CH2 involving a Trp residue. The interval of two Lys residues along the α-helix is made suitable for binding by adjacent phosphodiester groups of DNA. It is considered that CH2 is bound by DNA through electrostatic interaction of two Lys residues of one

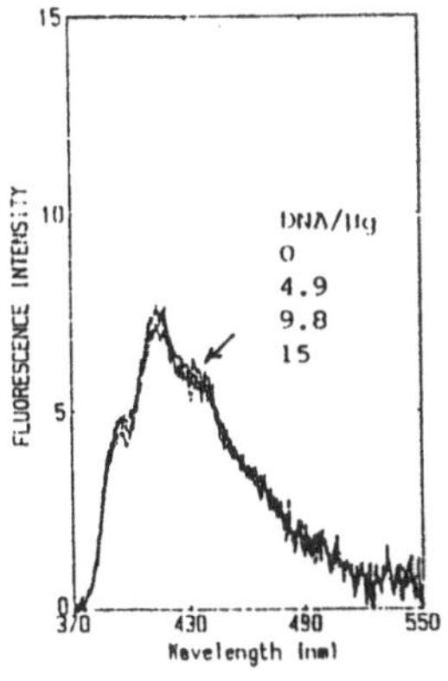

Fig. 10 Fluorescence spectra of CH1 in a buffer solution containing λ DNA of different concentrations.

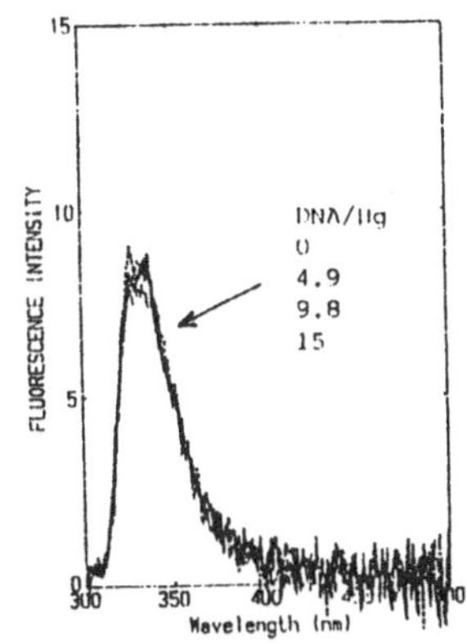

Fig. 11 Fluorescence spectra of F12-C8KL in a buffer solution containing λ DNA of different concentrations.

pendant chain with DNA backbone, and the anthryl group of the other pendant chain is intercalated into a base pair.

Fluorescence spectroscopy showed that F12-C8KL was not bound by λ DNA (Fig. 11). Although each pendant chain of F12-C8KL has two Lys residues, the spatial arrangement should not be suited for binding DNA, because the conformation of pendant chain in a buffer solution was seriously distorted from α-helical conformation as shown by CD measurement.

The cyclic moiety of CH2 can form complex with Ca^{2+}. When Ca^{2+} was added to a buffer solution of CH2, emission either from the Trp residue or from the anthryl group was quenched. CH2 might have formed a large aggregate upon complexation with Ca^{2+} to quench the emissions. However, the fluorescent emissions were restored by DNA addition (Fig. 12), suggesting that the aggregate of Ca^{2+} complex of CH2 dissociates upon binding of CH2 by λ DNA. However, the interaction of CH2 with DNA should be different in the presence of Ca^{2+}, since the emission from the Trp residue is observed in the presence of Ca^{2+}, while it is not observed in the absence of Ca^{2+}. The arrangement of two pendant chains of CH2 might have been changed by Ca^{2+} complexation at the cyclic moiety. Further investigation on this point is needed.

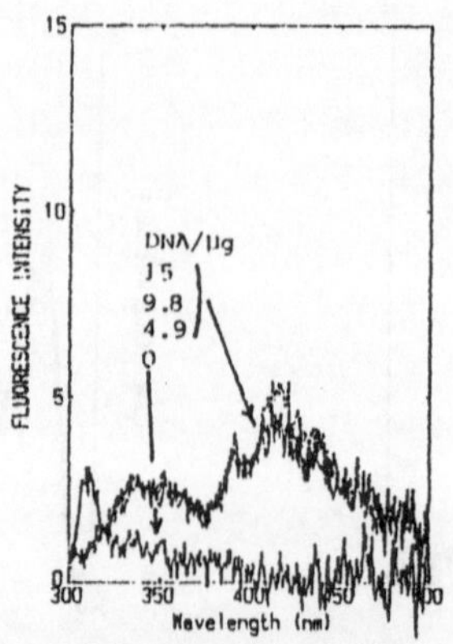

Fig. 12 Fluorescence spectra of CH2 in a buffer solution containing Ca^{2+} (50 mM) and λ DNA of different concentrations.

REFERENCES

1 Richardson JS, Richardson DC (1989) Trends in Biochem Sci 14:304

2 DeGrado WF, Wasserman ZR, Lear JD (1989) Science 243:622

3 Mutter M, Vuilleumier S (1989) Angew Chem Int Ed Engl **28**:535

4 Mutter M (1988) Trends in Biochem Sci 13:260

5 Imanishi Y, Kimura S, in *"Molecular Conformation and Biological Interactions"*, eds. Balaram P, Ramaseshan S, Indian Academy of Sciences, Bangalore, pp.645-659 (1991).

6 Landschulz WH, Johnson PF, McKnight SL (1988) Science 240:1759

7 Cuenoud B, Schepartz A (1993) Science 259:510

8 Lau SYM (1984) J Biol Chem 259:13253

9 Gibson NJ, Cassim JY (1988) Biochemistry 28:2134

A New Family of Water-Soluble Polymers: Polymers of N-Substituted Acrylamide Derivatives

Shouei FUJISHIGE/(Former Res.Inst. for Polymers & Textiles)
Tokyo Kasei University:Kaga 1-18,Itabashi,Tokyo,Japan 173

Abstract: Most polymers dissolve into solvents endothermically which means their solubility increases on heating. In contrast to this some polymers lose their solubility in water at higher temperatures above a critical point,but regain it on cooling the solvent below the critical temperature. In addition to well known poly(N-isopropyl)acrylamide,water-soluble polymers from a series of N-substituted acrylamide derivatives were synthesized and the characteristic solubity change induced by thermal stimulation was investigated by viscometry, light scattering, NMR and DSC.

1.INTRODUCTION

Many solid polymers are sensitive to heat stimulation and show characteristic REVERSIBLE PHASE TRANSITIONs at the GLASS TRANSITION TEMPERATURE and MELTING TEMPERATURE depending mainly on their own primary structure. This thermal sensitive property is being used in the fabrication of thermoplastic materials into any desired form in a similar means to that of fibers in the case of nylon, polyesters or films and other three dimensional products. In addition to the thermoplastic properties of polymers in the solid state, the molecular motion of polymers in solution is also achieved by thermal stimulation and has extensively been studied after pioneering works of Flory-Huggins. The effects of such thermal stimulation were detected experimantally as "sol-gel transformation","viscosity changes","phase separation" and so on. Egg-white and other aqueous solutions of natural proteins such as enzymes and hormones show such detectable behavior in the form of

K.P. Ghiggino (Ed.)
Progress in Pacific Polymer Science 3

an irreversible phenomenon by the word of "thermal denaturation" or "thermal deactivation". Detailed studies on such behavior as elemental process of biological species has been limited due to the lack of adequate methodology suitable for the measurement of such the phenomena. When aqueous solutions of poly-L-lysine and of poly-L-proline are heated, phase separation takes place whereas on subsequent cooling both precipitated polymers dissolve again and the behavior is reversible with a significant delay time to the cyclic thermal stimulations[1]. Aqueous solution of polyethylene-oxide shows similar thermal behavior but its phase transition temperature is observed above 100°C [2]. Similar behavior is also observed in[3,4] aqueous solutions of methylcellulose,hydroxyethyl-cellulose and polyvinylalcohol with a little amount of acetyl groups, however, we have to be careful because the solution properties,especially the thermodynamic properties of these polymer solutions depend not only on the molecular weight of the polymer samples but also on the degree of substitutions of the samples. It is well known that aqueous polyvinylmethylether solution shows similar thermal behavior[5] but in this case, the solubility of the polymer into water is being restricted mainly by its stereospecific structure depending on the preparative conditions of the polymer.

In contrast to those polymers described above,aqueous solutions of polyvinyloxazolidinon[6] and poly(N-isopropyl)acrylamide[7] have been studied in rather more detail, however, in spite of the fact that the solution properties including such a thermodynamic quantities should always be depending on the molecular weight of the polymer samples, those previous experimental works had been done with unfractionated polymer samples. Poly(N-isopropyl)-acrylamide (=PNiPAM) has the repeating unit[8] structure comparable to that of polypeptides as shown in Fig.1.

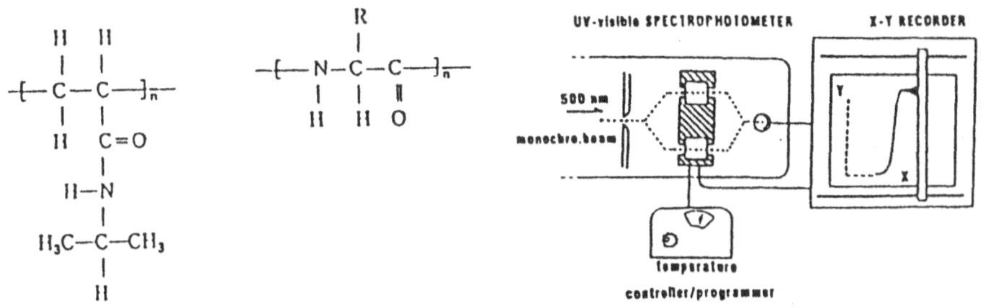

Fig. 1:Repeating unit of polypeptides as compared with those of polyacrylamide derivatives

Fig. 2:System to determine phase transition temperature of aqueous polymer solutions

In this paper, we would like to represent more detailed studies on such characteristic aqueous solution properties of PNiPAM and some additional results obtained with a series of polymers of N-substituted acrylamide derivatives.

2.EXPERIMENTALS and RESULTS

2-1) <u>Preparation of PNiPAM, other polyacrylamide derivatives and molecular weight fractionation</u>: PNiPAM and other polyacrylamide derivatives were prepared by conventional free-radical polymerization of purified monomer which was synthesized by the reaction of acrylic or methacrylic acid chloride with the equivalent amount of monoamine compound to give the corresponding N-substituted acryl- or methacryl-amide derivatives. Most of the polymers were fractionated from ca.1% dry acetone solution by n-hexane as the precipitant at room temperatures[9]).

2-2)<u>Viscometry and determination of Mark-Howink-Sakurada Eqs.</u> The following equations were obtained for fractionated PNiPAMs[9]) and fractionated poly(N-isopropyl)methacrylate[10]) (=PNiPMAM):

PNiPAM):$[\eta] = 14.5 \times 10^{-2}\ Mn^{0.5}$ in water at 20°C

$= 24.9 \times 10^{-2}\ Mn^{0.46}$ in water at 15°C

$= 9.59 \times 10^{-3}\ Mn^{0.65}$ in Tetrahydrofurane at 27°C

PNiPMAM):$[\eta] = 9.66 \times 10^{-3}\ Mw^{0.68}$ in water at 20°C

2-3)<u>Determination of Phase Transition Temperature:</u> In the previous pioneering work on PNiPAM, Guillet et al.[7]) used their naked eye to determine the transition temperatures because the transparency of the aqueous PNiPAM solutions is quite sensitive to the temperature.In this work, a more sensitive system composed of a spectrophotometer as shown in Fig.2 was constructed to measure the transparency at 500 nm even at very dilute solutions. As shown in Fig.3,the phase transition temperature of PNiPAM aq.solutions is independent of its concentration from very dilute,0.033% to 1(w/v)%. This observation is in contrast to the previous finding as a typical phase diagram of lower critical solution temperature[7]).

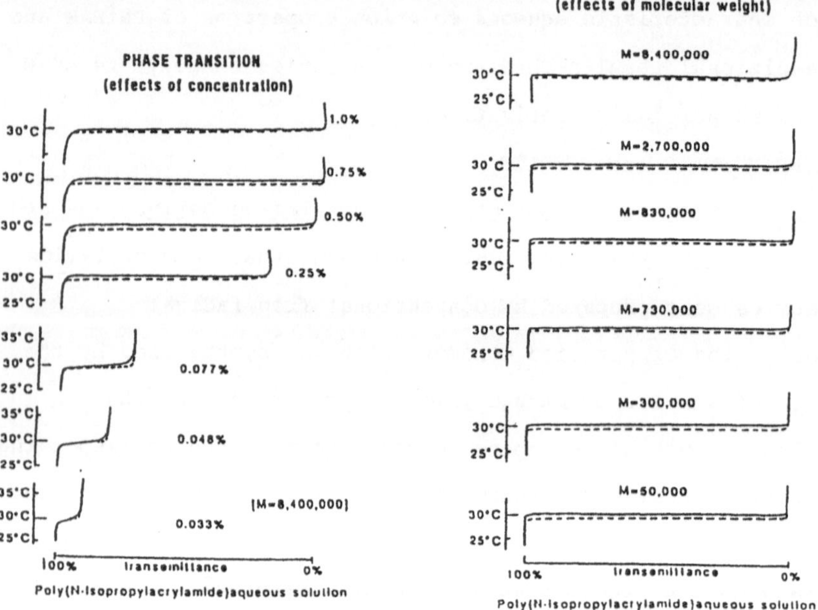

Fig. 3:Effects of concentrations of PNiPAM on transition temperature of aqueous solution

Fig. 4:Effects of molecular weight of PNiPAMs on transition temperature of 1(w/v)% aqueous solution

2-3)Effects of Chain Stiffness of Polymers on Phase Transition

As shown in Figs.3 and 4,aq.PNiPAM solutions lose their transparency quite sharply as a function of temperatures while the rate of heating(solid lines) and of cooling(dotted lines) has been kept constant (1°C/minute).Similar but slightly different behavior can be seen in Fig.5. A significant delay to recover the transparency is seen in the case of aq.PNiPMAM solution in contrast to that of aq.PNiPAM solution and a longer delay was observed in the case of aq.methylcellulose solution. These tendencies might be attributed to the chain stiffness of individual polymer samples in aq.solutions. Fig.6 is a typical example showing the effects of chain stiffness of PNiPAM compared with that of PNiPMAM where the average radius of gyration of PNiPMAMs per unit degree of polymerization is always higher than that of PNiPAMs[10]. This indicates PNiPMAM exist in a more expanded state in aq.solutions at 20°C than PNiPAM.

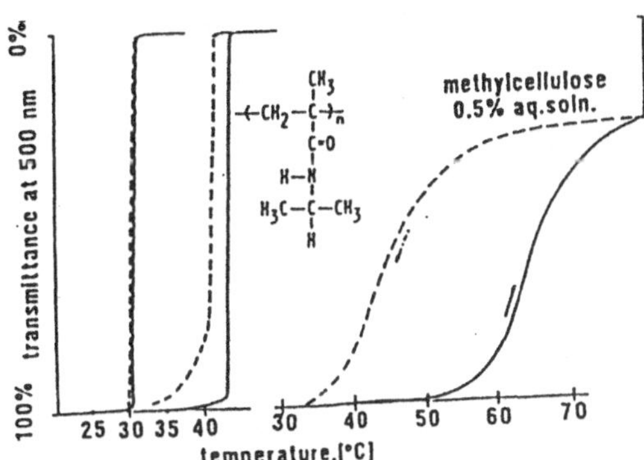

Fig. 5:Phase transition of aqueous 1(w/v)% solution of PNiPAM compared to those of 1(w/v)% PNiPMAM and 0.5(w/v)% methyl-cellulose

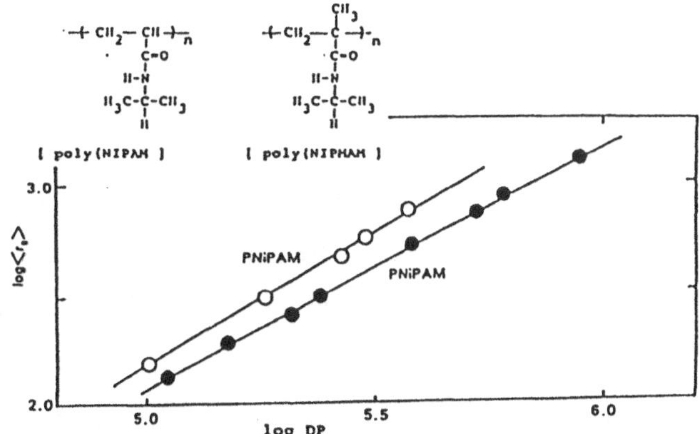

Fig. 6:Linear correlation between the average radius of gyration and DP of PNiPAMs in water at 20°C as compared to that of PNiPMAMs

2-4)Effects of temperature on a single PNiPAM in aq.solution: A home-made high precision type light scattering system was constructed by Professor Kubota,Gunma University and applied for an extremely dilute aq.solution as low as 0.001264(w/v)% of PNiPAM with the weight average molecular weight of 8,400,000. As it is seen in Fig.7,both the root mean square radius of gyration and the hydrodynamic radius of the single polymer chain vary drastically at a critical temperature corresponding to the phase transition

concentration:0.001264wt.%

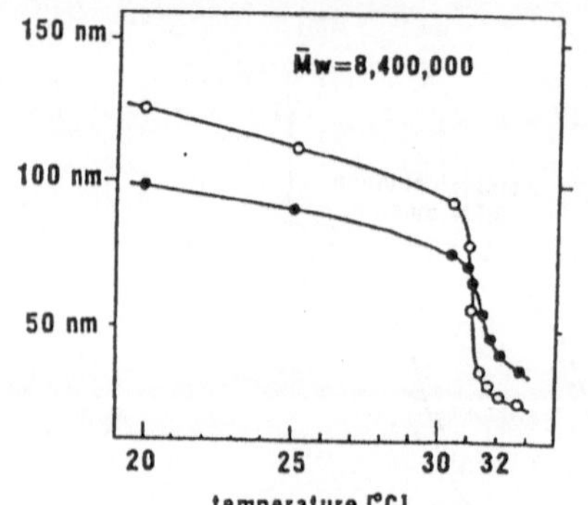

root-mean-square radius of gyration:

125 nm/20°C ⇆ 24 nm/32°C

hydrodynamic radius:

98 nm/20°C ⇆ 35.9 nm/32°C

reversible

COIL ⇆ GLOBULE

Fig. 7:Effects of temperature on the dimensional parameters of PNiPAM chain in an extremely diluted aqueous solution

temperature[11]. A series of experiments with different molecular weight samples confirmed[17] that, above the critical temperature, PNiPAM is dispersing as a collasped state (=globule) in aqueous medium and this conformational change is completely reversible with temperature as has been deduced as "COIL-GLOBULE TRANSITION" by Ptitzyn et al.[12]. An interpretation of this type of transition has been proposed by Matsuyama et al.[13]. The phase transition of dilute(less than ca.1%) aqueous solutions is detected as the solution becomes a milky suspension while the solution of higher concentration (more than ca.2%) gives rise to a "phase separation" which can be observed by naked eye.A polymer film prepared by heating such a dilute aquous PNiPAM solution is fragile but a film prepared by the same manner from the same PNiPAM but moderately concentrated aqueous solution behaves as an elastic material which might be attributed to the possible chain entanglements formed under such condition(Fig.9).

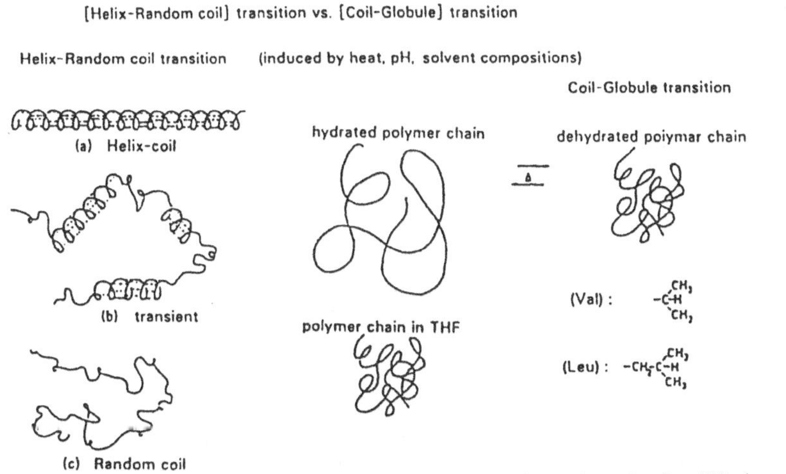

[Helix-Random coil] transition vs. [Coil-Globule] transition

Helix-Random coil transition (induced by heat, pH, solvent compositions)

Coil-Globule transition

(a) Helix-coil

hydrated polymer chain

dehydrated poiymar chain

(b) transient

(Val) : $-CH(CH_3)_2$

polymer chain in THF

(Leu) : $-CH_2-CH(CH_3)_2$

(c) Random coil

Fig. 8:Conformational transition of single chain molecule in dilute solution: [HELIX-RANDOM COIL] *vs.* [COIL-GLOBULE] transition

It is well known that an addition of small amount of ethylalcohol into aqueous polymer solution is effective to control the extent of hydration of solute polymer chains. In order to observe such an effect on the phase transition of aqueous PNiPAM solutions,even pure ethylalcohol is a good solvent for the polymer, however, neither concentration dependence nor molecular weight dependence were observed on their phase diagram (Figs.10).

Poly (N-isopropylacrylamide)

(non α-helical polyamides)

SOLUBLE in Water at low temp.	mod. conc. aq. soln.	dilute aq. soln.	extremely dil. aq. soln. coil
	↑ ↓	↑ ↓	↑ ↓
INSOLUBLE in Water at high temp.	phase separation	turbid	globule

Fig. 9:Correlation of thermally induced phase transition in aqueous solution of PNiPAM

2-5)<u>Differential Scanning Calorimetry on the phase transition:</u> As shown in Fig.11,thermally induced phase transition takes place and is observed as an endothermic peak for each of the polymer with different N-substituted side chains as:at 21.5°C with 11.6(ca/g) or ca.6.5(J/mol of monomer unit concentration) for Poly(N-n-isopropyl) acrylamide and at 31.9°C with 11.1(cal/g) or ca.4 to 5(J/mol of monomer unit concentration) for PNiPAM, and 45.5°C accompanied with less energy for poly(N-cyclopropyl)acrylamide. Further discussion appeared in Ref.[14]).

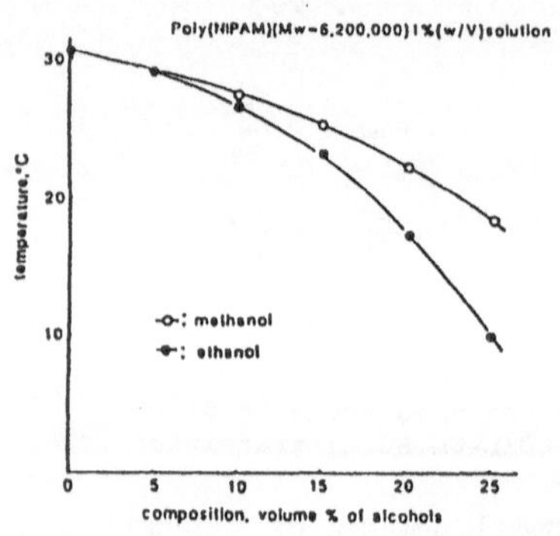

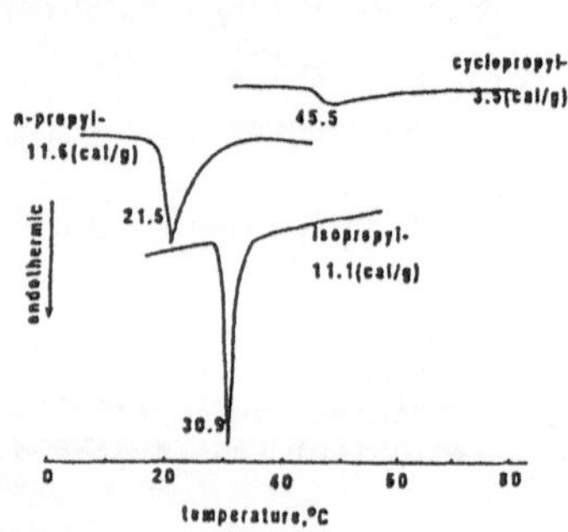

Fig.10:Effects of solvent compositions on the phase transition temperature of PNiPAM solutions in mixed solvents composed of water/methylalcohol and of water/ethylalcohol, respectively

Fig.11:Phase transition of polyacrylamide derivatives in aqueous solutions detected by DSC

2-6)<u>NMR applied for aqueous PNiPAM solutions at the vicinity of phase transition:</u> In order to elucidate the role of water molecules in aqueous PNiPAM solutions, the ^{1}H spin-lattice relaxation time, T_1 as well as the spin-spin relaxation time,T_2 were measured as a function of temperatures between 10°C and 50°C and it was found that the T_2 of water molecules in aqueous PNiPAM solutions increases as a function of temperature but decreases drastically at the vicinity of the phase transition temperature and then increases again while that of pure water increases linearly in this temperature range as shown in Figs.12.

2-7)<u>Thermally induced Volume Contraction of Nanometer-sized PNiPAM Beads in Aqueous Medium:</u> Fine mono-dispersed beads composed of a crosslinked PNiPAM were prepared in the presence of micellar structures of surfactants by free radical polymerization and the thermally induced volume contraction of the polymer beads was monitored by measuring the dynamic light scattering as a function of temperature from 25°C to 35°C and *vice versa*.

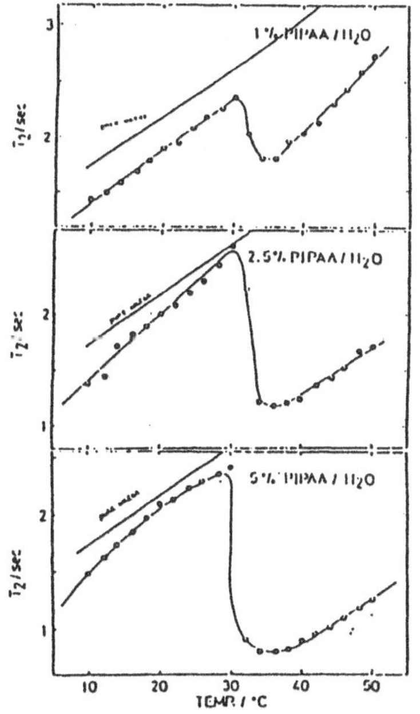

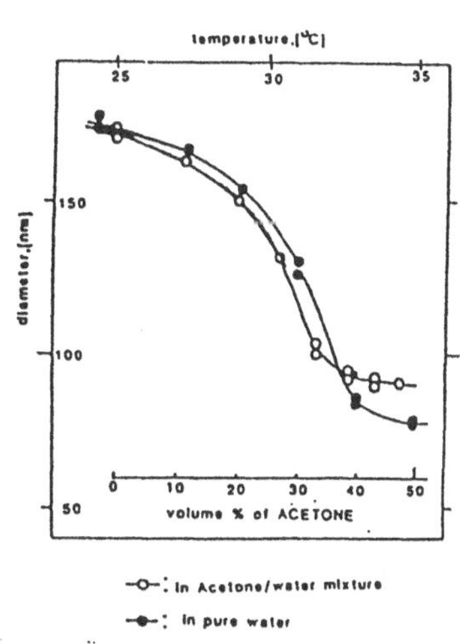

Fig.12:Temperature dependence of the SPIN-SPIN RELAXATION TIME,T_2 of 1-, 2.5- and 5-(w/v)% aqueous PNiPAM solutions compared to that of pure water

Fig.13:Effects of temperature and of solvent composition on the volume contraction of nano-meter sized beads of PNiPAM copolymerized with methylenebisacrylamide in the presence of micellar structures of surfactants

The diameter of the polymer beads of a typical example is found to be 175 nm at 25°C but changes abruptly at a critical temperature corresponding to the phase transition temperature of aqueous linear PNiPAM solution, that is 31°C, then attains to a steady state at about 35°C as of 75 nm. Similar behavior was observed even at a fixed temperature but with an aqueous medium mixed with a certain amount of acetone(Fig.13).

2-8)Syntheses of N-Substituted Acrylamide and Methacrylamide Derivatives and Phase Transition Temperature of Polymers Obtained

A series of polyacrylamides and polymethacrylamides having different alkyl-, alkoxyalkyl- and hydroxyalkyl-groups in their side chain structure was prepared *via* free radical polymerization of N-substituted-acrylamides or -methacrylamides which have been synthesized from corresponding alkylamines and acryloyl chlorides. The solubility of these polymers in water is classified into three types depending on the side chain structures as;1)soluble in water irrespective of temperature,2)soluble in cold water but insoluble at higher temperatures,3)insoluble into water at any temperatures. Almost of all polymers belong to the 2) group behaves quite similar to those of aqueous PNiPAM solutions described above and show that the phase transition temperatures depend mainly on their own primary structure as represented in Fig 14 [16].

3 Conclusion

According to the modern solution thermodynamics, most linear polymers dissolve into their solvent endothermically which means that the solubility of the polymer increases on heating the solvent and/or the polymer solution. On the contrary to this, however, some polymers lose their solubility into water at higher temperatures above a critical point but regain the solubility on cooling the solvents and or the polymer solutions below the critical condition. In this work, a series of N-substituted polyacrylamides and polymethacrylamides were prepared and, among those,PNiPAM and PNiPMAM were studied in more details as an typical example of such a new family of water soluble polymers.

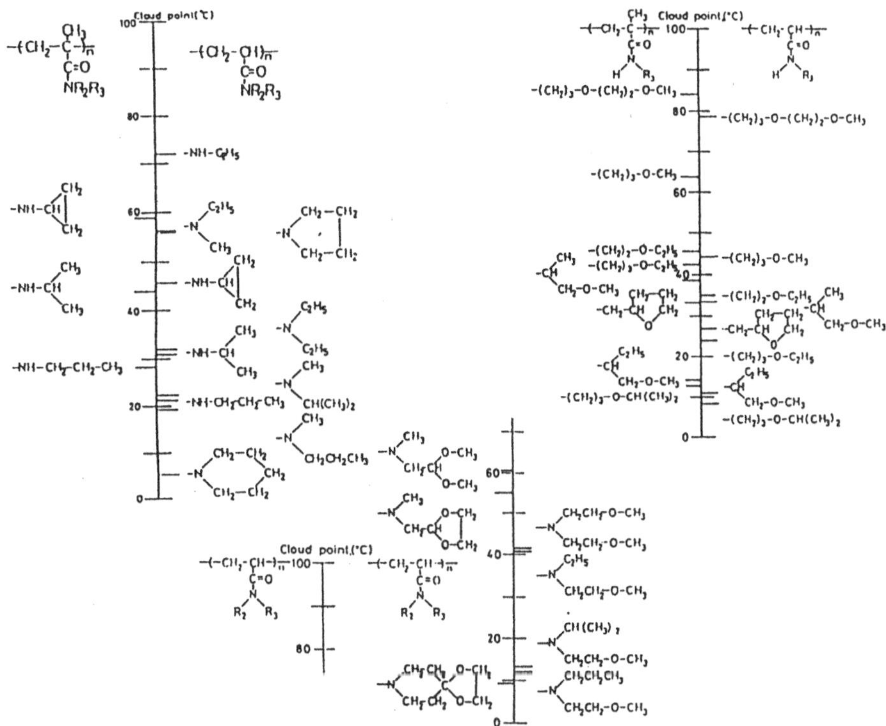

Fig.14:New family of water-soluble polymers of N-substituted acrylamide derivative

References

1 Mattice W L,Mandelkern L (1976) Macromolecules 4:271

2 Saiki S,Kuwahara N,Nakata M,Kaneko M (1976) Polymer 17:675

3 Frank F (1988) in: Finch C A (ed) Chemistry and Technology of Water-soluble Polymers pp.157-178 Plenum Press New York

4 Borchard W (1988) in: Finch C A (ed) Chemistry and Technology of Water-soluble Polymers pp.113-124 Plenum Press New York

5 Horn R A,Almeida J P,Day A F,Yu N-T (1971) J.Colloid Interface Sci. 35:77

6 Klotz I M (1965) Fed.Proc.24(Suppl.No.15) S24

7 Heskins M,Guillet J E (1968) J.Macromol.Sci.,Chem.A-2:1441

8 Scarpa J S,Mueller D D,Krotz I M (1967) J.Am.Chem.Soc. 89:602

9 Fujishige S (1987) Polymer J. 19:297

10 Kubota K,Hamano K,Kuwahara N,Ando I,Fujishige (1990) Polymer J 22:1050

11 Fujishige S,Kubota K,Ando I (1988) J.Phys.Chem. 93:3311

12 Ptitsyn O B,Kron A K,Eizner Yu Ye (1968) J.Polym.Sci.C-16:3509

13 Matsuyama A,Tanaka F (1991) J.Chem.Phs. 94:781

14 Schild H G,Tirrell D A (1990) J.Phs.Chem. 94:4352

15 Ohta H,Ando I,Fujishige S,Kubota K (1991) J.Polym.Sci.29:963 ; (1991) J.Mol.Structures 245:391

16 Ito S (1989) Kobunshi Ronbunsyu 46:437; (1990) Kobunshi Ronbunsyu 47:467

17 Kubota K,Fujishige S,Ando I (1990) J.Phys.Chem. 94:5154

Note: Quite recently,papers appeared with respect to POLY(N-ISOPROPYLACRYAMIDE) were reviewed by Dr.H.G.Schild in Progress in Polymer Science Vol.17,pp.163-249(1992).

The Measurement and Meaning of Rate Coefficients in Free-Radical Polymerizations

RADICAL POLYMERIZATIONS

Johan P A Heuts, Paul A Clay, David I Christie, Mark C Piton*, Jelica Hutovic, Scott H Kable, Robert G Gilbert

Chemistry School, Sydney University, NSW 2006, Australia

*current address: Department of Chemistry, Queen's University, Kingston, Ontario K7L 3N6, Canada.

Abstract: New experimental data and theory are reported for obtaining three types of rate parameters controlling kinetics and molecular weight distributions in free-radical polymerizations. (1) (a) Propagation rate coefficients (k_p) for butyl acrylate are found using pulsed-laser polymerization, with the novel variant of carrying out the procedure at very low temperatures (−10 to −40 °C); this obviates some problems hitherto found with the technique. (b) A means of calculating k_p from first principles from transition state theory is given, using accurate quantum mechanical computation of the transition state properties. These calculations give quantitative accord with experiment for ethylene, and reveal that the magnitude of the frequency factor for propagation is dominated by three hindered rotors in the transition state. Since these rotations are influenced by the penultimate unit, this suggests that significant penultimate unit effects are likely in copolymerizations. The effect of these rotations also explains the observed effect of deuteration on k_p. (2) A new technique for measuring the transfer rate coefficient (k_{tr}) is presented: initiation by pulsed laser, whence k_{tr} can be obtained from the high molecular weight slope of a plot of ln(number molecular weight distribution) against molecular weight. The method, which can be employed even when extensive termination is taking place, is tested with transfer from MMA to triethylamine. (3) The rate coefficient for termination can be successfully modelled from a knowledge of the diffusion coefficients for oligomeric species as functions of the degree of polymerization and the polymer fraction. New PFG NMR diffusion data for a series of model compounds in polystyrene/benzene give results for this dependence which are in accord with a semiempirical scaling law suggested previously [Piton MC, Gilbert RG, Chapman BE, Kuchel PW (1993) Macromolecules 26: 4472], suggesting that this scaling should be a useful basis for predicting termination rate coefficients.

INTRODUCTION

Properties such as the overall rate and the molecular weight distribution in a free-radical polymerization are controlled by the rate coefficients for radical creation from initiator, propagation, and chain-stoppage events: termination by combination and disproportionation, and transfer (to monomer, chain transfer agent or polymer). However, as illustrated in Fig. 1, literature compilations of rate coefficients [1] show a wide range of values under ostensibly the same conditions. While the scatter of values of rate coefficients in Fig. 1 might be seen in a single set of measurements and so represent statistical fluctuations, presumably each of the literature data shown in this figure are themselves proper averages over a number of individual measurements and should therefore show little variation between different workers using different techniques if the results were what they purport to be. Faced with such problems, an IUPAC Working Party has begun to establish some benchmark values and reliable measurement techniques for some of these quantities [2,3].

K.P. Ghiggino (Ed.)
Progress in Pacific Polymer Science 3

In the present paper, we apply some of these "established" techniques to obtain data for new systems; we also present a new technique for measuring the transfer rate coefficient which is more flexible than that conventionally used.

We also examine theoretical approaches to the *a priori* calculation of these rate coefficients, both for quantitative prediction and qualitative understanding. The type of question to which an answer is sought is: why is the value of the frequency factor for the propagation rate coefficient typically $10^6 - 10^7$ dm^3 mol^{-1} s^{-1}, and how can trends in this quantity (e.g., from acrylates to methacrylates) be qualitatively understood. The goal of calculating polymerization rate parameters from first principles may seem distant, but is by no means unattainable. The values of Arrhenius parameters for corresponding reactions in the gas phase are now routinely semi-quantitatively predictable (e.g., [4]), and in favourable cases may be found accurately by a combination of quantum mechanical calculations and appropriate application of transition state theory. Radical-molecule transition states, as occur in many of the reactions of interest in free-radical polymerizations, are known to be relatively unaffected by the presence of solvent (e.g., [5-7]) and indeed radical reactions are usually insensitive to solvent effects [8]. For this reason, one might suppose that quite good estimates of the rate coefficients for such reactions in condensed phases (i.e., in a polymerization medium) could be attained by the same means as have proved successful in the gas phase.

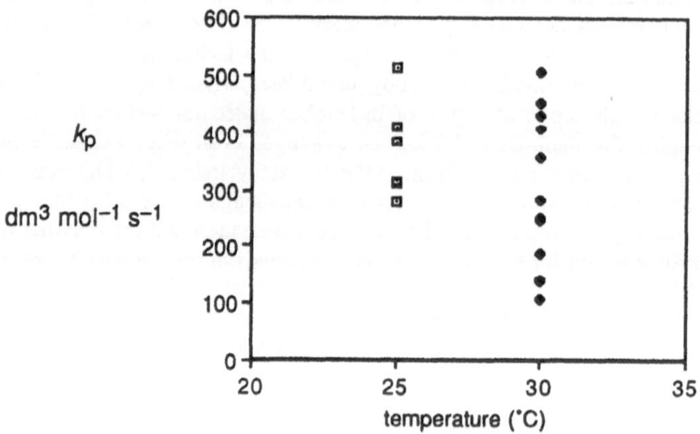

Figure 1 Propagation rate coefficients for methyl methacrylate at 25 and 30 °C, from the data tabulated by Brandrup and Immergut [1].

PROPAGATION

We here report new experimental data for the propagation rate coefficient, k_p, and new means of calculating this quantity from *ab initio* theory.

Experimental data

Pulsed-laser polymerization (PLP) [9-11] has emerged as a particularly reliable method for determining k_p [3,12]. Some (but by no means all) chains photo-initiated by one laser pulse are terminated "instantly" by short radicals formed from the subsequent one, as illustrated in Fig. 2.

The value of k_p can then be found from a knowledge of υ_p, the degree of polymerization of the chains so formed, the monomer concentration [M], and t_f, the time between pulses:

$$\upsilon_p = k_p[M]t_f \tag{1}$$

It is found that υ_p corresponds closely to the point of inflection on the GPC trace of the polymer formed in the process [12-16]. One test for reliability of data from this method is if this point of inflection is slightly below the maximum in this GPC trace, and higher harmonics are observed above this (the latter corresponding to termination by radicals formed two or more pulses after the initiating one). Because of the importance of the point of inflection in checking for consistency, it is useful when reporting PLP results to publish both typical GPC traces and their derivatives (wherein the lowest turning point makes the position of the corresponding point of inflection in the GPC trace particularly apparent). It should be noted that while the above criteria for reliability of PLP data are certainly sufficient, reliable k_p values may also be obtained from experiments which do not show all these features.

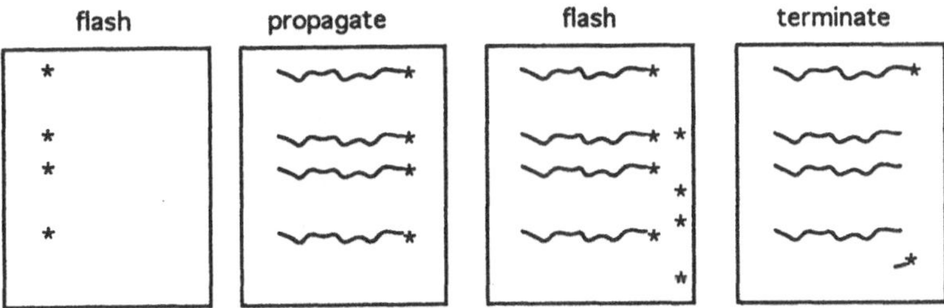

Figure 2. Illustrating the principle of pulsed-laser polymerization (PLP): k_p can be found from a knowledge of the degree of polymerization of chains terminated "instantly" by short radicals formed in the pulse subsequent to the initiating pulse.

Finding conditions in a PLP experiment so that the above criteria are obeyed is often difficult, and may involve a comparatively narrow window of experimental parameters. Those parameters which can be independently controlled are monomer concentration, concentration of photoinitiator, laser pulse frequency, laser power, and total illumination time. Optimization with regard to these conditions is constrained by having to ensure that laser illumination of the sample is uniform (including negligible attenuation of the laser beam over the length of the sample cell), and that [M] changes negligibly over the course of the experiment [11,17] (both requirements restrict the total amount of polymer that can be formed, which can lead to problems in GPC analysis). Previous attempts to measure k_p for acrylates had explored a wide range of conditions, but had failed [18] to find a GPC trace with a clear point of inflection, let alone one which exhibited the above desirable characteristics (although a PLP value for k_p for butyl acrylate has been reported [19], no GPC trace was published). We have also obtained PLP data for butyl acrylate at ambient temperatures over a very wide range of conditions, but were unable to find any results which showed a clear point of inflection.

What can be the origin of this problem in the application of PLP to acrylates? The point of inflection is a moderately subtle feature in a molecular weight distribution, and it is not unreasonable to assume that this could be readily masked by some sort of side reaction: for example, transfer to monomer or to some trace impurity (which would alter the molecular weight distribution), or the occurrence of backbiting (self-transfer), which would mean that the GPC trace would no longer be a simple transform of the number molecular weight distribution (transfer to polymer would cause a similar problem, but is unlikely at these high dilutions). Alternatively, the presence of a significant exotherm might smooth out a point of inflection. Now, such reactions as transfer have significantly higher activation energies than do propagation reactions, and hence would occur to a lesser extent at lower temperatures. Hence one might suppose that possible side reactions could be diminished by carrying out PLP at very low temperatures; similarly, an exotherm is diminished at low temperatures because propagation is slower. Polymerization can be carried out at very low temperatures with a photoinitiated system such as PLP. Since curvature of Arrhenius plots is minimal, and since solvent effects are generally unimportant for propagation reactions [5-7], Arrhenius parameters for k_p obtained at very low temperatures should be valid at for "ordinary" polymerization temperatures, as long as the low-temperature medium remains liquid, or at least does not become a glass. It therefore seems that the problems encountered with applying PLP to acrylates could be overcome by operating at very low temperatures.

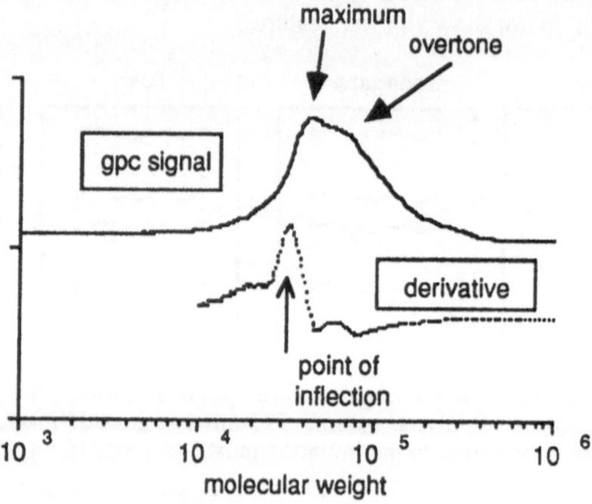

Fig. 3. GPC trace and derivative from PLP on butyl acrylate at –19 °C; note point of inflection, maximum and overtone (shoulder). Conditions: initiation by 308 nm Questek xenon chloride excimer laser operating at 35 mJ per pulse, pulsed at 30 Hz, 200 pulses, initiator = 1.77×10^{-4} mol dm^{-3} benzoin, [M] = 4.72 mol dm^{-3} in tetrahydrofuran.

We now give PLP data for butyl acrylate (BA) at very low temperatures (–40 to –10 °C) which indeed satisfy the criteria stated above as being sufficient for reliability. Butyl acrylate was distilled prior to use; GC analysis showed only a single peak. Typical results are shown in Fig. 3: note in particular the clear point of inflection, situated at a molecular weight slightly less than the maximum in the GPC trace and whose molecular weight is about a factor of 2 less than that of the overtone. GPC calibration was carried out by the "universal" method, with Mark-Houwink parameters $K = 8.57\times10^{-3}$ cm^3 g^{-1}, $a = 0.865$ [18]. The results of a series of such experiments are shown in Fig. 4. It should be noted that while the PLP data given here indeed conform to the stated reliability criteria, the scatter among the data at the higher temperatures is higher than is the norm for PLP; concomitantly, it is hard to reproduce runs which satisfy these criteria at the higher temperatures. The reasons for these problems are not known, and may well be connected to the possible reasons suggested above for the failure to obtain "reliable" PLP data at higher temperatures: a side reaction in acrylate polymerization due to backbiting or an adventitious impurity, perhaps in additon to an exotherm. In the latter context, direct measurement of the exotherm using a thermocouple under the conditions of Fig. 3 showed a rise of 4 °C.

$$k_p\ (\mathrm{dm^3\ mol^{-1}\ s^{-1}}) = 10^{7.4 \pm 0.4} \exp(-20 \pm 2\ \mathrm{kJ\ mol^{-1}}/RT)$$

Also shown in Fig. 4 are k_p values for BA reported by Buback and Degener [19]. The data of Buback and Degener are at a higher temperature than the present results, and were obtained by three techniques: rotating sector, time-dependent pulsed-laser polymerization, and PLP (although as stated no GPC traces were reported for the PLP result). Although all data appear consistent, the Arrhenius parameters for the present PLP data are not consistent with the data obtained by the alternative techniques at higher temperatures.

The PLP k_p data of Fig. 4 obey

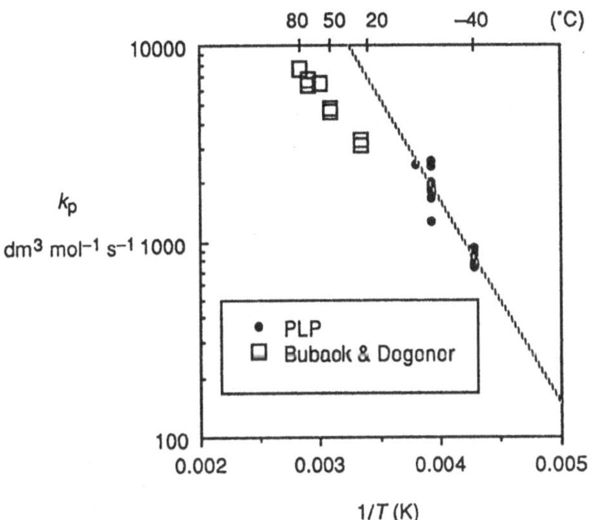

Fig. 4. Arrhenius plot for BA. Filled circles: PLP results (present work); open squares: results reported by Buback and Degener [19]. Broken line is Arrhenius fit to present data.

Theory: First Principles Calculation Of k_p

Transition state theory, combined with *ab initio* quantum mechanics, provides a means of calculating k_p (and other rate coefficients which are not diffusion-controlled) from first principles. The starting point is the transition state expression (e.g., [4]):

$$k = \frac{k_B T}{h} \frac{Q^\dagger}{Q} e^{-E_0/k_B T} \tag{2}$$

where k is the rate coefficient, $Q^\dagger$ and Q are the partition functions for transition state and reactant respectively, and E_0 the critical energy for reaction. The transition state is that slice through the complete potential function of the system such that it is assumed that any trajectory passing through this slice started in reactant space, finished in product space, and passed through only once [4]. For a reaction, such as propagation, with a barrier between reactant and product spaces, the transition state is simply located at that barrier (i.e., the saddle point), whose energy relative to reactants is E_0. The Hamiltonians of reactant and transition state are usually assumed separable, in which case the partition functions can be written as products of vibrational, rotational and translational terms (for a reaction such as propagation which occurs on a single electronic surface, the electronic factor is unity). Assuming that the vibrational components can be treated as harmonic oscillators and the rotors as rigid (free or hindered), there are simple analytic formulae for each component of the partition functions, which may be obtained from the geometry and energy derivatives of the reactant and transition state, which in turn can be located systematically by quantum chemistry methods (e.g., [20]). Moreover, eq 2 can be combined with these formulae to yield straightforward (although algebraically complicated) expressions for the Arrhenius frequency factor (A) and activation energy (e.g, [4]).

We here report results perfomed at a very high level of *ab initio* quantum theory; these calculations are very computationally demanding. In carrying out such calculations, careful checks must be carried out on the method used for finding the required properties of the transition state. While easier computations (e.g., molecular mechanics or semi-empirical quantum mechanical methods such as AM1, or low levels of *ab initio* theory) can be readily implemented through user-friendly, and readily accessible software, this can lead to misleading results unless the method of choice is properly tested. Tests include seeing if the *ab initio* calculation of an activation energy and frequency factor converges with increasing basis set and degree of inclusion of electron correlation; or, for semi-empirical methods, seeing if the chosen package reproduces known vibrational frequencies of analogous radicals and known activation energies of an analogous reaction. In the present case, appropriate tests were performed.

Calculations were performed for k_p using eq 2 and high-level *ab initio* computations (up to the QCISD(T)/6-311G** level of theory [21]) using Gaussian 92 [22]. The particular system chosen was the propagation of ethylene, which is sufficiently small that calculations for (say) a heptyl radical (the largest used in the present calculations) propagating with an ethylene monomer can be carried out at a relatively high level of theory. The results discussed here, and reported in detail elsewhere [23], go well beyond a mechanical application of the foregoing formulae. In addition to proper testing as just discussed, the most important innovation is to note that the lowest frequencies (say, ≤ 200 cm^{-1}) are in fact better treated as *hindered rotors* rather than as harmonic oscillators [4]. Moreover, since the partition function of such a rotor [24] depends significantly upon the hindrance angle and barrier height, these quantities in turn were also calculated quantum mechanically by scanning the energies of these rotations at the transition state as a function of rotational angle. Finally, the effect of mass, i.e., the fact that the propagation process of interest is for ethylene and an infinite rather than a finite macroradical, is taken into account through taking the proper limit of the translational and rotational partition functions.

It was found that the calculated *frequency factor* (A) readily converges with increasing level of sophistication of the quantum calculation, as shown in Table 1 for the propagation of a short species (an ethyl radical). However, it was found that the *activation energy* converged only at a very high level of theory (up to QCISD(T)/6-311G**[23,25]).

The reason that A converges so readily can be ascertained by examining the individual components of the partition functions. The mass component is trivial. The vibrational terms for frequencies above 200 cm^{-1} are all close to unity (recall $q_{\text{vib}} = \left[1 - \exp(-h\nu/k_BT)\right]^{-1}$) and their ratios, in $Q^\dagger/Q$, converges readily. Frequencies lower than 200 cm^{-1} are all treated as hindered internal rotors, whose partition function is determined largely by the geometry; it is well established that geometries can be calculated accurately at a relatively low level of quantum theory. The same holds for the external rotations (which comprise only those of the monomer for propagation of a macroradical).

The frequency factor for propagation of a macroradical calculated as above is in good accord with the best current experimental data: the transition state/quantum calculation yields 4×10^6

$dm^3\ mol^{-1}\ s^{-1}$, while experiment [26,27] gives $6\times10^6\ dm^3\ mol^{-1}\ s^{-1}$. This both suggests the reliability of the method used, and also (since convergence was obtained at a relatively low level of theory) that the method is applicable to monomers much larger than ethylene.

Table 1. Convergence of the frequency factor for propagation with level of quantum theory, for propagation of an ethyl radical ($C_2H_5^{\bullet} + C_2H_4$); calculated for $T = 350$ K.

method [20]	A ($dm^3\ mol^{-1}\ s^{-1}$)
HF/3-21G	2.5×10^7
HF/6-31G*	2.2×10^7
MP2/6-31G*	2.4×10^7

Inferences From Transition State Calculations

These calculations also give considerable physical insight into the propagation event. Since the vibrational partition functions are of order unity, it is the rotational and translational terms which dominate the value of A. For a macroradical, the only external rotation is that of monomer. The remaining terms are those of the hindered internal rotations, which are determined by their geometry and barrier to torsion. Our calculations show that there are three such modes for ethylene propagation, illustrated in Fig. 5. From this result, a number of important *general* implications for propagation can be drawn.

One very important conclusion is obtained by approximating these as being unhindered (free) rotations, to obtain an order-of-magnitude estimate of the frequency factor for propagation. A is then entirely determined by the geometry and the forms of these internal rotations; a back-of-envelope calculation, using the typical values of the appropriate moments of inertia (e.g., [4]) then gives $A \approx 10^6 - 10^7\ dm^3\ mol^{-1}\ s^{-1}$. This is indeed the observed order of magnitude for the frequency factor for propagation reactions. Hence it is seen that the size of such frequency factors is governed by the loss of rotational entropy as the monomer free rotation is converted into various hindered rotations in the transition state.

Another important conclusion comes about when it is noted that the hindrances of the internal rotations in Fig. 5 are influenced by the *penultimate unit*: for example, the third of the rotations in Fig. 5 brings the H atoms of the incoming ethylene and the penultimate unit into close proximity. This suggests that the propagation rate coefficient in a *copolymerization* might be influenced by the penultimate unit to the extent that the partition function of the hindered rotation is so influenced. The size of such an effect is, of the order of a factor of 2 (as estimated from the formula for the variation of the partition function of a hindered rotation with typical changes in hindrance angle; see [4], p 147). Now, there is considerable debate in the literature (e.g., [28]) on the importance of penultimate unit effects. The theory given here implies that the *penultimate unit could affect the value of k_p by a factor of approximately 2*.

Another implication of the importance of hindered rotations in the transition state is the effect of deuteration on k_p. Because the rotational partition function is dominated by the moment of inertia of the rotating moiety, and because Fig. 5 shows that these involve hydrogen atoms, the effect of hindered rotations can also at least partly explain the observation [29] that deuteration of MMA has a significant effect on the value of k_p.

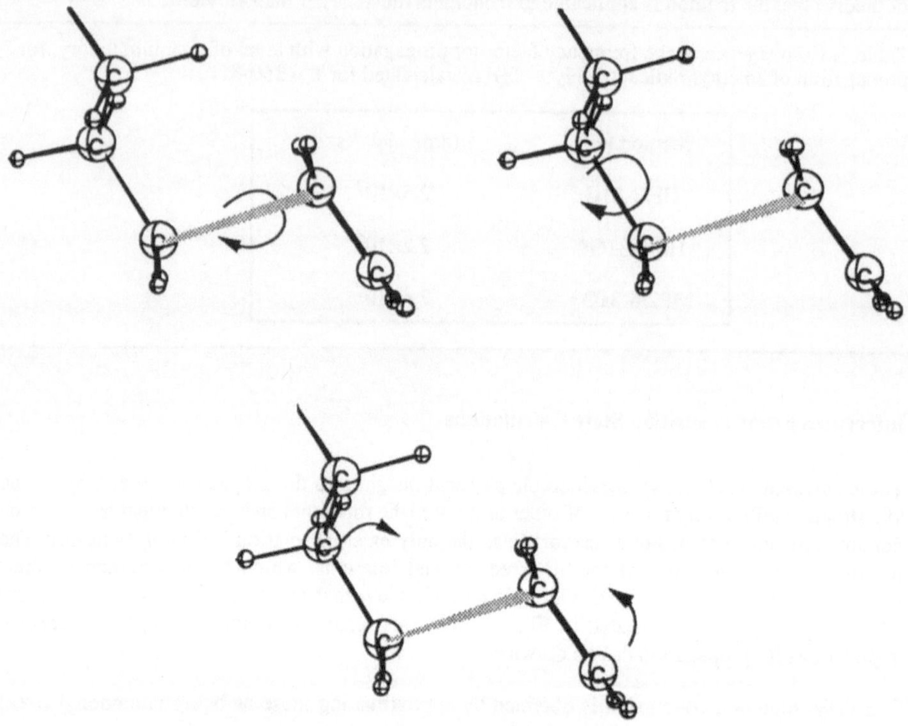

Fig. 5. The three hindered rotations which dominate the value of the frequency factor for ethylene propagation with a polyethylene radical [23]. For clarity, only the last three carbons of the ethylenic macroradical are shown. The forming bond is shaded.

A final inference can be obtained by comparing the value of the frequency factor obtained for butyl acrylate with those of other monomers. A for BA is close to the well-established value for styrene [2] and significantly greater than that for MMA [30]. This can be rationalized from the effect of the hindered rotations in the transition states. In MMA, the methyl group severely hinders such rotations, resulting in a much smaller frequency factor compared to styrene and BA, which are much less hindered. The activation energy for BA however is close to that for MMA: the effect of the acrylate group dominates $\Delta H^{\ddagger}$.

TRANSFER

The traditional means of obtaining the rate coefficient for transfer (k_{tr}) to monomer or to chain-transfer agent has been from the *average* molecular weight in a polymerizing system, *viz.*, the intercept of a Mayo plot of number-average molecular weight against $[I]/R_p$, where [I] is initiator concentration and R_p the polymerization rate (e.g., [31]). However, this method was devised before full molecular weight distributions became available, through gel-permeation chromatography. It has recently been noted that considerably more information on transfer is available from the complete number molecular weight distribution (MWD). It has been shown

[32-34] that accurate values of k_{tr} are found from the high molecular weight limit of the MWD. A summary of the theory [34] is now given, showing how k_{tr} can be obtained by a new method which promises greater flexibility than does a Mayo plot.

The number of chains with molecular weight M is defined as $P(M)$, the number MWD. The theory for the molecular weight distribution (which for simplicity we give here for a bulk or solution system) takes into account all events which can start and stop growth of polymer chains: initiation, termination by combination (k_{tc}) and by disproportionation (k_{td}), propagation and transfer of free radical activity from a growing chain to monomer ($k_{tr,M}$) and to added chain transfer agent A, $k_{tr,A}$; transfer to polymer is neglected in the present treatment. It is essential to take the chain-length dependence of k_t into account [35-38], for which a complete model has been given by Russell *et al.* [37,38], who present a means of determining the population of free radicals of degree of polymerization N, $T(N)$. Simple kinetic considerations lead to the following result [34] for the number of *dead* chains with degree of polymerization N at time t:

$$\frac{\partial P(N,t)}{\partial t} = (k_{tr,M}[M]+k_{tr,A}[A])T(N) + 2\int_1^{N-1} k_{tc}(N',N-N')\, T(N')\, T(N-N')\, dN' + 2T(N)\int_1^{\infty} k_{td}(N,N')\, T(N')\, dN' \tag{3}$$

where the chain-length dependence of k_t on the two degrees of polymerization N and N' is denoted by $k_t(N,N')$. Numerical solutions to the MWD equations can be obtained for any form of the component rate coefficients [34]. Approximate analytic solutions may also be deduced. After using the fact that termination is always diffusion controlled [36], and that the mutual diffusion coefficient involved is the sum of separate components from the two terminating chains (see eq 8 below), one finds [34]:

$$\lim_{M\to\infty} P(M) = \exp\left\{-\frac{k_{tr,M}[M] + k_{tr,A}[A] + \langle k_t\rangle [R^\bullet]}{k_p[M]}\,\frac{M}{M_0}\right\} \tag{4}$$

where $\langle k_t\rangle$ is $k_t(N,N')$ averaged over all chain lengths, M_0 the molecular weight of monomer, and $[R^\bullet]$ is the total radical concentration. The $\langle k_t\rangle[R^\bullet]$ term will become negligible at low initiator concentration. Hence one can state that for *any* system,

$$\lim_{M\to\infty,\ [I]\to 0} P(M) = \exp\left\{-\frac{k_{tr,M}[M] + k_{tr,A}[A]}{k_p[M]}\,\frac{M}{M_0}\right\} \tag{5}$$

The number MWD may in turn be found from the GPC signal as a function of elution volume, $G(V)$, as follows [34,39]. Let $\mathcal{V}(M)$ be the GPC calibration curve (i.e., $\mathcal{V}$ is the volume at which a monodisperse molecular weight sample of molecular weight M elutes). One then can show:

$$P(M(\mathcal{V})) = \frac{G(V)}{M\,\dfrac{dM}{d\mathcal{V}}} \tag{6}$$

If the calibration curve is linear, then $\log_{10}M$ is linear in $\mathcal{V}$. Eq 6 then takes the specific form:

$$P(M) = \frac{G(V)}{2.303\,M^2} \quad \text{(linear region of calibration curve)} \tag{7}$$

Hence eqs 5 and 7 show that $k_{tr,M}$ and $k_{tr,A}$ may always be determined from the high molecular weight limit of a GPC curve, given a value of k_p. It is important to note that this can be performed even in a system where extensive termination occurs. The limiting behaviour of eq 4 in fact can be readily reached, and this technique obviates the extrapolation which is required in the Mayo method.

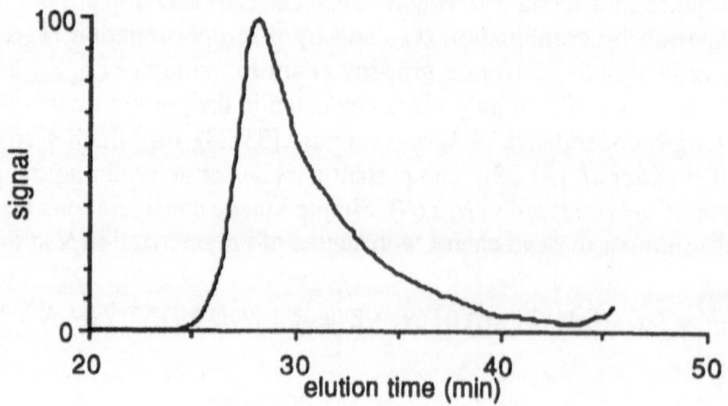

Figure 6. GPC trace, with baseline removed, for pulsed-laser initiated triethylamine/MMA system.

Fig. 6 shows the GPC trace from a pulsed-laser photoinitiated experiment with MMA monomer at 58 °C (with benzoin as photo-initiator) containing 2 mole % triethylamine as chain-transfer agent. Fig. 7 shows the same data plotted as ln$P(M)$. The straight-line region at higher M is most apparent, as predicted by eq 5. The behaviour at lower M can be ascribed to termination. The slope of the straight-line region can be obtained from a series of such data for different concentrations of chain-transfer agent, and the results plotted as a function of [A]/[M], as in Fig. 8. From this and eq 5, $k_{tr,A}$ is found to be 88±9 $dm^3 mol^{-1} s^{-1}$, in reasonable accord with the literature value [40] of 110 $dm^3 mol^{-1} s^{-1}$. Since the latter was obtained by the Mayo method, it is felt that the present result is in fact more accurate. The present technique also has the advantage that k_p can be measured in the same system, as can $k_{tr,M}$ (as apparent from Fig. 8).

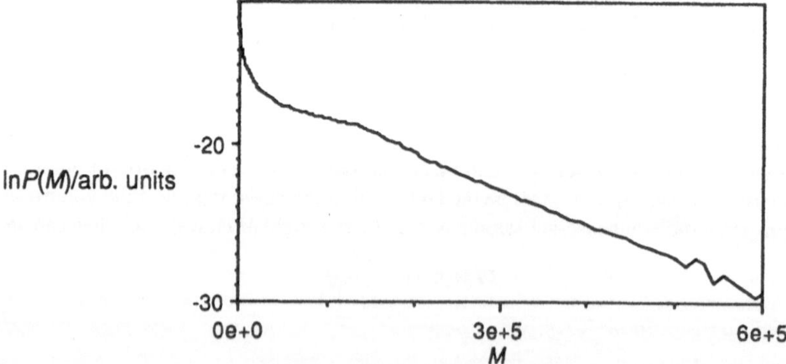

Figure 7. Data of Fig. 6 plotted as ln$P(M)$, showing straight-line region at higher M.

These results show that one can obtain transfer constants by the new technique even in systems such as these where termination is important. It is hoped that systematic pursuit of such studies will lead to predictive models for k_{tr} in the near future.

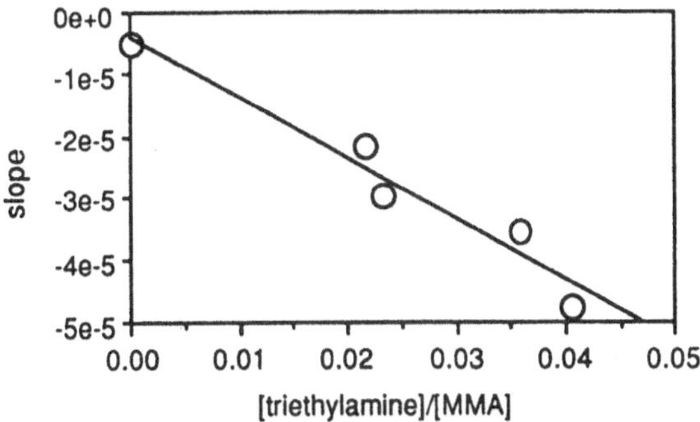

Figure 8. Slope from series of experiments as in Fig. 7 as function of [triethylamine]/[MMA monomer].

TERMINATION

Extensive *a priori* models for the diffusion-controlled rate coefficients which govern termination have now been developed [34,37,38,41]. An essential component of this treatment, noted above, is that the termination rate coefficient depends on the lengths of the two chains. The final expression for termination by combination is:

$$k_t(N,N') = 2\pi p[D(N) + D(N')]\sigma \tag{8}$$

Here σ is the van der Waals radius of a monomer unit, p is the probability that termination occurs upon diffusive encounter (this may be less than unity because of spin multiplicity), and the diffusion coefficient of an N-mer is given by the sum of centre-of-mass and reaction-diffusion terms:

$$D(N) = D_{cm}(N) + D_{rd} \tag{9}$$

A simple but accurate expression for the reaction-diffusion term (which is independent of chain length) has been given elsewhere [41].

The centre-of-mass term requires a knowledge of the diffusion coefficient as a function of degree of polymerization and of the weight-fraction polymer, w_p. It is essential to note that the species whose diffusion is significant for kinetics and MWDs are *oligomeric*: say, of degrees of polymerization ten or less. While there are extensive data and models for diffusion of much longer polymeric species in polymer/monomer systems (see [42] for a review), there is a paucity of data and models for diffusion of oligomers in polymer solutions above c^* and where the molecular weight of the polymer matrix corresponds to that typical of free-radical polymerizations. It has been found [38] that eqs 8 and 9 give accord with relaxation data which are highly sensitive to the termination process, using limited diffusion data for oligomers [43] which were fitted to an empirical equation:

$$D_{cm}(N) = D_{mon}(w_p)\left(\frac{M}{M_0}\right)^{-0.5-1.75w_p} \tag{10}$$

where $D_{mon}(w_p)$ is the diffusion coefficient of a monomeric species.

Some data for $D_{cm}(N)$ are not in accord with frequently accepted scaling laws (e.g., [43]). While improved models are gradually emerging (e.g., [44]), there is a need for more extensive data for species of the size of interest for modelling termination kinetics. As a contribution to this, we show in Fig. 9 new pulsed-field gradient NMR data obtained for a series of model compounds (hydrocarbon penetrants of a molecular weight in the range contributing to significant termination rates in typical systems), using a technique described in detail previously [43]. Also plotted are the *predictions* of the semi-empirical relation of eq 10. It can be seen that these predictions (which involve no adjustable parameters, but which were fitted to data on different penetrants) fit the new data very well. It is therefore suggested that eq 10 can be used in a wide range of situations for modelling k_t through eq 8. It is also hoped that data such as these can be used in the future to test and develop quantitative models for the process of diffusion as a function of polymer fraction and molecular weight.

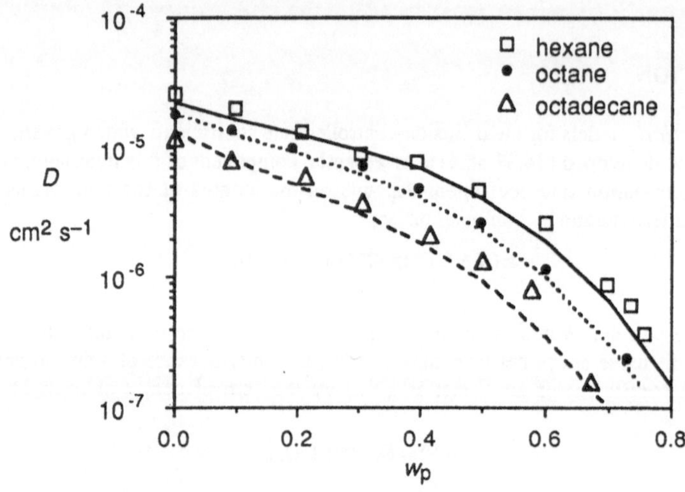

Figure 9. Diffusion coefficients for various trace penetrants in polystyrene/benzene solutions at 25 °C; w_p = weight fraction polymer in benzene/polystyrene. Points: measured by PFG NMR; lines: calculated using eq 10.

CONCLUSIONS

Modelling rates and molecular weight distributions of free radical polymerization requires a knowledge of the rate coefficients for propagation, transfer and termination. The present paper gives new means of obtaining such information experimentally and theoretically, and the use of theoretical models for qualitative understanding. The use of PLP at very low temperatures overcomes technical problems associated with this otherwise reliable technique, and widens its application considerably. Accurate quantum calculations of the transition state parameters, and proper use of these in calculation of k_p, have led to quantitative and qualitative understanding of the values and trends in the Arrhenius parameters for propagation. Especially important is the realization from these theoretical studies (1) that the partition function of the transition state for propagation is dominated by three hindered rotations, (2) that this can be used to show that a significant penultimate effect is to be expected in copolymerizations, and (3) to explain the effects of deuteration on k_p. The way is also open to use the long-chain behaviour of the number molecular weight distribution obtainable from conventional GPC to develop an equivalent data bank and understanding for transfer. New data on penetrant diffusion have supported a semi-empirical model for the dependence of the diffusion coefficient on degree of polymerization and weight fraction; this is of use in predicting termination rate coefficients, and for developing models for oligomer diffusion for termination processes.

Support from Rohm and Haas (US) and the Australian Research Council are gratefully acknowledged, as is the collaboration of Prof Phillip Kuchel and Dr Robert Chapman in the diffusion coefficent measurements.

(1) Polymer Handbook; 3 ed.; Brandrup A, Immergut EH, Ed.; Wiley Interscience: New York, 1989

(2) Buback M, Garcia-Rubio LH, Gilbert RG, Napper DH, Guillot J, Hamielec AE, Hill D, O'Driscoll KF, Olaj OF, Shen J, Solomon D, Moad G, Stickler M, Tirrell M, Winnik MA (1988) J. Polym. Sci., Polym. Letters Ed. 26: 293

(3) Buback M, Gilbert RG, Russell GT, Hill DJT, Moad G, O'Driscoll KF, Shen J, Winnik MA (1992) J. Polym. Sci., Polym. Chem. Edn. 30: 851

(4) Gilbert RG, Smith SC (1990) Theory of Unimolecular and Recombination Reactions; Blackwell Scientific, Oxford and Cambridge Mass.,

(5) Walling C (1957) Free Radicals in Solution; Wiley, New York,

(6) Mill T, Mayo F, Richardson H, Irwin K, Allara DL (1972) J. Am. Chem. Soc. 94: 6802

(7) Morrison BR, Piton MC, Winnik MA, Gilbert RG, Napper DH (1993) Macromolecules 26: 4368

(8) Reichardt C (1979) Solvent Effects in Organic Chemistry; Verlag Chemie, Weinheim, New York,

(9) Aleksandrov HP, Genkin VN, Kitai MS, Smirovna JM, Sokolov VV (1977) Kvantovaya Elektron. (Moscow) 4: 976

(10) Olaj OF, Bitai I (1987) Angew. Makromol. Chem. 155: 177

(11) Davis TP, O'Driscoll KF, Piton MC, Winnik MA (1989) Macromolecules 22: 2785

(12) Deady M, Mau AWH, Moad G, Spurling TH (1993) Makromol. Chem. 194: 1691

(13) Olaj OF, Bitai I, Hinkelmann F (1987) Makromol. Chem. 188: 1689

(14) Olaj OF, Zifferer G, Gleixner G (1987) Macromolecules 20: 839

(15) O'Driscoll KF, Kuindersma ME (1994) Makromol. Chem., Theory Simul. 3: in press

(16) Lu J, Zhang H, Yang Y (1993) Makromol. Chem., Theory Simul. 2: 747

(17) Davis TP, O'Driscoll KF, Piton MC, Winnik MA (1989) J. Polym. Sci., Polym. Letters Ed. 27: 181

(18) Davis TP, O'Driscoll KF, Piton MC, Winnik MA (1991) Polym. International 24: 65

(19) Buback M, Degener B (1993) Makromol. Chem. 194: 2875

(20) Hehre WJ, Radom L, Schleyer PvR, Pople JA (1986) Ab initio molecular orbital theory; Wiley, New York,

(21) Pople JA, Head-Gordon M, Raghavachari K (1987) J. Chem. Phys. 87: 5968

(22) Frisch MJ, Trucks GW, Head-Gordon M, Gill PMW, Wang MW, Foresman JB, Johnson BG, Schlegel HBR M.A., Replogle ES, Gomperts R, Andres JL, Ragavachari K, Binkley JS, Gonzales C, Martin RL, Fox DJ, Defrees DJ, Baker J, Stewart JJP, Pople JA, *Gaussian 92 (Revision A)* (1992); Pittsburgh, PA Gaussian, Inc.

(23) Heuts JPA, Radom L, Gilbert RG "A priori prediction of rate coefficients for propagation in free-radical polymerizations", manuscript in preparation

(24) Troe J (1977) J. Chem. Phys. 66: 4758

(25) Wong MW, Pross A, Radom L (in press) Isr. J. Chem.

(26) Schweer J, PhD thesis, Göttingen, 1988

(27) Russell GT, personal communication

(28) Fukuda T, Kubo K, Ma Y-D (1992) Prog. Polym. Sci. 17: 875

(29) Olaj OF, Schnöll-Bitai I (1990) Makromol. Chem., Rapid Commun. 11: 459

(30) Hutchinson RA, Aronson MT, Richards JR (1993) Macromolecules 26: 6410

(31) Elias HG (1977) Macromolecules: Structure and Properties; Plenum, New York,

(32) Ballard MJ, Napper DH, Gilbert RG (1984) J. Polym. Sci., Polym. Chem. Ed. 22: 3225

(33) Whang BCY, Ballard MJ, Napper DH, Gilbert RG (1991) Aust. J. Chem. 44: 1133

(34) Clay PA, Gilbert RG "Molecular weight distributions in free-radical polymerizations", manuscript in preparation

(35) Adams ME, Russell GT, Casey BS, Gilbert RG, Napper DH, Sangster DF (1990) Macromolecules 23: 4624

(36) Benson SW, North AM (1962) J. Am. Chem. Soc. 84: 935

(37) Russell GT, Gilbert RG, Napper DH (1992) Macromolecules 25: 2459

(38) Russell GT, Gilbert RG, Napper DH (1993) Macromolecules 26: 3538

(39) Shortt DW (1993) J. Liq. Chromat. 16: 3371

(40) Bamford CH, White EFT (1958) Trans. Faraday Soc. 54: 268

(41) Russell GT, Napper DH, Gilbert RG (1988) Macromolecules 21: 2133

(42) Lodge TP, Rotstein NA, Prager S (1990) In: I. Prigogine and S. A. Rice (ed) Advances in Chemical Physics. John Wiley New York Vol. 79; p 1

(43) Piton MC, Gilbert RG, Chapman BE, Kuchel PW (1993) Macromolecules 26: 4472

(44) Vrentas JS, Vrentas CM (1993) Macromolecules 26: 1277

Structure-Reactivity Relationships in Acyclic Diene Metathesis (ADMET) Polymerization

K. B. Wagener

Department of Chemistry, University of Florida
P. O. Box 117200, Gainesville, FL 32611-7200

Abstract: The structure/reactivity behavior for pure hydrocarbon diene monomers, and for dienes containing heteroatoms has been examined. Steric hindrance is the controlling factor for hydrocarbon monomers, and intramolecular electronic interactions determine the reactivity of dienes possessing heteroatom functionality. This electronic interaction phenomenon is termed the "Negative Neighboring Group Effect".

INTRODUCTION

Polymerization by means of metathesis is a subject spanning more than 30 years now, with the first reactions being focused upon ring-opening metathesis chemistry.[1] Ring opening polymerizations to produce elastic polymers are known and in fact used commercially today (Figure 1).

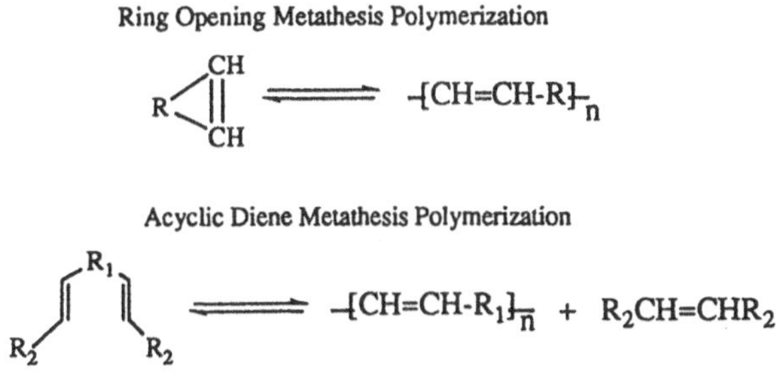

Ring Opening Metathesis Polymerization

$$R\langle CH=CH \rangle \rightleftharpoons -[CH=CH-R]_n-$$

Acyclic Diene Metathesis Polymerization

$$R_2CH=CH-R_1-CH=CHR_2 \rightleftharpoons -[CH=CH-R_1]_n- + R_2CH=CHR_2$$

Figure 1. Metathesis Polymerization Methods

While ROMP chemistry is well understood, another means of producing polymers by metathesis has received less attention and is best described as acyclic diene metathesis (ADMET) polymerization. The chemistry, which also is illustrated in Figure 1, requires the condensation of an acyclic diene and produces a repeat unit which is identical to that generated by ROMP chemistry. The polymerization is driven by the removal of a small molecule. Figure 2 illustrates the structural change that occurs in diene monomers where the internal sp^2 carbons form the link between growing monomer molecules, while the external sp^2 carbons are released as a small molecule. In this example ethylene is removed in order to drive this polymerization. The concept is not new, with results being published in the early 1970's, yet all

K.P. Ghiggino (Ed.)
Progress in Pacific Polymer Science 3

previous attempts had proven to be unsuccessful in generating high molecular weight, linear polymers.

In recent years we have been able to delineate the key to success in ADMET chemistry, that being the selection of a catalyst free of Lewis acids. Acidic catalysts permit the intervention of vinyl-addition chemistry which precludes formation of high polymers through step polymerization.[2] To prove the point, styrene was used as a model compound to explore possible mechanistic paths for reacting olefin systems. It was shown that if Lewis-acid containing catalyst systems are used, then vinyl-addition chemistry predominates, whereas the choice of a Lewis-acid free catalyst system (Shrock's tungsten catalyst[3] is used in this example) results in complete domination of metathesis chemistry instead.[4] These observations

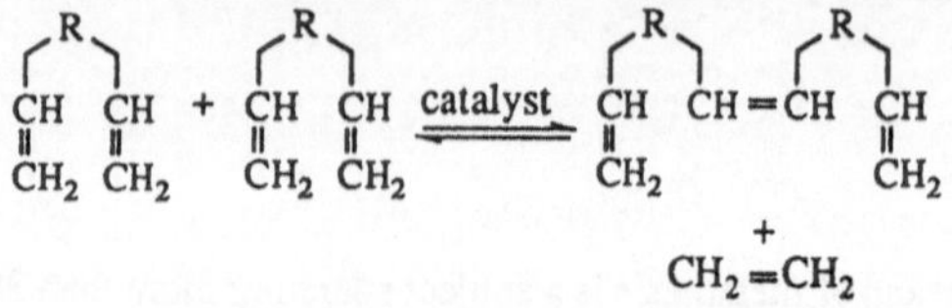

R R R R
CH CH + CH CH catalyst CH CH=CH CH
CH_2 CH_2 CH_2 CH_2 CH_2 CH_2
+
$CH_2=CH_2$

Figure 2. Monomer Structural Changes in ADMET Chemistry

are illustrated in Figure 3. Thus, if an alpha omega diene, such as 1,9-decadiene, is placed in contact with a Schrock alkylidene (either the tungsten or molybdenum version) in a bulk

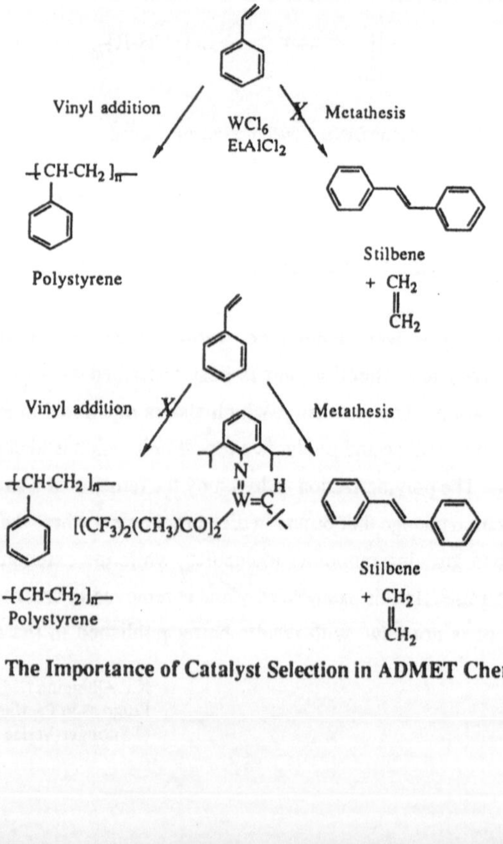

Vinyl addition
WCl_6
$EtAlCl_2$
Metathesis
$-[CH-CH_2]_n-$
Polystyrene
Stilbene
+ $CH_2=CH_2$

Vinyl addition
Metathesis
$-[CH-CH_2]_n-$
N
W=C H
$[(CF_3)_2(CH_3)CO]_2$
Stilbene
+ $CH_2=CH_2$
$-[CH-CH_2]_n-$
Polystyrene

Figure 3. The Importance of Catalyst Selection in ADMET Chemistry

polymerization scheme, then high molecular weight polyoctenamer is generated with no competing reactions being observed whatsoever (Figure 4). The polymer has a number average molecular weight in the range of 50,000, which means the step polymerization reaction has proceeded to the point of connecting more than 500 monomer units through ADMET chemistry. These results led us to examine the structure reactivity behavior for a series of ADMET monomers.

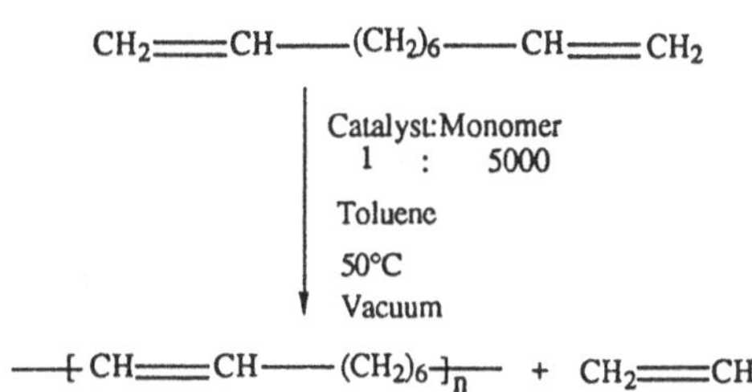

Figure 4. The First High Polymer Made by ADMET Chemistry.

RESULTS AND DISCUSSION

The ADMET Polymerization Mechanism. The mechanism responsible for ADMET chemistry is shown in Figure 5, where the alkylidene catalyst is represented using the shorthand notation $L_nM{=}CR_2$. The alkylidene combines with an olefin to form a metallacyclobutane ring, which then enters a polymerization cycle consuming monomer to form successive cyclobutane rings and evolve a small molecule (in this case, ethylene).[5] It should be noted that this mechanism generates polymers with well-defined endgroups, these being vinyl entities, and we have proven this to be the case. We have tested this mechanism, using both hydrocarbon dienes as well as dienes possessing functional groups, with the intent of examining both steric and electronic interactions which might influence polymerizability.

Structure Reactivity Relationships for Pure Hydrocarbon Monomers. A series of nonconjugated alpha omega dienes has been synthesized with methyl substituents placed in various positions, in order to define the structure reactivity behavior of these monomers.[6] All investigations were done using the same reaction procedure, i.e., combining 500 mols of monomer with 1 mol of catalyst on a high vacuum line apparatus (10^{-6} ml mercury) in the absence of solvent. The reactions are spontaneous, if in fact they do occur, and ethylene instantly evolves from the reactor as the ADMET chemistry proceeds. The results of this nonconjugated hydrocarbon diene monomer study are shown in Figure 6. It is clear that the position of the methyl group plays a profound role on the reactivity of the monomer itself. For example, terminal methyl groups result in very rapid spontaneous polymerization (as is the case for 2,6-

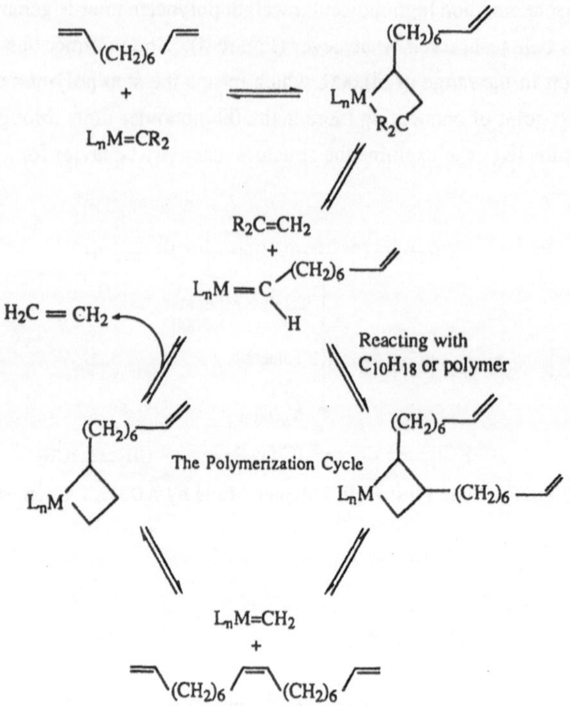

$(CH_2)_6$ + $L_nM{=}CR_2$

$R_2C{=}CH_2$

$H_2C = CH_2$

Reacting with $C_{10}H_{18}$ or polymer

The Polymerization Cycle

$L_nM{=}CH_2$

Figure 5. The ADMET Polymerization Mechanism.

octadiene) while movement of the methyl groups to the internal sp^2 carbon (as is the case in 2,5-dimethyl-1,5-hexadiene) terminates the reaction completely. Removing one of the methyls (i.e., 2-methyl-1,5-hexadiene) results in dimerization at the 5 position in a quantitative fashion.

Further, it is shown in Figure 6 that this methyl steric effect can be observed even when the methyls are placed alpha to the metathesizing olefin itself. Thus, 3,4-dimethyl-1,5-hexadiene does not polymerize, whereas 3-methyl-1,5-hexadiene dimerizes in the 5 position. These steric interactions are quite subtle, and by examining the ADMET polymerization cycle shown earlier, it becomes evident that the formation of a metallacyclobutane ring and its stability is influenced by simple steric interaction.

In order to demonstrate exactly how subtle these steric interactions can be, the methyl substituents were "tied back" in the form of a cyclohexane ring, as illustrated for the monomer 1,2-divinylcyclohexane. In this case, the diene polymerizes quite successfully. There is no question that steric interaction plays an important role in ADMET polymerization chemistry, and we have taken advantage of this observation in the synthesis of several unique unsaturated polymer systems.[6]

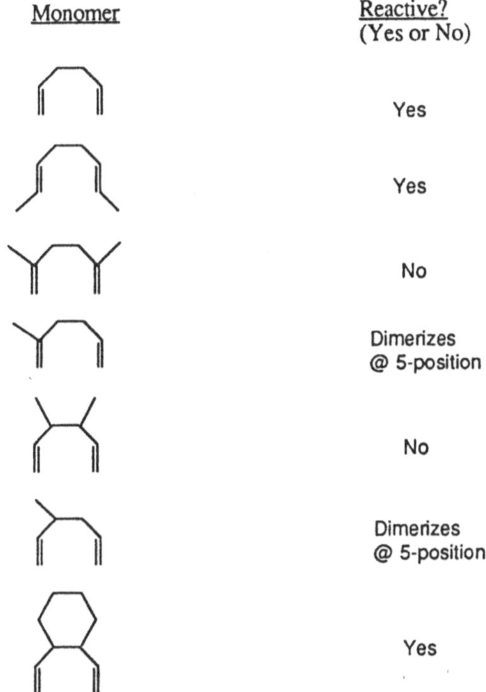

Figure 6. An ADMET Hydrocarbon Monomer Study.

<u>Structure Reactivity Relationships for Heteroatom-Containing Monomers</u>. A similar study has been undertaken using a series of dienes possessing heteroatoms in the spacing unit between the olefins themselves. Figure 7 illustrates our observations for ether-containing dienes.[7] While it is evident that divinyl ether and diallyl ether do not polymerize under the conditions described, bis(4-butenyl)ether and analogous monomers possessing greater numbers of methylene spacing units do condense by ADMET chemistry. In fact, with three or more methylene spacing units present between the ether oxygen and the olefin, there is no difference in apparent reactivity of the monomers as compared with hydrocarbon monomers. This observation has been substantiated for a large number of functional groups including esters,[8] ketones,[9,10] carbonates,[11] aromatic amines,[12] imides,[12] and thioethers,[13] in addition to the ether study illustrated above. We call this phenomenon the "Negative Neighboring Group Effect" where it is apparent that intramolecular complexation of the Lewis-based containing functional group with the catalyst obviates ADMET chemistry.

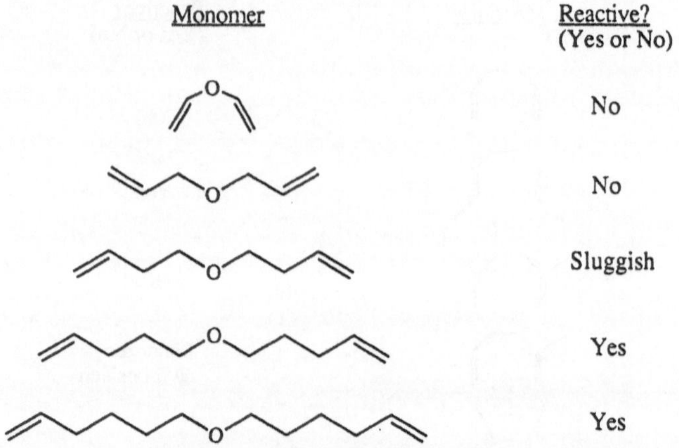

Monomer	Reactive? (Yes or No)
	No
	No
	Sluggish
	Yes
	Yes

Figure 7. A Reactivity Study for the Ether Functional Groups.

Thioethers are particularly interesting since sulfur is thought to be quite effective in complexation with metals such as molybdenum and tungsten. In fact, the reactivity observed for thioether monomers is no different from that of regular ethers, and with sufficient spacing between sulfur and the olefin, there is no apparent difference between the reactivity of heteroatom-containing and hydrocarbon dienes. For example, Figure 8 illustrates the polymerization of bis-4-pentenylsulfide where ^{13}C NMR spectra are given for the monomer and the "crude polymer" that is generated. It should be noted that these are very clean polymerizations. No other repeat units were observed besides that which is shown in Figure 8.

__Summary Statements__. These observations lead us to believe that the ADMET reaction is useful polymerization chemistry. It was recognized 20 years ago for its potential but has not been demonstrated until very recently. Obviation of competing vinyl-addition reactions is the key to the success of this reaction, and it is now evident that both pure homo- and copolymers can be made by this method. In the case of hydrocarbon monomers, steric hindrance controls the polymerizability of the diene under study. In the case of heteroatoms, intramolecular electronic interactions are the controlling factor, a phenomenon we term the "Negative Neighboring Group Effect". We are continuing these structure reactivity studies, extending the work to conjugated hydrocarbon dienes and to monomers possessing heteroatoms such as phosphorus and aliphatic nitrogen.

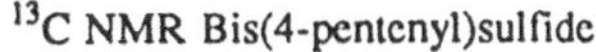

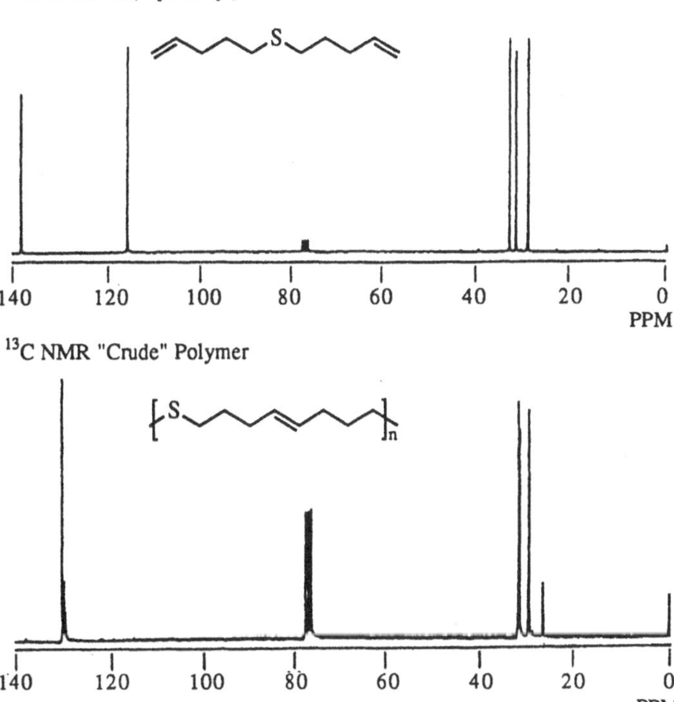

Figure 8. ^{13}C **NMR Data for the Polymerization of Bis(4-pentenyl)sulfide.**

ACKNOWLEDGMENTS

We would like to acknowledge the National Science Foundation, the Defense Advanced Research Projects Agency (DARPA), and the Shell Development Company for financial support of this work.

REFERENCES

[1] Ivin KJ (1983) Olefin Metathesis, Academic Press, New York

[2] Lindmark-Hamberg M, Wagener KB (1987) Macromol 20:2949

[3] Feldman J, Schrock RR (1991) Prog Inorg Chem 39:1

[4] Wagener KB, Boncella JM, Nel JG, Duttweiler RP, Hillmyer MA (1990) Makromol Chem 191:365

[5] Wagener KB, Boncella JM, Nel JG (1991) Macromol 24:2649

[6] Konzelman J, (May 1993) PhD Dissertation, University of Florida

[7] Brzezinska K, Wagener KB (1992) Macromol 25:2049

[8] Wagener KB, Patton JT (1992) Macromol 25:3862

[9] Wagener KB, Patton JT, Forbes M, Myers TL, Maynard HD (1993) Polym Intl 32:411

[10] Forbes ME, Wagener KB, et al (1992) J Am Chem Soc 114:10978

[11] Wagener KB, Patton JT (1993) Macromol 26:249

[12] Wagener KB, Portmess J, and Marmo C (1994) unpublished results

[13] O'Gara JE, Portmess JD, Wagener KB (1993) Macromol 26:2837

In-Situ Spectroscopy of Redox Reaction in PPY

D.Y. Kim, D.H. Kim+, J.Y. Lee and C.Y. Kim

Polymer Materials Lab., KIST, P.O. Box 131, Cheongryang, Seoul, KOREA
+ Spectroscopy Lab., KRISS, P.O. Box 3, Taeduok, Taejon, KOREA

ABSTRACT: In-situ Raman spectra of PPy are obtained with excitations of 514.5 and 1060nm. Strong Raman scattering observed with a deep reduction at 514.5nm excitation coincides with the degree of transparency of PPy of which color turns into pale yellow from dark on the reduction. The Raman scattering intensity of PPy with the deep reduction is 10 folds stronger than that of fresh PPy. The intensity enhancement of the Raman spectra with 1064nm excitation is insignificant but the spectra reveal much more information than those with 514.5nm excitation. Conjugation length generally shortens at the early stage of reduction but recovers on prolonged reduction with 1064nm excitation. Reduced PPy is easily oxidized after cutting-off the applied potential and the Raman spectrum of reduced PPy left in the system for 30min is almost the same as that of fresh PPy.

INTRODUCTION

Polypyrrole(PPy) formed by electrochemical polymerization has been scrutinized in depth in terms of possibilities of applications(1) or mechanisms of functionality such as conductivity and reactivity(2, 3). PPy is regarded as a stable material in electrical conductivity(4). However, the reactivity of the materials degrades on repeated redox reaction and the mechanism of the degradation on the reactivity is poorly understood to curtail it(5, 6).

Many attempts have been made to learn changes in chemical structures on redox reaction or chemical treatments. In-situ ultraviolet-visible spectroscopy has been employed to study the oxidation states of the partially oxidized PPy and the neutral one(7). The spectroscopy gives information on electron transitions from the lower energy state to the higher ones in the polymer. IR spectra reveal a featureless background absorption between 4000 and 1600cm^{-1} where the electron state of oxidized PPy with free charge carriers absorbs the energy around 1 eV(8). The region from 1600 to 700cm^{-1} contains spectra related to carbon-carbon stretching vibrations as well as bending vibrations of carbon-hydrogen or nitrogen-hydrogen

K.P. Ghiggino (Ed.)
Progress in Pacific Polymer Science 3

bonds(9, 10). However, specimens are dried for IR spectroscopy and some changes in chemical structures are expected on drying the samples.

Raman spectroscopy has an advantage over IR spectroscopy since H_2O is transparent in Raman spectroscopy and most of solvents leave little trace on compensation. In-situ Raman spectroscopy closely reveals changes in chemical structures during electrochemical reaction in PPy. The resonant Raman effect makes it possible that the specific segments of a polymer molecule is identified when the segments are excited by the Raman energy source(11, 12).

The present work attempts to trace chemical changes in the PPy structure on a doping/dedoping process in an aqueous system of an electrolyte by employing Raman and optical spectroscopy.

EXPERIMENTAL

Pyrrole(Aldrich, 99%) used as the monomer was dried with CaH_2 for 24 hours and distilled under reduced pressure. Sodium dodecylsulfate (NaDS)(Aldrich, 98%) was used as the electrolyte without further purification.

PPy films were prepared galvanostatically by applying a constant current of 5 mA/cm^2 to a platinum electrode with a working area of 1cm x 1cm in the reactant solution with 0.36mol of pyrrole and 0.036mol of the electrolyte in 40ml of distilled water. After purging the solution with nitrogen, the anodic current was imposed for 20 sec to make a film 0.3μm thick. The film was washed with distilled water to remove the monomer and soluble materials. The specimen was never dried.

The redox reaction was carried out in a cell which was designed to fit in both the instruments of optical and Raman spectroscopy. The reference electrode used was a Ag/AgCl, 4 mol KCl and the applied potential range for redox reaction was between 0.3 and -0.7V. Cyclovoltammograms were produced by using a set supplied by Hokuto Denko, model HA-301 and HB-104.

The optical spectrometer was assembled in the laboratory and the data were processed with a personal computer. Raman spectra excited at 514.5nm were obtained on an Ar^+ laser(Coherent Innova 95) with a double monochromator(Jobin Yvon U1000) and Raman spectra with 1064nm excitation were measured on a Perkin-Elmer System 2000 FT-Raman spectrometer. An argon ion laser and a Nd:YAG laser were operated at 514.5nm and 1064nm, respectively, for Raman excitation.

RESULTS AND DISCUSSION

In-situ Raman spectra of PPy with 514.5nm excitation after reduction stepwise at various potentials are shown in Figure 1. The bands are assigned as in Table 1 and seems not to change the positions with a progress of reduction except the one at ca 1575cm^{-1}. The band of the fresh PPy at 1575cm^{-1} moves downward by 20cm^{-1} with deep reduction indicating that the conjugation length becomes large on reduction(11). It is notable that the intensity of Raman spectrum of a fresh PPy increases 10 folds if it is reduced deeply. This effect has also been observed with 488nm excitation. The only band which is not grown proportionally on reduction is the one at ca. 1000cm^{-1}, which is assigned as bipolaron or polaron(12).

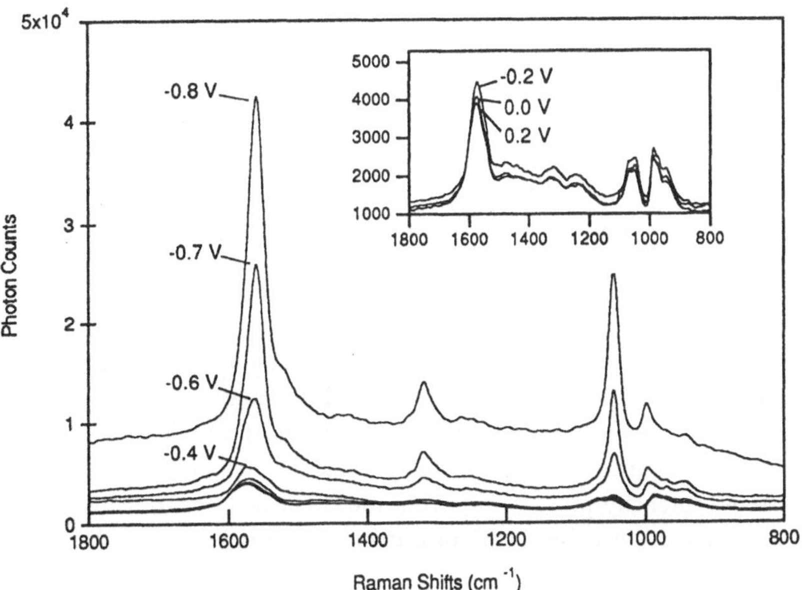

Figure 1. In-situ Raman spectra with 514.5 nm excitation during electrochemical reduction of PPy in 0.1 M NaDS aq. solution. Applied potential(V) vs. Ag/AgCl reference electrode.

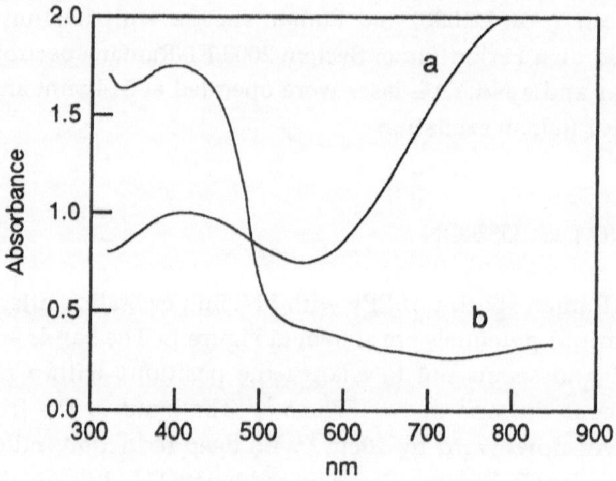

Figure 2. UV-VIS spectra of PPy in (a) oxidized state as grown and (b) electrochemically reduced state in 0.1 M NaDS aq. solution.

Electrochemically neutralized PPy has a fair absorption at 488nm and the fresh one with bipolarons shows some absorption at the same wave length as shown in Figure 2. It may be expected that the chemical segments in the neutralized state become resonant with the Raman excitation energy of 488nm and display high absorption intensities. However, there is a weak absorption at 514.5nm for both the neutralized or the fresh one and no Raman resonance is expected to give high absorption intensities. It is clear that the resonant Raman effect is not the sole factor to intensify the spectra. It is suggested that the intensity of the Raman shifted bands of fresh PPy becomes weak due to the absorption of not only the incident beam but the Raman scattered beam by the dark polymer itself. The dark PPy film changes the color into yellow on reduction and the incident light is absorbed less by the reduced film to enhance the intensity of the Raman spectra. The Raman shifted bands are also absorbed less by the reduced sample than by the dark fresh one to enhance the band intensity. There is, however, a possibility that the 10-fold intensity-jump comes from the enhanced π-π^* transition with the resonance effect since the absorbance even at 514.5nm is noticeable.

Figure 3 shows in-situ Raman spectra of PPy with 1064nm excitation after reduction stepwise at various potentials. There are more bands than in Figure 1 and each band is assigned as in Table 1. It is apparent that there are two sets of spectra with one intermediate. The spectra up to reduction potential of 0.1V are almost the same as that of the fresh sample indicating little change in the structure at the early stage of reduction. The spectra of PPy at reduction potentials between -0.4~0.7V look alike with each other but different from those up to 0.1V. The spectra of PPy at reduction potentials between 0.0~0.3V seem to be a combination of the two sets.

Table 1. Raman band frequencies of PPy excited by Ar ion and Nd:YAG laser source for 514.5 and 1064 nm, respectively.

Virbrational modes	Raman shifts (cm^{-1}) excited by	
	514.5 nm	1064 nm
ν(C=C)	1582	1580
ν(C-N) + ν(C-C), ring	1462	1473
ν(C-C), ring	1378	1370
δ(C-H), in-plane bending	1233 and 1087	1230 and 1080
δ(C-H), out-of-plane bending	976, 937	935

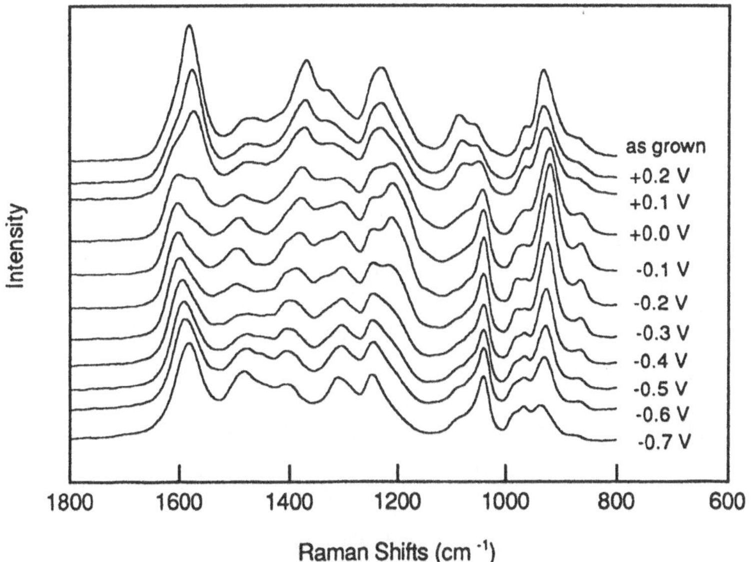

Figure 3. In-situ FT-Raman spectra with 1064 nm excitation during electrochemical reduction of PPy in 0.1 M NaDS aq. solution. Applied potential(V) vs. Ag/AgCl reference electrode.

There seems to be twelve Raman shifted bands in the region between 800 ~ 1650cm^{-1} when the Raman spectra are deconvoluted as shown in Figure 4. The change in the band positions during reduction is observed through the whole region. The band at 1580cm^{-1} produces a shoulder at a long wave number on the reduction process, which dominates at the reduction potential of 0.1V but shrinks with the deep reduction. It is interesting to note that the conjugation length shortens on early reduction, which is not observed with 514.5nm excitation. The probable explanation is that the conjugation length of PPy with polarons is detected as a short size. The band at 830cm^{-1} gradually decreases in the intensity with a progress of the reduction. The bands at 1380, 1240 and 1100cm^{-1} lose the intensity on reduction and shift either to higher or lower energy. It is obvious that the chemical structure of PPy changes with the reduction potential. Detailed assignment of the absorption bands will be reported in the future but the intensity and position of the bands related to -C-H bending vibration at ca 1240, 1100 and 930cm^{-1} indicates the structural changes during the reduction process.

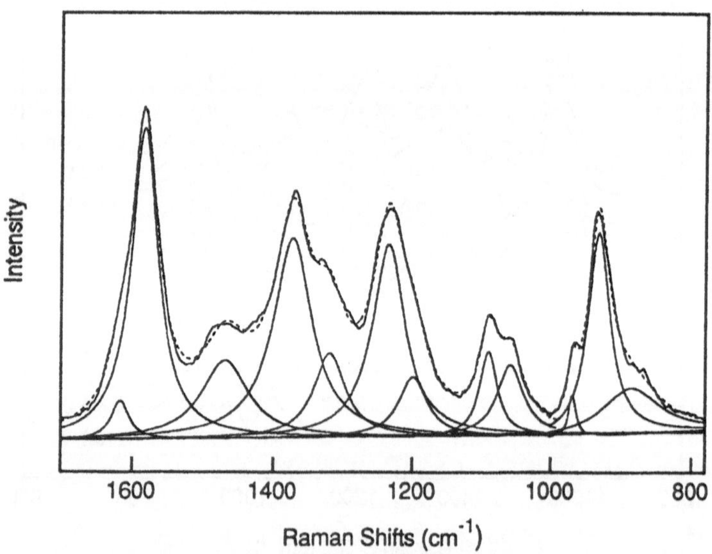

Figure 4. Peak deconvolution of FT-Raman spectrum of PPy as grown. Dashed line for a simulated spectrum.

It is known that the fresh PPy contains bipolarons mostly and polarons a little based on the visible light spectrum as shown in Figure 2 (8). The bipolarons absorb the energy of 1064nm and the Raman spectra with excitation of 1064nm should experience the resonant Raman effect(13). The integrated intensity at ca 1580cm^{-1} deduced from the deconvoluted Raman spectra is plotted against the reduction potential as shown in Figure 5. The peak area of the band at 1580cm^{-1} is related to the concentration of the long conjugation. The area decreases with reduction up to the potential of -0.1V and changes a little with further reduction. The peak area of the band at 1610cm^{-1} increases with reduction up to the potential of -0.1V and becomes stabilized with further reduction. It is demonstrated that the longer the conjugation length is, the larger the shifted wave number becomes(11). The first three spectra from the top in Figure 3 may be regarded as the resonant Raman spectra with localized excitation while the concentration of bipolaron is high. The Raman intensity of PPy reduced at the potentials lower than 0.0V is lost no more. The little loss of the intensity may be due to either the fact that the deep reduction still leave enough bipolarons in PPy or the one that the resonant Raman effect is not observed.

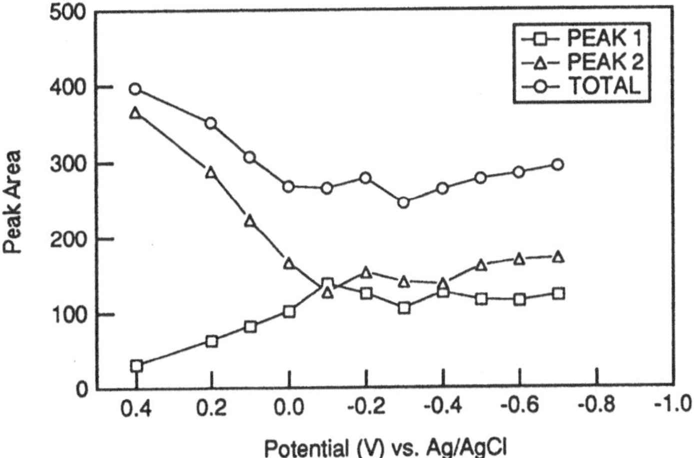

Figure 5. The change of peak area vs. applied potential during electrochemical reduction of PPy in 0.1 M NaDS aq. solution. Peak 1 and 2 for 1610 and 1580 cm^{-1}, respectively.

However, it is inconceivable that a significant amount of bipolarons remains in the polymer after a deep reduction which makes the dark film pale yellow as shown in Figure 2. The absorbances at 520 and 800nm are traced against the reduction potential as shown in Figure 6. The absorbance at 800nm associated strongly with bipolarons gradually decreases up to the reduction potential of -0.1V and momentarily increases up to the potential of -0.3V to become nil at the full reduction. The absorbance at 520nm shows some increase from the reduction potential of -0.1V on expense of the concentration of bipolarons. These changes may be correlated with the result of cyclovoltammogram as shown in Figure 7. A strong reduction current is observed at the potential of -0.2V. It is speculated that the reduction rate at -0.2V is high to increase abruptly the concentration of polarons, the intermediate on a full reduction, which couple with each other to form bipolarons again. At the same state, the polaron concentration shows a decrease. The decent intensity of Raman spectra with 1064nm excitation at any reduction stage of PPy is obtained because of little loss of the incident or shifted beam intensity of 1064nm through the film.

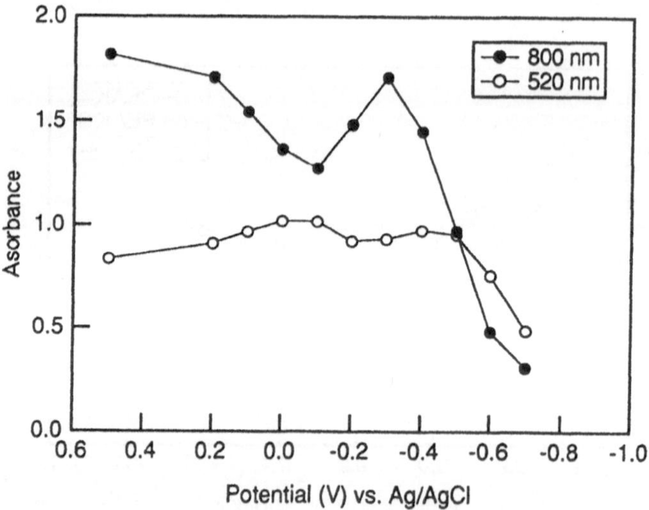

Figure 6. Absorbance changes vs. applied potential at 520 and 800 nm in UV-VIS spectra of PPy during electrochemial reduction in 0.1 M NaDS solution.

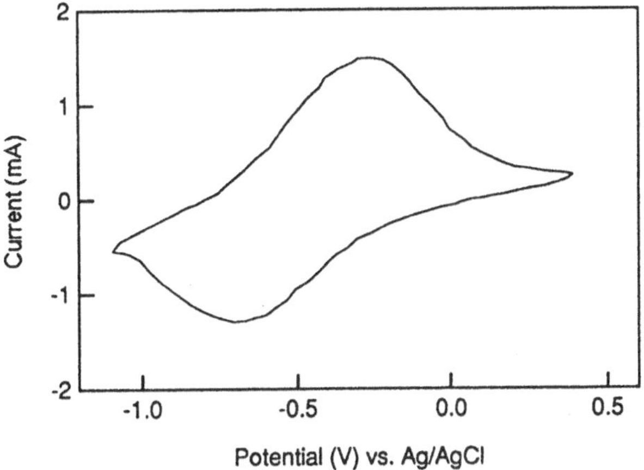

Figure 7. Cyclovoltammogram of PPy in 0.1 M NaDS aq. solution. Applied potential(V) vs. Ag/AgCl reference electrode. Scan rate of 100 mV/sec.

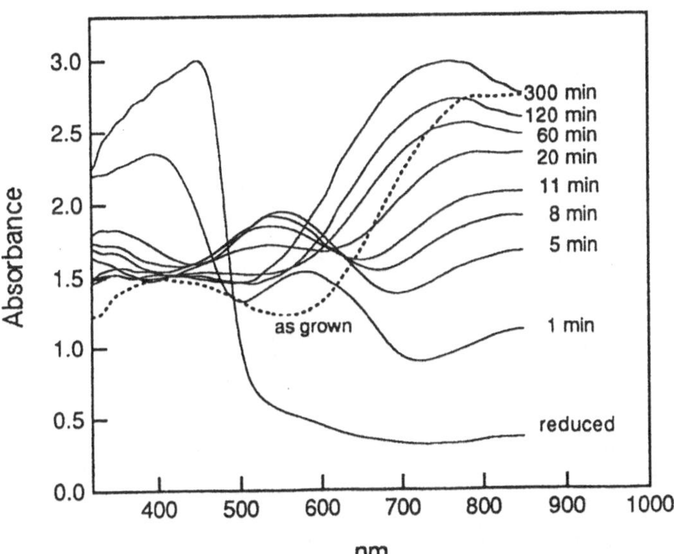

Figure 8. UV-VIS spectra of fully reduced PPy with time in wet state without applied potential in 0.1 M NaDS solution. Dotted line for PPy film as grown.

Electrochemically polymerized PPy is regarded as a stable material but electrochemically reduced PPy is very unstable as shown in Figure 8. The polymer is reduced at -0.8V to show an absorption at 470nm associated with π-π^* transition. The deeply reduced film left in the system after cutting-off the potential produces an absorption peak at 580nm in 1min while the peak at 470nm shifts to 400nm. The film shows a very strong absorption at 560nm in 5 min, of which spectrum looks identical with that of the film reduced at -0.2V. The absorbance beyond 700nm shows the peak position moving toward shorter waves with time and to 760nm at 300min. The optical spectrum at 300min has the absorption peak at 400nm as high as that of fresh PPy but the absorbance beyond 400 nm is larger than that of the fresh one to reach the peak at 760nm. The reduced PPy seems to be oxidized but has different structures from the fresh PPy. It is known that an electrochemically reduced PPy is easily doped with oxygen to become a good electrical conductive materials(14). It becomes important to understand if PPy doped with oxygen is reduced intact in the chemical structure.

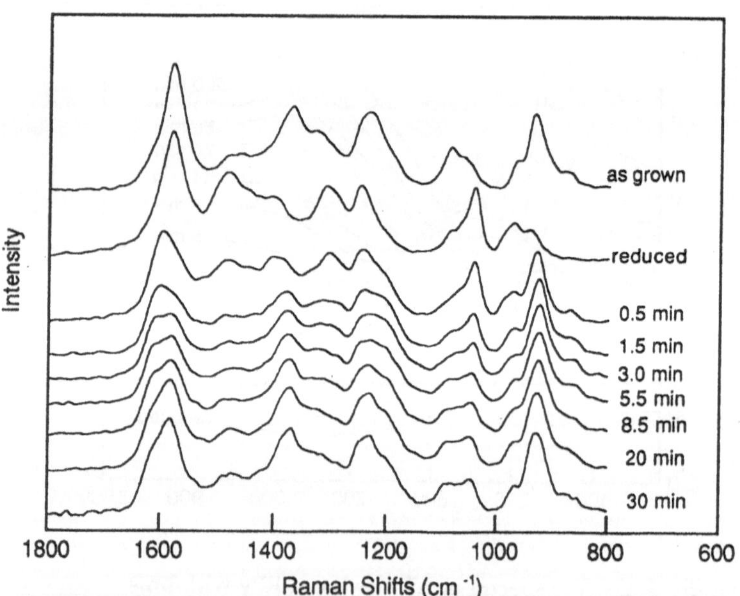

Figure 9. FT-raman spectra of fully reduced PPy with time in wet state without applied potential.

Figure 9 shows Raman spectra with 1064nm excitation of reduced PPy left in the system. The spectrum of PPy left for 0.5min is different from that of the fully reduced polymer. The conjugation length at ca 1610cm^{-1} changes quickly and other changes are also noticed at several wave numbers. When the reduced polymer is left in the system for 30min after cutting-off the applied potential, the spectrum is much different from that of the fully reduced one. It is interesting to note that the spectrum left in the system for 30 min is almost the same as that of fresh PPy. It is speculated that the Raman active species in both the fresh polymer and the exposed one for 30 min are almost the same. However, the reactivity of the species in redox reaction is hardly guaranteed to be the same with each other. Cyclovoltammetry proved that a stable cyclability is observed in the nitrogen-purged system while the redox reactivity degrades quickly in the system without the nitrogen-purging.

CONCLUSION

In-situ Raman spectra of PPy with excitation of 1064nm reveals changes in the chemical structure on reduction. The conjugation length based on the intensity as well as position of the band at ca 1600cm^{-1} becomes short at an early stage of reduction but recovers with prolonged reduction. The spectra with Raman excitation of 514.5nm shows no conjugation length shortening but lengthening. The absorption intensity of Raman spectra with excitation of 514.5nm grows remarkably large with a progress of reduction. It is suggested that the 10 fold increase in the absorption intensity on a deep reduction is associated with the transparency of the film on reduction at a given wave length. The dark PPy film prepared electrochemically becomes pale yellow on reduction. The resonance Raman effect is observed on fresh PPy or at the early stage of reduction when the specimen is excited with 1064nm.

When the fully reduced film is left in the system for 30 min without applied potential, the film turns color easily from pale yellow into dark indicating the polymer is oxidized to contain bipolarons. The naturally oxidized film shows almost an identical Raman spectrum with the fresh one although the optical spectrum looks close to that of the film reduced at -0.2V. The reduced PPy film is very unstable and may be damaged permanently on a reduction process. The redox reactivity decreases with a prolonged cycling because of irreversible structural changes in the polymer molecules.

REFERENCES

1 Miller JS (1993) Adv Mat 5:587, 671
2 Tan KL, Tan BGT, Kang ET, Neoh KG (1991) J Chem Phys 94:5382
3 Ko JM, Rhee HW, Kim CY (1992) Makromol Chem Macromol Symp 33:353
4 Stotheim TA (ed) (1986) Handbook of Conducting Polymers. Marcel Dekker, New York
5 Ko JM, Rhee HW, Park SM, Kim CY (1990) J Electrochem Soc 137:905
6 Ko JM, Rhee HW, Kim CY (1992) Makromol Chem Macromol Symp 53:81
7 Yakushi K, Lauchlan LJ, Clarke TC, Street GB (1983) J Chem Phys 79:4774
8 Brédas JL, Scott JC, Yakushi K, Street GB (1984) Phys Rev 30:1023
9 Lei J, Liang W, Martin CR (1992) Synth Met 48:301
10 Novák P, Rasch B, Vielstich W (1991) J Electrochem Soc 138:3300
11 Tian B, Zerbi G (1990) J Chem Phys 92:3892
12 Frukawa Y, Tazawa S, Fujii Y, Harada I (1988) Synth Met 24:329
13 Frukawa Y, Ohta H, Sakamoto A, Tasumi M (1991) Spectrochem Acta 47A:1367
14 Street GB, Clarke TC, Krounbi M, Kanazawa K, Lee V, Pfluger P, Scott JC, Weiser G (1982) Mol Cryst Liq Cryst 83:253

Disubstituted Poly (Bithiophene) s: Syntheses and Applications

H.S.O. Chan*, S.C. Ng, H.H. Huang, S.H. Seow

Department of Chemistry, Faculty of Science, National University of Singapore, Kent Ridge, Singapore 0511.

ABSTRACT

The properties of electrochemically prepared poly(3,3'-dichloro-2,2'-bithiophene) (PDCBT) and poly(3,3'-dibutoxy-2,2'-bithiophene) (PDBOBT) were studied and compared with polybithiophene (PBT) in order to investigate the influence of different substituents. Smaller band gap and enhanced neutral-to-doped transition were observed in PDBOBT, whereas PDCBT was characterized with larger band gap and relatively unstable doped state.

Polybithiophene film coated on indium tin-oxide (ITO) glass plate was used to detect low levels of ascorbic acid based on changes in the UV-visible absorption characteristics of the doped polymer. To further investigate the influences of substituents on polybithiophene, both PDCBT and PDBOBT were used to detect ascorbic acid and the performance of these sensors were evaluated.

INTRODUCTION

Polythiophene has a rigid polymeric backbones which renders it insoluble in common solvents thereby limiting their processability and consequent bulk-scale application. Polythiophene functionalized with appropiate pendant substituents will introduce structural alterations in the polymers making them more soluble. Since the introduction of substituents in monomers can also impart a significant influence on the other properties of the resulting polymers, the mechanical, electrical, electrochemical and optical properties of various π-conjugated polymers can be controlled by incorporation of substituents with appropriate structural and electronic properties in the respective monomers.

Substitution of the thiophene rings with electron-releasing alkyl groups was reported to afford polymers with smaller band gap and lower redox potential [1, 2], with these effects being accentuated upon replacement of the alkyl groups with alkoxy substituents [3]. Conversely, it was found that electron-withdrawing halogen moieties on polythiophenes shows totally opposite effects [4].

The study of structure-property relationship of polymer [5] is often hampered by the random occurrence of " head-to-head" and "head-to-tail " coupling in polymerizing

K.P. Ghiggino (Ed.)
Progress in Pacific Polymer Science 3

functionalized thiophenes which result in non-stereoregular polymers. However, symmetrically disubstituted bithiophenes provide structural regularity that is required to render a high degree of chain regularity in the resulting polymers [6]. Such stereoregularity in polymer backbone is important to control various tunable properties of any potentially useful polymer as well as to minimize complications in the study of its structure-property relation [7].

Conducting polymer-modified electrodes have potential usage in the fabrication of chemical detector for analytical applications [8]. As many of such research activities have focused extensively on the application of polypyrrole and its derivatives in the fabrication of biosensor especially glucose sensors [9], we have worked on the development of analytical probes using polythiophenes and polybithiophenes. An amperometric glucose sensors fabricated using enzyme/polybithiophene modified electrode has recently been reported [10].

Apart from glucose, ascorbic acid is another commonly found biochemical in food and clinical samples. Whilst analytical measurements for ascorbic acid have made use of spectrophotometric methods, amperometric detection or volumetric titration, we have herein relied on colour changes of the polymer matrix during the reduction of doped polybithiophenes by ascorbic acid as a mean of analytical detection and measurement. This approach utilizes the redox property and the electrochromatic nature of polybithiophenes upon reduction by ascorbic acid.

EXPERIMENTAL

Materials

Ascorbic acid (Merck), tetrabutylammonium perchlorate (Bu_4NClO_4) (TCI), hydrazine (Merck) and $FeCl_3$ (Merck) were used as purchased. Acetonitrile (J.T.Baker) was refluxed over calcium hydride and distilled prior to experiments.

Instrumentation

Electrochemical polymerization were performed using an EG&G 273A potentiostat/galvanostat. A single-compartment electrochemical cell consisting of a working electrode, a platinum foil as counter electrode and a reference electrode was employed for all experiments. All potentials were measured against an Ag/0.1 mol dm^{-3} $AgNO_3$(acetonitrile) reference electrode (0.34 V *vs.* SCE). All potential values reported are corrected with respect to SCE. UV-visible absorption spectroscopy was performed using a Hewlett Packard Model 8452A diode array spectrophotometer. The cell temperature was maintained at 25°C by a Hewlett Packard Model 8909A temperature control unit. Elemental analyses were performed by the Microanalysis Laboratory in our Department, using the Perkin-Elmer C, H, N analyzer (Model 240-C). Melting points (mp) were obtained on a Thomas Hoover capillary melting apparatus and are uncorrected. Mass spectra were measured on a VG 7035E micromass mass spectrometer at a source temperature of 200°C and at 70 eV electron energy. ^1H NMR spectra were determined at 300 MHz with a Brucker ACF300 spectrometer using $CDCl_3$ as solvent and $SiMe_4$ as internal reference.

Syntheses of Monomers

Synthesis of 2,2'-bithiophene (BT) was performed using Grignard coupling of 2-thienylmagnesium bromide with 2-bromothiophene in the presence of $NiCl_2dppp$ [dppp=$Ph_2P(CH_2)_3PPh_2$] as catalyst. Direct bromination of BT was effectively accomplished by treatment with 4 mol equiv. of bromine in chloroform to afford 3,3',5,5'-tetrabromo-2,2'-bithiophene (TBBT). The symmetrically disubstituted 3,3'-dibromo-2,2'-bithiophene was prepared by selective debromination of TBBT at the 5,5'-positions using powdered zinc and *n*-propanol-acetic acid mixture heated under reflux as previously reported [11].

<u>3,3'-Dichloro-2,2'-bithiophene (DCBT):</u> 3,3'-Dibromo-2,2'-bithiophene (6.9g, 21.4mmol) was dissolved in DMF (15ml) at 120°C under N_2 atmosphere before addition of CuCl (6.4g,64.2mmol) in DMF (15 ml). After 24 hours, the reaction mixture was allowed to cool to room temperature before the solvent was removed under high vacuum. The product was then dissolved in ether and the solution was dried using anhydrous $MgSO_4$. After filtration, the ether was removed and the product was purified dy distillation under high vacuum (4.18g, 84%). The fraction boiling at 130-140°C / 0.1 mm Hg was collected to afford a crytalline product; mp 33-35°C; ^1H nmr $\delta(CDCl_3)$ 7.41 (2H, d, J = 5.37 Hz), 7.04 (2H, d, J = 5.62 Hz); m/z 234/236 (M^+: 100%/33%). Anal. Calcd. for $C_8H_4Cl_2S_2$: C, 41.03; H, 1.71; Cl, 29.91; S, 27.35. Found: C, 40.88; H,1.69; Cl, 29.21; S, 27.50.

<u>3,3'-Dibutoxy-2,2'-bithiophene (DBOBT):</u> Sodium metal (1.99g, 0.087moles) was completely reacted with 50ml of *n*-butanol under nitrogen atmosphere. After addition of CuO (1.27g, 0.016moles) and potassium iodide (0.063g, 0.38mmol), 3,3'-Dibromo-2,2'-bithiophene (5.01 g, 0.0156mmol) dissolved in DMSO was transferred into the reaction mixture and was allowed to stirred at 120°C for 6 days. After addition of deionized water, the product was extracted from the aqueous layer using ether and the organic solution was dried using anhydrous $MgSO_4$. After filtration, the ether was removed and the product was purified by distillation under high vacuum to afford product (2.45g, 49%), mp 98-99°C; ^1H nmr $\delta(CDCl_3)$ 7.07 (2H, d, J = 5.60 Hz), 6.83 (2H, d, J = 5.60 Hz), 4.10 (4H, t, J = 6.50

Hz), 1.83 (4H, m), 1.56 (4H, m), 0.98 (6H, t, J = 7.40 Hz). Anal. Calcd. for $C_{16}H_{22}S_2$: C, 61.90; H, 7.14; S, 20.65. Found: C, 61.57; H,7.14; S, 19.86.

Polymerization

All polymer films were grown galvanostatically from 0.05 mol dm^{-3} monomer solutions in acetonitrile containing 0.1 mol dm^{-3} Bu_4NClO_4 using a current density of 0.5 mA cm^{-2} unless otherwise stated. The polymers were deposited on either a platinum plate or ITO-coated glass as stated in different experiments. Having rinsed with acetonitrile to remove residue monomer and soluble oligomers, the polymer-coated electrodes were then air-dried. Using freshly prepared 2% hydrazine solution, the as-synthesized polymers coated on conducting glass electrodes were dedoped and stored in a clean environment before use. The conducting glass electrodes coated with PBT, PDCBT and PDBOBT (of *ca.* 10 μm thickness each) were designated as PBT-ITO, PDCBT-ITO and PDBOBT-ITO respectively.

Spectrophotometric Detection of Ascorbic Acid

Before the experiments, neutral PBT-ITO and PDCBT-ITO were re-doped using freshly prepared 0.5% $FeCl_3$ solution in acetonitrile containing 0.1 mol dm^{-3} Bu_4NClO_4, whereas neutral PDBOBT-ITO was re-doped using freshly prepared 0.5% $FeCl_3$ solution in acetonitrile/water (1:1) mixture containing 0.1 mol dm^{-3} Bu_4NClO_4. Ascorbic acid solutions were prepared immediately prior to experiments using degassed Alpha-Q water. The time-based absorption of freshly re-doped polymer-coated ITO-glass electrode immersed in standard ascorbic acid solution was recorded at constant wavelength.

RESULTS AND DISCUSSION

Table 1 Electrochemical and optical properties of polybithiophenes

Polymer	Neutral			Doped		
	E_{pa} (V)	λ_{max} (nm)	colour	E_{pc} (V)	λ_{max} (nm)	colour
PBT	1.01	488	red	0.88	480 732	blue
PDCBT	1.20	470	brown-red	1.07	466 700	blue
PDBOBT	0.07	526	violet-blue	-0.17	<400 >700	blue

UV-Visible Spectroscopy

The extent of conjugation in π-conjugated polymers can be estimated from the maximum absorption wavelength (λ_{max}) of the neutral polymers because the longer conjugation length is indicated by lower π-π* transition energy. The results of solid-state UV-visible absorption data are shown in Table 1 and the spectra are shown in Figure 1. Comparison of λ_{max} shows that the conjugation length of the poly(bithiophene)s follows the trend: PDCBT < PBT < PDBOBT.

It has been reported that the steric interaction between the sulphur atoms and the chlorine substituents in solid-state PDCBT impart slight torsion between adjacent thiophene rings [7]. Therefore coplanarity of the thiophene rings, which is essential for inter-ring conjugation, is hindered in PDCBT as compared with that of PBT.

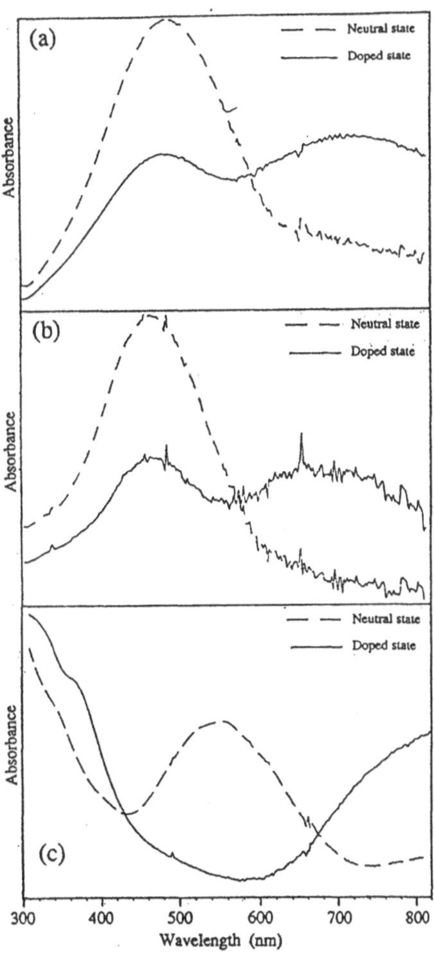

Figure 1 UV-visible spectra of polymer film coated on ITO glass. (a) PBT; (b) PDCBT; (c) PDBOBT.

In contrast to chlorine substituent in PDCBT, incorporation of butoxy groups at the 3- and 3'- positions in PDBOBT reduce the steric hindrance between the first atom in the side chain (i.e O atom) and the sulphur atom in thiophene ring. As deduced from the λ_{max} observed, the degree of coplanarity in PDBOBT is higher than that of PBT despite the smaller Van der Waal's radius of the H atom (1.20 A) than that of the O atom (1.40 A). McCullough et al. [12] reported that interaction between the flexible alkyl chains renders planar comformation along the structurally homogeneous polythiophene. The exceptionally small band gap of PDBOBT can therefore be correlated with the presence of flexible butoxy chains at 3,3'-position which improved coplanarity and structural homogeinity of the polymer backbone.

The spectra of doped polymers are featured by blue-shift and decreases in intensity of the original absorption peaks and also the appearence of polaron/bipolaron absorption peaks at the region of lower energies (above 700nm) [13, 14]. From these spectral changes shown in Figure 1, the degree of doping for the respective polymers can be estimated. The presence of predominant polaron/bipolaron absorption and remarkedly low intensity of the original absorption peak in the spectra of doped PDBOBT indicate an extremely high doping level in the polymer. However, such a high doping level is not observed in PBT since the intensity of the polaron/bipolaron absorption is slightly higher than that of the original adsorption peak.

Electrochemical Characterizations

During cyclic voltametry experiments, the colour of polymers were observed to change during the anodic/cathodic cycles in which the polymers were switched from undoped state to doped state repeatedly. Such colour change during the redox transition are illustrated in Table 1. From the cyclic voltamograms (CV), the anodic peak potentials (E_{pa}) of PBT, PDCBT and PDBOBT were determined to be 1.01V, 1.20V and 0.07V respectively, whereas the cathodic peak potentials (E_{pc}) were found to be 0.88V, 1.07V and -0.17V respectively. Hence the redox potentials of the polymers were found to decrease in the order PDCBT > PBT > PDBOBT suggesting that electronic nature of the substituents imparts a considerable effect to the redox transition of disubstituted polybithiophenes.

As indicated by the higher E_{pa} of PDCBT than that of PBT, the electron-withdrawing substituent supresses the oxidation of PDCBT by reducing the π-electron density of thiophene ring. On the contrary, the electron-donating substituent imparts an opposite effect as the E_{pa} of PDBOBT is much lower. Therefore, a highly doped state can be obtained when PDBOBT is re-doped as shown in Figure 1(c).

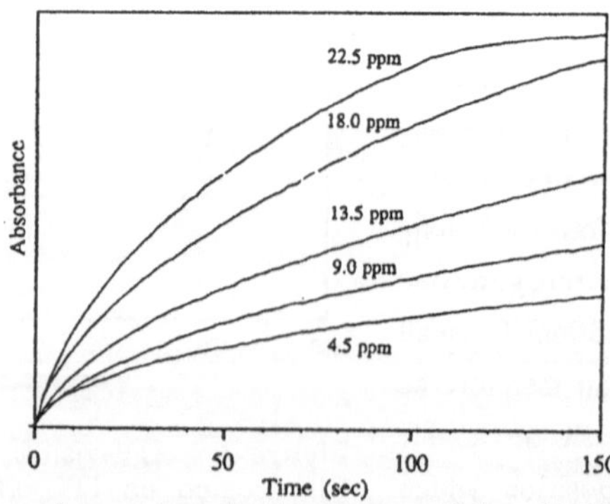

Figure 2 Time-based spectroscopic absorbances of doped PBT film in ascorbic acid of various concentration with measurement wavelength fixed at 488 nm.

Spectrophotometric Detection of Ascorbic Acid

As the oxidation potential of ascorbic acid is approximately -0.104 V (*vs* SCE) which is lower than that of PBT (as shown in Table 1), the doped polymer therefore can be reduced by ascorbic acid to the neutral state. In a preliminary experiment, thin film of blue-doped PBT turned red within seconds upon immersion in a 500 ppm ascorbic acid solution. Such colour change was recorded by UV-visible spectroscopy. Similar to the spectra shown in Figure 1(a), both absorption peaks of doped PBT-ITO were displaced to a single peak with an λ_{max} at 488 nm upon reduction by ascorbic acid solution.

Figure 2 shows the time-based absorbance of doped PBT-ITO at constant wavelength upon reduction in ascorbic acid solutions of various concentration. When the concentration of ascorbic acid solution is above 20 ppm, the absorbance rose sharply and a maximim absorbance was attained almost immediately. For ascorbic acid solutions of lower concentration, the absorbance-time curve rose in a more gradual manner, suggesting that the time-based spectrophotometric responses of doped PBT coated on ITO electrode could be used, in principle, as a measurement of ascorbic acid concentration.

Using (A - A_0) values taken at different time, plots of log(A - A_0) against concentration gave straight lines with regression coefficient $R^2 = 0.99$ as shown in Figure 3. The absorbances (A) were taken at a specified time whereas the initial absorbance (A_0) corresponded to the absorbance at the zeroth second. In view of the linearity shown between log(A - A_0) and concentration, the plots in Figure 3 can be used as calibration curve to determine the level of ascorbic acid in samples of unknown concentration. The linear range of PBT-ITO for spectrophotometric detection of ascorbic acid is up to about 22 ppm and the optimum detection time falls in the range of 15 s to 100 s since the linearity between log(A - A_0) and concentration degenerates rapidly after the 100th second as shown in Figure 4.

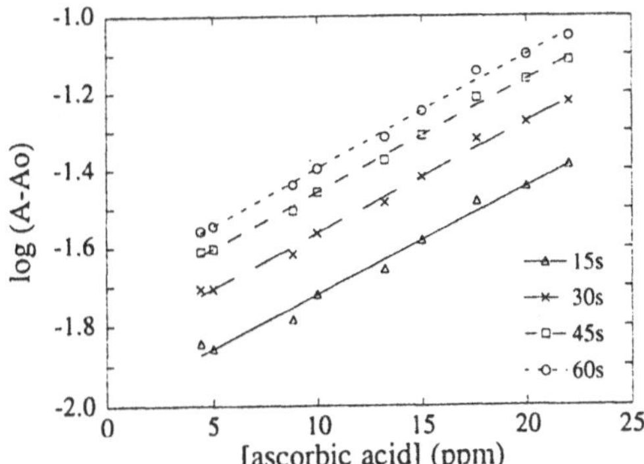

Figure 3 Calibration curve of log(A - Ao) against ascorbic acid concentration for doped PBT film.

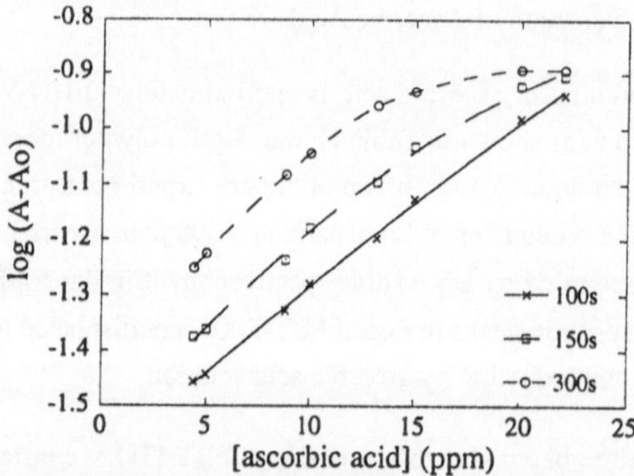

Figure 4 Plots of log(A - Ao) against ascorbic acid concentration for doped PBT film.

Figure 5 shows the spectrophotometric responses of various doped polymers in blank solution. The doped PDCBT-ITO is very unstable in water as indicated by its rapid conversion to undoped/neutral state as quantified by the absorbance at 470 nm and this instability can be rationalized by the high E_{pa} of PDCBT. Both doped PBT-ITO and doped PDBOBT-ITO exhibit satisfactory stability within the period of time required for spectrophotometric detection of ascorbic acid.

Using PDBOBT-ITO as detection probe for ascorbic acid, the linear range of spectrophotometric response between the 15th and 60th second has gone up to 75 ppm as shown in Figure 6. The higher capacity can be rationalized by the fact that highly doped PDBOBT contains larger amount of doped component per unit volume of polymer matrix than PBT as deduced from the UV-visible absorption spectra.

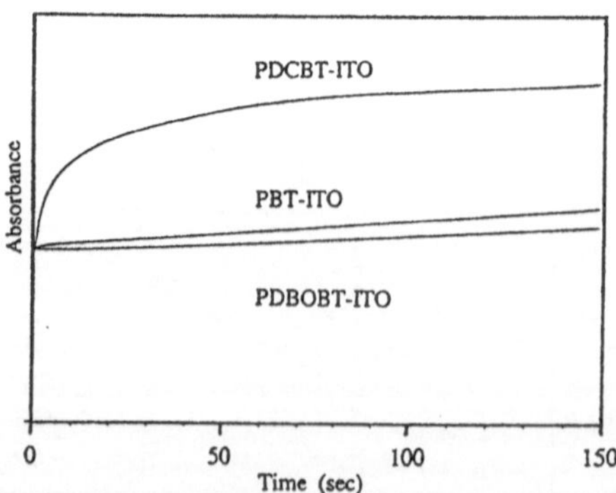

Figure 5 Time-based spectroscopic absorbances of doped polymer film in blank solution at constant wavelength. PBT: 488 nm; PDCBT: 470nm; PDBOBT: 526nm.

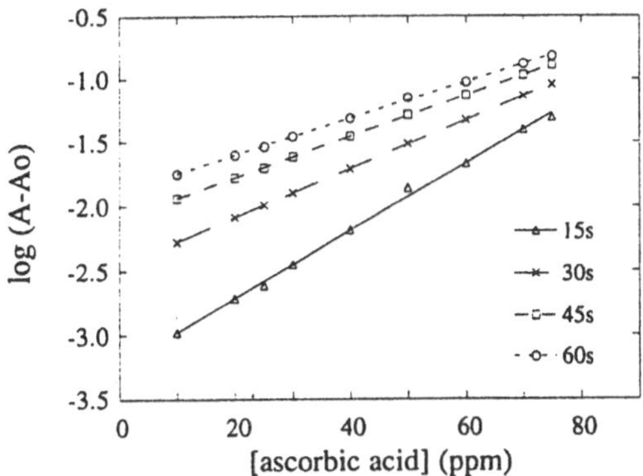

Figure 6 Calibration curve of log(A - Ao) against ascorbic acid concentration for doped PDBOBT film.

CONCLUSION

Using symmetrically disubstituted monomers, structurally homogeneous PDCBT and PDBOBT were prepared electrochemically and studied in comparison with PBT. PDCBT exhibited higher E_{pa} and less extended π-conjugation due to the chlorine substituent in the polymer backbone. On the contrary, the electron donating and flexible butoxy side chain caused much lower E_{pa} and longer conjugation length in PDBOBT. Highly doped PDBOBT therefore could be obtained and the doped state of the polymer was very stable as compared to the doped PDCBT.

Using the electrochromatic changes of these doped polybithiophenes upon reduction by ascorbic acid, a spectrophotometric detector was devised and tested. The PBT-based detector showed linear response to ascorbic acid up to 22 ppm. PDCBT was not used to fabricate the detector due to instability of the doped state. The linear range of PDBOBT-ITO for spectrophotometric detection of ascorbic acid was up to about 75 ppm and this was attributed to the highly doped state attained.

REFERENCES

1. Sato M, Tanaka S, Kaeriyama K (1986) J Chem Soc, Chem Commun 873
2. Jen KY, Miller GG, Elsenbaumer RL (1986) J Chem Soc, Chem Commun 1346
3. Leclerc M, Daoust G (1991) Synth Met 41:529
4. Waltman RJ, Bargon J, Diaz AF (1983) J Phys Chem 87:1459
5. Mao H, Xu B, Hocroft S (1993) Macromolecules 26:1163
6. Tanguy J, Pron A, Zagorska M, Kulszewics-Bajer I (1991) Synth Met 45:81
7. Chan HSO, Ng SC, Seow SH, Moderscheim MJG (1992) J Mater Chem 2:11:1135
8. Bartlett PN, Whitaker RG (1987/88) Biosensors 3:359

9. Lyons MEG, Lyons CH, Fitzgerald C, Bannon T (1993) Analyst 118:361
10. Chan HSO, Ng SC, Seow SH, Chi H (1993) In: Prasad PN (ed) Frontiers of Polymers and Advanced Materials. Plenum Publishing Corporation, New York
11. Novak I, Ng SC, Huang HH, Mok CY, Khor E, Kovac B (1991) J Phys Org Chem 4:675
12. McCullough R, Lowe RD, Jayaraman M, Anderson DL (1993) J Org Chem 58:904
13. Chung TC, Kaufman JH, Heeger AJ, Wudl F (1984) Phys Rev B 30:702
14. Ferraris JP, Hanlon TR (1989) Polymer 30:1319

Fluorescence Studies of Polymer Diffusion Across Interface

Mitchell A. Winnik,* Haeng-Boo Kim,[1] and Seigou Kawaguchi[2]

Department of Chemistry, University of Toronto
Toronto, Ontario, Canada M5S 1A1

ABSTRACT

We investigate the influence of surface properties on the kinetics of polymer diffusion across interfaces of films prepared from several different types of poly(*n*-butyl methacrylate) [PBMA] core-shell latex particles. A series of core-shell PBMA microspheres containing different amounts of methacrylic acid groups in their shell were prepared by three-stage emulsion polymerization. PBMA latex with poly(ethylene oxide) [PEO] chains at the surface was prepared by dispersion copolymerization of BMA with PEO macromonomer. Each pair of latex, one labeled with a donor (phenanthrene), the other, with an acceptor (anthracene), allows us to use direct non-radiative energy transfer experiments to follow polymer interdiffusion in their latex films. The interdiffusion was found to be significantly retarded but not suppressed, either by the presence of the acid groups in the latex shell, or by their salts. Neutralization of the acid groups by Ba^{2+} had a larger effect than Na^{+} on slowing down the polymer diffusion. On the other hand, PEO chains at the latex surface was found to promote interdiffusion in the early stages of film formation and aging.

INTRODUCTION

There has been a substantial interest over the past decade in the diffusion of polymer molecules across interfaces. The key features of interest for homopolymer interdiffusion are the influence of the interface on the kinetics of interdiffusion, and the connection between the extent of interdiffusion and the growth of strength of the joint.[3-7] One of the reasons for the broad interest in this topic is that polymer diffusion across interfaces represents the essential feature of a number of technologically important processes, such as welding of polymer slabs, sintering or compression molding of polymer powders, and the formation and aging of latex films.

One feature which distinguishes latex film formation from other polymer sintering processes is that latex microspheres have polar groups at the particle surface. These groups are introduced during the emulsion polymerization process used to prepare the particles, and they serve to provide colloidal stability for the particle dispersion. When the dispersion

K.P. Ghiggino (Ed.)
Progress in Pacific Polymer Science 3

is applied to a substrate and the solvent evaporates, compressional forces deform the spherical particles into space filling Wigner-Seitz cells,[8] and the polar groups originally at the particle surface now sit at the interface between adjacent cells. When these groups are few in number and well separated, they presumably remain relatively isolated in the interface. If present in larger amounts, as in core-shell latex materials, they then form an interconnected membrane in the film. This hydrophilic membrane gives the nascent film a morphology similar to that of a foam.[9]

Physicists used to thinking about foams see this membrane as an absolute barrier to interdiffusion of the latex polymer. From this perspective, one can *define* coalescence in such films as the rupture of the membrane which then allows intimate contact between the core polymer molecules of adjacent cells. According to this relatively simple model, the ruptured membrane retracts into phase-separated droplets, and then polymer interdiffusion occurs in much the same way as if there were no barrier. In many core-shell latex films, the membrane is composed of a copolymer or graft copolymer with a composition not too dissimilar from that of the latex core. Under these circumstances, polymer diffusion across the interparticle membrane will either involve the membrane polymer, or the core polymer will be obliged to diffuse through the copolymer membrane phase. This leads in principle to two interesting diffusion scenarios: the membrane acts as a barrier which retards but does not prevent polymer interdiffusion; or, alternatively, the membrane mixes with and plasticizes the core polymer and facilitates the rate of interdiffusion.

Here we report experimental results in which both types of phenomena are observed. These experiments involve poly(*n*-butyl methacrylate) [PBMA] latex films prepared from core-shell latex particles in which PBMA is the core polymer. One set of particles has a shell containing methacrylic acid [MAA] as a comonomer [P(BMA-*co*-MAA)]. Films prepared from the ion-exchanged latex have a carboxylic-acid-group-rich phase as an interparticle membrane, whereas films prepared from the same particles at high pH form an ionomer phase in the membrane. These structures retard but do not prevent interparticle polymer diffusion. A second type of PBMA, prepared from a poly(ethylene oxide) [PEO] macromonomer, has a shell rich in PEO chains. PEO is miscible with PBMA and has the potential to act as a plasticizer. Here we find that the membrane polymer actually accelerates interdiffusion in the early stages of the interdiffusion process.

To measure the rate and extent of polymer diffusion across a boundary, one needs a source of contrast. Contrast is achieved by labeling the polymer, for example, employing a small amount of deuterated latex in conjunction with small angle neutron scattering experiments,[10] or with fluorescent labels in conjunction with direct non-radiative energy transfer [DET] measurements.[11] In our laboratory, we prefer the DET methodology because of our ready access to beam time. Thus the latex we prepare are labeled with ca 1 mol% donor groups (D = phenanthrene, Phe) or acceptor groups (A = anthracene, An). Films are prepared from dispersions containing a 1:1 mixture of these particles.

EXPERIMENTAL

Materials: A series of core-shell PBMA latex particles containing different amount of MAA groups in their shell were prepared by three-stage emulsion polymerization under monomer starved condition (1ml/hr) at 80 ^{o}C.[12] The latex dispersions were purified using a mixed bed ion-exchanged resin to remove low molecular weight electrolytes including surfactant. When appropriately purified, the dispersion exhibited a tendency to form a colloidal crystalline phase, implying the complete removal of salts. The surface charge density (Q_s) was determined by potentiometric titration of freshly ion-exchanged latex.

PBMA latex with PEO chains at their surface was prepared by dispersion copolymerization of BMA with the PEO macromonomer(1) (0.97 mol% of BMA) in methanol-water media (20 vol% water) at 70 ^{o}C.[13] The complete consumption of PEO macromonomer was confirmed by gel permeation chromatography (GPC) and by ^{1}H-NMR. After polymerization, methanol was removed completely from the dispersion by dialysis against deionized water.

CH_3-O-$(CH_2CH_2O)_n$-$(CH_2)_7$-C_6H_4-CH=CH_2 (1)

n=53
M_w/M_n=1.06
Functionality=98%

The particle diameters were determined by dynamic light scattering (Brookhaven Instruments, Model BI-90 particle sizer); and molecular weights, by GPC (Waters Styragel columns, THF as eluent, PMMA samples as standards).

Film Formation: Latex films were prepared on small quartz plates from dispersions containing an equal number of Phe- and An-labelled particles. Samples were annealed for various periods of time in a temperature-controlled oven, removed and cooled to room temperature for fluorescence measurements, and then returned to the oven. Samples for fluorescence decay measurements were placed in small quartz test tubes and flushed with argon. Decay profile measurements were carried out using the single photon timing technique as described previously.[11] Phenanthrene decay profiles were monitored at 366 nm, with λ_{ex} = 300 nm. Decays were first fitted to the equation

$$I(t') = A_1\left[\exp\left(-\frac{t'}{\tau_D} - P\left(\frac{t'}{\tau_D}\right)^{\beta}\right)\right] + A_2\exp\left(-\frac{t'}{\tau_D}\right) \quad (1)$$

with $\beta = 0.5$, where first term approximates the contribution to the donor decay of energy transfer in regions of the film where polymer interdiffusion has taken place. A better measure of the extent of interdiffusion is obtained from the areas under the decay profiles, calculated by integrating $I_D(t')$, using parameters obtained in eq.(1) after normalizing at t'=0.

$$\text{Area} = I = \int_0^\infty I_D(t')dt' \cdot \qquad (2)$$

The decrease in area is proportional to the increase in extent of energy transfer, which in turn is related to the volume fraction of mixing, f_m.

$$f_m = \frac{\text{Area}(t) - \text{Area}(0)}{\text{Area}(\infty) - \text{Area}(0)} \qquad (3)$$

RESULTS AND DISCUSSION

Effect of Surface Carboxyl Groups on Interdiffusion

The characteristics of the latex particles and their constituent polymers are presented in Table 1. For each sample we were able to prepare latex particle pairs, one labeled with phenanthrene and the other with anthracene, which are similar in size, in molecular weight and molecular weight distribution, but differ only in the content of carboxyl groups.

Films were prepared from mixtures of donor- and acceptor-labeled latex, and, once dry, were annealed in an oven. The films were removed periodically, subjected to fluorescence decay analysis, and returned to the oven. From the data, the extent of polymer interdiffusion was calculated. In Figure 1 we present f_m values, plotted against square-root of annealing time, for this series of latex films. Two features of this graph are important. First, we note that the rate of polymer diffusion depends sensitively on the carboxyl group content of the latex: f_m values at the same annealing time decrease significantly in films prepared from latex with increasing carboxyl group content in the shell.

Table 1
Characteristics of PBMA Latex with Surface Carboxyl Groups

Latexes	Diameter [nm]	$10^{-5}M_w$	M_w/M_n	10^5Qs 1) [eq/g]	S. A 2) [A^2/acid]
An-MA0	129	4.8	3.6	1.0	730
Phe-MA0	128	4.1	3.2	0.99	750
An-MA2	152	4.8	2.9	4.52	140
Phe-MA2	147	4.4	3.2	4.98	130
An-MA4	150	4.2	2.9	7.98	79
Phe-MA4	150	4.0	2.7	6.54	96
An-MA6	149	4.9	3.0	11.6	55
Phe-MA6	146	4.3	2.6	12.1	54

1) Total surface charge density of $-OSO_3H$ and -COOH groups.

2) The average surface area occupied by an acid group.

The second important observation for the data in Figure 1 is that interdiffusion occurs even at the early stages of annealing time, in the all films. This result is very different from that reported by Joanicot et al.[9b] for a different latex film. They found that polyacrylic acid [PAA] at the surface of a poly(styrene-co-butyl acrylate) latex effectively suppressed interdiffusion until the film temperature exceeded the T_g of the PAA, at which point the polar membranes ruptured. The essential difference in the two systems is one of carboxyl content in shell. In their system, the shell polymer was extremely rich in PAA. In the latex particles we prepared, the composition of the shell-polymer did not exceed 9 mol% MAA, but this phase comprised about one-third of the polymer in latex. Thus our particles have a thicker shell composed of a materials likely to be more miscible with the core polymer.

Diffusion Coefficients

To proceed with the analysis, we need to be able to compare rates of interdiffusion among the various experiments. We calculate diffusion coefficients by fitting the data to a spherical diffusion model which satisfies Fick's laws of diffusion, eq (4),

$$\frac{\partial C(r,t)}{\partial t} = \frac{1}{r^2}\frac{\partial}{\partial r}\left(Dr^2\frac{\partial C(r,t)}{\partial r}\right) \tag{4}$$

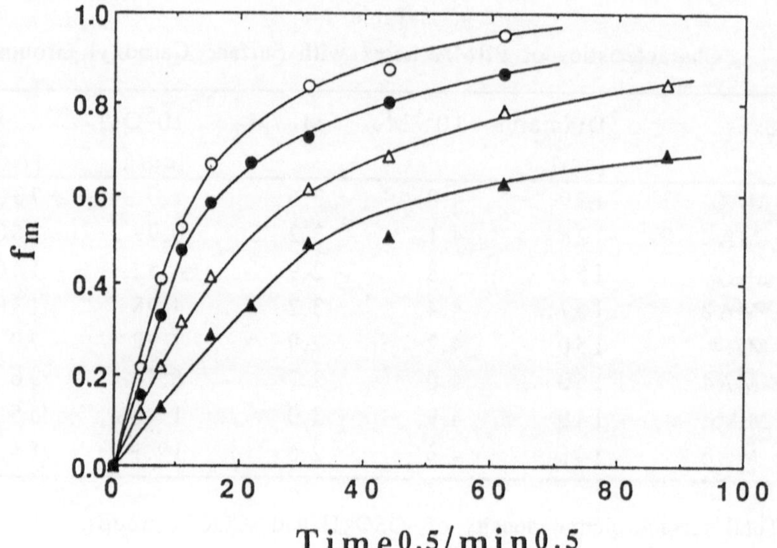

Figure 1. Plot of f_m vs the square-root of annealing time for latex films of MA0(○), MA2(●), MA4(Δ), and MA6(▲) at 90°C.

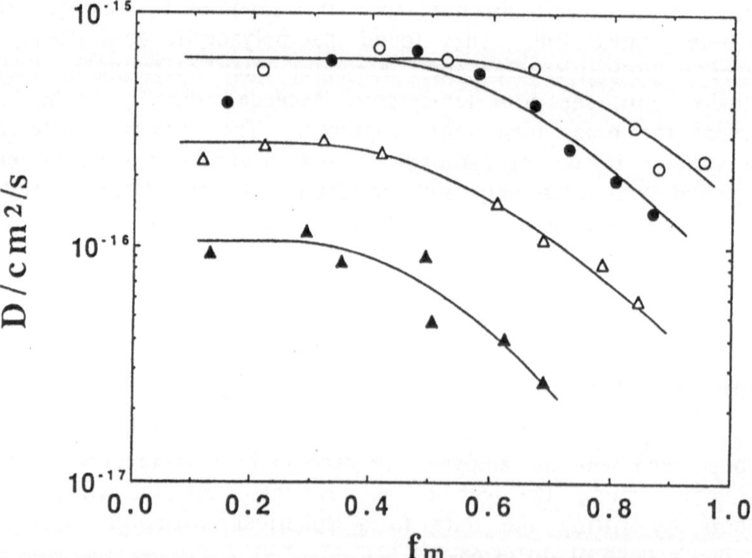

Figure 2. Plot of D values calculated from the data in Figure 1 as a function of f_m. The symbols are the same as in Figure 1.

so that at time t we have for the concentration profile of the donor-labeled polymer

$$C(r,t) = \frac{C_0}{2}\left\{\operatorname{erf}\left(\frac{R+r}{2\sqrt{Dt}}\right) + \operatorname{erf}\left(\frac{R-r}{2\sqrt{Dt}}\right)\right\} - \frac{C_0}{r}\sqrt{\frac{Dt}{\pi}}$$

$$\times\left\{\exp\left(\frac{(R-r)^2}{4Dt}\right) - \exp\left(-\frac{(R+r)^2}{4Dt}\right)\right\} \qquad (5)$$

where D, R, t, C_o are the diffusion coefficient, the radius of the latex particle, the annealing time, and the initial polymer segment concentration, respectively. D values are calculated by equating f_m with the fractional mass which has diffused across the interface, (M_t/M_∞), where $M_\infty = (4/3)\,\pi R^3 C_o$, and carrying out a numerical integration to find the best D value which satisfies the equation[11]

$$M_t = M_\infty - \int_0^R C(r)\, 4\,\pi r^2\, dr \qquad (6)$$

D values calculated from the data in Figure 1 are presented in Figure 2 as a function of f_m.

For each of the samples, we see that the D values are initially constant, and then decrease sharply once a certain extent of interdiffusion is attained. This crossover point where the diffusion begins to slow down shifts to lower f_m for films richer in -COOH groups. Since the D values are cumulative diffusion coefficients, the instantaneous decrease in D must be even more pronounced. Another feature of importance is the decrease in the initial D value with increasing acid group content. We interpret these results to mean that the most mobile chains in the system dominate diffusion at early times. In this system, the most mobile chains are the lowest molecular weight PBMA homopolymers which were originally a part of the PBMA seed. This polymer must diffuse through an acid-rich region to cross the interparticle interface. As the acid group content of the boundary layer increases, the local T_g increases, contributing to the decrease in initial diffusion rates. At longer times, the MAA copolymer contributes to the interdiffusion, and its D values decrease with increasing molecular weight and with increasing MAA composition.

Effect of Neutralization of Surface Groups

Films were prepared from the MA6 latex dispersion after first neutralizing the acid groups with inorganic base. Neutralization was performed by using one equivalent of base corresponding to the number of titratable acid groups in the latex, so that the degree of neutralization of the surface ionizable groups is unity. The D values obtained from these latex films are shown in Figure 3, where we compare the behavior of an unneutralized MA6 film with those prepared from dispersions that were neutralized with NH_4OH, NaOH, and $Ba(OH)_2$. The first feature of interest here is that the diffusion coefficients of the polymers with acid groups present as the ammonium salt are essentially identical at early annealing times with those with their acid groups in the protonated form. Some differences are apparent at later times, but the absence of a significant effect of NH_3 on polymer D value points to a tendency for the ammonium groups to dissociate upon drying or annealing of the film.[14]

When the MA6 film is neutralized by sodium hydroxide, one observes not only a substantially smaller initial value of D, but also a very pronounced decrease in D throughout the interdiffusion process. Moreover, this diffusion is retarded even further in the case of the barium carboxylate film. Here the initial D value is ca 10^{-17} cm^2/s, and appears to increase before decreasing. One possible reason for the initial increase of D with f_m may be related to the mutual attraction of ion pairs to form a segregated, ion-rich microphase. This kind of phase in ionomer materials has been detected by small-angle X-ray scattering.[15] The major conclusion to be drawn from this data is that neutralization of the carboxylic acid groups at the latex surface with NaOH and $Ba(OH)_2$ induces a pronounced decrease in the interdiffusion rate, and that the divalent Ba^{2+} salt is more effective than Na^+ at retarding interdiffusion.

Effect of PEO Chains at the Latex Surface

PBMA latex particles with a surface rich in PEO were prepared by dispersion polymerization in methanol-water and then transferred to a purely aqueous phase. The characteristics of these microspheres are listed in Table 2. The diameters of the two PBMA/PEO particles are very similar to one another and are only sightly larger than those of the PEO-free PBMA sample composed of similar molecular weight polymer. The PBMA/PEO particles contain 14 wt% PEO. This PEO is present in the form of a PBMA copolymer with PEO branches. In the GPC analysis of the latex components, the hydrodynamic volume of the branched copolymer may well be smaller than that of the linear polymer PBMA with the same molecular weight. Thus it is possible that the true molecular weight of the PBMA grafted PEO chain is greater than that of PEO-free linear PBMA.

Table 2
Characteristics of PBMA Latex with PEO Surface Chains

Latexes	Diameter/nm	$10^{-5}M_w$	M_w/M_n
An-PBMA[1)]	115	2.44	3.6
Phe-PBMA[1)]	117	2.86	3.4
An-PBMA/PEO[2)]	129	2.64	2.9
Phe-PBMA/PEO[2)]	132	2.35	2.7

1) SDS surfactant was removed by ionic exchange.

2) PEO macromonomer content is 14.4 wt%.

Films were prepared from the two sets of samples described in Table 2. One immediately apparent difference is that the PBMA/PEO latex formed transparent films at 20°C, whereas the PBMA dispersion required higher temperature (~30°C). The surface PEO chains appear to lower the minimum film forming temperature [MFT] of the latex. Interdiffusion measurements were carried out on these two films, and f_m values were calculated, as is shown in Figure 4. These results are quite surprising. One notices first that the surface PEO chains promote interdiffusion at early times. This result is consistent with the PEO also lowering the MFT of the system, and suggests that PEO acts as a plasticizer for the particle surface.

At later stages of the interdiffusion, the two curves cross, and the PEO-containing latex film experiences a smaller diffusion rate. These experiments are still in their very early stages, and thus it is premature to draw firm conclusions. Our current explanation for this retardation is that it corresponds to the diffusion of the PBMA-PEO copolymer. Its diffusion is slower either because of its enhanced molecular weight or because the branches interfere with the diffusion.

ACKNOWLEDGMENTS

The authors thank The Glidden Company, ICI, and ICI Canada, as well as NSERC Canada for their support of this research. S.K. thanks the Ministry of Education, Science and Culture of Japan for a fellowship.

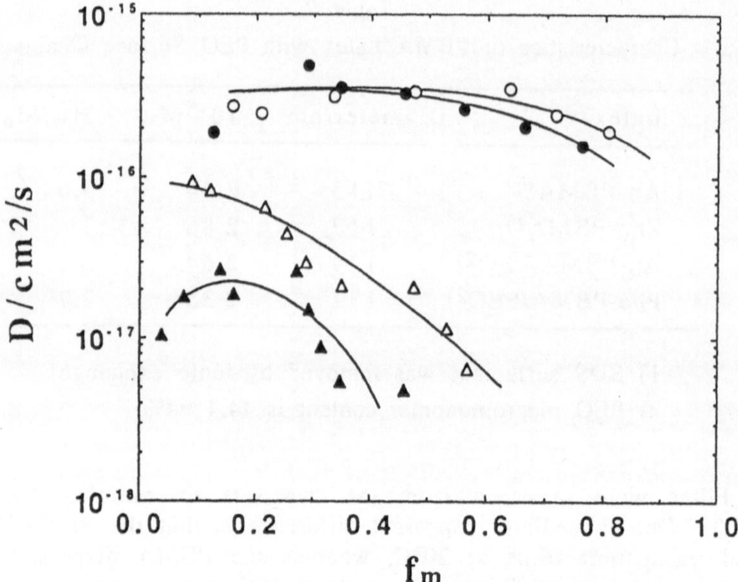

Figure 3. Plot of D values of latex films of MA6 annealed at 100 °C as a function of f_m. The surface ionizable groups of MA6 latex particles were neutralized by NH_4OH(●), NaOH(Δ), and $Ba(OH)_2$(▲). The top curve (O) refers to a films prepared from unneutralized MA6, plotted for comparison.

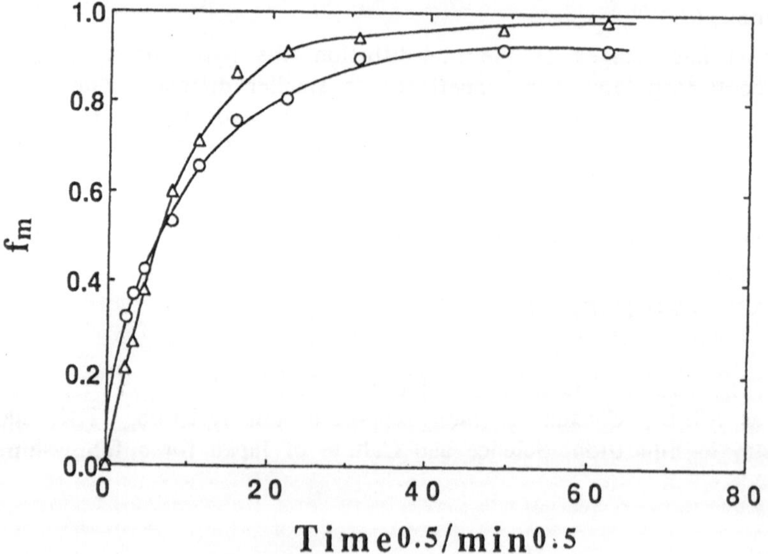

Figure 4. Plot of f_m vs the square-root of annealing time for PBMA(●) and PBMA/PEO (O) latex films at 90 °C.

REFERENCES

1 Permanent address, Faculty of Engineering., Hokkaido University, Sapporo, 060 Japan
2 Permanent address, Department of Materials Science, Toyohashi University of Technology, 1-1 Tempaku-cho, Toyohashi 441 Japan
3 Kausch HH, Tirrell M (1989) Annu Rev Mater Sci 19:341
4 Wool RP, Yuan BL, McGarel OJ (1989) Polym Eng Sci 29:1340
5 de Gennes PG (1980) C R Acad Sci Ser B 291: 219
6 (a) Prager S, Tirrell M (1981) J Chem Phys 75:5194 (b) Prager S, Adolf D, Tirrell M (1983) J Chem Phys 78:7015 (c) Prager S, Adolf D, Tirrell M (1986) J Chem Phys 84: 5152
7 (a) Wool RP, O'Conner KM (1981) J Appl Phys 52:5194 and 5953 (b) Kim YH, Wool RP (1983) Macromolecules 16:1115
8 Zallen R (1983) The Physics of Amorphous Solids, Wiley-Interscience, New York, pp 54 ff
9 (a) Chevalier Y, Pichot C, Graillat C, Joanicot M, Wong K, Lindner P, Cabane B (1992) Colloid Polym Sci 270: 806 (b) Joanicot M, Wong K, Richard J, Maquet J, Cabane B (1993) Macromolecules 26: 3168
10 Yoo JN, Sperling LH, Glinka CJ, Klein A (1990) Macromolecules 23:3962
11 (a) Zhao CL, Wang Y, Hruska Z, Winnik MA (1990) Macromolecules 23:4082 (b) Wang Y, Winnik MA, Haley F (1992) J Coatings Technol 64(811):51. (c) Wang Y, Winnik MA (1993) J Phys Chem 97:2507
12 Kim H-B, Wang Y, Winnik MA (1994) Polymer, in press
13 Kawaguchi S, Winnik MA, Ito K manuscript in preparation
14 Kim H-B, Winnik MA (1994) Macromolecules, in press
15 Eisenberg A (1977) Ion-Containing Polymers. Physical Chemistry and Structure, Academic Press, NY

Recent Studies on β-Crystalline Form of Isotactic Polypropylene

Guanyi Shi

Shanghai Institute of Organic Chemistry, Academia Sinica,
354 Fenglin Lu, Shanghai, 200032, P. R. China

CONTENT

1. Introduction
2. Crystallization Conditions
3. Mechanical Properties
4. Processing
    4.1. Compression Molding
    4.2. Extrusion
    4.3. Injection Molding
5. Elastomer Modified β PP Blends

1. INTRODUCTION

Isotactic polypropylene shows four distinguishable chain conformation with each being a 2 × 1/3 helix[1]. Different packing geometries lead to three well known crystalline structures (polymorphs), namely the monoclinic α-form, the hexagonal β-form and the triclinic γ-form. The appearance of these structures is critically dependent upon the crystallization conditions[1–5]. Among these three crystalline structures, the monoclinic α-form is by far the most common, being found in normal melt-crystallized or solution-crystallized IPP samples. In the industrial practice, therefore, β- and γ-form have been considered as laboratory curiosities[2]. The practical way for obtaining β-form with relative high purity is to use nucleating agent. Only a few materials are known to nucleate β-form spherulites[6, 7].

In 1980 we have found accidentally that in the presence of pimelic acid special grade isotactic polypropylene, containing substantial quantities of calcium stearate, crystallized predominantly to β-crystalline form[8]. Then a systematic research work has been conducted at this laboratory leading to the first discovery of a group of bicomponent β-nucleators, consisting of an organic dibasic acid, such as pimelic acid, azelaic acid etc. and an oxide, a hydroxide or a salt of a Group II A metal, such as magnesium, calcium, stronium and barium[9, 10].

By judicious selection of appropriate chemical and structure parameters, an effective β-nucleator tailored to increase the interaction with polypropylene was obtained. Such nucleator

K.P. Ghiggino (Ed.)
Progress in Pacific Polymer Science 3

has a variety of interesting features worthy of study. This new class of β-nucleator is interesting from scientific view point because it is formed by mixing a carboxylic acid type α-nucleator of IPP with inert inorganic filler or lubricator of polymers. Furthermore, the crystallization of IPP in the metastable β-form highlights the ability of nucleators to induce preferred orientation and arrangement of the helical polymer chains. Such nucleator may be also effective to promote crystallization and crystal transformation in other helical polymers. For instance, it is an effective additive which promotes the transformation of tetragonal phase to hexgonal phase in polybutene-1[11].

From a technological point of view, a bicomponent β-nucleator for polypropylene has various advantages over β-nucleator, reported previously in the literature, such as the quinacridone dye permanent red E3B (E3B)[6]. Firstly, the bicomponent β-nucleator is colorless, while the E3B is red in color. Secondly, the β-form of isotactic polypropylene (β PP) obtained by the melt-crystallization of commercial polypropylene in the presence of a bicomponent β-nucleator has much higher impact resistance[12] than that obtained by other β-nucleator[13]. Thirdly, such β PP has the tendency to the formation of numerous microvoids during tensile deformation[14]. Therefore, it can be utilized for preparing microporous film.

The present article describes the recent studies on the β-form of IPP with the use of the bicomponent nucleator. Particular attension was paid to the crystallization conditions, mechanical properties, processing and potential uses of the β PP.

2. CRYSTALLIZATION CONDITIONS

It has been reported that β-spherulites are formed at crystallization temperature between 80° C and 130° C with the optimum range being 100 to 125° C[2]. In order to determine the temperature range for the formation of β-crystals in our β-nucleated PP, the samples were crystallized by quenching in silicon oil and in water, thermostated at a selected temperature between 20 and 140° C. The relative amount of β-form in each of the samples, as expressed by the K value[2]

$$K = \frac{H(300)}{H(110) + H(040) + H(130) + H(300)} \qquad (1)$$

was determined by X-ray diffraction, where H(110), H(040) and H(130) are the heights of the three strong, equatorial α-form peaks (110), (040) and (130), and H(300) is the heights of the strong β-form peak (300).

Fig. 1 shows the K value as a function of Tc for samples crystallized by quenching in silicon oil and in water[15, 16]. Since in our experiments the samples were sandwiched between two copper foils, they were not in contact with liquid medium. Futhermore, the half-time of crystallization $t_{1/2}$ of the β-nucleated PP is less than a few seconds for Tc below 100° C as estimated from extrapolated $t_{1/2}$ vs. Tc curve, measured by differential scanning calorimetry[17].

It is most probable that the crystallization proceeds in the quenching period from temperature of melt to the pre-determined Tc for Tc < 100° C before the isothermal temperature is reached. The liquid with higher thermal diffusivity e.g. water cools the sample more rapidly than the liquid with lower diffusivity e.g. silicon oil. Consequently the difference between the actual and pre-determined Tc for samples quenched in water is lower than that for samples quenched in silicon oil. It is clear from Fig. 1 that a rapid decrease of the K value with decreasing Tc started at 80° C for samples quenched in water. Considering the results of Varga[18], the estimated lower critical temperature limit lies between 80 and 100°C.

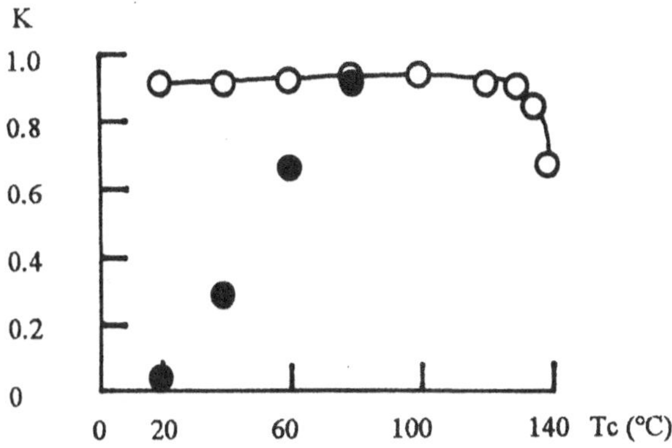

Fig. 1. Effect of crystallization temperature on the K value for β-nucleated polypropylene quenched in silicon oil (○) and in water (●)

We have subsequently shown (see section 4.3.) that injection molded specimens, which were chilled in air, have mixed polymorphic composition where the skin consists of α-crystals while the core comprises β-spherulites[16]. This fact supports the existence of a lower critical temperature limit for the formation of β PP.

The existence of an upper critical temperature limit was directly demonstrated by Fig. 1, being 140° C. Our radial growth rates data show that above 140° C the growth rate of α-spherulites is higher than that of β-spherulites, while it is lower for temperature below 140° C[10]. Therefore, above 140°C no production of β-form is possible.

3. MECHANICAL PROPERTIES

The properties of isotactic polypropylene in its β-crystalline form can be compared with commercial IPP of same grade in its α-crystalline form. The differences show up most clearly in impact resistance. The characteristic properties of two commercial grade polypropylene HGX330 and GP186 having melt flow index of 33 and 3 g / 10 min, respectively, supplied by Phillips Petroleum Company in its α- and β-form at room temperature are reported in Tabs. 1–3.

Table 1. Tensile Properties of Polypropylene (ASTM D638)

Tensile Properties	HGX330, 33MF		GP186, 3MF	
	α-form	β-form	α-form	β-form
at Yield				
Strength (MPa)	38.08	31.26	39.46	31.19
Elongation (%)	11.09	11.32	12.55	13.03
at Break				
Strength (MPa)	38.08	25.23	18.24	26.77
Elongation (%)	11.09	68.85	30.36	93.41

Table 2. Flexural Properties of Polypropylene (ASTM D790)

Flexural Properties	HGX330, 33MF		GP186, 3MF	
	α-form	β-form	α-form	β-form
Modulus (MPa)	1,597	1,516	1,626	1,577
Standard Deviation	17.29	9.21	20.25	6.52
Strength at Yield (MPa)	47.11	38.61	46.92	39.03

Table 3. Impact Resistance of Polypropylene (MTS High Rate Impact Test)

Impact Speed (in / sec)	Impact Energy to Failure (in-lb)			
	HGX330, 33MF		GP186, 3MF	
	α-form	β-form	α-form	β-form
10	16.5	45.3	23.9	72.2
50	13.4	40.3	21.4	73.3
100	11.6	22.1	16.8	64.6
300	17.3	17.1	6.8	43.3

It is clear from Tabs. 1–3 that the β PP has equal stiffness, slightly lower yield strength, much higher ductile-brittle transition and much higher impact resistance than α PP. The β PP has also lower moldings volumetric shrinkage than α PP as its density (0.889 g / cm^3) is lower than α PP (0.902 g / cm^3).

β PP has a lower temperature of melting than α PP: 150° C as compared with 165° C. It is rather strange that the β PP has higher heat deflection temperature (ASTM D648) than α PP:

123° C as compared with 113° C, which is inconsistent with the lower melting point of β-form. This is understandable if one determines the temperature dependence of the flexure modulus of α and β PP by measuring deflection as a function of temperature using Perkin-Elmer TMS-2 Thermomechanical Analyser. Fig. 2 shows the deflection of α and β PP films under a standard loading condition (264 PSI). It can be seen from Fig. 2 that a less pronounced increase of deflection with temperature was observed for β PP rather than for α PP. The reason for the less pronounced decrease of flexure modulus with temperature in β PP is not known but may be related to the β-β recrystallization after the initial melting of the unstable structure[19]. The β PP has better resistance to many chemicals at temperature up to 100° C than α PP. Chemical resistance tested covers 30 % HCl, concentrated formic acid, concentrated acetic acid, aldehydes, brines, and alkaline solutions.

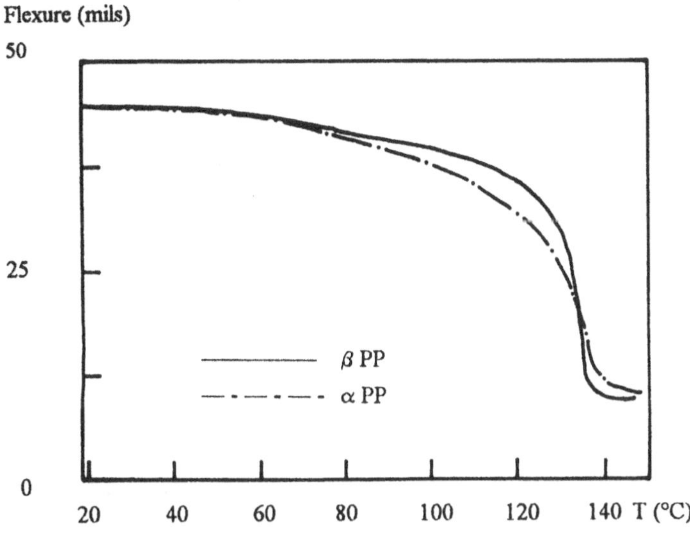

Fig. 2. The deflection of α and β PP sheets under a load of 264 PSI as a function of temperature at a heating rate of 5°C/min

4. PROCESSING

As it was demonstrated above, pure or predominantly β PP can be formed by melt-crystallization in temperature range of 100–135° C. It is crucial for the production of β PP based end-products if these crystallization conditions can be provided during processing.

Under the processing conditions, crystallization usually proceeds non-isothermally. In fact, the thermal environment may be widely variable in the residence space of the material. However, it was also determined that the optimum cooling rate was reported to be about 5° C to about 40° C per minute with significant amounts of β-spherulites produced at a cooling rate of 80°C/min.

It is demonstrated in the following sections how β PP based end-product can be produced by various processing techniques under the appropriate thermal and mechanical conditions.

4.1. COMPRESSION MOLDING

Sheets or films of different thickness can be formed by compression molding. It is easy to maintain mold temperature at temperature within 100 to 135° C for crystallization. Especially for thick moldings, due to the thermal inertia of the polymer, the mold can be chilled in air. Under this condition, the cooling rate is usually 40° C/min or less and the corresponding peak temperature of crystallization measured by DSC is 109° C or more. The typical melting curve of a molded film, cooling to room temperature by chilled in air after molding is shown in Fig. 3.

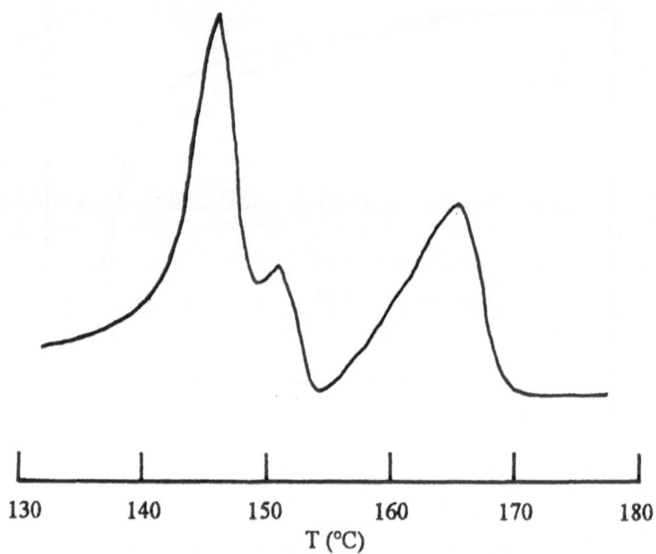

Fig. 3. Melting curve of sheet compression molded by chilled in air

It can be seen from Fig. 3 that the DSC trace, consisting of three partially overlapping process: a melting of original β_1-phase, a melting of stabler β_2-phase and melting of the α-phase is typical for β PP[19].

The most important mechanical properties of β PP were determined for molded sheets and compared to those of α PP manufactured in a similar manner as shown in Tabs. 1–3 (see section 3).

4.2. EXTRUSION

β PP extruded films can be prepared by the blown-bubble process or the slit-die extrusion process. In the blown-bubble process, the film is preferably cooled with a stream of air. In the slit-die extrusion process the cooling can be accomplished with a liquid cooling bath, chill roll or stream of air. The chill-roll is commonly used for the cast film process. It was found by us that the thermal condition of crystallization for β-nucleated PP can be provided in the slit-die extrusion process using commercial available equipments. The process for forming the microporous film invented by us comprises the steps of forming a film containing β-crystals and stretching the film for developing micropores.

In our process the β PP pellets are fed into a single screw extruder operated at about 240–260° C. The polymer melt is continuously extruded through a slit-die, slightly stretched in air and cooled on the metallic surface of a thermoregulated chill-roll. Temperature of the rolls can be adjusted by separate oil thermostats. If the roll temperatures keep in the temperature range of 100–130° C, the molten polypropylene crystallizes predominantly to the β-form. The micropores are formed simply by stretching the cast film uniaxially or biaxially. In uniaxial stretching the cast film was stretched in the machine direction with a roll type stretching machine at temperature below the melting point of the β-crystals. Lower temperature favors the formation of micropores, however, reduces the homogeneity of films. The pore volume of the film as measured from the apparent density increase rapidly at the first stage of drawing, pass a maximum then decreases gradually with further increasing draw ratio. The stretching induced microvoid formation and whitening makes this drawn film opaque, acceptable to printing ink and gives it the appearance of a fine paper[14]. In fact it is a new type of synthetic paper which does not contain any filler.

The films produced by the instant process have poor vapor porosity and non-wetting characteristics. The pore size, pore size distribution and pore volume were determined for drawn films with low and high draw ratio. These pore measurements were made by standard high pressure mercury intrusion porosimetry. The plots of the pore volume vs. pore radius for films with draw ratio 3 and 8 are shown in Fig. 4. It can be noticed from Fig. 4 that the pore size is rather small and less than a few thousands angstrons. However, that measured by scanning electron microscope is much larger, as shown in Fig. 5. This inconsistency can be explained as follows:

The chill-roll touched surface of the cast film consists of α-rich layer which is air-tight after stretching in spite of the inner layer being porous. Consequently, the whole film behave as less porous film. During the electron-microscopic observation electron beam etched the air-tight layer, then the porous structure of film can be viewed.

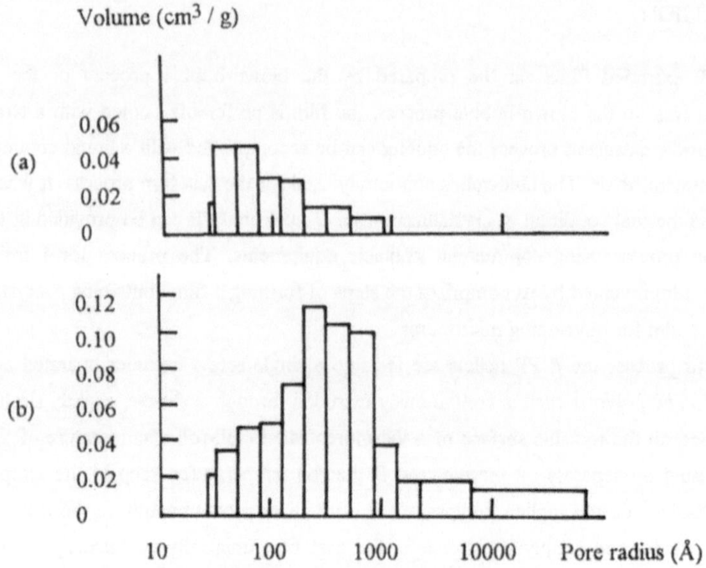

Fig. 4. Pore radius distribution of uniaxial drawn film from β PP with draw ratio of 3 (a) and 8 (b)

(a) (b)

Fig. 5. Scanning electron micrographs of sections from β PP drawn film

By biaxial drawing micro-porous film with enhanced vapor permeability and larger pores are obtained. The nitrogen permeability at 20° C for the film with thickness of 70 μm is 0.003 $cm^3 \cdot cm$ (STP) / $cm^2 \cdot sec \cdot cm$ Hg, which is comparable with the commercial available Celgard film. In order to modify the film surface for hydrophilicity and biocompatibility the plasma-grafting polymerization technique are utilized. The resulting film has high water uptake (50 %), excellent dimension stability and high permeability towards Na^+, Mg^{++} and urea being 3.2×10^{-6}, 5.8×10^{-8} and 1.75×10^{-6} cm^2 / min, respectively[20].

4.3. INJECTION MOLDING

The injection molding of β-nucleated PP were performed at the similar conditions using same machine as that of commercial PP. The results obtained are as follows:

1. At low mold temperatures (20 to 60° C) in the thinner (1 mm) plates, always α-modification formed regardless of the molding conditions.

2. At low mold temperatures (20 to 60° C) the thicker (3 mm or more) plates have mixed polymorphic composition where the skin consists of α-crystals while the core comprises β-spherulites. The charpy and Izod impact strength of thicker specimens are summarized in Tabs. 4, 5.

Table 4. Notched Charpy Impact Strength of Polypropylene
(Chinese Standard GB-1043-79)

Sample	Melt Flow Index (g / 10 min)	Impact Strength (Kg•cm / cm^2) at 23°C		at -20°C	
		α-form	β-form	α-form	β-form
SOF	0.47	10.8	36.6	3.7	4.5
1300	1.7	6.1	15.6	3.4	3.2
1400	5.6	3.4	10.1	2.1	3.4
F103-EA	3.1	4.3	8.2	2.4	2.6
Moplen-EP	1.5	18.9	31.0	3.8	7.9

Table 5. Izod Impact Strength of Polypropylene

Izod Impact (ft-lb / in)	GP186, 3MF		HGX330, 33MF	
	α-form	β-form	α-form	β-form
Notched 73°F	0.6	1.1	0.5	0.7
Unnotched 73°F	19	26	15	13
Unnotched 0°F	2.7	2.9	2.4	2.3

Based on the visual observation of broken specimens the stress whitening mechanism is proposed to explain the much higher impact strength of β PP. During the matrix deformed on impact numerous voids are created, by which additional energy was dissipated.

X-ray diffraction data show that the β-form transforms to the smectic or α-form during tensile deformation[14, 21]. Therefore, it is proposed that the microvoids formation originates from volume contraction during deformation since the crystalline density of the β-form (0.921 g / cm^3) is lower than that of the α-form (0.936 g / cm^3)[14]. However, the enhancement of Izod impact of PP by using E3B as β-nucleator[13] is lower than that by using a bicomponent β-nucleator: 50 % as compared with 100–330 %. Tab. 6 illustrates the effect of β-nucleator content on the impact strength of PP. The increase of impact strength with increasing β-nucleator content, together with the dependence of impact strength on the nature of the β-nucleator has added a further complication. It seems that the nucleation sites on the surface of chemi-adsorbed inorganic solid interact with polymer chains. This results in transient cross-links which increase the impact resistance of β PP.

Table 6. Effect of β-Nucleator Content on the Charpy Impact Strength of β PP

β-Nucleator Content (wt.%)	Impact Strength at 23°C (Kg•cm / cm ^{2})
0	13.0
0.0001	22.6
0.0005	24.2
0.001	28.5
0.01	34.1
0.05	45.7
0.5	42.7

5. ELASTOMER MODIFIED β PP BLENDS

The impact modification by the addition of elastomer traditionally decreases the stiffness of a polymer system resulting in lower flexural modulus. Loss in flexural modulus with an increase in elastomer level was observed in all elastomer modified PP blends. The impact modification of PP by the addition of β-nucleator does not reduce the stiffness and ultimate strength of PP. However, their impact strength at low temperature is unsatisfatory, as the glass transition temperature of α and β PP which determine the low temperature brittleness are identical.

Table 7. Mechanical Properties of E/P copolymer

Temperature (°C)	Yield Strength (MPa)		Ultimate Strength (MPa)		Elongation at Break (%)		Izod Impact (J / M)	
	α-form	β-form	α-form	β-form	α-form	β-form[b]	α-form	β-form
23	27.2	23.0	24.0	24.6	10~140	200	61.9	106.8
70	17.4	13.3	18.7	22.0	880	700		
110	10.7	8.1	> 10.7[a]	> 12.9[a]	> 1350[a]	> 1180[a]		

a) > means the sample was not broken at a gauge length of 15 mm, because the sample length at break is beyond the limited height of the environmental chamber.
b) Without necking.

Combining both methods of impact modification elastomer modified β PP blends with improved mechanical properties at high and low temperature are developed. If only small amount of E/P elastomer or ethylene unit is incorporated, the blend retains the stiffness and the tensile strength of PP, but has improved mechanical properties at low temperature. The addition of β-nucleator converts the PP from the α-form to β-form. Tab. 7 shows the comparison of the mechanical properties for the original and β-nucleated E/P block copolymer having ethylene content of 5.6 wt.% and MFI of 9.7 g / 10 min.

The E/P block copolymer used in our study consists of a multiple phase system of the PP homopolymer and ethylene propylene rubber. This blend exhibits a relatively high stiffness along with a markedly improved impact resistance at 23° C. This method of impact modification may be used for extruder blends as well.

ACKNOWLEDGEMENT

This work was supported by National Natural Science Foundation of China and Academia Sinica Selected Research Program.

REFERENCES

1 Natta G, Corradini P (1960) Nuovo Cim, Suppl 15: 40
2 Turner-Jones A, Aizlewood JM, Beckett DR (1964) Makromol Chem 75: 134
3 Keith HD, Padden FJ Jr, Walker NM, Wyckoff HW (1959) J Appl Phys 30: 1485
4 Lotz B, Wittmann JC (1986) J Polym Sci, Polym Phys Ed 24: 1541, 1559
5 Lotz B, Graff S, Wittmann JC (1986) J Polym Sci, Polym Phys Ed 24: 2017
6 Leugering HJ (1967) Makromol Chem 109: 204
7 Morrow DR (1969) J Macromol Sci, Phys Ed B3: 53
8 Shi G, Zhang J (1982) K'o Hsueh T'ung Pao (Foreign Lang Ed) 27: 290
9 Shi G, Zhang J, Jin H (1989) Chinese Patent: CN 1004076B; (1993) US Patent: 5,231,126
10 Shi G, Zhang X, Qiu Z (1992) Makromol Chem 193: 583
11 Zhang X, Zhang X, Shi G (1992) Thermochimica Acta 205: 245
12 Zhang J, Shi G, Cao Y, Wang H (1986) Acta Polym Sin: 241
13 Jacoby P, Bersted BH, Kissel WJ, Smith CE (1986) J Polym Sci, Polym Phys Ed 24: 461
14 Shi G, Chu F, Zhou G, Han Z (1989) Makromol Chem 190: 907
15 Zhou G, He Z, Yu J, Han Z, Shi G (1986) Makromol Chem 187: 633
16 Shi G, Cao Y, Zhang X, Hong J, Hua X (1992) Chinese J Polym Sci 10: 319
17 Shi G, Zhang X (1992) Thermochimica Acta 205: 235
18 Varga J (1986) J Thermal Anal 31: 165
19 Shi G, Zhang X, Cao Y, Hong J (1993) Makromol Chem 194: 269
20 Zheng J, Chu F, Zhou G, Shi G (1991) Functional Polymer (in Chinese) 4: 207
21 Asano T, Fujiwara Y (1978) Polymer 19: 99

Controlling Sequence Distribution in Copolyesters

C.W. Potter, J.C. Lim, G. Serpe, J. Economy

University of Illinois, Urbana, IL 61801

Abstract: Recent progress on elucidating the role of interchain transesterification reactions in the copolyesters is described. New data are presented supporting the chemical nature of the ordering process. An unexpected degradation reaction in the p-hydroxybenzoic acid/2,6-hydroxynaphthoic acid copolyester system is discussed.

INTRODUCTION

During the past several years we have been exploring the use of interchain transesterification reactions (ITR) as a tool for facilitating processing of high performance polyesters. We have recently reported on the potential of liquid crystalline copolyesters (LCP's) based on p-hydroxybenzoic acid (HBA) and 2, 6-hydroxynaphthoic acid (HNA) to form good high temperature adhesive bonds between metal substrates such as A1, Ti and steel.[1,2] We have also described the use of LCP materials to provide protective coatings for the above metals against corrosive liquids and gases and against abrasion.[3] We are now exploring the design of crosslinkable oligomeric polyesters that can be cured and laminated either as adhesives or matrices for advanced composites.[4]

Critical to the successful implementation of the above technologies is the need for a fundamental understanding of the high temperature chemical processes that take place in these polyesters. In this paper we briefly review some of our earlier work demonstrating the potential for randomization reactions via ITR.[5] We then consider some recent data that argues strongly that ITR processes can occur in crystalline regions near the T_{CN} of the aromatic copolyesters leading to partial ordering of the random sequence distribution and resulting in a sharply increased melting point and use temperature. An unusual degradation process has also been identified in the partially ordered HBA/HNA copolyester systems. This process appears to occur rapidly at temperatures well below the original polymerization temperature used in preparing these copolyesters.

K.P. Ghiggino (Ed.)
Progress in Pacific Polymer Science 3

BACKGROUND

In the past several years considerable progress has been made in the understanding of the microstructure of the liquid crystalline aromatic copolyesters. The behavior of the nematic melt has been given much attention because of its importance to the processing of LCP's. At first, the microstructure in the melt was thought to be stable [6], but later it was argued that chemical reactions occur causing crystallization induced ordering in some cases [7], and randomization in others [8,9,10]. This issue has for the most part been clarified with a recent publication by Economy and coworkers [8], demonstrating the potential for rapid randomization processes at elevated temperatures (above Tcn). When a 50/50 mixture of homopolymers of pHBA and HNA was compression molded at 450°C, the mixture reacted in less than 60 seconds to produce a 50/50 copolyester of pHBA/HNA with a random sequence distribution. These results, backed by NMR analysis, led to the conclusion that in the nematic melt, randomization through transesterification reactions occurs very rapidly at elevated temperatures and is entropically driven.

It is well accepted that upon heating the HBA/HNA copolyesters well below the crystal to nematic transition, that further crystallization occurs but essentially by a physical annealing process [11,12,13,14,15]. Differential scanning calorimetry results show a 4x increase in the transition enthalpy, i.e. an increase in the percent crystallinity, but with no appreciable change in the melting temperature of the annealed samples. This would indicate that no sequence ordering within the chains occurred. This was further substantiated by ^{13}C NMR showing that there was indeed no change in the microstructure [16].

Annealing close to the crystal nematic transition produces very different results, namely, a significant increase in melting point of 40-60°C [11,12,14]. Heating above the new crystal nematic transition to 400°C causes reversion back to the lower melting, random structure. These processes were shown to occur throughout the entire compositional range of the HBA/HNA system [17] (**Figure 1**).

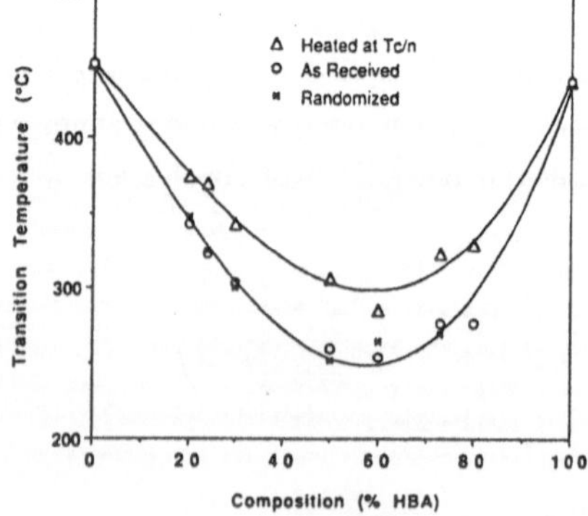

Figure 1 Transition Temperatures versus Composition in the 20/80, 24/76, 50/50, 60/40 and 73/27 PHBA/HNA Copolyesters

ORDERING AS A GENERAL PROCESS IN POLYESTERS

In order to establish the breadth of the high temperature randomization and ordering reactions, three compositions of the Xydar® copolyesters were studied in depth; SRT300, SRT700, and SRT900. It seemed reasonable to explore conversion of the as received Xydar® to a more random sequence distribution since there were early reports on Xydar® SRT300 indicating a blocky structure in the as received polymer [18]. SRT300 which consists of two part HBA and are part biphenol terephthalate (BPT) was heated at 460°C for 10 minutes followed by annealing at 270°C for 5 hours. A much lower melting material with a crystal nematic transition of 359°C was isolated. Presumably, this lower melting material is more random in nature as compared to the sequence distribution of the as received polymer. Interestingly, we observed that upon heating just below this new, lower crystal nematic transition, ordering took place resulting in a material with a transition of 444°C, far above the 408° melting point of the as received copolymer. **Figure 2** shows the randomization and ordering behavior for the SRT300. This same behavior was seen in the SRT700 and SRT900 materials as well. On the other hand based on simulations of X-ray diffraction patterns, Blackwell [19] has reported that melt spun fibers of Xydar® copolyesters show a random sequence. From what we have recently observed, we conclude that the as prepared copolymer is indeed blocky, but upon heating at the high temperatures needed for fiber formation, randomization of the blocky units very likely occurs.

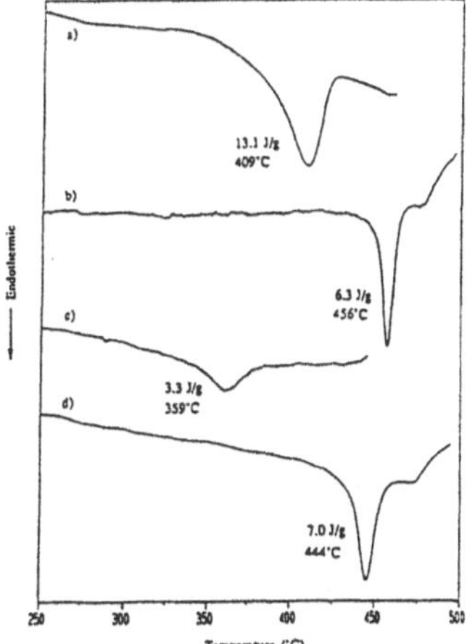

Figure 2 a) as received, b) (a) annealed 5 hours at 400°C, c) (a) heated to 460°C for 10 minutes, followed by annealing 5 hours at 270°C d) (c) annealed 20 hours at 349°C

These ordering processes appear to be far more general and have been observed in a non-liquid crystalline copolyester (LPE-8) derived from a mixture of sebacic acid, biphenol and hydroquinone (62/38 BP/HQ) (See **Figure 3**), as well as in a thermoplastic elastomer based on copolyester units and recently reported on by Fischer [20].

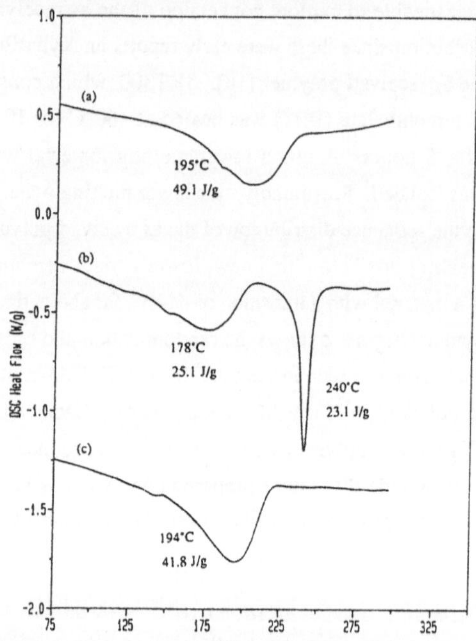

Figure 3 Ordering and Randomization Processes in LPE-8, a) (as prepared) annealed at 220°C for 4 hours, b) (a) annealed at 200°C for 62 hours, c) (b) heated to 300°C

PROPOSED MECHANISMS FOR ORDERING

The mechanism by which the ordering process occurs has been highly debated. The argument for a physical process proposes that ordering occurs through the melting of crystallites, followed by nucleation and growth by diffusion processes. This mechanism requires the preexistence of a significant concentration of ordered sequences. With a chemical process, ordering is described as occurring within existing crystallites by ITR on heating near the melting point [5, 17]. The driving force for these chemical ordering reactions arises from the improved packing and correspondingly higher density associated with a more highly ordered structure, as well as higher dipole interactions between chains.

Additional differences in going from the lower melting to the higher melting phase include a change from a hexagonal to orthorhombic crystal structure, an increase in density of 6% [21], and greatly increased insolubility of the copolyester [17]. There is practically no change in transition enthalpy and the ordered phase will revert back to the lower temperature phase on heating well above the new melting point. Interestingly, heating the random structure for a few minutes to at least 30°C above its Tcn a memory of the rod-like cystalline structure persists; however, heating above the Tcn of the ordered structure, no memory of this order is observed. It has also recently been indicated that the crystal structure of the ordered phase converts to a smectic melt [22]. Presumably the smectic phase on heating well above this transition would undergo a randomization process as indicated in **Figure 1**.

Three possible interpretations can be invoked to explain the ordering occurring on heating this system near its crystal nematic transition, namely, (1) a crystal to crystal transition, (2) a physical ordering process [12], and (3) a chemical ordering process [5]. In a crystal to crystal transition, such as that reported in the pHBA homopolymer [23], a structure of higher order (orthorhombic) converts to a structure of lesser order (hexagonal), which is the opposite of what is observed here. Regardless, it is clear that a more ordered phase could not arise spontaneously in a strictly thermal process starting with a random copolyester. Likewise, a substantial amount of experimental evidence has been collected which argues against a physical process as the major mechanism for ordering near the crystal nematic transition. Part of this evidence includes statistical analysis of the probability for pre-existing long sequences of ordered units within a random copolymer. The probability of having such a sequence in a two-dimensional array of chains, was found to approach a probability of less than 1% (See **Figure 4**). On the other hand, it would seem reasonable that near the crystal nematic transition, the endgroups have increased mobility within the crystallites facilitating further ordering of the microstructure through interchain transesterification reactions. It is argued that these reactions can only occur in the existing crystallites since in the coexisting nematic (non-crystalline) phase, randomizing reactions are occurring [8].

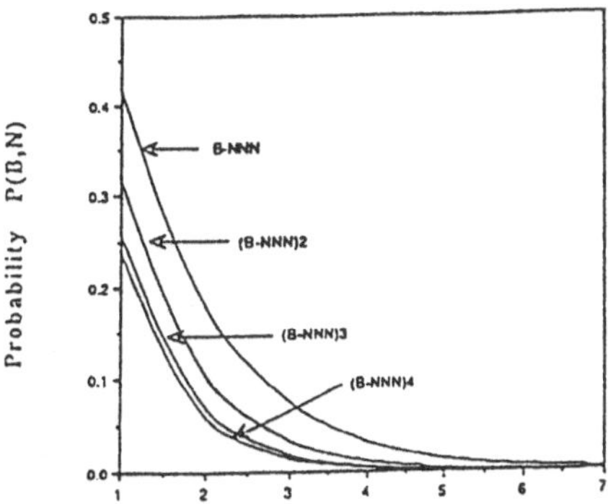

Figure 4 Curves for the Probabilities P(B,N) of (B-N-N-N) in Sequences for n = 1,2,3,4 in a Two Dimensional Crystallite as a Function of the Number of Chains in the Crystallite

RECENT INSIGHTS ON ORDERING

Because of the circumstantial nature of the above arguments, it seemed logical to pursue experiments which would provide more direct evidence as to the nature of the ordering process on heating near the Tcn.. In one approach, the effects of higher temperature annealing on the ordered structure of the copolyester systems were investigated. It seemed plausible that after some ordering at the original Tcn had occurred the increase in the transition would decrease the chain mobility in the crystallite, thus terminating further ordering. If indeed only partial ordering occurred then one might expect that further ordering might be induced by annealing at somewhat higher temperatures. In fact, this kind of process has now been clearly observed in the Xydar® series. Conversely, in the HBA/HNA system a completely unexpected degradation process has been identified (See next section).

A second approach to examine the nature of the ordered phase was to prepare a low M_N copolyester which could be ordered and then dissolved to permit use of solution techniques for more direct analysis of the mechanism for ordering.

DSC results of stepwise anneals for the SRT900 are shown in **Figure 5**. By increasing the second anneal temperature for the SRT900 sample from 280°C to 330°C, the resulting melting transition was 399°C, nearly 20°C higher than previously achieved for that system with one long anneal. Previous experiments showed that, even after annealing 60 hours at 290°C, the transition temperature never reached more than 380°C. With stepwise annealing, mobility of the chains was maintained within the crystallite, thus enabling the material to continue to order. A similar pattern was observed for the other Xydar® series copolyesters.

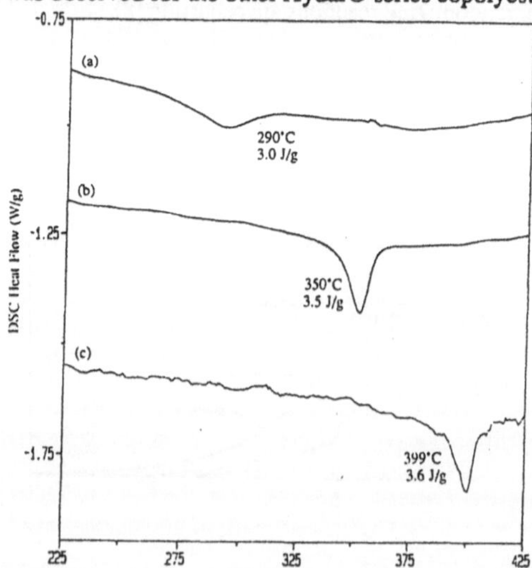

Figure 5 a) Randomized at 400°C for 15 minutes and annealed at 230°C for 10 hours, b) Sample "a" annealed at 280°C for 5 hours, c) Sample "b" annealed at 330°C for 10 hours

Based on these results we now believe that the increased melting point in these copolymers arises from some improved ordering of the sequences but not necessarily to the point where complete ordering occurs. During annealing near Tcn, as order within the crystalline regions increases, and the chains become better packed, it is reasonable to expect that chain mobility becomes suppressed. As mobility of the end groups is essential for the interchain transesterification reactions, the material cannot continue ordering at the original crystal nematic transition. It follows that one should be able to induce further ordering by increasing the anneal temperature close to the new transition.

A second approach to address the controversy surrounding the nature of the ordering process focused on synthesizing a soluble, low molecular weight form of the 50/50 HBA/HNA system using isophthalic acid as a capping agent (Mn ~ 4500). This oligomer exhibited the same ordering capability upon annealing near its crystal nematic transition, although the Tcn of the random copolymer was 236°C and the ordered copolymer had a transition at 279°C (See **Figure 6**). Fortunately the random and ordered oligomers were soluble in hot PFP.

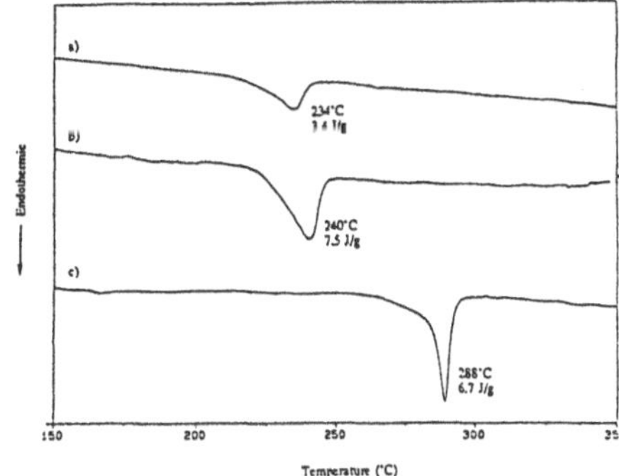

Figure 6 a) As prepared, b) (a) annealed 48 hours at 190°C, c) (b) annealed 48 hours at 230°C

An important insight into the nature of the ordered phase was obtained by precipitating the random and ordered oligomers from solution. DSC traces of the sample immediately following this procedure showed that the precipitated form of the control and the ordered sampler are significantly different, with transitions of 235°C and 250°C, respectively. Both specimens were then annealed 25°C below Tcn for a period of time adequate for diffusion processes, but not long enough for chemical processes to occur at these low temperatures. The control displayed a crystal nematic transition of 236°C, identical to the starting material. On the other hand, the ordered phase displayed a transition of 258°C (See **Figure 7**), a higher transition temperature, but not as great as in the original sample. These results provide further substantiation that ordering does not occur by a physical process but most likely by an ITR process.

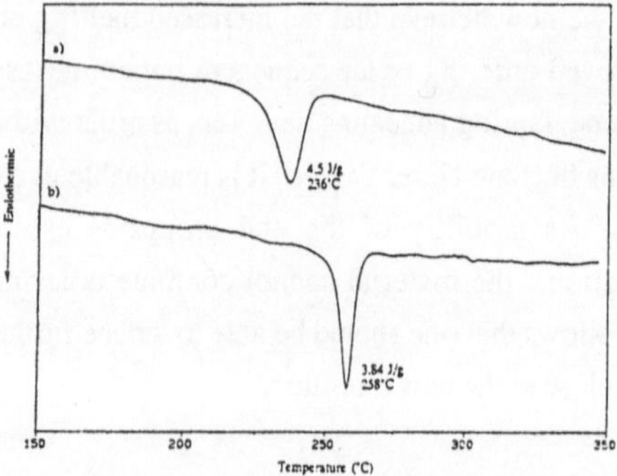

Figure 7 a) Random, Precipitated oligomer annealed 12 hours at 210°C, b) Ordered, precipitated oligomer annealed 12 hours at 225°C

THERMAL DEGRADATION OF THE HBA/HNA COPOLYESTER

An unexpected degradation process was observed in attempting to step-order the HBA/HNA copolyester system. Stepwise heating of the HBA/HNA 73/27, 50/50 and 24/76 compositions resulted in a complete loss of crystallinity, even though the annealing temperatures were well below the melting transitions and well below the original polymerization temperatures at 350°C.

A set of experiments were run to establish the length of time required for the disappearance of the endotherm upon annealing the ordered structures. In the case of the 50/50 HBA/HNA oligomer, it was found that this degradation occurred at 300°C in one hour (See **Figure 8**). When annealing the ordered 24/76 composition 20°C below its transition of 370°C, the transition enthalpy slowly decreases to zero over a period of 5 hours, although half of the endotherm is gone after only 1 hour (See **Figure 9**).

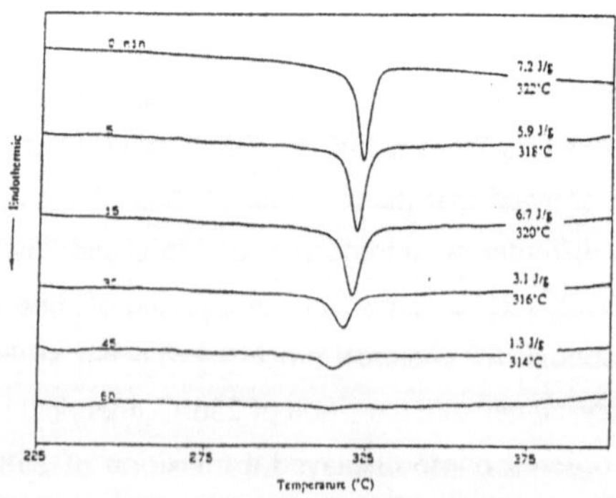

Figure 8 As received 50/50 HBA/HNA copolyester annealed at 200°C for 24 hours followed by 234°C for 24 hours, 270°C for 24 hours and then heated at 300°C for the indicated times

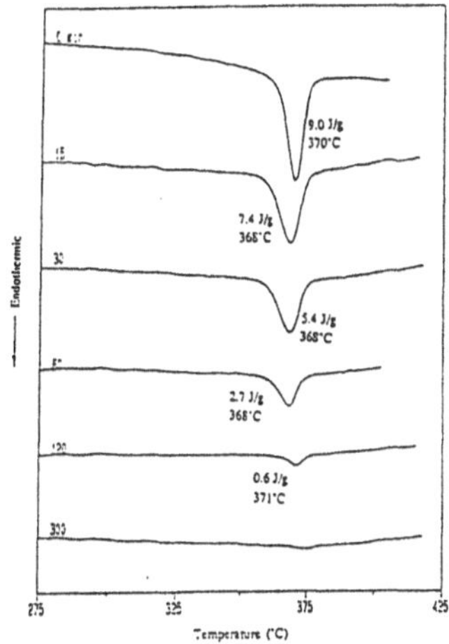

Figure 9 As received annealed in tube furnace: (1) 250°C for 24 hours, (2) 310°C for 24 hours. (3) 330°C for 18 hours, then annealed in the DSC at 350°C for the indicated times

Further analysis of this degradation phenomenum showed that these degradation reactions occur only within the highly ordered crystalline structure and not in the melt phase. A DSC scan of the as-received 24/76 sample first heated in the melt (at 350°C) for 10 hours, then annealed at 250°C for 24 hours shows the development of the random crystalline phase with an endotherm at 319°C. These results can be compared to the ordered material, which upon heating at 350°C for five hours and then 250°C again for 24 hours did not form the random structure. Additional characteristics of the degradation process are enumerated in Table 1.

TABLE 1
Characteristics of the Degraded HBA/HNA Copolyester

- Non-melting
- Lower molecular weight forms may soften >400°C
- All forms insoluble in pentafluorophenol
- Amorphous as indicated by X-ray and DSC
- No weight loss (350°C for 10 hours)

One possible interpretation of this behavior is that a thermal Fries reaction could occur in the aromatic copolyesters as a result of the closer chain packing in the ordered crystallite structure (orthorhombic). The Fries rearrangement would promote branching and crosslinking between chains, either by formation of a carbonyl unit with an adjacent chain or by ester interchange with the phenolic hydroxyl. One possible indication that thermal Fries reactions have occurred is a sharp discoloration of the polymer powder from a yellowish shade to a dark brown shade typical of formation of phenolic hydroxyls. This discoloration was observed with both compositions of the copolymer. Based on the preliminary data to date, it is reasonable to speculate that the non-melting character of these copolyesters arises from a modest amount of crosslinking resulting from Fries rearrangements at temperatures well below the conditions where these polymers normally decompose. The fact that this process is observed only in the HBA/HNA system suggests that the naphthalene unit may be more prone to these kinds of reactions.

ACKNOWLEDGEMENTS

We would like to thank NSF/DMR (MRL grant 89-20538) for providing the funding for this study. Thanks are also extended to Amoco and Hoechst for providing the samples used in this study.

REFERENCES

1 J. Economy, T. Gogeva, and V.J. Habbu, (1992) J. Adhesion, 37, 215

2 J. Economy and A. Andreopoulos, (1993), J. Adhesion, 40, 115

3 D. Frich and J. Economy, (1993) ACS Polymer Preprints, 69, 438

4 J. Economy, C. Goranov, D. Frich, J.S. Lim, PMSE PREPRINTS, Spring 94

5 J. Economy, R.D. Johnson, J.R. Lyerla, and A. Mühlebach, (1990), ACS Symposium Series No. 465, Editors R.A. Weiss and C.K. Ober, Chapter 10, 129

6 M.T. DeMeuse and M. Jaffe, (1988) Mol. Cryst. Liq. Cryst. Inc. Nonlin. Opt., 157, 535

7 R.W. Lenz, J.I. Jin, and K.A. Feichtinger, (1983) Polymer, 24, 327

8 A Mühlebach, J. Economy, R.D. Johnson, T. Karis, and J. Lyerla, (1990) Macromolecules, 23, 1803

9 G. Chen and R.W. Lenz, (1985) Polymer 26

10 D. W. Wiswe, H.G. Zachmann, K. Hahn, and E.W. Fischer, (1987) Macromolecules, 20, 1116

11 S.Z.D. Cheng, (1991) Macromolecules, 24, 2877

12 Y.G. Lin and H.H. Winter, (1991) Macromolecules, 24, 2877

13 Y.G. Lin and H.H. Winter, (1988) Macromolecules, 21, 2439

14 A. Kaito, M. Kyotani, and K. Nakayatama, (1990) Macromolecules, 23, 1035

15 G.R. Mitchell and A.H. Windle, (1985) Colloid. Polym. Sci. 263, 230

16 J. Economy and K. Goranov. "Thermotropic Liquid Crystalline Polymers for High Performance Application," submitted for publication August 1993, P. Hergenrother, ed. High Performance Polymers, Springer-Verlag.

17 J. Kachidza, (1991) "Ordering and Randomization Processes in the Copolyesters of 4-Hydroxybenzoic Acid and 2,6-Hydroxynapthoic Acid," Thesis, University of Illinois at Urbana-Champaign

18 R.D. Johnson, J. Economy, J. Lyerla, A. Mühlebach, (1989) Presented at the APS Mtg. in St. Louis, MO

19 J. Blackwell, H.M. Cheng, and A. Biswas, (1988) Macromolecules, 21, 39

20 Fischer and J. Economy, (1994) Pol. for Adv. Tech., in press

21 D.J. Wilson, A.H. Windle, and H.G. Zachmann, (1993) "crystallization Behavior of a Thermotropic Random Copolyester at Elevated Temperatures," International Conference on Advanced Polymer Materials, Dresden, Germany

22 D. Yoon, Private Communication

23 J. Economy, W. Volksen, C. Viney, R. Geiss, R. Siemens, and T. Karis, (1988) Macromolecules, 21, 2777

Performance on Demand - A New Era in Polymer Science (A Case Study Using Conducting Polymers)

Prof. G.G. Wallace

Intelligent Polymer Research Laboratory, Department of Chemistry, University of Wollongong, Northfields Avenue, Wollongong, NSW, 2522, Australia

The notion of **Performance on Demand** is not alien to polymer research scientists. Usually however, the demand has come from engineers and has been aimed at the polymer scientist with a view to achieving a desired level of performance from both researcher and material. In most instances the time lag associated with the demand and the response has been much longer than either the engineer or the researcher would like.

A new era is emerging. Numerous technologies searching for new generation materials now request that the performance specifications of polymers are such that properties (chemical and/or physical) can be changed on demand, in-situ with an appropriate (short) response time. Nature has mastered this approach. Nature uses dynamic polymer systems to ensure that basic functions such as sensing, movement, growth and repair are possible. However, as molecular engineers, we have yet to come to grips with how we can assemble synthetic structures capable of such complex behaviour.

How can we create structures capable of performance on demand?

At least two approaches are obvious. The first involves the design and development of material systems capable of responding to stimuli that are regulated and imposed by human intervention. The second approach involves the production of materials capable of sensing, processing information and actuating an appropriate response without human intervention. In this latter case the material is said to assume some level of intelligence.

This article, will deal with the former (simpler) approach which requires the polymer to have sufficient dynamic character to perform on demand in response to application of external stimuli by us humans. We see this as the first step in the creation of truly intelligent materials. We also see that this simpler approach will in itself have major ramifications on numerous technologies. For example:

Controlled Chemical Release Systems: The controlled release of drugs or agricultural chemicals from appropriate polymer formulations is of interest to numerous researchers and pharmaceutical companies (1-2). The ability to program release profiles by the application of external stimuli enables the design of more effective and more efficient systems.

Separation Technologies: The use of polymer membranes in chemical and biological processing systems has increased dramatically in the past 20 years (3, 4). Traditionally separations were based on physical differences eg., particle/molecular size. More recently, chemically active membranes have been employed. It is

K.P. Ghiggino (Ed.)
Progress in Pacific Polymer Science 3

recognised that the ability to effect a more complicated chemical separation process relies on the use of multifunctional materials. The ability to introduce and control such multifunctional characteristics in-situ adds another powerful dimension to separation for chemical and biotechnological applications. For example in the minerals processing area the use of membranes where selectivity and transport characteristics can be controlled in-situ enables multistream processing to be implemented in a single unit. In the protein area the separation of complex macromolecules may be possible if different modes of interaction can be introduced at predetermined intervals on an appropriate time scale.

Cell Culturing: The development of controllable dynamic polymer surfaces should revolutionise our approach to cell culturing both for biotechnological and biomedical purposes. The ability to trigger cellular events in-situ can lead to improved efficiency in culturing or removal/release of products generated by cells. It may also be possible to spatially direct cell growth enabling complex repair processes (eg. nerve repair) to be initiated and controlled in-situ.

Sensing Technologies: The development and use of polymer sensors has increased dramatically in recent years (5, 6). Polymers have proven particularly useful in providing complicated recognition capabilities. As for membranes (see above) the development of sensing systems whose performance can be tuned in-situ provides the basis of more generic and powerful sensing systems. The adjustment of properties at a molecular level enables the possibility of tuning into particular analytes. The oscillation of chemical properties, even during a recognition process can be used to provide more sophisticated and more powerful analyte recognition systems.

If one accepts that controllable dynamic behaviour is in fact desirable then groups of materials that can be monitored and controlled as required using tools currently at our disposal must be identified.

Polymers being the most versatile and dynamic of all materials have a crucial role to play in this new area of science and technology. Obviously, even within this class of materials certain groups will emerge as being most appropriate. We believe conducting electroactive polymers (CEPs) is one of these groups of materials. Polymers such as polypyrroles (I), polythiophenes (II) and polyanilines (III) have unique properties that enable their dynamic chemical properties to be studied, manipulated and controlled.

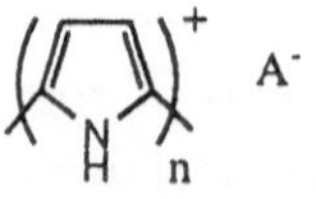

(I) 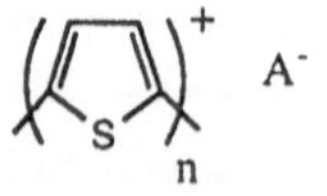 (II) 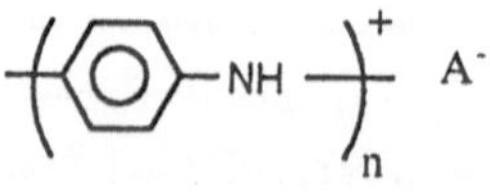 (III)

[A^- is a counterion incorporated during synthesis. n is between 2-4].

In this article we report on some recent advances in our laboratories in using conductive electroactive polymers as materials capable of performance on demand. The basis of the ability of these materials to respond to electrical stimuli is the fact that they are conductive and electroactive. The conductivity allows electrical stimuli to be delivered throughout the bulk of the material. The response mechanism is the actual redox reaction induced by the imposition of an appropriate electrical stimulus. One such response involves release of counterions, according to:

$$\left[\text{pyrrole}\right]_n^{+} A^- \rightleftharpoons \left[\text{pyrrole}\right]_n + A^- \quad \ldots (1)$$

The electrodynamic (performance) parameters will be determined by ;

- the energy required to induce a transition.
- the kinetics of the transition, and
- the chemical nature of the transitions.

Each of these may be manipulated during synthesis. Several of the variables available are discussed below:

The synthesis and characterisation of these materials is discussed, and some advances in application areas being developed in our laboratories are considered. In essence three critical areas are addressed, these are the ability to:

- **Create,**
- **Communicate, and**
- **Control.**

the structure.

CREATION - SYNTHESIS AND PROCESSING

Conducting electroactive polymers (CEPs) such as polypyrrole can be synthesised chemically or electrochemically, according to:

$$\text{pyrrole} \xrightarrow[\text{OXIDISE}]{A^-} \left[\text{pyrrole}\right]_n^{+} A^- \quad \ldots (2)$$

This simple yet elegant polymerisation process allows the dynamic properties (and hence the ability to obtain performance on demand) to be (pre)determined at the time of assembly.

There are several key variables that can be manipulated during polymer assembly in order to alter the properties of the system. For example:

The Monomer: The chemical properties such as the ion exchange behaviour or the ability to undergo hydrophobic interactions are influenced by the choice of the monomer. For example polyaniline has been shown to be less hydrophobic then polypyrrole (7, 8). The electrochemical properties are also affected. Both

PPy and PTh undergo a one electron transition according to the reaction denoted in equation 1. However polyaniline is known to undergo two reversible electrochemical transitions (8).

Even with PPys or polythiophenes the presence of other functional groups on the 3 or 4 position can be used to adjust the chemical properties and the electrochemical switching properties. For example the addition of alkyl groups results in polymers that are more hydrophobic (9) and also more soluble in common organic solvents. We have also shown that the addition of carboxy groups introduces self doping to the polymer and therefore cation exchange properties (10, 11).

The Counterion: The counterion incorporated into the polymer during synthesis influences both the chemical and electrochemical properties of the material. For example with polypyrroles the incorporation of simple anions such as chloride produces an anion exchange material (12). However, incorporation of larger counterions such as dodecylsulfate (13) or polyvinylsulfate (14, 15) produce cation exchange materials.

Physical Form: The physical form in which conducting polymers may be produced is important in that for most applications they must be integrated with other materials in order to a mechanically stable and practically viable structure. Most CEP's are not soluble in common solvents. Alkylated polypyrroles can be dissolved in common solvents such as CH_2Cl_2, $CHCl_3$ or xylene. Most polyanilines are soluble in N-methyl pyrrolidone and when some counterions are used eg. camphor sulfonic acid then solubility in cresol can be induced.

In our laboratories we have been interested in developing innovative methods of CEP processing. Two of these studies involve:

- Preparation of conducting colloids.
- Preparation of conducting polymer - hydrogel composites.

Previously conducting polymer colloids have been produced using chemical means (16). This involves oxidation of the monomer using a soluble oxidant such as $FeCl_3$ in the presence of a steric stabiliser such as polyethylene oxide. Using this approach spherical monodisperse colloids can be produced. However with the use of chemical oxidants the number of counterions (A^- equation 2) that can be incorporated is limited and polymers produced using chemical oxidants are usually less conductive than those produced electrochemically. We have now developed a method by which conducting polymers can be produced electrochemically (17). Using a flow through cell (Figure 1) and with stabilisers present in the electrolyte conducting polymer colloids can be produced electrochemically. A range of conducting polypyrrole colloid with different counterions (eg. NO_3^-, Cl^-, paratoluene sulfonate) have been produced. In addition polyaniline colloids have been synthesised using this flow through electrochemical approach.

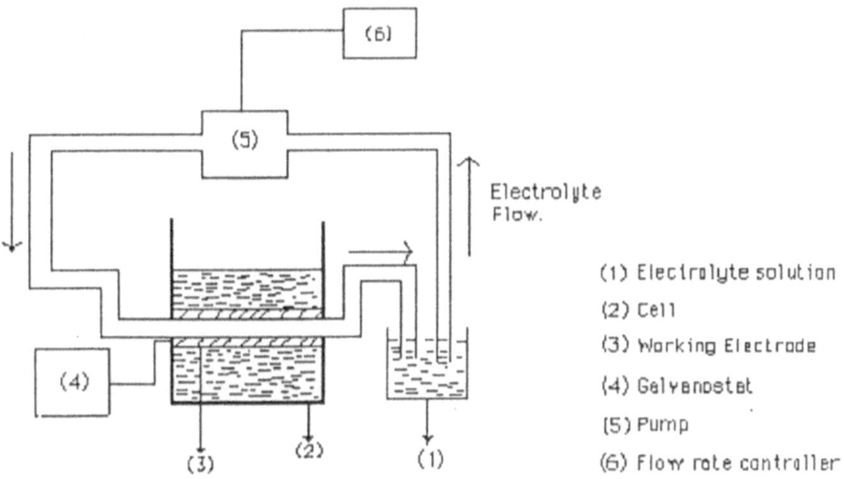

Figure 1: Flow through polymerisation cell for colloid production.

We have also discovered that CP-hydrogel composites can be sythesised electrochemically (18). This involves casting electrodes into the hydrogel during gelation. The gel containing electrodes is then placed in a monomer solution and polymerisation can be initiated electrochemically. The high water content of the gel allows electropolymerisation to occur within the structure (Figure 2). These structures retain the properties of hydrogels and conducting polymers and this is an excellent example of conducting polymer integration.

Figure 2: Hydrogel cell set up.

LINES OF COMMUNICATION: The fact that they are electroactive polymers (Equation 1) provides a direct means by which their chemical and electrical properties can be controlled. This direct line of communication coupled with novel in-situ techniques provide a means by which the polymer properties can be studied and manipulated both during and after growth.

However, the process described in equation 1 is worthy of further investigation. The fact that we can communicate directly with CEPs using electronic means gives us some insight into the behaviour of these materials and a closer examination of cyclic voltammograms (plots of current flow versus potential) indicates that the above equation is simplistic and even misleading.

Consider the voltammogram (plot of potential applied versus current flow) showing the oxidation/reduction of simple polypyrrole in Figure 3. This voltammogram was obtained using a conventional sized electrode and appears to reveal a single oxidation/reduction process. It was such information that led to the simplistic description in equation 1. However using these conventionally sized electrodes, high concentrations of electrolyte were used and as a consequence important information lost. For example, Figure 3b shows the cyclic voltammogram obtained when a microelectrode is used and hence lower concentrations of electrolyte can be employed (19). Two distinct oxidation/reduction couples are now obvious and the potential shift observed with varying electrolyte concentration indicates that one is related to anion incorporation/expulsion the other due to cations (19). Another interesting feature is obvious on both voltammograms. As soon as the potential is swept in a negative direction substantial current flow commences. Initially this was attributed to charging current, simply movement of ions with no direct polymer reduction occurring. An alternative explanation is that the polymer is reduced sequentially from +0.70 V as we scan negative. That is the polymer has a number of sites with a series of E^{o} values and different sites are activated at different potentials - causing ion movement - at different potentials.

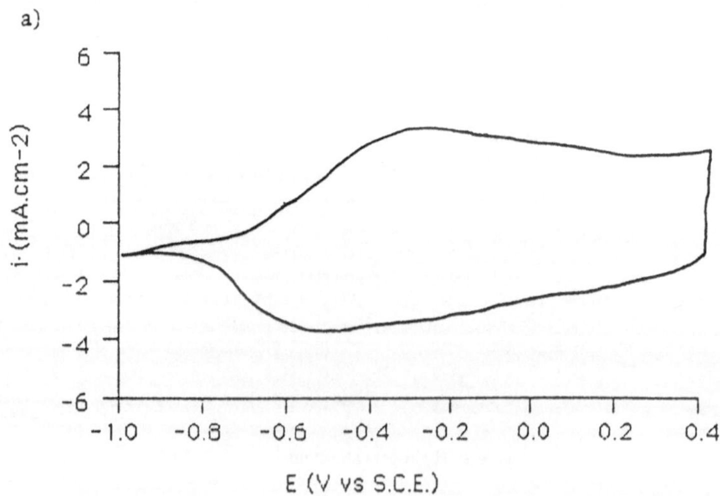

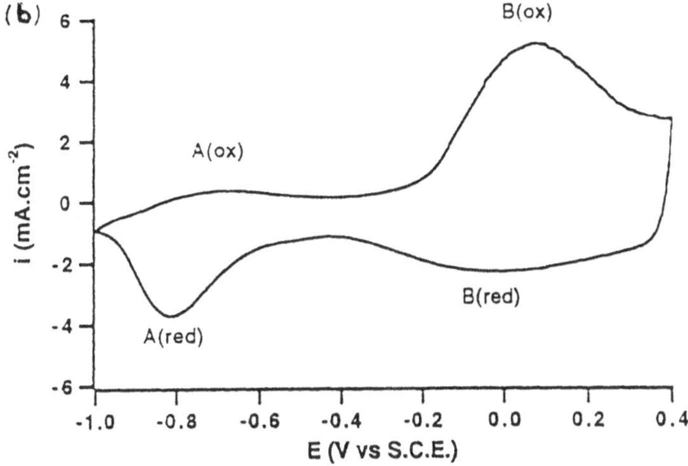

Figure 3. Cyclic voltammogram showing oxidation/reduction of polypyrrole (a) macroelectrode, (b) micorelectrode

Some new communication tools help shed more light on this dynamic behaviour. Among these new techniques are Electrochemical Quartz Crystal Microbalance (EQCM) and resistometry.

EQCM - This technique can be used in-situ to allow small changes in the mass of the polymer to be monitored as electrical stimuli are applied. The system is based on monitoring the changes in the frequency of oscillation of a quartz crystal as the mass of the polymer deposited on the crystal changes (14). This enables us to now monitor both current flow and changes in mass (Figure 4) as the polymer is oxidised and reduced. Note, that the continuous reduction theory is now verified. Mass changes are observed as soon as the potential travels in a negative direction and this continues, interestingly, almost linearly. As the polymer is reoxidised it increases almost linearly from the onset of oxidation.

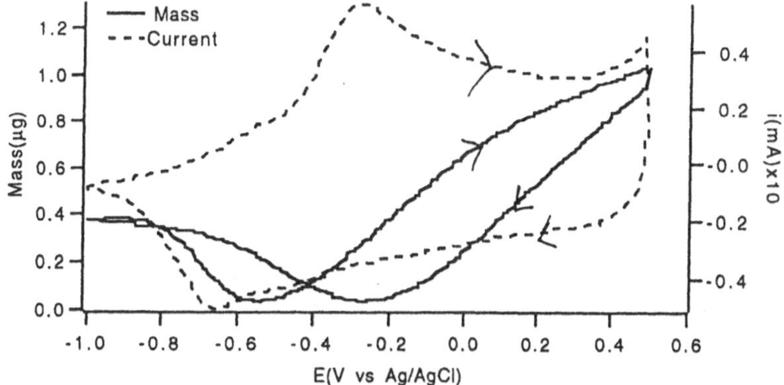

Figure 4: QCM - Simultaneous Mass-Current profile for PPy/PTS (exposed to a potential ramp) in 0.2M NaCl, Scan Rate =10 mVs^{-1}

The competitive nature of anion/cation movement is also obvious even in simple polymer counterion systems. If the potential is scanned more quickly from -1.00 to +0.70 V the anion, with now less time to move, is not ejected so efficiently - a potential is then reached where charge compensation is achieved by cation inclusion and expulsion.

The use of resistometry, another in-situ technique, provides further information on this process. Resistometry allows changes in the resistance of the bulk polymer to be monitored in-situ as electrical stimuli are applied (20). From voltammetric data it was originally assumed that *i* flow and resistance changes were synchronised processes. Resistometric experiments that allow both *i* and R to be monitored at the same time reveal that this is not the case (Figure 5). There is an obvious time lag between the current flow and the resistance transition. Upon reflection this is understandable since the current flow is merely the process that instigates changes in bulk polymer properties and it is these changes that are reflected in the change in resistance.

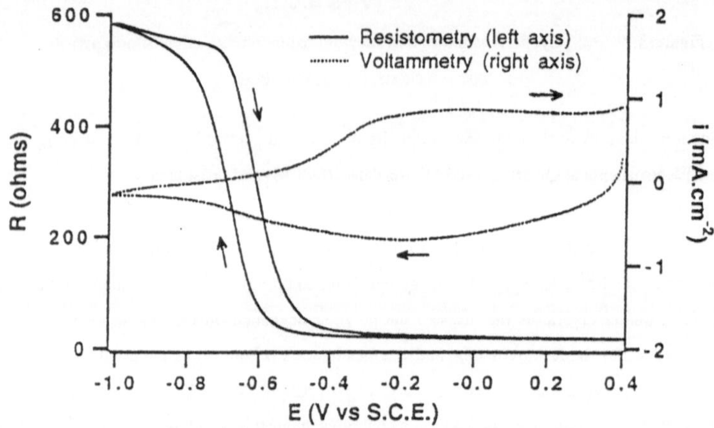

Figure 5: Cyclic voltammogram and corresponding cyclic resistogram recorded in 1.0 $molL^{-1}$ KCl, Scan Rate=20 mV s^{-1}

We have now developed a Simultaneous Multidimensional Analyser for Conductors (SMAC) that allows current flow, mass changes and resistance transitions that occur upon application of electrical stimuli to be monitored (21). This enables the system to be monitored in-situ in real time.

Another interesting technique developed by us in recent years is Inverse Liquid Chromatography (ILC). Using carefully selected molecular probes - the nature of the molecular interactions (eg. ionic, hydrophobic, etc) occuring at CEP/solution interfaces and the effect of electrical stimuli on them can be determined. For example, a simple series of molecules based on benzene with different functional groups attached has been used to probe hydrophobic, ionic, H bonding, and electron donor interactions (22). A series of polyaromatic hydrocarbons has been used to probe the effect of molecular geometry (eg. the length/breadth of the molecule) on the ability to undergo interactions (23). It is even possible to monitor complex bio molecular interactions such as antibody - antigen interactions and to study the effect of applied potential on this using chromatographic methods (24).

CONTROL

Having created conducting polymer structures and developed appropriate lines of communication. We can assemble systems whereby the material's properties can be controlled in-situ. Then performance on demand is possible. The degree of control available and how this can be utilised is best illustrated by considering some areas of application currently being developed in our laboratories.

Separations - A new membrane separations technology wherein the transport properties of the material can be manipulated in-situ has been developed. This technology utilises the ability that transport can be initiated and stopped (Figure 6). This is achieved by using the electrical stimuli to load up and then to expel ions from the polymer membrane (25, 26). The selectivity of transport can also be tuned in-situ by adjusting the magnitude of the potential and the frequency of pulses applied. For example in the experiment involving transport of K^+ and Na^+.

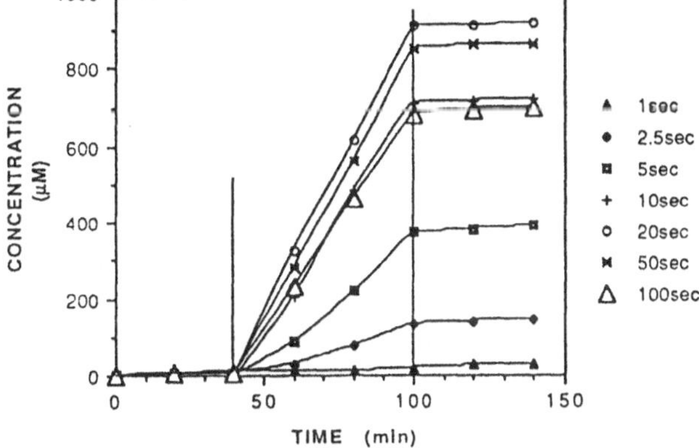

Figure 6: Transport of K+ across PPy/PTS membrane. Source solution: 0.2M KCl, Receiving solution: deionised water. Symmetric pulsed potential waveform with different pulse widths were applied. Potential Range: +0.7V to -1.0V (vs Ag/AgCl)

This ability to change transport properties will have implications in areas as diverse as chemical processing to biotechnology. In the former area, electromembrane systems that can be used to carry out complex chemical reaction and separation processes. Increased efficiency will be achieved since a single system can be modified in-situ to enable multistep processes to be carried out. In the biotechnology area we have recently discovered that these dynamic polymers are biocompatible and that true cellular communications are possible (27). Mammalian cells can be cultured directly on conducting polymer surfaces and then the material can be stimulated to initiate and regulate cell growth and differentiation.

Sensors

The same dynamic properties that enable performance on demand have been used in the devlelopment of a new chemical sensing technology (28, 29). The use of potential stimuli to control molecular interactions in-situ has been used as the basis of sensors suitable for environmental monitoring of species such as phosphate and nitrate to the monitoring of proteins of biomedical significance (29). In this latter area the ability to control biomolecular interactions in-situ has been particularly imporant - resulting in the development of a rapid - powerful - new biosensing technology. Systems for detection of medium sized proteins such as human serum albumin, small proteins, thaumitin, an artificial sweetener and even small molecules, p-cresol, have been developed. In this latter case the antibodies are raised against the p-cresol / bovine serum albumin conjugate. Some flow injection analysis responses to this conjugate are shown in Figure 7. Sensitive reproducible and rapid responses are obtained.

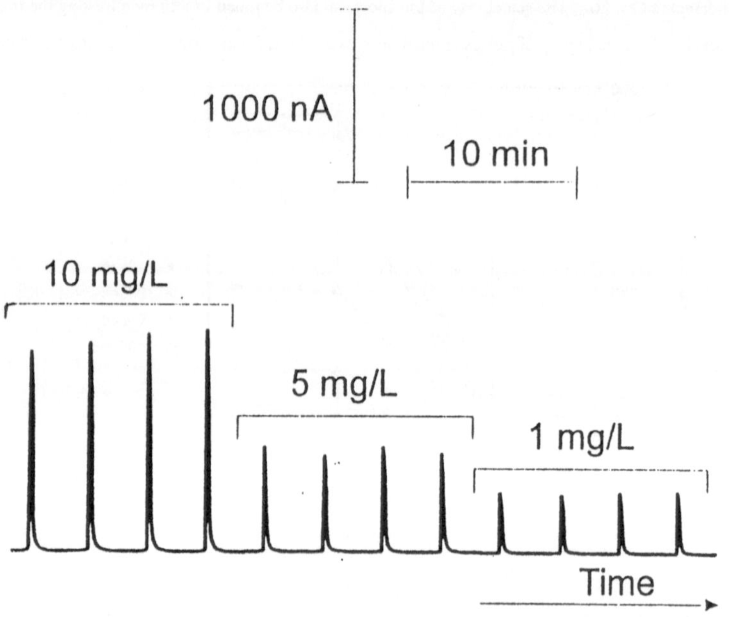

Figure 7: Flow injection analysis responses obtained for p-cresol/Bovine serum albumin conjugate

CONCLUSIONS

There is much to be gained from the use of dynamic systems whose properties can be controlled to produce performance on demand. There is a need for continued refinement of our approach and the tools we use to create, communicate and control these structures.

REFERENCES

1) Illum L, Davis SS, (Eds) (1987) Polymers in Controlled Drug Delivery Systems, IOP, Bristol.

2) Anderson JM, Kim SW, Krutson K (1990) Advances in Drug Delivery Systems, 4, Elsevier, Amsterdam.

3) Mearcs P, (1976) Membrane Separation Processes, Elsevier, New York.

4) Other membrane book - several missing from my office.

5) Imisides MD, John R, Riley PJ, Wallace GG (1991) Electroanal., 3:879-889.

6) Wallace GG, (1993) Analyst, 118:329-334.

7) Teasdale PR, Wallace GG, (1994) Reactive Polymers, Submitted.

8) Teasdale PR, Wallace GG, (1993) PhD Thesis, University of Wollongong.

9) Ge, H, Gilmore, KG, Wallace, GG (1994) J.Liq. Chrom., In Press

10) Ge, H, Ashraf, S, Gilmore, KG, Too, CO, Wallace, GG (1993) J. Liq.Chrom., 16, 7.

11) Ge, H, Wallace, GG (1992) Reactive Polymers, 18, 133.

13) Zhao, H, Price, WE, Wallace, GG, React. Polym., Submitted.

14) Lien, M, Smyrl, WH, Morita, M (1991) J. Electroanal. Chem. 309, 333.

15) Mirohseni, A, Price, WE, Wallace, GG (1993) Polymers Gels Networks, 1, 61.

16) Eisazadeh, h, Spinks, G, Wallace, GG (1992) Mat. For., 16, 341.

17) Eisazadeh H, Spinks G, Wallace GG. (1994) Polym. Comm., Submitted.

18) Gilmore, KG, Hodgson, AJ, Wallace, GG (1994) Polymer Gels and Networks, Submitted

19) John, R, Wallace, GG (1993) J. Electroanal. Chem., 354, 154.

20) John, R, Talaie, A, Fletcher, S, Wallace, GG (1991) J. Electroanal. Chem., 319, 365.

21) Talaie, A, Wallace, GG (1994) Proc. Interfaces II, In Press

22) Ge, H, Wallace, GG (1991) J. Chrom., 588, 25.

23) Hawick, C, Wallace, GG (1994) Chromatographia, In Press.

24) Hodgson, AJ, Lewis, TW, Maxwell, KE, Spencer, MJ, Wallace, GG (1990) J. Liq. Chrom., 13, 3091.

25) Talaie, A, Sadik, O, Wallace, GG (1993) J. Int. Mat. Sys. Structures, 4, 43.

26) Price, WE, Wallace, GG, Zhao, H (1992) J. Electroanal. Chem., 334, 111.

27) Aoki, T, Hodgson, AJ, Gilmore, KG, Wallace, GG (1994) Mater. Chem., Submitted.

28) Wallace, GG (1993) Analyst, 118, 329.

29) Wallace, GG (1993) Chem. Brit., Nov. 967.

Reversible Optical Storage in Azo Polymers

A. Natansohn and P. Rochon

Department of Chemistry, Queen's University, Kingston, Ontario K7L 3N6 and Department of Physics, Royal Military College, Kingston, Ontario K7K 5L0, CANADA

Abstract: Amorphous high-Tg azobenzene-containing polymers can be used as reversible information storage materials in form of films. The procedure takes advantage of the photochemically activated trans-cis-trans isomerization of the azobenzene groups. The written information is stable below the polymer glass transition temperature. Writing and erasing can be performed up to 10^5 times on the same spot of the polymer film. The reorientation mechanism depends on the type of substituents bound to the azobenzene group, on the type of the neighboring group on the polymer chain and on the tendency of the polymer to self-organize. The efficiency of the process depends on the maximum absorbance of the azobenzene group and can be tailored to reach unity. This property of azobenzene-containing polymers can be used in optical and holographic memory, waveguides and sensors, and photonic devices.

INTRODUCTION

It is well known that azobenzenes isomerize from trans to cis and from cis to trans when subjected to light of appropriate wavelength (1). This property has been the basis of a series of applications in polymer science, such as the use of azobenzene groups as probes for various types of motion within polymer chains. A review of azobenzene photochemistry and some of its applications in polymer chemistry appeared in 1989 (2). More recently, some novel potential applications involving azobenzene-containing polymers were reviewed (3). They are based on supramolecular orientation of the rigid azobenzene groups. For example, azobenzene mesogens form various liquid crystalline phases and their order can be destroyed by illumination, because the cis form of the azobenzene obtained by photoisomerization has a bent shape and cannot form ordered domains. In another example, the azobenzene mesogens can be aligned by poling in an electric film to give materials with second order nonlinear properties. Photoisomerization can also be exploited in switching the ordering of Langmuir-Blodgett films containing azobenzene groups.

Our research is using high glass transition temperature amorphous polymers in which substituted azobenzenes are bound on the side chain or in the main chain, as thin films. The films are subjected to polarized light, which induces trans-cis-trans isomerization cycles and subsequent rearrangement of the azobenzene groups preferentially perpendicular to the polarization direction. This preferential director of the azobenzene groups creates a dichroism

K.P. Ghiggino (Ed.)
Progress in Pacific Polymer Science 3

which translates into birefringence. Original disorder can be easily restored by switching the polarization of the light from planar to circular (4-6).

Background And Previous Literature

The reorientation of azobenzene groups subjected to polarized light was first observed in 1983 by Todorov et al (7) using methyl orange dissolved in poly(vinyl alcohol). The phenomenon was very unstable, decaying almost immediately as the laser was switched off. Independently, in 1987, Ringsdorf et al (8) reported a similar phenomenon in liquid crystalline side-chain polymers. The polymer sample was a film which was initially subjected to an electric field to obtain a general director for the nematic or smectic organization of the azobenzene groups. After this preorientation, the film was subjected to light of polarization parallel to that of the director. The azobenzene groups orientation was thus changed at 90° from its initial director. A series of papers followed from Wendorff's laboratory in which the phenomenon was shown to be applicable to a variety of liquid crystalline polymers containing substituted azobenzene groups. Erasing the orientation was performed by heating the polymer film above its Tg or (more efficiently) above its clearing temperature.

Our initial reports (4-6) concentrated on **amorphous high-Tg** polymers, for a few reasons. First, if an amorphous polymer is used, there is no need (or possibility) to preorient the azobenzene groups in an electric field. Second, the orientation induced by the laser is stable below the polymer glass transition temperature. If **no spacers** are present between the azo groups and the polymer backbone, the polymer can be designed to have a very high Tg, hence a very good stability of the "written" orientation. A very important point is that azobenzenes which are **bound within the polymer** will keep their orientation much better than azobenzenes simply dissolved in the polymer. Our oldest sample has kept its "written" birefringence for almost three years now. Probably the most important improvement in our system is the use of **circularly polarized light** to erase the signal. This allows for local erasure, as opposed to complete erasure of the whole film in the case of the liquid crystalline polymer system.

Brief Description of the Phenomenon

The rate of trans-cis isomerization depends on the angle between the laser polarization and the azobenzene group orientation by a $\cos^2\Theta$ factor. Azobenzenes perpendicular to the laser polarization direction are not activated. While undergoing multiple trans-cis-trans isomerization cycles, the azobenzene groups change their orientation. Any group falling perpendicular to the polarization direction will remain in that position, while all others will be reactivated. In the end, there will be a slight increase in the concentration of azobenzene groups perpendicular to the polarization direction, which will give the dichroism and birefringence. When the laser light is turned off, dissipation of energy by heat reduces the birefringence by ca. 20%, but the remainder is frozen in the film for a long time. Circularly polarized light activates all azobenzene groups, restoring the initial disorder of azobenzene orientations. A schematics of the process is presented on the next page:

Random orientation (disorder)	planarly polarized light → ← circularly polarized light	**Supramolecular order**

REORIENTATION MECHANISM

Parameters

There are a few important parameters governing the extent and rate of reorientation of the azobenzene groups in an amorphous polymer film.

The nature of the azo group: The most important structural feature of an azobenzene group is the type of substituents it possesses. If there is no amino substituent, or if there are no electron-donor - electron-acceptor substituents to confer a push-pull electron movement, the cis isomer is relatively long lived. If such substituents are present (aminoazobenzenes or stilbene-like azobenzenes), the cis isomer reverts back to trans in a few seconds at room temperature, which means that its absorbance and other characteristics cannot be directly measured. Typically, the maximum absorbances of cis and trans isomers of an aminoazobenzene or of a stilbene-like azobenzene are superimposed, while nonsubstituted cis azobenzenes have absorbances at higher wavelengths than the trans isomers. Thus in the nonsubstituted azobenzenes, the trans-cis and cis-trans isomerization processes can be separately activated.

An example on how the reorientation process is affected by the type of azobenzene used is given below. Poly[4'-[[2-(acryloyloxy)ethyl]ethylamino]4-nitroazobenzene] (pDR1A) (5) and poly[4-(2-methacryloyloxy)ethyl-azobenzene] (pMEA) (9) are subjected to irradiation with a polarized argon laser at 514 nm.

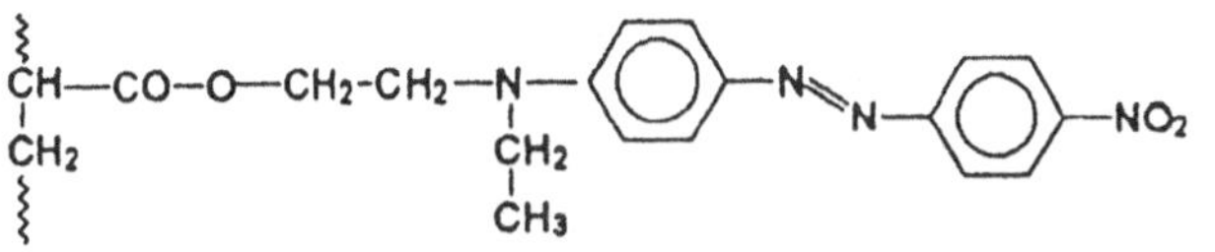

CH—CO—O—CH$_2$-CH$_2$—N(CH$_2$CH$_3$)—C$_6$H$_4$—N=N—C$_6$H$_4$—NO$_2$ (backbone CH$_2$) pDRIA

CH$_3$—C(CH$_2$)—CO—O—CH$_2$—CH$_2$—C$_6$H$_4$—N=N—C$_6$H$_5$ pMEA

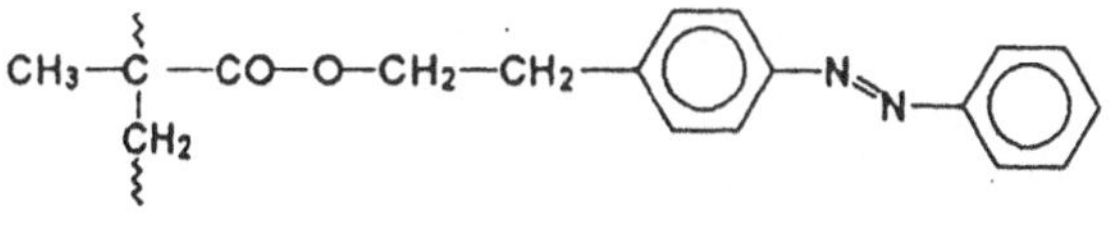

The writing/erasing curve is very similar in both cases, however there are significant differences in the level of birefringence achieved and the time required to achieve it. pDR1A has a maximum absorbance at 490 nm, very close to the laser wavelength. pMEA has a maximum absorbance at 330 nm, while its cis isomer absorbs at 450 nm. pDR1A absorbs the laser light more efficiently, thus the saturation level of birefringence is 0.08 and is achieved, depending on the film thickness, in about 100 ms. pMEA has only a tail of its absorbance at 514 nm. Its maximum level of birefringence is 0.009 and it takes over 10 seconds to achieve it. This experiment is done with a laser power of 3 mW/cm^2. pMEA can also be "written" with UV light.

Another example in which the nature of the azobenzene determines different "writing" behavior has been published previously (6).

The polymer structure: This is obviously very important, as it determines the potential mobility of the azobenzene group, the stability of the "writing" (by its glass transition temperature) and the overall thermal stability of the material.

The film thickness: Depending the absorbance of the azo group and on the wavelength of the laser, thicker films may present a nonlinear behavior which was described in another publication (10). There is a maximum workable thickness of the film for a certain wavelength: $4/\alpha_o$, where α_o is the absorption coefficient of the film. Obviously, pMEA films will have a greater useful thickness than pDR1A films at 514 nm, due to the great difference in absorption at this wavelength.

The azo group content: Intuitively, the more azo groups are present in a sample, the greater will the achievable birefringence be. This was confirmed in our experiments and in others found in the literature. We have also observed a neighboring group effect associated with increasing azo content, and this effect will be described in a following subheading.

The self-organizing tendency of the polymer: The literature contains examples of semicrystalline polymers (11,12) in which much higher levels of birefringence (0.2-0.3) can be attained. This is achieved because "writing" on such films reinforce the thermodynamic tendency of these polymers to form organized domains. We will show in a following subheading that in the process of reorientation, some other rigid groups present in the polymer are moved by "sympathy", and that this cooperative motion enhances the effect created by the laser.

Efficiency and Stability

Figure 1 illustrates the result of an experiment with a pulsed laser. The aim of the experiment was to determine how many times one can "write" on the same spot of the film before the film is damaged. It is clear from Figure 1 that the determining factor is not the number of times one can "write and erase", but the overall power input into the film. With a power of 5mW, the film is damaged earlier than with a power of 1 mW, even though the number of cycles at 10Hz and at 100 Hz is quite different. For 100 Hz, one can see that 10^5 cycles can be attained (with 1 mW power) which indicates good long term usage of such a material. There probably is a correlation between the overall thermal stability of these films and the resistance under a laser light, but this correlation is still to be established. A typical pDR1A film starts decomposing when heated above 180°C.

The overall thermal stability of the induced reorientation (of the "writing") is illustrated in Figure 2. It is clear that heating of a "written" sample destroys the induced orientation, and as one gets closer to the glass transition temperature, all reorientation is lost.

When trying to "write" at different temperatures, the amount of birefringence remaining after the laser is switched off clearly depends on how far from Tg is the writing temperature. This experiment illustrates how important is to synthesize polymers with high glass transition temperatures to be used in this reversible optical storage process.

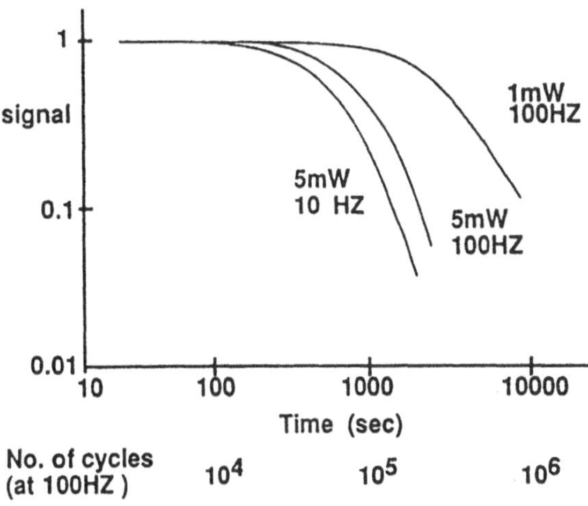

Figure 1. Normalized signal (proportional to the square of the birefringence) as a function of the time the pulsed laser is kept ON on the same spot on a film of pDR1A.

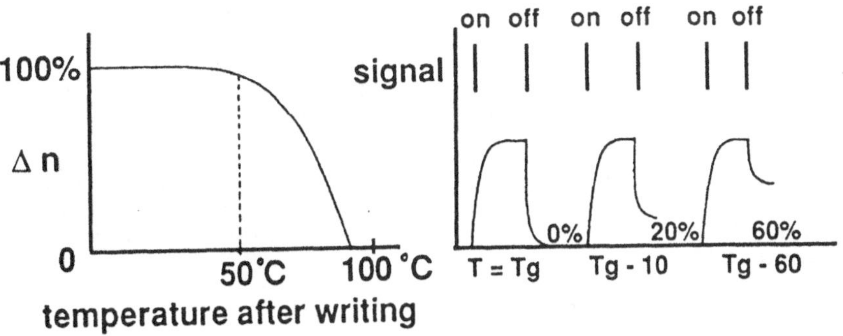

Figure 2. Thermal stability of "writing" in a pDR1A film. a. Birefringence as a function of temperature. b. Normalized signal in a sample written at about Tg, about 10 degrees below Tg and at about 60 degrees below Tg (room temperature for pDR1A).

The efficiency of the process can be best described by the diffraction efficiency obtained when inscribing a phase grating (as in holographic storage). This is defined as the ratio between the amount of probe light which is diffracted by the grating and the amount of incident light (13). The efficiencies for pDR1A at 514 nm are in the range of 10^{-4} for films of 0.3 nm thickness. This type of efficiency depends on the film thickness (the thicker the film the more efficient the diffraction), and one has to take into account the fact that at this wavelength pDR1A is absorbing, hence its useful thickness is very low. The best photorefractive polymer reported in the literature has an efficiency of 1.2% (14) for a 125μ thick film. When normalizing, the square root of efficiency per cm is 8.8 for the photorefractive polymer and 745 for pDR1A (15).

These results suggest that for holographic storage use, nonabsorbing films may be much better.

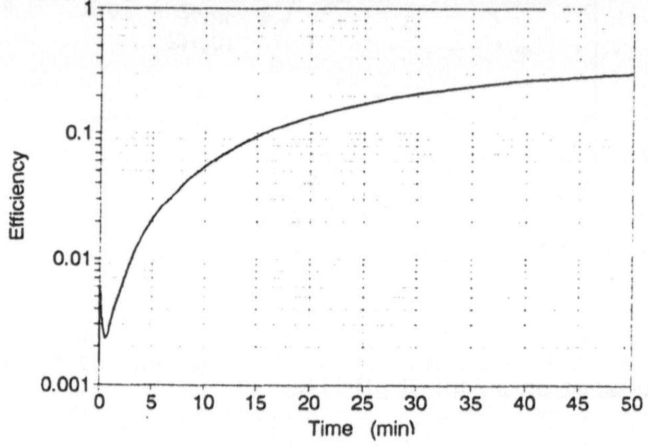

Figure 3. Diffraction efficiency as a function of time for a 12μ film of pMEA irradiated at 514 nm.

Figure 3 shows how the efficiency, expressed as diffracting efficiency, increases with time for a pMEA film (12μ thick) at 514 nm. The process is very slow, because pMEA barely absorbs this wavelength, but the efficiency levels are very high : up to ca. 20%. Thicker films should show even better efficiencies. Thus, if pMEA is not such a good candidate for reversible fast writing as described previously, it may be the best material for holographic storage use.

Neighboring Group Effect

As it was mentioned above, the level of birefringence that can be achieved in a sample depends - among other factors - on the azo group content in the sample. Various references in the literature report induced dichroism and/or birefringence in polymers containing different amounts of azobenzene groups. We have decided to systematically study the effect of composition on the reorientation phenomenon. To this end, we have synthesized and characterized two series of copolymers and one series of blends. The copolymers are poly(DR1A-co-MMA) and poly(DR1M-co-MMA) [DR1M is the methacrylate equivalent of DR1A] and the sequence distribution has been analyzed by solution NMR and found to be fairly different between the two series (16). This means that for two samples of the same composition, poly(DR1M-co-MMA) will have a different distribution of units than poly(DR1A-co-MMA). The blends were prepared by mixing various amounts of pDR1A and pMMA. With a proper thermal pre-treatment, the two polymers are miscible at the molecular level in any proportion (17).

Figure 4 illustrates the birefringence of all these sample as a function of the azo group content (by weight). It is very clear in the figure that, apart form the overall azo group content, there is some other factor influencing the saturation level of birefringence. All films reported in Figure 4 have similar thicknesses and similar Tgs.

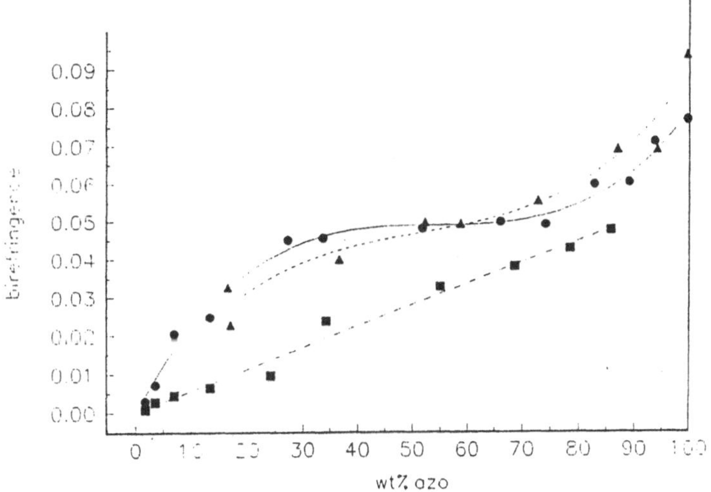

Figure 4. Saturation birefringence level as a function of weight percent azo in the copolymer or blend. (▲) poly(DR1A-co-MMA); (•) poly(DR1M-co-MMA); (▪) blends of pDR1A with pMMA.

For the blends, there is a linear increase in the birefringence level as the azo group content increases. For the copolymers, however, there is a rapid increase at low azo content, followed by a plateau region (between 25 and 75 weight% azo), and then another rapid increase at very high azo content. The two copolymers also behave slightly differently from each other (18). Work is in progress to correlate this behavior with the difference in sequence distribution between the two copolymer series.

More evidence for the influence of the next neighbor on the reorientation of an azo group is provided by the level of relaxation after the "writing" laser is turned off, which depends on composition for the copolymers and has a different shape for each series of copolymers. Copolymers with high azo content retain more of the signal than copolymers with low azo content. This level of relaxation does not depend on composition for the blends (18). Also, the cis-trans thermal relaxation apparent rate constant can be modeled very well using the triad distribution in these two series of copolymers. Again, the cis-trans thermal isomerization is about twice as fast for azo groups with azo neighbors than for azo groups with methyl methacrylate neighbors (19).

Cooperative Motion

From the above described results it is very clear that there is an "amplification" effect in reorientation if two azo groups are next to each other. The term amplification has been used to describe a similar phenomenon of reorientation of liquid crystals triggered by azo-containing Langmuir-Blodgett monolayers (20). For the phenomenon described in Figure 4, an enhancement of about 33% can be calculated for instances where azo groups neighbor other azo groups and tend to reorient together.

This phenomenon is also known for liquid crystalline copolymers containing azo mesogens and mesogens of the same shape and situated at the same site on the copolymers, but containing ester or amide bonds instead of azo. These mesogens are not responding to the laser light, but they do reorient by "sympathy" when the azo mesogens are moved by the laser light (21).

As we mentioned in a previous section, if the polymers are semicrystalline, there may be a three-fold amplification of the induced birefringence. We have synthesized pDR19T:

$$-CO-CH=CH-C_6H_4-CH=CH-CO-O-CH_2-CH_2-N(C_6H_4-N=N-C_6H_4-NO_2)-CH_2-CH_2-O-$$

following the literature procedure (12) and investigated this unexpected increase in birefringence using our "writing" procedure and infrared and UV-Vis dichroism. A typical writing curve for pDR19T is shown in Figure 5.

The "writing" curve in Figure 5 confirms that the level of birefringence achievable in an amorphous pDR19T film is much higher than in pDR1A. However, it takes tens of seconds to achieve saturation in a film of comparable thickness with pDR1A, where it takes about 100 ms. The difference appears from the fact that - as the laser start reorienting the azo groups - whole crystalline domains are formed and oriented in the preferred direction. Indeed, at point B, where the laser is turned off, the birefringence **increases**, instead of decreasing as it does in amorphous polymers. The heat which has to be dissipated through the sample reinforces the thermodynamic tendency of the rigid groups to organize in crystalline domains. Erasing is very slow and rewriting never gives the same level of birefringence that was initially achieved in the amorphous film, because the disorder achieved by erasing is composed now of organized domains randomly oriented.

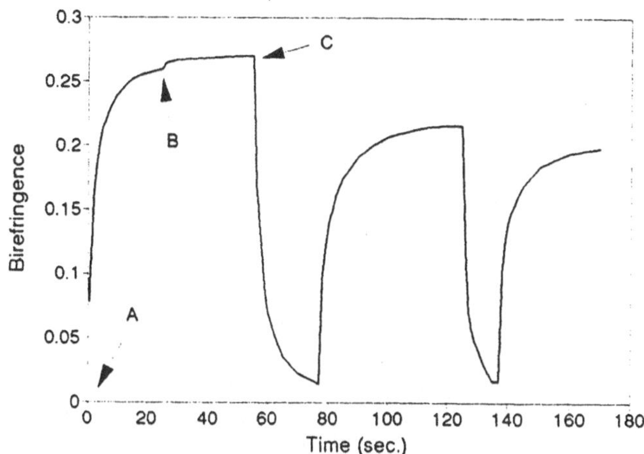

Figure 5. "Writing" curve on a freshly prepared pDR19T film. A: 514 nm planarly polarized laser ON; B: laser OFF; C: circularly polarized laser ON. Same procedure is repeated at 80 seconds and at 135 seconds.

An IR dichroism study (22) confirms that the dichroism of the azo groups is much greater in pDR19T than in pDR1A, but that - surprisingly - the other rigid groups in the polymer structure (-CH=CH-Ph-CH=CH-) also show a significant dichroism in the same direction as the azo groups. Thus, this reorientation by "sympathy" which has been reported in Langmuir-Blodgett films and liquid crystalline polymers is a much more general phenomenon and may involve rigid groups which are of a completely different shape and situated at a completely different site in a polymer structure.

POTENTIAL APPLICATIONS

Because of the potential of repetitive usage, an obvious application of the amorphous azo polymers is in erasable optical memory disks or cards. The rate of "recording" is in the milliseconds range. For read only memory, or write once read many times, liquid crystalline or semicrystalline polymers may be preferred, because the time to "write" the information is not critical, while the "contrast" (birefringence) is greater than in amorphous polymers.

Holographic storage also seems a fairly good potential use for these polymers. They are probably the only materials which can be reversibly recorded and which do not require any development of the image after recording. One advantage of these materials is that one could define a certain direction and store an image in that direction, then define another direction, at a fairly close angle to the first one, and store another image in the new direction. A multitude of directions could be defined this way, allowing a very high density of storage.

Some research groups have proven the waveguide properties of the laser-induced "writing" on polymer films (23). One could imagine a "printed circuit" of waveguides on a polymer film having the main advantage that any connection could be erased or added at any time during the lifetime of such a device. Simplified versions of such waveguides can act as sensors in buried places. Second harmonic generation can also be obtained in these polymer films by using a combination of laser-induced reorientation and electric field poling (24). The advantage is that heating above Tg (as in "regular" poling) is not necessary.

CONCLUSIONS AND FUTURE WORK

A reversible order-disorder transition can be induced photochemically in amorphous high-Tg azobenzene-containing polymers. The mechanism of the reorientation process is still under study, but the most interesting aspect are the neighboring group effects and the cooperative motion between like and unlike rigid groups present in the polymer structure. The reorientation depends on the maximum absorbance of the polymer and its relation with the "writing" laser.

Our research group is investigating this phenomenon along a few lines. We are synthesizing new azobenzene-containing polymers with various absorbances in order to obtain high Tg and thermally stable polymers. We are studying the neighboring group effects on the efficiency of this process and are investigating new polymer structures which are liquid crystalline of a lower orientational order, or can become liquid crystalline by being subjected to polarized laser light.

ACKNOWLEDGMENTS

The students who performed this work are Shuang Xie, Darryl Brown, Chris Barrett, Soi To, Todd Jones, Chloë Stuart and Jessy Mao. We thank prof. Michel Pézolet's for his IR dichroism work.

Financial support from the Office of Naval Research, U. S., as well as from NSERC Canada and the Department of Defence Canada is gratefully acknowledged.

References:

1 Rau H (1990) in Rabek JF (ed) Photochemistry and Photophysics Vol II, CRC Press, Boca Raton, Florida, chapter 4
2 Kumar G, Neckers D (1989) Chem Rev 89:1915
3 Xie S, Natansohn A, Rochon P (1993) Chem Mater 5:403
4 Rochon P, Gosselin J, Natansohn A, Xie S (1992) Appl Phys Lett 60:4
5 Natansohn A, Rochon P, Gosselin J, Xie S (1992) Macromolecules 25:2268
6 Natansohn A, Xie S, Rochon P (1992) Macromolecules 25:5531
7 Todorov T, Tomova N, Nikolova L (1983) Opt Commun 47:123
8 Eich M, Wendorff JH, Reck B, Ringsdorf H (1987) Makromol Chem Rapid Commun 8:59
9 Natansohn A, Rochon P, Mao J, Xie S (1993) OSA Technical Digest "Organic Thin Films for Photonic Applications" 17:147
10 Rochon P, Bissonnette D, Natansohn A, Xie S (1993) Applied Optics 32:xxx
11 Tregold RH, Allen RA, Hodge P, Khosdehl E (1987) J Phys D Appl Phys 20:1385

12 Chen M, Yu L, Dalton L, Shi Y, Steier W (1991) Macromolecules 24:5421
13 Gunter P, Huignard JP (1988) in Photorefractive Materials and Their Applications, Springer Verlag 1:30
14 Donkers MCJM, Silence SM, Walsh CA, Hache F, Burland DM, Moerner WE, Twieg RJ (1993) Opt Lett 18:1044
15 Rochon P, Natansohn A, Mao J, Barrett C, Xie S (submitted) Phys Rev Lett
16 Xie S, Natansohn A, Rochon P (submitted) Macromolecules
17 Xie S, Natansohn A, Rochon P (submitted) Macromolecules
18 Brown D, Natansohn A, Rochon P, Xie S (1993) OSA Technical Digest "Organic Thin Films for Photonic Applications" 17:248
19 Barrett C, Natansohn A, Rochon P, Xie S (1993) OSA Technical Digest "Organic Thin Films for Photonic Applications" 17:151
20 Seki T, Sakuragi M, Kawanishi Y, Suzuki Y, Tamaki T, Fukuda R, Ichimura K (1993) Langmuir 9:211
21 Wiesner U, Reynolds N, Boeffel C, Spiess HW (1991) Makromol Chem Rapid Commun 12:457
22 Natansohn A, Rochon P, Pezolet M, Audet P, Brown D, To S (submitted) J Am Chem Soc
23 Sekkat Z, Dumont M (1992) Appl Phys B 54:486; Mol Cryst Liq Cryst. Tech B Nonlinear Opt 2:359
24 Blanchard PM, Mitchell GR (1993) Appl Phys Lett 63:2038

Highly Electrically Conductive Polymer Composites and Blends

T. J. Kang*, Y. Miyaki**, J. H. Han, T. Motobe, Y. E. Whang and S. Miyata

Division of Chemical and Biological Science and Technology, Tokyo University of Agriculture and Technology, Nakamachi 2-24-16, Koganei, Tokyo 184 , Japan
**Department of Fiber and Polymer Science, Seoul National University, Seoul 151-742, Korea*
***Polymers Research Laboratory, Tosoh Corporation, Yokkaichi, Mie, Japan*

Introduction

Conducting polymers have received a great deal of attention during the last decade because these polymers are considered as promising new materials of many potential applications, in which the polymeric materials have not been considered to be applicable before. Among the various conducting polymers polypyrrole has recently attracted much interests because of its high electrical conductivity combined with its good overall environmental stability.

In recent years significant researches have been reported on highly conductive polypyrrole synthesized by methods of electrochemical polymerization, chemical polymerization, chemical vapor polymerization, and photo-electrochemical polymerization[1-8]. All these polymerization methods readily yield conductive polypyrrole and the polymer shows overall stability under ambient conditions.

The polypyrrole film prepared chemically or electrochemically in organic solvent containing electrolyte exhibits conductivity up to 500 S/cm, which is a sufficient electrical conductivity for the most of the applications. Unfortunately, however, these polymers show poor mechanical properties as well as the lack of processability. This is one of the reason for their limited utilization despite of promising new materials of potential applications.

Several attempts to overcome these drawbacks have been made, in particular a method of blending with conventional polymers[9-18]. However, mechanical blending in melt state is not desirable because the conductivity of the polymer decreases due to the thermal degradation during processing. Furthermore, a high concentrarion of conducting polymer is required to form a conducting path.

In this study novel methods of preparing highly electrically conductive polymer composites and blends have been studied. A chemical diffusive polymerization process for preparing highly electrically conductive poly(vinylchloride)-polypyrrole composite has been developed. And a method of preparing conductive poly(vinylacetate)-polypyrrole composite film by casting on a substrate has also been developed.

Experimental

Preparation of Poly(vinylchloride)-polypyrrole Composite Film

Poly (vinylchloride)-polypyrrole composite film was prepared by allowing the oxidative polymerization to take place in PVC matrix with diffused pyrrole. The PVC film was immersed in a swelling solution containing pyrrole so that the pyrrole monomer could diffuse into the polymer matrix and then subsequently polymerized in an oxidative solutions containing oxidant in an binary solvent system of similar solubility coefficients with PVC film. Mixture of acetone, n-hexane and pyrrole was chosen as a diffusion solution because it yields highly conductive and transparent composite film with good processability.

The typical composition of diffusion solution for the processing of highly electrically conductive composite films consists of n-hexan, acetone, and pyrrole with the volume percentage of 87.8, 9.8, and 2.4, respectively.

K.P. Ghiggino (Ed.)
Progress in Pacific Polymer Science 3

After the immersion of the PVC film into the difusion solution the oxidative polymerization of pyrrole takes place in a 1.0 mole $FeCl_3$ solution of binary solvent system with acectonitrile and methanol. The oxidation solution requires the capability of swelling PVC for pormoting the deeper penetration of the oxidant into the polymer matrix. For the mixture of acetonitrile and methanol with 1.0 mole $FeCl_3$ oxidant, the mole fraction of acetonitrile determines the compatibility with the swollen PVC as well as the level of the oxidation potential. The mole ratio of 85/15 of acetonitrile/methanol resulted high electrical conductivity of the composite film with the solvent system of n-hexane, acetone, and pyrrole when polymerized at 0°C. The oxidation potential also varies with the amount of oxidant($FeCl_3$) as well as the reaction temperature.

The oxidative polymerization was carried out in the ice bath to keep the polymerization temparcture close to zero degree. The polymerization temperature affects the molecular structure of polypyrrole and conjugation length as well as the dopant level. The nature and extent of chemical defect sites and charge distribution characteristics of polypyrrole will be affected with the temperature of the polymerization. It is well known that polypyrroles synthesized under lower temperature below 0°C has longer conjugated chain length with well defined oriented structures of less defects which results in high electrical conductivity.

Preparation of Poly(vinylacetate)-polypyrrole Composite Film

A mixed solution of ferric chloride and poly(vinylacetate) was first prepared by dissolving them in methanol. The oxidation potential of the solution was controlled within the optimum range by changing the amount of ferric chloride. Then, pyrrole was added to the mixed solution with stirring. A conducting polymer composite film was prepared by casting the mixture on a substrate. After sufficient drying, it was rinsed with methanol to remove the remaining ferric chloride and the ferrous chloride which was formed from the polymerization of pyrrole.

When the mixture was casted on the substrate, its color changed gradually from lightly yellow to dark green. This indicates that the polymerization of pyrrole progresses under the increased oxidation potentioal driven by the evaporation of the solvent from the casting solution. The mixture of pyrrole and ferric chloride-poly(vinylacetate) solution gradually transforms to poly(vinylacetate)-polypyrrole composite as the polymerization of pyrrole progresses.

Results and Discussion

Morphology of the Composite Films

The poly(vinylchloride)-polypyrrole composite film was examined with SEM and TEM microscopes. It is clear with the TEM examination that the conducting layer exists on the surface of the PVC film with the thickness of about 0.2~1.0 μm on both sides(Fig.1) and the thickness of the conducting layer varies with the processing conditions such as diffusion and oxidation time as well as oxidation potential. The amount of polypyrrole content in the composite also varies with the processing conditions from as low as 37% to up to 86%. The conducting layer in Fig.1 consists with the composite of polypyrrole and PVC with the polypyrrole mole ratio of 64% determined from the atomic ratio of N with the X-ray photoelectron spectroscopy. It is observed under the microscope that the conducting layer exists as a homogeneous mixture of polypyrrole and PVC polymer. The polypyrrole particles are uniformly dispersed in the conducting layer and form a network of conducting path.

The morphology of poly(vinylacetate)-polypyrrole composite film is very different from that of polymer blend prepared from the solution polymerization. Polypyrrole aggregates are connected with each other in the former case, which results in a polypyrrole network throughout the composite, while polypyrrole aggregates are separated from each other in the latter case. It is expected in the former case that a spinodal decomposition occurs during the phase separation process because of sudden changes in the concentration of pyrrole and ferric chloride as well as the viscosity of poly(vinylacetate). These sudden changes are caused by the evaporation of solvent.

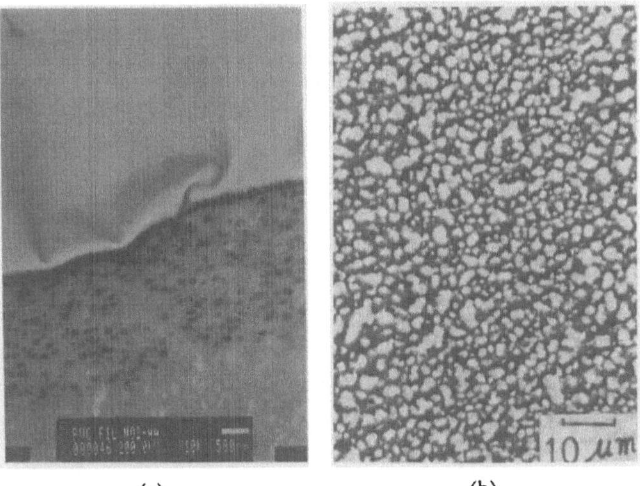

(a) (b)

Fig. 1 Morphology of highly electrical conductive composite films of poly(vinylchloride)-polypyrrole (a) and poly(vinylacetate)-polypyrrole (b)

Electrical Conductivity of the Composite Film

The electrical conductivity of the composite films were measured with the standard four probe method. The electrical conductivity is given as;

$$\sigma = 0.2206 \times \frac{i}{v \times d}$$

where, s is the conductivity in S/cm.
v is the applied potential in volt.
d is the thickness of the conducting layer in cm.
i is the current in ampere.

The typical conductivity of the poly(vinylchloride)-polypyrrole composite film shows about 20 S/cm with the 50~60% transparency at 500 nm. The pure polypyrrole layer which is synthesized under the same conditions as that of the poly(vinylchloride)-polypyrrole composite measured with the thin disk pressed from the precipitate of the polypyrrole showed the electrical conductivity of about 250 S/cm.

The poly(vinylacetate)-polypyrrole composite film also showed high electrical conductivity of about 10 S/cm which resulted from the polypyrrole network structure formed from the spinodal decomposition during the casting process.

The electrical conductivity of solution-cast film depends on the initial oxidation potential of the solution as well as on the holding time of the solution state before casting. If the initial oxidation potential of the solution is either too low or too high, no highly conductive film can be obtained. It is expected that a pyrrole monomer will be evaporated together with solvent when the oxidation potential is too low, however, phase separation has already progressed to a considerable extent before casting when the oxidation potential is too high.

The optimum range of initial oxidation potential of the solution is 480-560 mV (versus SCE). The holding time of solution before casting may be associated with the degree of polymerization of pyrrole and the extent of phase separation.

It is surprising that the electrical conductivity can be obtained as high as 10 S/cm with only 5 wt.% of pyrrole monomer being incorporated, while the incorporation of 40 wt.% of pyrrole leads to 0.01 S/cm in the solution polymerization (Fig.2). The network structure of Polypyrrole is maintained even when the ratio of pyrrole monomer is decreased. Therefore, a network structure of polypyrrole can be obtained with only 5 wt.% of pyrrole and this brings about the highest conductivity.

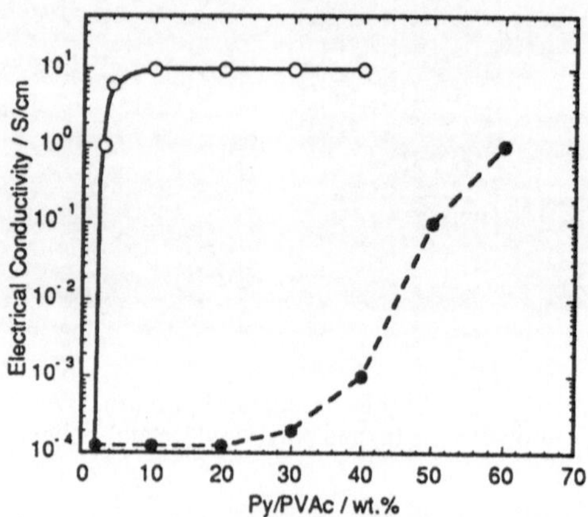

Fig. 2 Electrical conductivity vs. the content of polypyrrole: (○) ; poly(vinylacetate)-polypyrrole composite film prepared from the solvent evaporation method, (●) ; poly(vinylacetate)-polypyrrole composite prepared from the solution polymerization.

Conclusions

The process of preparing highly electrically conductive composite films of poly(vinylchloride)-polypyrrole as well as poly(vinylacetate)-polypyrrole have been studied. The solvent system of n-hexane, acetone, and pyrrole mixture offers effective swelling medium of PVC film for preparing the highly electrically conductive poly(vinylchloride)-polypyrrole composite film. The poly(vinylacetate)-polypyrrole composite film has been prepared from the solvent evaporation method.

The oxidative polymerization of the diffused pyrrole in the binary solvent system of acetonitrile and methanol gives high conductivity of the polypyrrole as well as the good penetration of the oxidant into the PVC polymer matrix. The mole ratio of acetonitrile to methanol in the binary solvent system with 1.0 mole of $FeCl_3$ showed optimum in the range of 0.80 to 0.85.

In the poly(vinylacetate)-polypyrrole composite film polypyrrole forms a network structure by spinodal decomposition and poly(vinylacetate) is cross-linked by ferric chloride during the polymerization process which results in excellent chemical stability of the composite.

The thickness of the conducting layer in the PVC film was measured to be about 0.2~1.0 μm. The mole ratio of polypyrrole to the PVC matrix in this conducting layer was estimated to be about 64%. The analytical testing of the composite film shows the formation of homogeneous mixture of conductive layer within the 0.2~1.0 μm of thickness on the film surface.

The electrical conductivity of the composite films of poly(vinylchloride)-polypyrrole and poly(vinylacetate)-polypyrrole were measured as high as 20 and 10 S/cm, respectively. These highly electrically conductive composite films showed excellent environmental stability under ambient conditions.

References

1. S. Machida, S. Miyata, and A. Techagumpuch (1989) *Synth. Met.* **31** : 311-318.
2. S. P. Armes (1987) *Synth. Met.* **20** : 365-371.
3. M. M. Castillo-Ortega, M. B. Inoue, and M. Lnoue (1989) *Symth. Met.* **28** : C65-C70.
4. T. H. Chao, and J. March (1988) *J. Polym. Sci. : Polym. Chem.* **26** : 743-753.
5. R. B. Bjorklund (1987) *J. Chem. Sc., Faraday Trans. I* **83** : 1507-1514.
6. R. E. Myers (1986) *J. Electron. Mater.* **15** : 61-69.
7. J. A. Walker, L. F. Waltmann, and E. F. Witucki (1988) *J. Polym. Sci. : Polymer Chemistry* **26** : 1285-1294.
8. B. Zinger, P. Shaier, and A. Zemel (1991) *Synth. Met.* **40** : 283-297.
9. M. A. Depaoli, R. J. Waltmann, A. F. Diaz, and J. Bargon (1985) *J. Polym. Sci. : Polym. Chem. Ed.* **23** : 1687-1698.
10. X. Bi, and Q. Pei (1987) *Synth. Met.* **22** : 145-156.
11. G. B. Street, S. E. Lindsey A. I. Nazzal, and K. J. Wynne (1985) *Mol. Cryst. Liq. Cryst.* **118** : 137-148.
12. O. Niwa, M/ Mikita, and T. Tamamura (1985) *Makromol. Chem., Rapid Commum.* **6** : 375-379.
13. T. Ojio, and S. Miyata (1986) *Polym. J.* **18** : 95-
14. O. Niwa, and T. Tamamura (1984) *J. Chem. Soc. Chem. Commun.* : 817-
15. B. Wessling, and H. Volk (1986) *Synth. Met.* **15** : 183-
16. T. A. Ezquerra, F. Kremer, M. Mohammadi, J. Ruhe, G. Wegner, and B.Wessling (1989) *Synth. Met.* **28** : C83-
17. J. H. Han, T. Motobe, Y. E. Whang, and S. Miyata (1991) *Synth. Met.* **45** : 261-264.
18. T. J. Kang, S. Miyata, and Y. Miyaki (1993) *Polym. Bulltin* : 31; 593-599

Single-Chain Single Crystal

Haishan Bu, Erqiang Chen, and Shengyong Xu

Department of Materials Science, Fudan University
Shanghai 200433, People's Republic of China

Abstract: A new method that makes use of monomolecular patches on a liquid surface has been developed for the crystallization of single chain, and single-chain single crystals of poly(ethylene-oxide) and isotactic polystyrene were investigated. The molecular mass and molecular mass distribution of samples could be found to match a statistical analysis of single-chain particles. The observations of morphology and electron diffraction indicated that single-chain particles are crystals, even single crystals. The observed single-chain crystals have very regular morphology, and several new kinds of morphologies were found for the first time. Proof was brought that the single-chain single crystals do not reach equilibrium state on crystallization, rather assume the lamellar morphology with chain-folded macroconformation, also known from crystallization of polychain crystals.

INTRODUCTION

Review of the literature on crystallization and annealing of flexible, linear macromolecules[1], revealed observations that the properties of a chain within a crystal may deviate from the properties of the crystal as a whole. Other evidence indicates that it is possible to isolate single chain in a solution, and it seems that a starting point to elucidating the behavior of a chain in a crystal must begin with a crystal that consists of only one chain. Recently a new method that makes use of monomolecular patches on a liquid surface has been developed for crystallization of single chain, and single-chain single crystals of poly(ethylene oxide) (PEO)[2] and isotactic polystyrene (i-PS)[3,4] have been investigated.

The separation and observation of amorphous single chain of polychlorostyrene was first made by Boyer et al.[5], further experiment on other polymers followed[6,7]. The purpose of the experiments was to estimate the molecular mass and obviously reliable molecular mass measurement could be achieved. But the method developed for amorphous polymers did not give proper molecular mass data for crystallizing polymers. Two earlier observations point to the existence of single-chain single crystals and single-chain domains within a larger crystal, one is that of self-nucleation[8], the other, the special morphological features seen on annealing of lamellar, chain-folded crystals[9]. The most direct evidence of the growth of single-chain single crystals was given by Bittiger[10] on crystallization of dilute solution of cellulose tricarbanilate. It must be noted that this polymer is a rather stiff macromolecule with intramolecular hydrogen bonds and may possess an ideal structure to preserve monochain

K.P. Ghiggino (Ed.)
Progress in Pacific Polymer Science 3

domains. Our efforts here are directed toward duplicating such crystallization with flexible, linear macromolecules. In this paper, the experimental results of single-chain single crystals of PEO and i-PS are presented, the molecular mass and molecular mass distribution of samples could be found to match with a statistical analysis of single-chain particles. The observations of morphology and electron diffraction indicated that particles are crystals, even single crystals. The observed single-chain single crystals have very regular morphology and do not reach equilibrium state, rather assume the lamellar morphology with chain-folded macroconforma-tion, as known from crystallization of polychain crystals.

SINGLE-CHAIN SINGLE CRYSTALS

According to polymer solution theory, there is a critical concentration C^* where molecular chains begin to overlap. Below concentration C^*, polymer chains should exist as separated, single chains. Thus a dilute solution with concentration less than C^* was used to prepare single molecular chains by using Langmuir film balance. Such solution was then spread on a water surface, isolated single-chain particles formed after the solvent had evaporated. Next, the surface was compressed properly to concentrate the particles, but avoiding aggregation. The particles were then transferred to copper grids for electron microscopy. The copper grids were covered with collodion film, reinforced with evaporated carbon. A detailed description of preparation of these particles of single chains was given in a previous paper[2]. The "bio-spread method" was also attempted to prepare single chains. A needle was inserted into the water at an angle of 45°, then a drop of the dilute solution was placed on the needle, it slipped down the needle and spread quickly on the water surface. The resulted particles were transferred to copper grids by touching the grids to the water surface.

The coated copper grids were put in a thermostat under a nitrogen atmosphere to crystallize the particles at a certain temperature. The copper grids were then observed either with a Hitachi H-500H transmission electron microscope (TEM), Philips EM-420 TEM, or Opton EM-902 TEM.

In the investigation of i-PS single-chain single crystals, an i-PS fraction of high molecular mass with narrow molecular mass distribution was used, the number-average molecular mass, M_n, and polydispersity, dw, were 1.34×10^6 and 1.25, respectively; the concentration of solution of i-PS in benzene was 1×10^{-3} wt-%. Fig. 1 shows the single-chain single crystals grown on crystallizing single-chain particles at 448.2±0.1 K for 3 h. Clearly, most of the particles have polygonal shapes instead of being irregular spheres in amorphous state. The particles become more transparent to the electron, indicating a thinner structure. The uniformity of the contrast suggests a lamellar morphology. When the samples were crystallized for longer time, the regularity increased. these facts, together with the electron diffraction results, described below, indicate that the particles are crystals, even single crystals. Furthermore, the crystalline c-axis prefers to orient in a direction normal to the substrate.

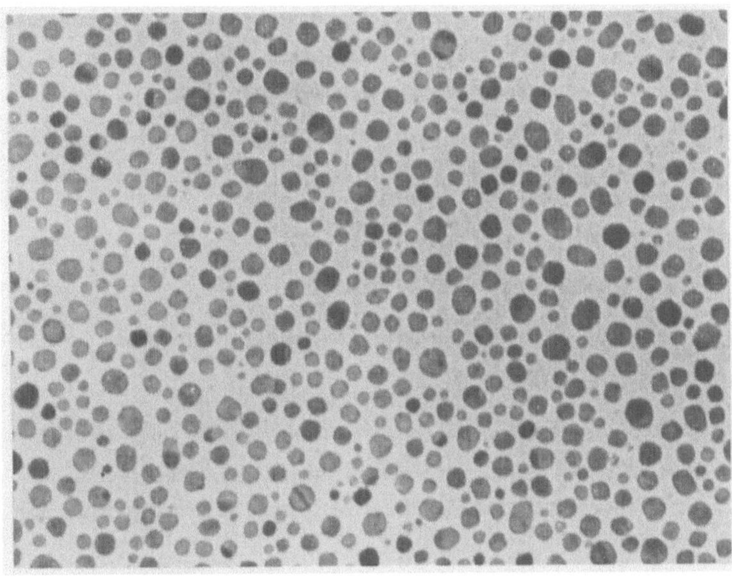

Fig. 1 Single-chain single crystals of i-PS fraction crystallized at 448.2 K for 3 h. The magnification bar indicates 100 nm.

Assuming the chains form a small equilibrium crystal, their dimensions can be calculated in the case of a minimum free energy (Ref. 1, Sects. 3.1 and 7.1.2),

$$L_e = (\gamma_e^2 V / \gamma^2)^{1/3} \tag{1}$$

where γ_e is the (001) surface free energy[11], reported to be 31mJ/m^2; γ, the side-surface free energy, usually assumed to be about 1/5 of γ_e ; and V is the total crystal volume. Using the number-average molecular mass of i-PS fraction and the X-ray density of crystalline polystyrene(1.11Mg/m^3)[1], one can compute L_e for i-PS fraction to be 36.8 nm. The corresponding lateral dimensions (assumed to give a quadratic cross-section) is 7.4 nm.

The thickness of single-chain single crystals was measured by shadowing at an angle of 12°, and it was between 10 and 20 nm with an average of 14 nm, far below the equilibrium value. Even single-chain droplets of polystyrene can thus not reach equilibrium on crystallization, but form metastable lamellae with a much reduced fold length. Assuming classical theory of crystallization can be applied to the crystallization of a single chain, the free enthalpy of formation of a primary homogenous nucleus is (Ref. 1, Sect. 5.1.1):

$$\Delta G = 4LAN^{0.5}\gamma + 2NA^2\gamma_e - NA^2L\Delta g \tag{2}$$

where Δg represents the bulk free enthalpy of fusion; L is the dimension of the nucleus in the chain direction; N is the number of stems in the nucleus with a cross-section A^2 ; γ and γ_e represent lateral and end surface free energies, respectively. The critical nucleus dimensions are calculated by differentiation of ΔG with respect to L and N and setting the derivatives equal to zero and substituting Δg by $\Delta h_f \Delta T / T_m^0$:

$$L^* = 4\gamma_e T_m^0 / (\Delta h_f \Delta T) \tag{3}$$

and

$$N^* = 16\gamma^2 T_m^{02} / (\Delta h_f \Delta T A)^2 \tag{4}$$

where T_m^0 is the equilibrium melting temperature; Δh_f, the heat of fusion per cm³; and ΔT , the supercooling.

With the value of γ_e =31 mJ/m² , T_m^0 =516.2 K, and Δh_f=108.5MJ/m³,[12] L^* can be calculated to be 8.7 nm at the crystallization temperature of 448.2 K. On the basis of fluctuation theory, the thickness of the crystal grown from the nucleus should be somewhat larger than L^*, and

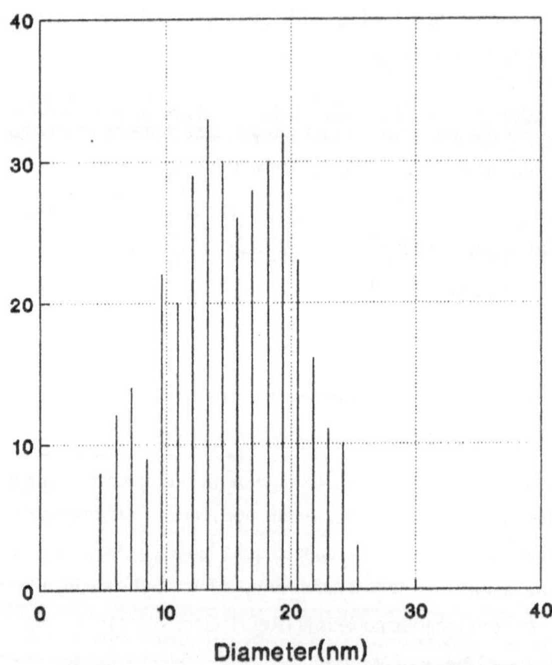

Fig. 2 Particle diameter distribution of i-PS single-chain crystals

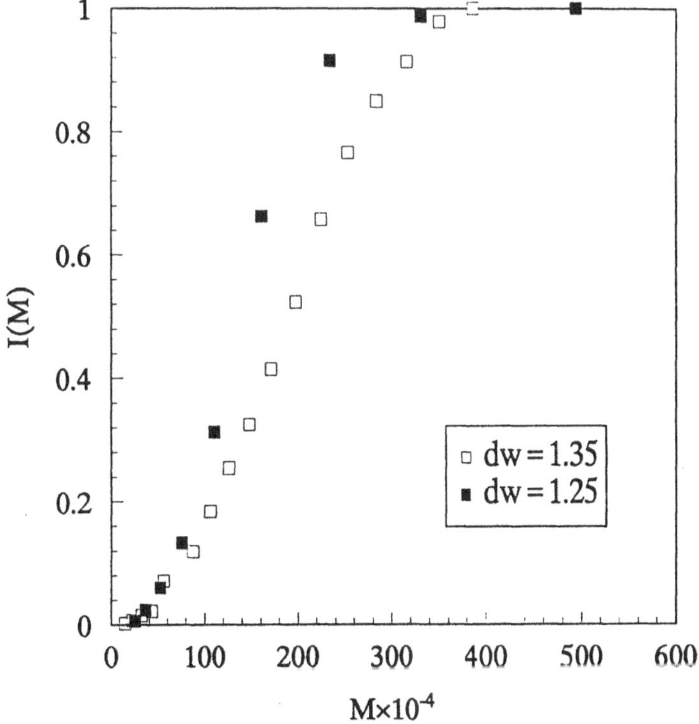

Fig. 3 Molecular mass distribution of i-PS fraction: Open squares from particle distribution, filled squares from the GPC curve.

an increase is also possible from annealing at the crystallization temperature. Taking account of the errors in the measurement of shadow-length, L^* is closer to the measured value of 14 nm than the equilibrium length L_e. The larger experimental thickness, as well as a wider distribution than assumed by the classical theory is most likely an indication of annealing subsequent to growth.

The size of particles, shown in Fig. 1, was digitized with a computer using a scanner for electron micrograph. The cross-sections of the particles could be obtained by image analysis, based on the principle of mathematical morphology and were converted to particle diameters. The particle diameter distribution is shown in Fig. 2, and it is transformed to the molecular mass distribution, as shown in Fig. 3, assuming the thickness of the crystals to be 14 nm and their density to be 1.11 Mg/m^3. The number-average molecular mass can be obtained to be 1.5×10^6 with a polydispersity index of 1.34, in reasonable agreement with the results from GPC. Most of the analyzed particles must thus have been of single chain size.

The largest molecular mass, corresponding to the onset of GPC curve, is 4.8×10^6 for i-PS fraction. This chain would form a particle with a diameter of about 28 nm, thus a particle larger than that can not be considered to be a single chain. The largest particles considered in the image analysis was 25.4 nm.

Similarly, the single-chain single crystals of PEO were obtained by collecting single-chain droplets of melt from a hot water surface at 353 K on microscope grids and quenching to room temperature. PEO is soluble in cold water, but not in hot water. It was also found that the molecular mass of PEO sample could be matched with the statistical analysis of single-chain particles

MORPHOLOGY

An i-PS crystal grown from dilute solution has usually regular, hexagonal appearance. The morphology of the single-chain single crystals grown in this research were, however, sometimes of quite different morphology. When single-chain single crystals were grown at 448.2 K for a short time, e.g. 3 h, polygonal crystals, as show in Fig. 1, resulted. Even close to circular shape were seen. Single-chain single crystals grown at the same temperature, but for a longer time, e.g. 8 h, showed much more regular-shaped crystals, as shown in Fig. 4. By close observation typical morphologies as displayed in Fig. 5 could be seen. The left row in Fig. 5 collects crystals which are close to the conventional hexagonal shape, the others contain modifications which are new morphologies observed for the first time. The morphology of all crystals has the common feature of the characteristic angle of 120°. Obviously the occurrence of the new morphologies is related to different growth rate at different growth face. Besides, it is interesting to note that the single-chain single crystals did not nucleate heterogeneously on the carbon film surface, but rather homogeneously. Only in the latter case could they settle flat on the substrate, as observed.

Tent-like crystals can also be recognized, revealing distinct sectors and sectorization boundaries as shown in Fig. 6. As is well known, the tent-like morphology and sectorization of polyethylene crystals grown from solution indicate the different ordering of the fold structure within each sector of the specific crystal growth face[13]. Tent-like crystals of polyethylene can only be seen when the crystals float in the solution, usually a planar lamellae is observed due to the collapse of crystals on a substrate by slip along the chains axis. The here presented tent-like single-chain single crystals are so small, that they can avoid the collapse. It may be assumed that the tent-like single-chain single crystals of i-PS have the same origin as the polyethylene crystals grown from solution.

For the single-chain single crystals of PEO, polygonal crystals were found on quenching single-chain droplets of melt to room temperature, as reported in a previous paper[2], recently

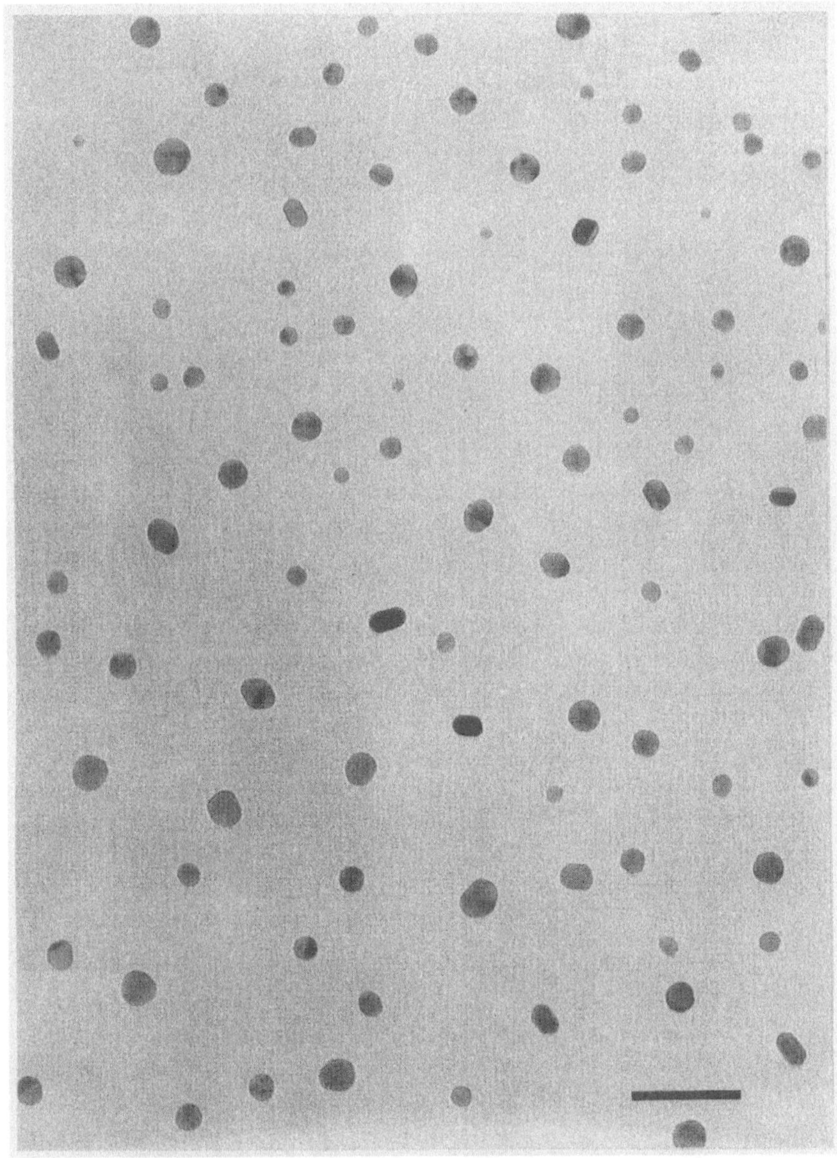

Fig. 4 Regular-shaped single-chain single crystals of i-PS crystallized at 448.2 K for 8 h. The magnification bar indicates 50 nm.

regular square crystals could be observed when crystallizing single-chain particles at 318 K for 5 h.

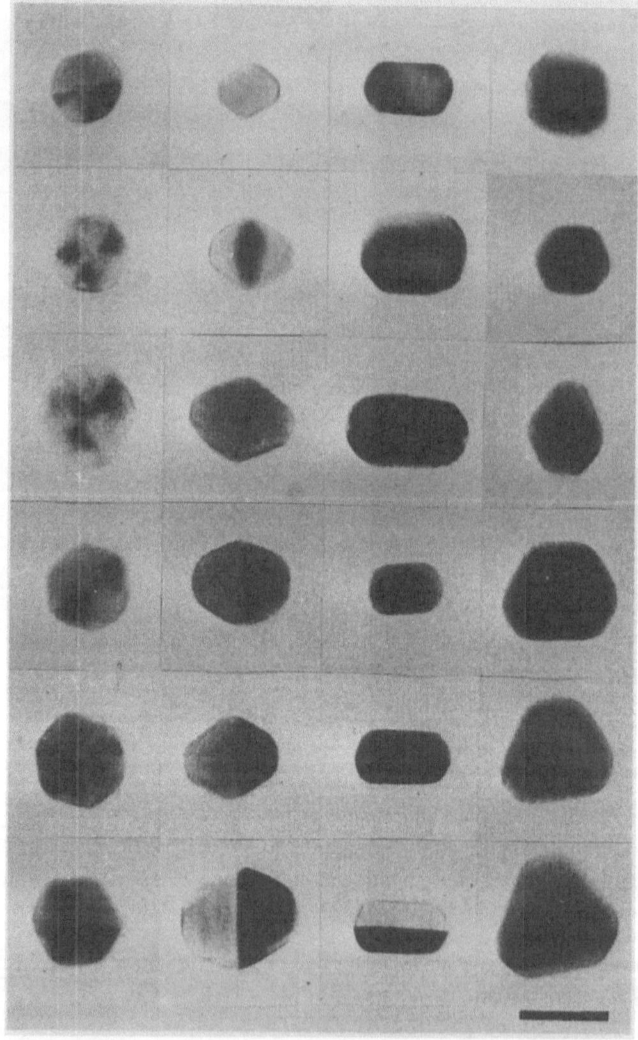

Fig. 5 Typical morphology of single-chain single crystals of i-PS. The magnification bar indicates 20 nm.

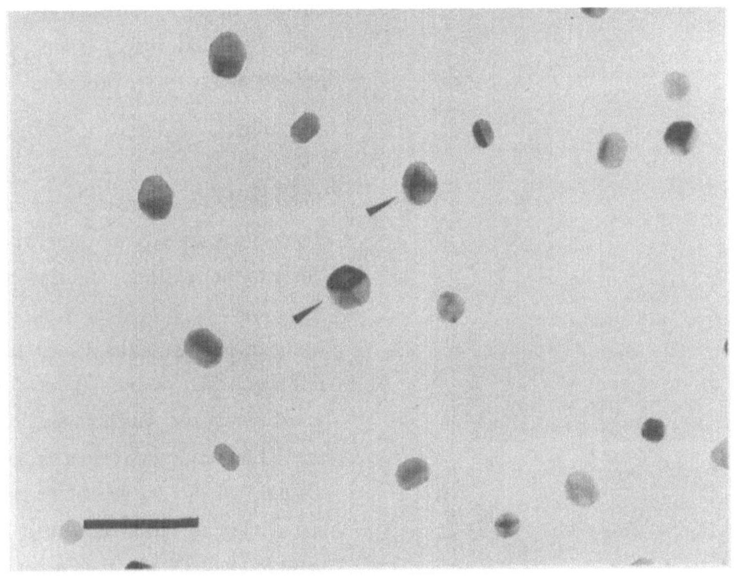

Fig. 6 Tent-like morphology of single-chain single crystals of i-PS. The magnification bar indicates 50 nm.

CRYSTAL STRUCTURE

Selected-area electron diffraction was carried out with the Philips EM-420 for the larger crystal which may contain several tens of chains and for simultaneous analysis of a large number of single-chain crystals. Micro-beam electron diffraction was performed on isolated single-chain crystal. For the latter, the electron beam was concentrated to a spot with a diameter of 30-40 nm, to cover no more than a single-chain crystal.

It was reported that i-PS crystal has a trigonal unit cell[14] and the *hk0* electron diffraction from an i-PS single crystal has hexagonal symmetry[15]. Fig. 7(a) shows a typical hexagonal selected-area electron diffraction pattern from a larger i-PS crystal which may contain several tens of chains with lamellar morphology, and the c-axis normal to the substrate and parallel to the incident electron beam. Fig. 7(b) shows a selected-area electron diffraction pattern from a large number of single-chain crystals. Because the c-axis of all crystals preferred to orientate at right angle to substrate, and the a-axis and b-axis orientated randomly, the *hk0* reflec-tion rings were produced.

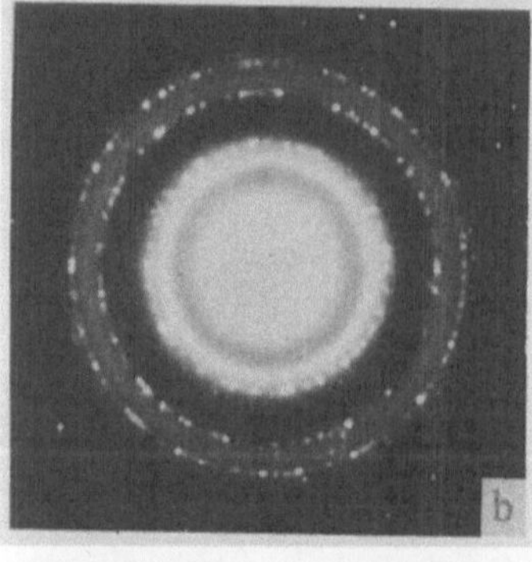

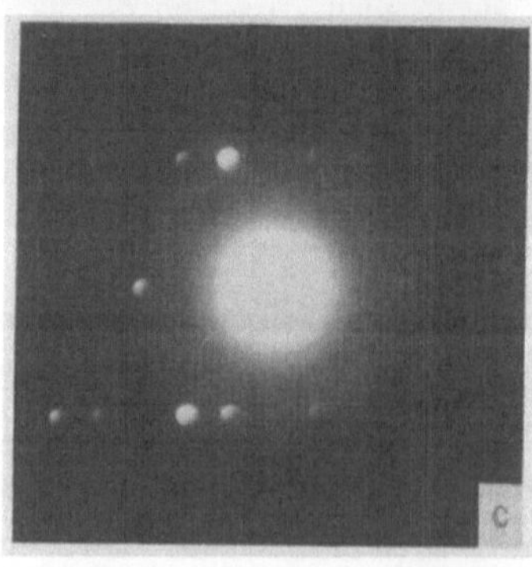

Fig. 7 Electron diffraction patterns of i-PS crystals
(a) Selected-area electron diffraction pattern from a large i-PS single crystal (multi-chain).
(b) Selected-area electron diffraction pattern from a large number of i-PS single-chain single crystals.
(c) Micro-beam electron diffraction pattern from one i-PS single-chain single crystal.

Micro-beam electron diffraction pattern from a single-chain single crystal is shown in Fig.7 (c). Usually the c-axis of these single-chain single crystals are not exactly at right angle to the substrate, and since the micro-beam electron diffraction is rather sensitive to the orientation of the crystal, the electron diffraction pattern do not have full hexagonal symmetry. Furthermore, because of the large beam current used, reflections disappeared after 60-90 s of irradiation, so it is not easy to obtain complete diffraction pattern. Undoubtedly, it is necessary to do more work, to elucidate the structure of single-chain single crystal.

Selected-area electron diffraction pattern of PEO was also obtained from a single-chain single crystal, and the six reflections around the center correspond to (022), (0$\bar{2}\bar{2}$), ($\bar{1}$21), (1$\bar{2}\bar{1}$), (101) and ($\bar{1}$0$\bar{1}$), with spacing of 0.503 nm according to the known monoclinic unit cell of PEO. However, the electron diffraction pattern is different from the diffraction pattern reported in the literature, could a new crystal structure form in the case of crystallization of single chain? the investigation of structure of single-chain crystal is still in progress.

FOLDING SURFACE

Periodic strips were found on the top surface of single-chain crystal with an average periodic length of 0.52 nm between strips, by using scanning tunnelling microscopy[3]. Based on the hexagonal shape of crystal, the top surface is considered to be the folding surface, thus each strip stands for a single molecular layer lamella viewed in the direction of the c-axis. The observed periodic length between strips is approximately in agreement with (220) spacing of i-PS crystal. The folding can also be supposed to be regular and adjacent on the basis of the uniformity of the strips.

STATISTICAL THERMODYNAMICS THEORY

A statistical thermodynamics theory is developed in order to give an analysis of the equilibrium state of single-chain single crystal based on a chain-folded lattice model[16]. Assuming the thickness of single-chain crystal to be l (in units of skeletal c-c bond) in chain direction and the number of loops to be ν and neglecting two chain-end cilia on the fold surface, then the following equation can be found for the single-chain crystal composed of M skeletal bonds

$$M = l(\nu+1) + \sum_{n=n^*} n\nu_n \tag{5}$$

and

$$\nu = \sum_{n=n^*} \nu_n \tag{6}$$

where $\nu+1$ is the number of stems in crystal; ν_n the number of loops containing n bonds; and the summation index n goes from n^*, the minimum number of bonds required to form a loop, to M.

The partition function and free energy of the single-chain crystal are calculated. From the calculation of free energy, the equilibrium thickness l_e of the single-chain crystal and melting temperature $T_{m,e}$ corresponding to the equilibrium state of the minimum free energy can be obtained.

$$l_e = M^{1/3}\left(\frac{-2kT\ln\lambda}{f_1 f_2 \Delta h_f}\right)^{2/3} \tag{7}$$

and

$$T_{m,e} = T_m^0\left(1 - \frac{3(f_1 f_2 \Delta h_f)^{2/3}(-2kT_{m,e}\ln\lambda_{m,e})^{1/3}}{2M^{1/3}\Delta h_f}\right) \tag{8}$$

where Δh_f is the heat of fusion (per bond) of the equilibrium crystal; T_m^0, equilibrium melting temperature; T, annealing temperature; k, Boltzmann constant; f_1 , f_2 and λ , the related parameters. Clearly l_e is related to the molecular mass of single chain and annealing temperature, while $T_{m,e}$ depends on molecular mass. Similar results can also be derived from thermodynamics analysis of single-chain crystal.

ACKNOWLEDGEMENTS

This work was supported by National Science Foundation of China and Chinese National Basic Research Project-Macromolecular Condensed State, and was in cooperation with Professor B. Wunderlich. We would like to thank Professor Renyuan Qian for helpful discussions.

REFERENCES

1. B. Wunderlich, *Macromolecular Physics*, Vols. 1-3, Academic Press, New York, 1973, 1976, 1980.
2. H. S. Bu, Y. W. Pang, D. D. Song, T. Y. Yu, T. M. Voll, G. Czornyj, and B. Wunderlich, *J. Polym. Sci., Polym Phys. Ed.*, 29, 139 (1991).
3. H. S. Bu, E. Chen, J. Yao, S. Xu, and Y.Kuang, *Polym. Eng. Sci.*, 32, 1209 (1992).
4. H. S. Bu, E. Chen, S. Xu, K. Guo, and B. Wunderlich, *J. Polym. Sci., Polym. Phys. Ed.*, to be published.
5. R. F. Boyer and R. D. Heidenreich, *J. Appl. Phys.*, 16, 62 (1945).
6. D. U. Quale, *Polymer*, 8, 217 (1967).
7. J. Kumaki, *Macromolecules*, 19, 2258 (1986).
8. D. Blundell, A. Keller, and A. J. Kovacs, *J. Polym. Sci., Part B*, 4, 481 (1966).
9. W. O. Statton, and P. H. Geil, *J. Appl. Polym. Sci.*, 3, 357 (1960).
10. B. Bittiger and E. Husemann, *Makromol. Chemie*, 96, 92 (1966).
11. N. Overbergh, H. Berghmans, and H. Reynaers, *J. Polym. Sci., Polym. Phys. Ed.*, 14, 1177 (1976).
12. *ATHAS Data Bank*, see for example, *Thermal Analysis*, Academic Press, Boston, 1990.
13. D. C. Bassett, *Principles of Polymer Morphology*, Cambridge University Press, Cambridge, 1981.
14. G. Natta and P. Corrandini, *Macromol. Chem.*, 16, 77 (1955).
15. Masaki Tsuji, Saroj K. Roy, and R. St. John Manley, *Polymer*, 25. 1573 (1984); *J. Polym. Sci., Polym. Phys. Ed.*, 23, 1127 (1985).
16. E. Chen, X. Hu, and H. S. Bu, *Makromol. Chemie*, to be published.

Polymers in New Water Treatment Technologies

B. A. Bolto

CSIRO Chemicals and Polymers
Private Bag 10, Rosebank MDC
Clayton, Vic 3169, Australia

Abstract: New methods of water and wastewater treatment are emerging from CSIRO which have high rates of both reaction and separation. The role of polymers in novel adsorption and coagulation technologies and the influence of polymer structure on performance are reviewed.

INTRODUCTION

CSIRO has for many years had active research under way on improved methods of cleaning up water and wastewater by means of processes based on adsorption and coagulation phenomena. The initial work on ion exchange for desalination, using resins which could be regenerated with hot water rather than chemicals, led to an emphasis on high-rate systems, both for the reaction step and for the separation of the adsorbed or accumulated impurities from the purified water, and the development of a range of new purification processes. All of these physicochemical methods, whether based on adsorption or on coagulation followed by sedimentation or flotation, have polymers as one of the necessary ingredients. The role of polymers in the processes will be outlined, together with the rationale for determining the optimum polymer structure for a particular function.

ADSORPTION PROCESSES

SIROTHERM Desalination

In a novel adsorption process for the removal of dissolved salt from brackish water, a mixture of weakly basic and weakly acidic ion exchangers is regenerated by a mere hot water wash, rather than by separation of the two resins and regeneration with chemicals in the usual manner (1):

$$R_b + R_aH + Na^+ + Cl^- \underset{90^o}{\overset{20^o}{\rightleftharpoons}} R_bH^+Cl^- + R_a^-Na^+$$

The driving force for the regeneration is the 30-fold increase in the dissociation of water at the higher temperature. The actual regenerants are the additional protons and hydroxyl ions formed on heating. The vital parameters are the buffering characteristics of the resins, the levels of basicity and acidity of the component resins and the way they change with temperature, the oxidation resistance of the resins (especially the basic resin) and the salinity level. The process is well suited to the partial demineralisation of brackish waters.

Buffering properties: To obtain the most effective utilisation of the exchange sites, it is essential that the ionic loading of the resins be insensitive to pH in the region of interest. The resins should be loaded as much as possible in the cold state, which means that they both should be highly ionised at near neutral pH. On heating both resins behave as weaker electrolytes, but

K.P. Ghiggino (Ed.)
Progress in Pacific Polymer Science 3

which means that they should both be mainly in the non-ionised form. It is difficult to achieve this for both resin types at the appropriate pH level.

Such a buffering characteristic is strongly dependent on the polymer structure, and is readily achievable in the case of the basic resin where the requirements, summarised in an earlier review (2), include:

- one type of group: secondary or tertiary amino
- one or more alkyl substituents on the nitrogen
- at least four carbon atoms separating the amino groups
- a non-polar polymer backbone

The phenomenon is best illustrated by the plots obtained when the basic resin, suspended in salt solution of constant concentration, is titrated at constant temperature against mineral acid. With resins made by aminating a crosslinked chloromethylated polystyrene with the simple multifunctional aliphatic polyamines $NH_2C_2H_4(NHC_2H_4)_mNH_2$, where m = 0 to 3, very sloping curves are produced. The pH change from 20 to 80% neutralisation, ΔpH, is a measure of the buffering properties of the resins, with a low value indicating strong buffering behaviour. For these polymers it ranges from 3.3 to 5.7. If the amination is done with diethylamine a mixture of tertiary amino and quaternary ammonium groups can be obtained, in the ratio of 74 : 26, whereupon ΔpH is 2.0. If the resin is boiled in alkali to dealkylate most of the quaternary ammonium sites to tertiary amino types, ΔpH becomes 0.4, which shows that the desired buffering effect has been attained.

Homofunctionality alone is not the predominant factor, since crosslinked or otherwise insoluble forms of polymers containing only primary amino groups as the basic sites have unsatisfactorily high ΔpH values. The values obtained were polyvinylamine, 3.8, polyvinylbenzylamine [**1a**], 3.2 and aminoethyl cellulose, 2.1.

For resins having tertiary amino sites, ΔpH is dependent on the nature of the substituents on the nitrogen. If hydroxyethyl rather than ethyl groups are present on the nitrogens in the polystyrene resin, the value is raised from 0.4 to 1.6.

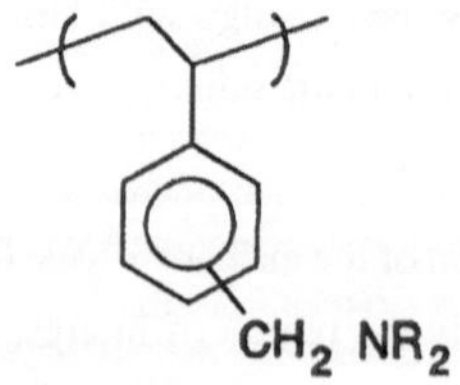

$CH_2\ NR_2$

1a: R = H
1b: R = Et
1c: R = Me

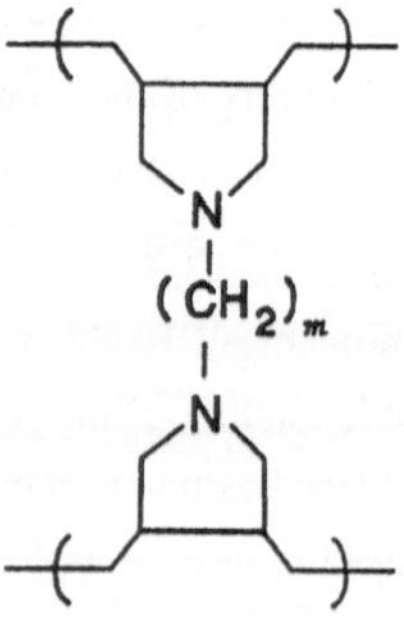

N
$(CH_2)_m$
N

2

The amino groups must not be in close proximity, as shown in the titration of the self crosslinked resins **2** made from the bis-diallylamines $R_2N(CH_2)_mNR_2$, where R is allyl (3). For m values of 2, 3 and 4 the ΔpH data are 3.5, 1.0 and 0.6 respectively, while for m of 5 to 10 the level remains at 0.2 to 0.3.

The polymer backbone must be essentially non-polar, as can be seen from the ΔpH values for resins containing appropriate basic sites, but attached to structures as below:

polyepichlorohydrin/piperidine [3]	1.6
poly(t-butylaminoethyl methacrylate)	0.5
polyvinylbenzyldiethylamine [**1b**]	0.4
polytriallylamine [**4a**]	0.5

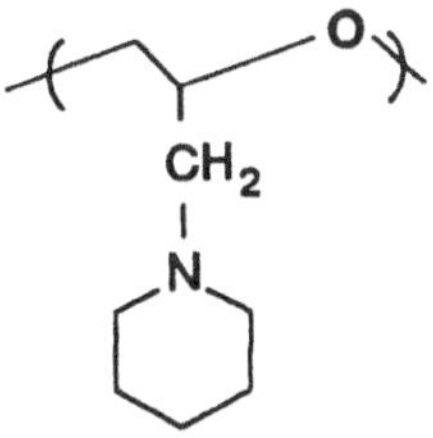

O
CH2
N
3

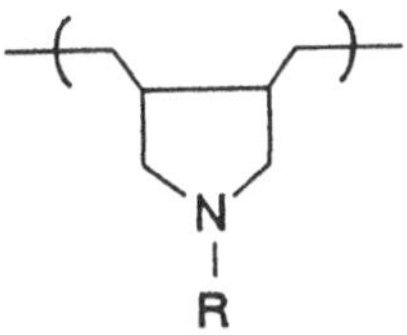

N
R
4a: R = allyl
4b: R = alkyl

It is clear that to obtain plateau behavior and the accompanying enhanced buffering capacity a non-polar molecular environment is essental. To obtain plateau behaviour in weakly acidic resins is much more difficult, in that they are inherently more polar structures because of the higher concentration of exchange sites (2). A similar curve shape is obtained for crosslinked polyacrylic and polymethacrylic acids, and for a resin made by reacting thioglcolic acid with chloromethylated polystyrene. However, a noticeable effect is produced by incorporating non-polar entities into a carboxylic acid resin. Thus introducing methyl methacrylate or vinyl isobutyl ether units into a poly(acrylic acid) resin causes ΔpH changes from -2.2 to -1.2 and -1.4 respectively, but any improvement made is negated by the severe lowering of the concentration of acidic groups. It is more practical to merely employ an acrylic acid resin and a higher ratio of acidic to basic sites. The reverse effect pertains when vinyl alcohol residues are present, the level then becoming -3.6.

Basicity levels of polyamine resins: Two structures which exhibit marked buffering properties are the polyvinylbenzyldialkylamines **1** crosslinked with divinylbenzene and polyalkyldiallylamines of structure **4b** crosslinked with bis-1,6-diallylaminohexane (2). With larger alkyl groups there is a decrease in the basicity of the polymers, because of the less polar environment around the nitrogen, and in the capacity or concentration of active sites, expressed as meq/g. For a 7-9% crosslinked styrenic resin the pH at half neutralisation in 1760 mg/L saline at 20° varies as follows:

Alkyl groups in **1**	Me	Et	n-Pr	i-Pr	n-Bu	i-Bu
pH at half neutralisation	6.7	5.6	5.3	5.7	4.6	3.0
Capacity, meq/g	5.0	3.7	3.2	2.4	2.8	2.8

For a 10% crosslinked diallylamine resin the corresponding data are:

Alkyl group in **4b**	Me	Et	n-Pr
pH at half neutralisation	9.0	9.1	8.5
Capacity, meq/g	8.4	7.8	6.7

The degree of substitution likewise influences these properties, as can be seen when the content of amino groups is increased in the amination of a chloromethylated resin with dimethylamine to produce **1c**, with more basic species being formed as the degree of substitution rises:

pH at half neutralisation	4.4	5.9	6.2	6.7	6.8
Capacity, meq/g	1.5	3.7	4.5	5.0	5.1

The variation of basicity with temperature in the polyamine resins described so far is mainly governed by the change in pK_W, but the effect is greater for plateau-type resins. Resins having the formula **5** have been synthesised by quaternising a crosslinked poly(4-vinylpyridine) with 1-bromo-6-aminooctane (4). They are capable of forming micellar units which are markedly temperature sensitive. Titration of the primary amino groups at 20° and 80° gives mid-point pH differences of 2.8 units, or almost double the amount ascribable to the change in pK_W. Unfortunately, because of the long hydrocarbon chain the total base capacity of the resin is about half that obtained with the preferred propyl version of the polyalkyldiallylamine resins **4b**, so there is no overall advantage.

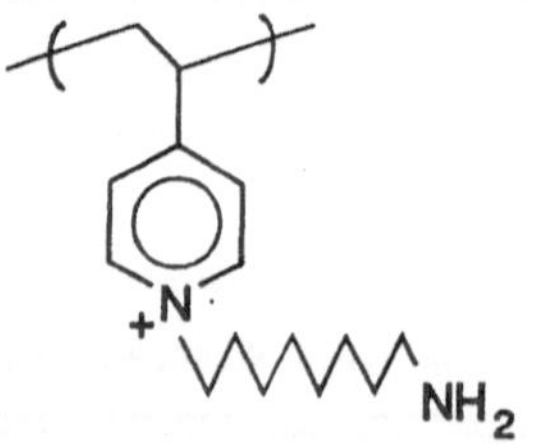

+N NH2

5

<u>Oxidative stability:</u> Both preferred resin types are thermally stable under the process conditions, and while oxidation is not a problem with the crosslinked poly(acrylic acid) resin, a slow oxidative degradation of the polyamine resin can occur during the hot regeneration stage via reaction with oxygen dissolved in the water. Some interesting features have arisen from oxidation studies of basic resins, linking stability with structure (5). The more basic polymers are more readily oxidised in general, and primary amino sites are susceptible, as are hyroxyethyl substituents. For polymers containing the one type of amino group, and which have similar basicity levels, a change in the backbone is significant. Crosslinked polyacrylamide, polyether

and polystyrene derivatives are least affected, and polytriallylamine are the most sensitive, losing 28% of their basic sites in an accelerated oxidation test, versus 1% for a polyvinylbenzyldimethylamine. Amide or ester links are not hydrolytically stable, and polystyrene resins have the disadvantage of low capacities, so attention was focussed on polyethers.

As already outlined, the plateau phenomenon seems to disappear when ether links are present. By aminating crosslinked polyepichlorohydrin with secondary amines to produce structures akin to 3, it was found that the less polar products showed an improved titration curve shape as regards the presence of a plateau. The best results were obtained with homopiperidine or mixes of piperidine with dipropylamine or dibutylamine, to give ΔpH values of about 0.6, whereas with dimethylamine, dipropylamine or piperidine the levels were 1.0 to 1.5. A very interesting feature of these two categories of polyether resins was that those exhibiting the better plateau were four times more sensitive to oxidation. A further fascinating effect was that, in contrast with all earlier observations, increasing the size of the alkyl substituent gave different results under hot conditions versus cold. For a resin made with a mix of dipropylamine and dimethylamine, ΔpH was 1.0 at 20°, but only 0.2 at 80°. This outcome is in accord with the increased hydrophobic interactions which occur with a more bulky substituent, and the increase in this effect on heating. The practical implication is that although the apparent thermal variation of basicity is zero at 70% resin loading, it is 1.1 pH units at 20% loading.

Rates of salt adsorption: Since the rections involved depend upon a proton transfer step at near neutral pH levels, the rate of salt uptake is very slow when resins of conventional size (300-1200 μm diameter) are used. The rate needs to be 100 times faster for the process to be competitive. To achieve this the basic and acidic sites must be in close proximity. There are several options: putting both groups in the one monomeric unit, putting them both within the same polymer particle in a 'snake-cage' format, or utilising microparticles. The amphoteric resins defined by the first two systems are unsatisfactory because of self neutralisation effects, in which internal salt structures are formed between the carboxylate and protonated amino groups, rendering them inaccessible to the inorganic ions. However, reducing the particle size of the two homofunctional resins to 1-5 μm gives the desired result (5).

Handling such finely divided material presents difficulties which may be overcome by forming them into composite beads of normal size, with or without an inert binder. A composite resin has been made from mixtures of the basic and acidic microparticles bound together by poly(vinyl alcohol) crosslinked with glutaraldehyde, a matrix which is permeable to water and salt. The resin was used for some years in batch operation of the process in a packed column, desalting 600 kL/day of a poor quality town supply for industrial purposes. Continuous operation offers significant cost advantages, so conventional continuous contactors have been explored on the pilot scale, but they are complex and difficult to control, which highlights the need for a new approach to continuous operation.

The ideal format for the resin is a porous bead of mosaic structure which is free of inactive matrix, and is composed of domains of one active species bonded to a domain of the other active type. The major obstacle to this approach is achieving an effective segregation of the active materials so that internal neutralisation does not occur. The perfect preparative route would be a one step procedure involving a solution of two monomers, with unwanted interactions minimised by blocking procedures. These may utilise counterions, especially multi-charged or bulky ions, precipitants and low pH conditions to ensure that the acidic monomer is uncharged, or neutral precursors such as amides or esters (5). The products actually consist of one species embedded in a matrix of the other. It is advantageous in fact to have a system which is initially heterogeneous, in which the amine monomer is polymerised around microparticles of the acidic polymer, either preformed or exist initially as droplets of monomer in a separate solution. Blocking of sites is still essential, but a 50% improvement in salt uptake capacity is obtained, indicating that virtually all the sites present in the product are available.

Selective and Readily Regenerable Ion Exchangers

The nature of the substituent group on the nitrogen in quaternary ammonium polymers has a profound influence on the selectivity of the adsorbent for anions, as does the nature of the polymer backbone. Thus resins such as **6a** with a polystyrene backbone and tributyl substituents on the quaternary nitrogen prefer nitrate over sulfate (6), contrary to conventional resins **6b** containing trimethylammonium sites, and the quaternary ammonium resin **7** made from polydimethyldiallylamine or those obtained by aminating polyepichlorohydrin with tertiary amines to give **8**. Regeneration is achieved with brine.

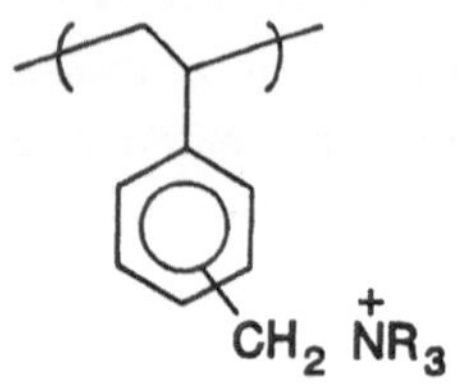

$CH_2\ \overset{+}{N}R_3$ | $\overset{+}{N}$ Me Me | CH_2 $^{+}NR_3$

6a: R = Bu
6b: R = Me

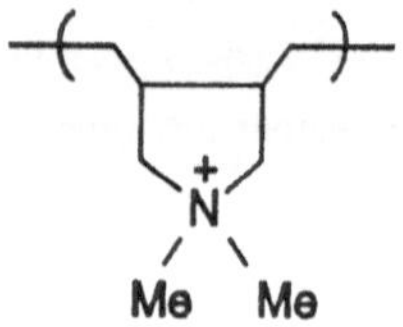

7

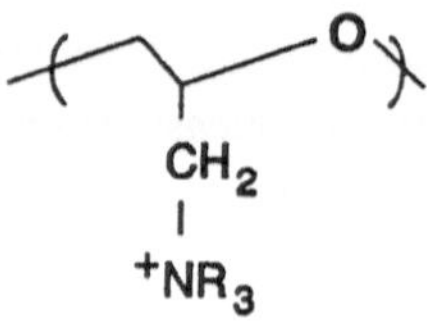

8

Sulfur-containing resins specific for mercury uptake have been made from polyepichlorohydrin and thioglycolic acid or thiourea, but they are difficult to regenerate, just as are known selective resins of the thiol type (7). However, cationic mercury is readily adsorbed by commercial iminodiacetic acid resins, from which it can be removed as the anionic chloro complex by a brine wash; alternatively, when there is already chloride present in the raw water the mercury can be adsorbed as the chloro complex on a commercial quaternary ammonium resin, from which it can be recovered by a complexing agent such as sodium sulfite (8).

Reacting polyepichlorohydrin with sulfite introduces sulfonic acid sites, but some chain scission occurs at ether linkages, depending on the nature of the crosslinker used (9). Cleavage is more pronounced when a dithiol provides the crosslinking, while substitution predominates when a diamine is used. As with all strong electrolyte resins, regeneration requires considerably more than the stoicheometric amount of alkali or acid, highlighting the advantages of weak electrolyte systems, where only about a 15% excess of regenerant is necessary. Thus a chromate selective but oxidatively stable weakly basic resin has been made from crosslinked polyacrylonitrile beads by reaction with diethylenetriamine (10). It is selective for Cr(VI) over sulfate and chloride, and is readily regenerated with alkali.

Magnetic Ion Exchangers

Another way of handling microparticles and one which allows continuous operation is to incorporate a magnetic filler within the beads (11). Resin beads of size 20-200 μm which contain 30% w/w of gamma Fe_2O_3 flocculate strongly when magnetised, so that rapid settling results. Because of the high voidage, the magnetised beads may be directly pumped without great attrition, which is not possible with conventional resins. This makes a truly continuous system feasible. The resin is used in the magnetised state since on agitation the flocs disperse so that rapid reaction can occur with the single beads. When passed to a quiescent zone the flocs reform. Composite magnetic resins, made up of a mixture of weakly basic and acidic resins and the magnetic filler embedded in a crosslinked PVA matrix, have been tested on flows of 1 ML/day to confirm the large-scale practicability of a continuous version of SIROTHERM desalination in a novel fluidised bed system.

A magnetic resin having weakly acidic properties only has been synthesised by grafting acrylic acid polymers onto a magnetic core of crosslinked PVA containing a magnetic iron oxide (12). It was successful in pilot tests of calcium bicarbonate removal from ground water, and in a full-scale demonstration of nickel salt recovery from electroplating rinse water, where reuse of the nickel and water was shown to be a viable proposition (11). Other shell format resins were made by similarly grafting 2-hydroxy-3-methacryloyloxypropyltrimethylammonium chloride to produce a strongly basic material. Polyacrylamide grafts have been converted to primary and tertiary amino forms by Hofman and Mannich reactions respectively (5). Another weakly basic resin has been prepared from poly(vinyl chloride) or polyepichlorohydrin beads containing an iron oxide by aminating them with polyethyleneimine (13).

The advantages of continuous contactors based on magnetic resins are that the amount of resin needed is only 25-33% of that required in other systems, the equipment is smaller and simpler, and there is no necessity for a prior filtration step. Magnetic resins can also be used to detoxify sludges and slurries, when employed in large bead form so that separation from the gelatinous sludge is facilitated (11).

A selective magnetic adsorbent made from activated carbon and a magnetic filler enclosed within a cage of crosslinked PVA removes small nuisance molecules from food processing slurries, but leaves untouched macromolecules which have a high nutritional value (5).

NON-CHARGED MAGNETIC POLYMER PROCESSES

Magnetic polymers have been developed for applications that exploit the high voidage of the magnetised particles. The conglomerate network formed by linkages of particles along flux lines persists in the settled state, leaving tortuous channels and large cavities. If the particles are irregular in shape the effect can be accentuated, and a bed can be formed which is difficult to compress.

A magnetic polymer lacking functional groups has been prepared in hydrophilic form from a vinyl chloride/vinyl acetate copolymer for use as a reusable filter aid in small-scale water treatment (11). Styled on diatomaceous earth, which must be disposed of after the filter bed is saturated, the particles can be separated from the impurities with a hydrocyclone and reused.

Floating vesicular magnetic particles made from a hydrophobic polymer such as polystyrene have been used for recovering oil from harbour spills (11). The unmagnetised material is sprayed on the spill and recovered on rotating magnetic discs, trapping the oil within the voids of the magnetic floc. Recovery of the oil can be achieved by centrifugation.

COAGULATION AND SEDIMENTATION

SIROFLOC Clarification

Insoluble suspended matter can be removed from water by means of positively charged particles. This is normally done with flocs of the insoluble hydroxides of aluminium or iron, but to accelerate the reaction and separation stages by magnetic particle technology it is possible to employ magnetic shell graft resins containing quaternary ammonium sites, with brine as the regenerant. However, it has been found to be far more practical and economic to employ finely divided iron oxide in the form of 1-10 μm sized particles of magnetite, Fe_3O_4. Its surface is positively charged at pH levels below 6.5, when the magnetite will adsorb and coagulate negatively-charged colour bodies and colloidal material (11). If there are high levels of impurities present, a coagulant aid in the form of a **soluble** cationic polyelectrolyte is required to bind the excess to the already loaded and now negatively charged magnetite. The system is regenerated with alkali, whereupon the charge on the magnetite is reversed and the impurities are repelled from its surface.

The role of the polyelectrolyte has been extensively studied (14). Logically, a weakly basic polymer would seem be the most appropriate, as it would be protonated and hence positively charged under acidic conditions, and neutral as the free base form in alkali, which would encourage its release from the negatively charged species. The polyamines studied fell into two categories: short chain polymers of molecular weight 10^4 to 10^5 daltons, such as polyethyleneimine, the condensation polymer **9** obtained from epichlorohydrin and a primary amine such as methylamine, polymers of structure **10** made by aminating polyepichlorohydrin with secondary amines, polyvinylamine, and uncrosslinked versions of the polyalkyldiallylamines **4b**. None performed satisfactorily in coagulation tests at pH 6 because of the weak binding of the impurities to the magnetite. Turbidity shearing occurred with the impurity particles separating from the magnetite during agitation and pumping.

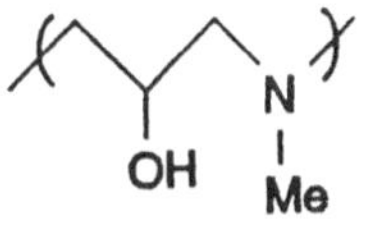

9 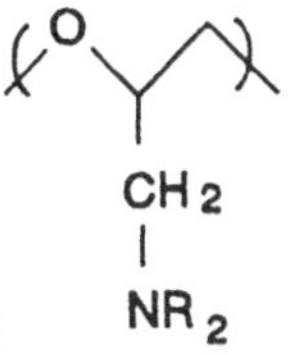 10

This was alleviated by using the second category of polymers: those of much greater chain length, by a factor of 100, one example being **11**, copolymers of acrylamide with dimethylaminoethyl acrylate. However, the lower charge density mitigates against better performance. In all cases the release of impurities in alkali was excellent. The best polyamine overall was polydiallylamine, but the clarification performance was not adequate.

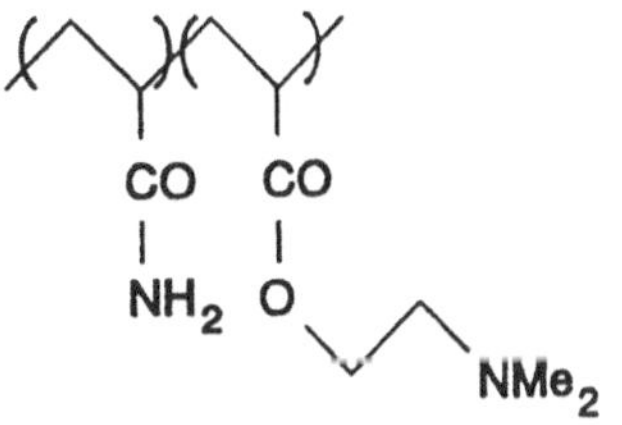

11

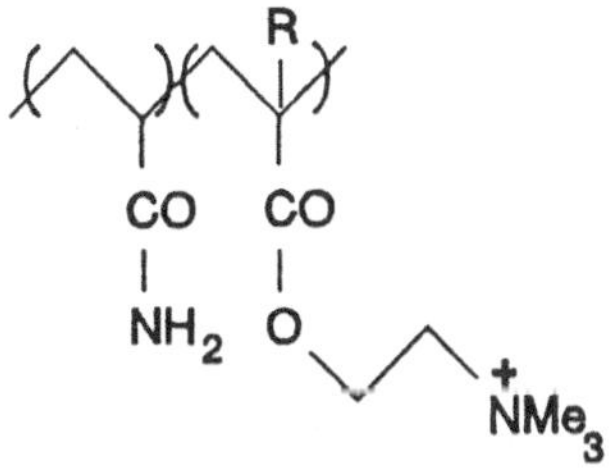

12a: R = H
12b: R = Me

The quaternary ammonium polymers investigated included quaternised forms of some of the above polyamines, as in the products from the reaction of epichlorohydrin with dimethylamine and from the amination of polyepichlorohydrin with trimethylamine. Others were the products from the quaternisation of poly(4-vinylpyridine) with methyl bromide, from the amination of chloromethylated polystyrene with trimethylamine to give the soluble version of **6b**, and from the polymerisation of dimethyldiallylamine to yield a soluble version of **7**. These were all short chain types; as before long chain versions were made, this time by copolymerising acrylamide with the chlorides of trimethylammoniumethyl acrylate or methacrylate to produce **12a** and **b** respectively. The highly charged short chain products were the most successful at water clean up, the best being polydimethyldiallylamine and the condensation polymer from epichlorohydrin and dimethylamine. The former is preferred on health clearance grounds. The regeneration characteristics were a function of chain length, with the shorter chains preferred. With long chain species there can be a problem with removal of the flocculant in the regeneration stage. Six full-scale SIROFLOC water treament plants of capacities up to 70 ML/day have been constructed or are being built which utilise a cationic polyelectrolyte.

In an attempt to improve upon the ease of regeneration, novel polyampholytes have been synthesised which are cationic at slightly acidic pH levels, but neutral or anionic under alkaline conditions (15). The rationale is depicted in the equation

$$—B^+—AH— \; + \; OH^- \rightleftharpoons —B^+—A^-— \; + \; H_2O$$

A range of structures was explored, including copolymers of methacrylic acid with monomers such as trimethylammoniumethyl acrylate or dimethyldiallylammonium chlorides to give **13** and **14** respectively, terpolymers of these with methyl methacrylate, and polymers containing phenolic or amino acid moieties as well as quaternary ammonium groups. Methacrylic acid moieties are somewhat too acidic for the purpose, as in homopolymers they are more than half ionised at pH 6, and with cationic groups present the ionisation will be much greater. Hence the majority of the cationic sites will be neutralised, rendering the system only 10 to 20% as effective in clarification as a purely cationic polymer, although it regenerates five times more readily. This is also the case for the terpolymers containing methyl methacrylate, even though they are slightly weaker acids because of the less polar environment.

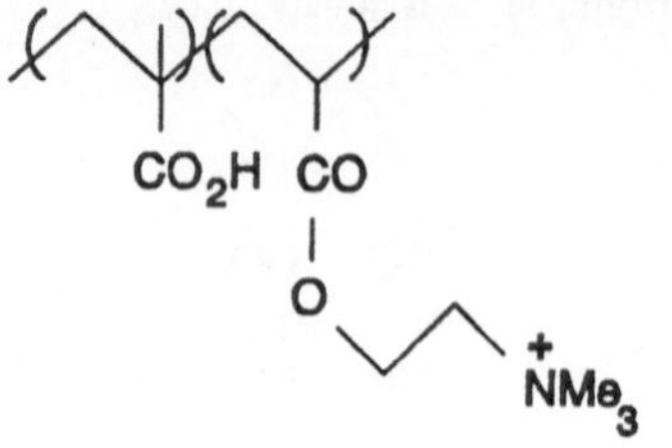

13 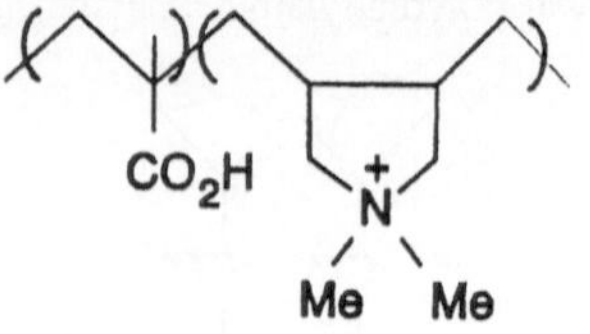 **14**

Phenolic groups, as in phloroglucinol residues introduced into chloromethylated polystyrene after the amination of half the chloro groups with trimethylamine to give **15**, would be expected to be only about 3% ionised at pH 6. A polymer of this type is 80% as effective as a purely cationic one, but is four times more facile in regeneration. A further route to the desired charge reversal behavior is via the introduction of amphoteric groups into a cationic polymer as in **16**.

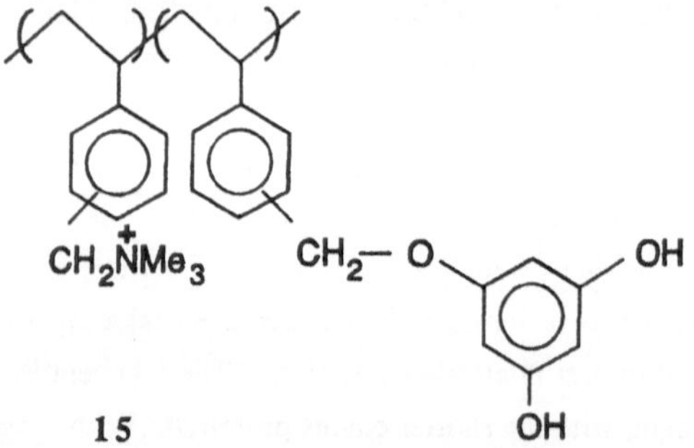

15 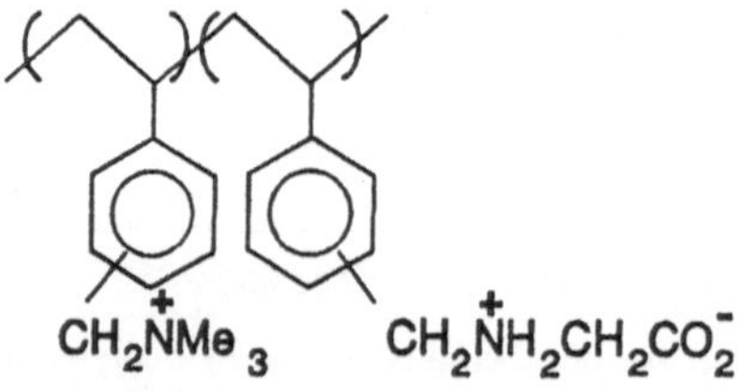 **16**

An aminoacid substituent such as glycine which, although zwitterionic, is overall neutral at pH 6, will become deprotonated and anionic at pH 11. These products are 90% as effective versus cationic homopolymers and regenerate two to three times more easily. Thus there is a consistant trend: as clarification improves, regeneration becomes more difficult. So far no product has been obtained which will match the cationic homopolymer in clarification performance yet give a system which is more readily regenerable.

COAGULATION AND FLOTATION

Aqueous industrial effluents containing organic contaminants as emulsified or colloidal material are amenable to separation by flotation because of the inherent hydrophobicity of the impurities (16). Unless there is much dissolved coloured material present, when a metal salt coagulant may have to be added, the only additive necessary is a polymeric flocculant to coalesce the contaminants. A high-rate system is then possible by employing large air bubbles to create a foam; the organics are recovered in the froth, leaving a clear subnatant solution. The large energy requirement of competing processes such as dissolved air flotation (which employs small bubbles) is avoided, but at the expense of the flocculant addition. Cationic polyelectrolytes perform this function very satisfactorily, and pilot plant work has shown that for oily wastes from steel mills and greasy wastes from wool scouring the removal of oil and grease can be better than 90%. There is further scope for improving performance by designing more appropriate flocculants.

CONCLUSIONS

The demands of adequate performance in proposed new water treatment processes have required a thorough understanding of the influence of molecular structure on equilibria and of physical format on reaction and separation rates. Non-polar substituents have a profound effect on ion binding involving insoluble adsorbents containing amino, carboxylic acid or quaternary ammonium sites. By the introduction of such substituents, proton release from amino groups on heating is facilitated, and the binding of monovalent rather than divalent anions is preferred by quaternary ammonium groups. In the field of soluble polyelectrolytes the selection of the optimum structure is less straightforward. Assisting in the heterocoagulation of colloids with magnetite by adding cationic polymers which will allow reuse of the magnetite presents a formidable challenge, with the requirement of maximum binding for the working stage and of minimum binding for the regeneration stage. Amphoteric species are probably the answer here. In foam flotation performance can be enhanced by tailoring the polyelectrolyte to suit the properties of the material to be flocculated.

Solid reagents in finely divided form are ideal for accelerating reaction rates, and can be conveniently handled by a composite particle or magnetic filler approach. The need to improve upon the inherently slow reactions in the SIROTHERM process has spawned a series of continuous versions of adsorption and coagulation processes, which radiate from the common magnetic particle axis. Of the ones described, the SIROFLOC clarification process has been the most successful, with the installation of full-scale plants in three countries.

REFERENCES

1 Bolto BA, Weiss DE (1977) In: Marinsky JA, Marcus Y (eds) Ion exchange and solvent extraction, vol 7. Dekker, New York

2 Bolto BA (1980) In: Goethals EJ (ed) Polymeric amines and ammonium salts. Pergamon, Oxford

3 McLean CD, Ong AK, Solomon DH (1976) J Macromol Sci - Chem A10: 857

4 Joyce DE, Kurucsev, T (1980) Polymer 21: 1457

5 Bolto BA (1983) Progress Polymer Sci 9: 89

6 Jackson MB,and Bolto BA (1990) Reactive Polymers 12: 277

7 Jackson MB, Vickers LJ (1988) Reactive Polymers 8: 11

8 Becker NSC, Eldridge RJ (1992) Water 19(5): 33

9 Bolto BA, Jackson MB, Vickers LJ (1986) Reactive Polymers 4: 181

10 Eldridge RJ, Vickers S (1992) In: Slater MJ (ed) Ion exchange advances. Elsevier, London

11 Bolto BA (1990) Waste Management 10: 1

12 Clemence LJ, Eldridge RJ, Lydiate J (1984) Reactive Polymers 2: 197

13 Clemence LJ, Eldridge RJ (1988) Reactive Polymers 8: 27

14 Anderson NJ, Blesing NV, Bolto BA, Jackson MB (1987) Reactive Polymers 7: 47

15 Anderson NJ, Bolto BA, Eldridge RJ, Jackson MB (1993) Reactive Polymers 19: 87

16 Beeby JP, Dixon DR, Gray SR, Smitham JB (1993) Proc 15th Fed Conv. Australian Water and Wastewater Association, Gold Coast

Novel Application of Natural Rubber Latex as Natural Flocculant for Tin Tailings Slurry

C. C. HO[*+], K. C. LEE and E.B. Yeap[**]

[*]Department of Chemistry and [**]Department of Geology, University of Malaya, 59100 Kuala Lumpur, Malaysia

Abstract: Natural rubber (NR) latex concentrate is essentially a stable dispersion of polyisoprene latex particles in an aqueous serum phase containing some minor constituents of soluble non-rubbers whereas tin tailings slurry is a very stable dispersion of mainly submicron size clay minerals with some heavier silt particles which normally settle out relatively fast. Under normal pH conditions, the negatively charged clay particles in the slurry would remain disperse for an extended period of time. There has not been much success in destabilizing this slurry with most of the common synthetic polymeric flocculants. It was found that when used in combination with an inorganic salt (e.g. calcium chloride) under optimum conditions, the negatively charged natural rubber latex particles can function as a polymeric flocculant in bringing about a complete heterocoagulation of the mixed dispersions whereby producing a clear supernatant and a porous sediment. The effect of pH, slime and latex concentrations on the heterocoagulation behaviour of the latex-slurry mixed dispersion was investigated. The influence of these factors on the efficiency of destabilization, the floc size of the sediment and the settling rate was compared with those which use calcium chloride only as coagulant. The floc size of the heterocoagulated latex-slurry mixtures decreased with increasing pH of the dispersion and slime concentration. The slurry could not be destabilised effectively using either latex or coagulant alone.

INTRODUCTION

Polymeric flocculants are water-soluble, high molecular weight macromolecules that can function by a bridging mechanism to link together the disperse particles of a colloidal dispersion. Synthetic polymers such as polyacrylamides are widely used in water clarification, particulates removal from waste-waters in effluent treatment and in the control of the state of aggregation of the disperse phase in many industrial processes.

It has been reported previously that coagulation and flocculation have been employed with various degree of success in treating slimes or slurries left behind by mining activities [1,2]. Both inorganic electrolyte and synthetic polymer are used jointly to effect destabilization of these slurries. Natural products such as corn starch and hydrolysed Cassava have also been successfully empolyed in the flocculation of tar sand tailings [3]. However in our search for an effective flocculant for tin tailings slurry, essentially a stable dispersion of clay minerals, we have not had much success with any of the commercial

[+] Corresponding author

K.P. Ghiggino (Ed.)
Progress in Pacific Polymer Science 3

synthetic polymeric flocculant available [4]. Neither did natural macromolecules such as corn starch and alginates work. It was with a combination of natural rubber latex particles and an inorganic electrolyte that an effective flocculation system was found for these slurries.

Indeed natural rubber latex concentrate is a dispersion of polyisoprene latex particles in an aqueous serum phase containing a small amount of soluble non-rubber constituents. These latex particles are stabilised by adsorbed proteins and long chain fatty acid soaps derived from the hydrolysis of phospholipids originally present in the latex from the *Hevea Brasiliensis* tree [5]. Thus the latex particles are amphoteric in nature and become negatively charged at alkaline pH. On the other hand, the overall nett charge on the submicron size plate-like clay particles in the slurry is strongly dependent on pH : the edges are positively charged at acidic pH whereas the basal surface is negatively charged. At alkaline pH the clay particles are negatively charged. Thus the interaction between the NR latex particles and the clay particles in the slurry, both carrying negative charges at neutral pH and comparable in size, is expected to be complicated with respect to their chemical compositions and polydispersity in particle size compared to well-defined model systems. It is envisaged a heterocoagulation process was involved between the rubber latex particles and the clay particles. Heterocoagulation phenomena of mixed systems have been observed on well-defined systems since the Sixties but the process has seldom been employed for the destabilization of mining slimes as such. It has also been known that adsorption or adhesion of small particles onto larger ones could destabilize a dispersion.

In the present study, slurry from three ex-tin mining ponds with different mineral compositions were heterocoagulated with NR latex in the presence of calcium chloride salt. The effect of pH, latex and slurry concentrations and nature of slime types on the stability of the mixed dispersions were studied by monitoring the transmittance of the supernatant, the settling rates, aggregate size and morphology of the sediment of the mixed dispersion. An explanation of the observed phenomenon through an electrostatic bridging mechanism of the latex particles to the clay particles of the slurry via the calcium ions is proposed.

MATERIALS AND EXPERIMENTAL

Samples And Chemicals

Slimes collected from three different ex-mining slurry ponds : Malim Nawar (MN), Pengkalan (PN) and Serendah (SH), Malaysia were air-dried, pulverised, sieved (60 mesh BS) and stored at ambient temperature before use. The clay mineral compositions of the slurries have been reported previously [6] and are given in Table 1. Commercial high ammonia NR latex concentrate (HA) prepared by centrifugation was used as received. The dry rubber content (DRC) of the HA latex was about 60.5% w/w. The particle size, determined by electron microscopy ranged from 0.05 to 0.6 micron with an average size of 0.24±0.13 micron.

Other analytical grade chemicals were used as received.

Table 1. Mineral Composition, Physical and Chemical Properties of Slimes

	Samples		
	MN	PN	SH
Kaolinite (%, w/w)	65.1	71.2	32.3
Illite (%, w/w)	10.2	8.2	9.7
Amorphous clay (%, w/w)	10.0	nd	24.7
Montmorillonite (%, w/w)	nd	7.8	nd
Quartz (%, w/w)	15.3	12.0	33.3
Amorphous* content	5.6	1.4	21.0
Total organic content(%, w/w)	1.46	1.97	-
Average Particle Size and Size Range (m)	0.32 ± 0.18 0.07 ± 0.96	0.14 ± 0.08 0.03 ± 0.86	- -

nd - not detected
* - determined by chemical analysis

Heterocoagulation Procedure

8.0 g of air-dried slurry were dispersed in 50.0 cm^3 of distilled water in a 100 cm^3 measuring cylinder. The appropriate volume of calcium chloride solution was then added followed by distilled water to give a volume of 70.0 cm^3. The cylinder was inverted 15 times to effect mixing of slurry. 0.16 g of HA latex concentrate dispersed in 30.0 cm^3 of distilled water were added to the measuring cylinder to give a final volume of 100.0 cm^3. Thus the final dispersion contained a latex concentration of 0.1% w/v. The cylinder was immediately inverted 15 times to ensure complete mixing of the contents. The contents of the cylinder were allowed to settle under gravity without further disturbance. The sediment height i.e. sedimentation volume was then recorded at regular intervals until no further change was observed. The settling rate was obtained from the initial slope of the plot of sediment height versus time. After the dispersion was allowed to settle for 20 minutes, 5.0 cm^3 of the supernatant was pipetted out (at a point about 3.0 cm below the liquid surface in the cylinder) and diluted to 25.0 cm^3 with distilled water. The transmittance of this was measured at a wavelength of 450 nm. High transmittance corresponds to clear supernatant and effective flocculation whereas low transmittance corresponds to ineffective flocculation and cloudy supernatant. A visually clear supernatant would give transmittance in excess of 85%. The floc size of the sediment was measured using optical microscopy after there was no further change in the sediment height. Optical micrographs taken under the same magnification showed the floc structure of the sediment.

RESULTS AND DISCUSSION

Effect Of pH

In the mixed dispersion, the surface charges of both the clay and HA latex particles are pH dependent. The electrophoretic mobilities of both the clay and latex particles attained maximum (constant) values above pH about 7 [7] reflecting complete ionisation of the surface groupings on these particles. The solution chemistry of the inorganic coagulant, usually metal ions, is also influenced by pH. Thus the coagulation of clay dispersion is expected to be a function of pH of the mixed dispersion. Edge-to-edge (EE) and edge-to-face (EF) associations of clay particles prevail over acidic and neutral pH range at low electrolyte concentration, while face-to-face (FF) association predominates under alkaline pH at high electrolyte concentration. Visually the photomicrographs (Fig 1) for the MN-HA mixed dispersion at 8% w/v slime and 0.1% w/v latex concentrations shows very clearly the decrease in floc size with increasing pH. At pH 2 and 3, the sediment consisted of solely large flocs resulting in effective solid separation and high transmittance. At pH 6, large flocs and many small aggregates were formed in the sediment. On further increase of pH to alkaline conditions, all large flocs were absent and the sediment consists of very small and well-dispersed clay particles only. Thus it is clear that the latex-slurry mixed dispersion would remain stable at pH 10 giving cloudy supernatant in the absence of any added inorganic electrolyte (e.g. calcium chloride). In addition 0.1% w/v HA latex itself is stable in the presence of even up to 1.81×10^{-3} mol dm^{-3} (200 ppm) $CaCl_2$.

The effect of added calcium chloride can be seen in Fig. 2 for the MN-HA mixed dispersion. There was a three-fold increase in the optimum concentration of calcium chloride (defined

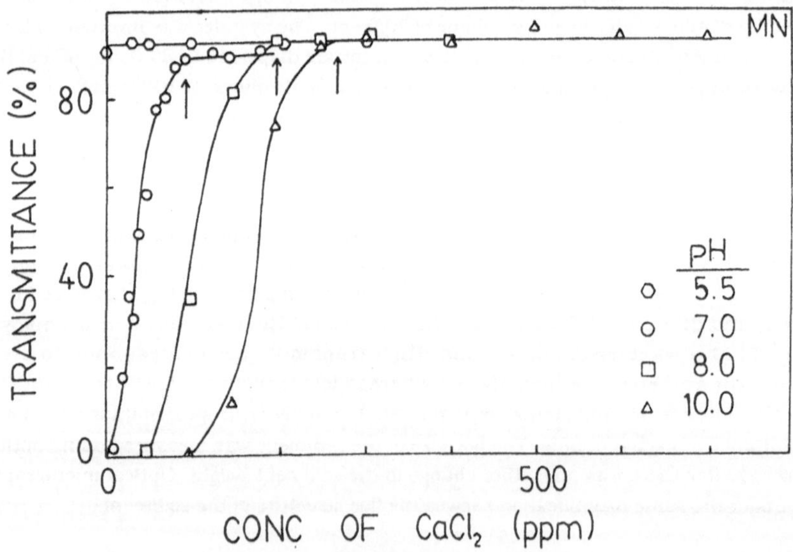

Fig. 2. Effects of pH on the heterocoagulation of MN slime-latex mixed dispersion (8.0% w/v slime and 0.1% w/v latex) studied as a function of $CaCl_2$ concentration.

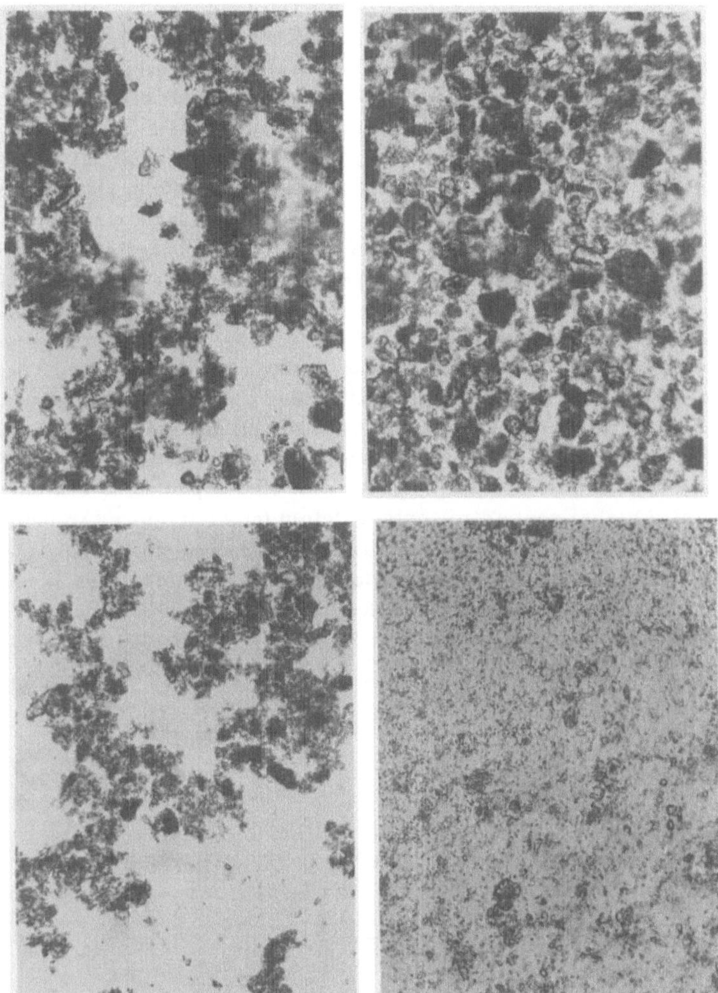

Fig. 1. Photomicrographs show the effect of pH on the state of aggregation of 8% w/v MN slime-latex mixed dispersion in the presence of 0.1% w/v latex: (a) pH 2; (b) pH 3; (c) pH 6 and (d) pH 10.

as the concentration at which the transmittance of the supernatant achieved maximum value i.e. > 80%) needed to bring about heterocoagulation when the pH was raised from 7 to 10. At pH 5.5 the mixed dispersion of HA latex and clay were completely flocculated even in the absence of calcium chloride. Thus addition of more calcium chloride to the mixture had no effect on the transmittance of the supernatant which remained high throughout. The edge of the clay particles is positively charged in acidic pH, whereas at pH 5.5 the negative charge on the HA latex particles is low but above its isoelectric point [8]. Subsequently the weakly negatively charged latex particles could interact with the positive sites on the edges of the clay particles, leading to heterocoagulation of the mixed dispersion. At pH 7.0 residual positive charge on the edges of the kaolinite platelet could still link up with the basal surface of another kaolinite platelet via the negatively charged latex particles, if sufficient calcium ions are present on the basal surface to serve as 'electrostatic bridges'. Above pH 7, the edges of the clay platelets were rendered negative as a result of specifically adsorbed OH^-. The net negative charges on both the clay and latex particles are much higher compared to those at acidic pH and strong repulsion keeps the particles apart resulting in a cloudy dispersion. Under this condition, interaction between the clay and latex particles would be by 'electrostatic bridges' provided by calcium ions and significantly higher concentration of calcium ions is required to bring about heterocoagulation as evidenced by the higher optimum concentration of calcium chloride required at pH 10. The function of Ca^{2+} ions as 'electrostatic bridges' in enhancing adsorption of anionic bioflocculant on river sediment [9] and anionic surfactant on kaolinite [10] has been reported recently.

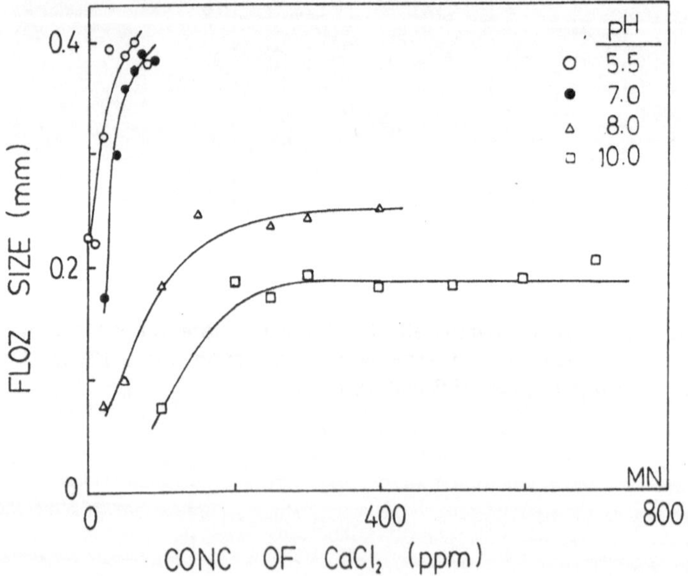

Fig. 3. Effect of pH on the floc size of MN slime-latex mixed dispersion (8% w/v slime and 0.1% w/v latex) studied as a function of $CaCl_2$ concentration.

The sedimentation volume of the heterocoagulated sediment of a 8.0% w/v MN slurry and 0.1% w/v latex was 22 cm^3 at pH 5.5. This was decreased to 13 cm^{-3} when the pH was increased to 10 and in the presence of 350 ppm calcium chloride . This clearly shows that the more open 'card-house' structure of EF association results in a much larger sedimentation volume of the flocs than the 'card-pack' structure of FF asociation brought about by high electrolyte concentration [11]. This is well correlated with a decrease in floc size of the sediment with increase in pH at any particular concentration of calcium chloride as shown in Fig 3.

Nature Of Slurry

The extent of heterocoagulation of the mixed dispersion of the slurries from three ponds each containing 0.1% w/v of HA latex and plotted in the form of transmittance of the supernatant of the mixed dispersion versus pH is shown in Fig. 4. A sharp drop in transmittance was observed at pH 6-7 for MN-HA and PN-HA systems but at higher pH (7-9) for SH-HA system. This means that below these pHs the respective mixed dispersions were completely destabilised. Thus for example the supernatant of the MN-HA system was clear in acidic pH, whitish at pH 7 but appeared brownish at pH > 7. In the presence of $6.5x10^{-4}$ mol dm^{-3} (72 ppm) calcium chloride, MN slurry itself and MN-HA mixed dispersion were found to be stabe at a higher pH (7-9) compared to cases when they were absent. This difference in stability could be traced to the nature of the slurry itself. SH dipersion was acidic (pH 4.1) while the natural pH of MN and PN dispersions were 6.7 and 7.8 respectively. Under acidic condition, hetero-coagulation via interaction of positively charged edges of the clay particles with latex particles would occur as outlined above.

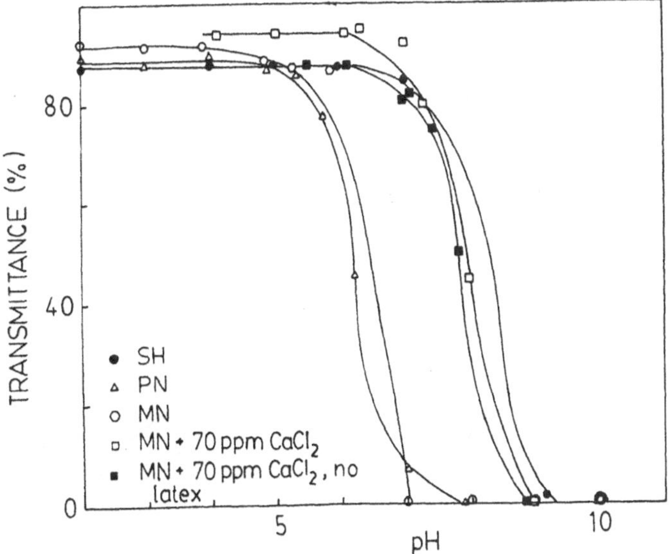

Fig. 4. The variation of transmittance of supernatant with pH of various slime-latex mixed dispersions (8% w/v slime and 0.1% w/v latex). For comparison MN slime dispersion in the presence of 70 ppm $CaCl_2$ was included.

Thus the transmittance of the supernatant of MN-HA and PN-HA systems remained clear and constant at > 90% at pH 2-6. For the acidic slurry from SH pond, hydrolysis of its amorphous content (Fe_2O_3) results in the formation of hydrolysed ionic species [6] that could behave as electrostatic bridges and allow adsorption of the negatively charged latex particles on the clay particles to take place leading to destabilisation of the SH-HA dispersion. At about pH 7 in the absence of added electrolyte, the weak interaction between the residual positive charges on the edges of the clay platelets and the HA latex particles resulted in excess of unaggregated free latex particles in the supernatant as revealed by its whitish appearance. The whitish supernatant is also indicative of the absence of dispersed clay particles. The occurrence of EE and EF associations among clay particles resulted in their complete removal from the aqueous phase. Under alkaline condition, both edge and basal surface of the clay particles as well as the latex particles are all strongly negatively charged and hence would remain well-dispersed and so the supernatant appeared brownish. Thus heterocoagulation can only take place through the 'electrostatic bridge' mechanism in the presence of metal ions at high pH. In the case of the acidic SH slurry, formation of ferric hydroxide under weakly alkaline pH destabilised the otherwise stable clay and latex particles through co-precipitation mechanism giving rise to a slightly higher pH at which zero transmittance of the supernatant was obtained.

Fig 5 shows the floc size of the three types of slurry as a function of pH in the presence of 6.31×10^{-4}mol dm^{-3} calcium chloride (70 ppm). The general trend of decreasing floc size of the sediment with increasing pH was again noted for all the mixed dispersions. Some minor difference in floc size between the three different types of slurries especially at low pH was evident. In addition, the floc size of the sediment for the PN-HA mixed dispersion was much smaller than that for the MN-HA system at all pH leading to a slower settling rate for the former.

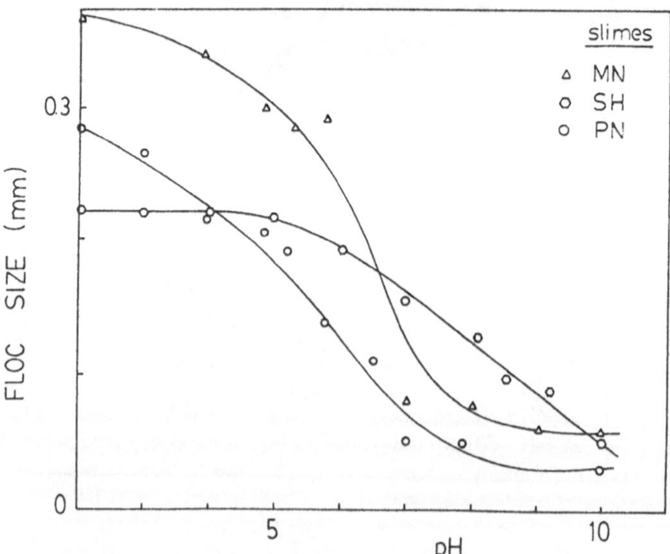

Fig. 5. The variation of floc size with pH of various slime-latex mixed dispersions (8% w/v slime and 0.1% w/v latex).

Effect Of Latex Concentration

At optimum pH 7 and a slurry concentration of 8% w/v, the transmittance of the supernatant decreased as the latex concentration increased for any particular calcium chloride dosage as shown in Fig. 6. Thus the transmittance for 0.5% w/v latex system was the lowest over the entire calcium chloride concentration range studied. A minimum concentration of 2.7×10^{-4} mol dm^{-3} calcium chloride (30 ppm) was needed before any heterocoagulation was noted (i.e. transmittance of supernatant commenced to increase from 0% latex), whereas for 0.1% w/v latex, a lower calcium chloride concentration of 10 ppm was found. The optimum concentration of calcium chloride to cause destabilisation of the mixed dispersion was about 80 ppm for the low latex concentration but at 120 ppm at 0.5% w/v latex. The floc size of the mixed dispersion, containing 0.1% w/v latex (see Fig. 7) was the largest and it increased very rapidly with calcium chloride concentration and prior to attainment of a constant size at the optimum concentration of calcium chloride. There was a four-fold increase in floc size. The aggregate size of MN slurry itself was very much smaller and this only increased slightly with calcium chloride concentration. Thus the settling rate was the highest for the mixed dispersion containing 0.1% w/v latex. When excess latex was present as the case of mixed dispersion containing 0.5% w/v latex, not only a higher optimum concentration of calcium chloride was needed but a network structure of flocs was also formed resulting in lower settling rate. Thus at optimum dosage of latex (0.1% w/v) the largest flocs and the highest settling rate were observed. The above results clearly demonstrates that the dosages of calcium chloride and latex, the pH of the mixed dispersion must be carefully evaluated to achieve optimum destabilisation through the bridging mechanism outlined above.

Effect Of Slurry Concentration

The solid concentration of slurry that could be pumped practically from mining ponds ranges from 8 - 15% w/v. Fig. 8 shows the effect of slurry concentration on the heterocoagulation behaviour of the slurry at pH 10 in the presence of 0.1% w/v HA latex. It can be seen the optimum concentrations of calcium chloride for heterocoagulation of the mixed dispersion were essentially the same when the slurry concentration was increased from 2 to 8 % w/v. The critical coagulation concentration of a particular electrolyte is known to increase with decreasing solid concentration in dilute dispersions due to a large distance of separation between particles at low solid concentration. In the present system, the different particles were similarly charged and electrostatic attraction between them would be minimum. Thus collision between particles leading to heterocoagulation is ineffective when the slurry concentration is low. Thus a higher electrolyte concentration ($CaCl_2$) was required for compressing the electrical double layers to enable closer approach of particles so that the bridging mechanism mentioned above can become operative. It would appear that the slurry concentration used was too high for this effect to be observed here.

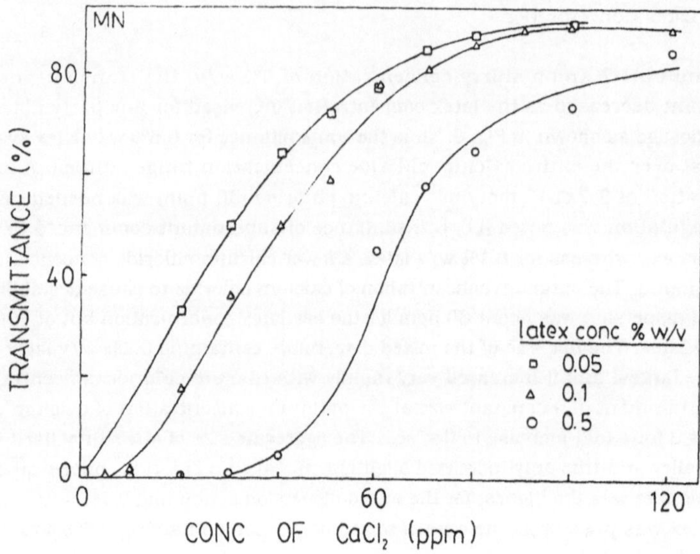

Fig. 6. Effect of latex concentration on the heterocoagulation of MN slime-latex mixed dispersions (8% w/v slime and 0.1% w/v latex at pH 7) studied as a function of $CaCl_2$ concentration.

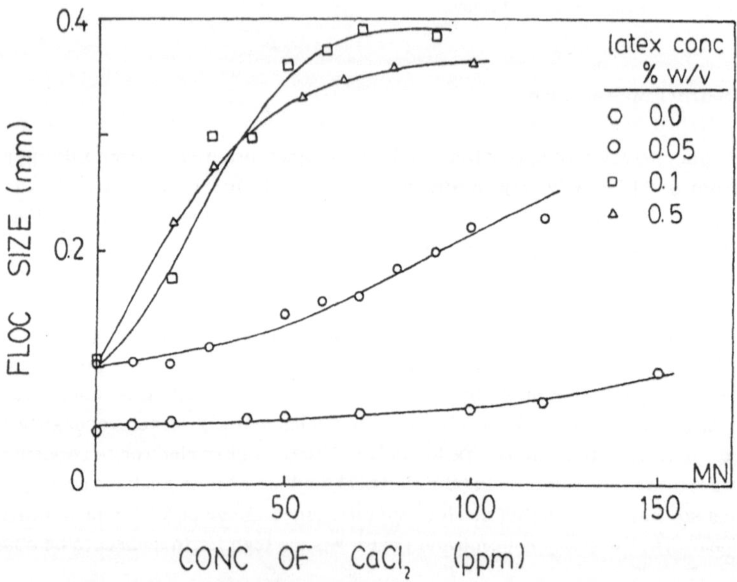

Fig. 7. Effect of latex concentration on the floc size of MN slime-latex mixed dispersions (8% w/v slime and 0.1% w/v latex, pH 7) studied as a function of $CaCl_2$ concentration.

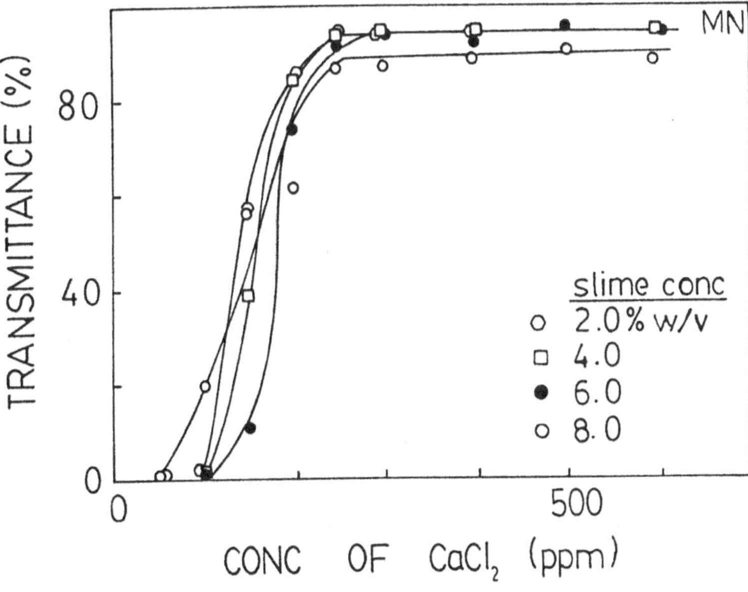

Fig. 8. Effect of slime concentration on the heterocoagulation of MN slime-latex mixed dispersion (0.1% latex at pH 10) studied as a function of $CaCl_2$ concentration.

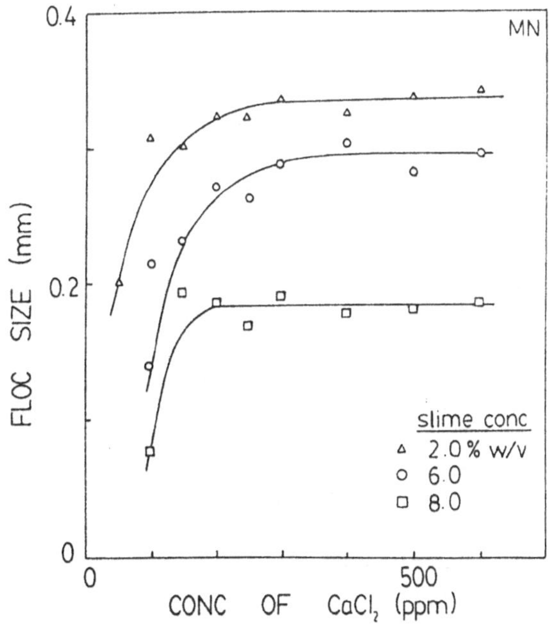

Fig. 9. Effect of slime concentration on the floc size of MN slime-latex dispersions (0.1% w/v latex at pH 10) studied as a function of $CaCl_2$ concentration.

The effect of slurry concentration on the heterocoagulation is more obvious through floc size measurement of the sediment as shown in Fig. 9. At a particular slurry concentration, the floc size increased rapidly with increasing calcium chloride concentration until a maximum floc size was reached at about the optimum concentration of calcium chloride beyond which the floc size remained constant. However when the floc sizes at various slurry concentrations were compared, it was found that the floc size decreased with an increase in slurry concentration. In fact the floc size at 8% w/v slurry concentration was only about half the floc size at 2% w/v slurry concentration at the same calcium chloride concentration. The photomicrographs in Fig. 10 shows another series of experiment where the variation of floc size of the sediment with much higher slurry concentrations (up to 25% w/v) was determined at pH 7 in the presence of 0.1% w/v HA latex and 90 ppm $CaCl_2$. Again the same trend was observed. The present results seem to be at variance with previous observation [12] that floc size should increase with increasing solid concentration. It should be noted that as the slime concentration increased, the particle number ratio of HA latex particles to clay particles involved in floc formation decreased. It was also observed that the floc size of the **heterocoagulated** latex-slurry sediment was substantially larger than the **homocoagulated** slurry. In the presence of **excess** latex, **heterogeneous** sediment (HA latex coagulum intermingled with heterocoagulated latex-slurry) would occur. As far as we are aware, no other flocculant behaves in this way, especially in the formation of heterogeneous sediment. In the case of polymeric flocculant, restabilisation usually occurs instead in the presence of excess polymer [13]. This demonstrates that for the present system, a critical assessment of the optimum conditions is essential for effective heterocoagulation, any overdosing or underdosing of latex would nullify the dramatic effect of the HA latex.

CONCLUSION

In contrast to the water-soluble synthetic polymeric flocculants used in flocculation of colloidal dispersions, the present system of mixed dispersion of latex-slurry operates via a hetero-coagulation process in the presence of metallic ions acting as electrostatic bridges linking the dissimilar particles together. Intrinsic to this behaviour is the anisotropic character of the clay particles: the surface charges on the basal surface and on the edge of the plate-like clay particles are different and strongly pH-dependent. In addition both the latex and clay particles are polydisperse in size and the dispersions complex in chemical compositions. And yet a total heterocoagulation of the latex and clay particles with complete removal of disperse particles from the liquid phase can be achieved at optimum dosage of latex and electrolyte at the natural pH of the slurry (near neutral). Underdosing of latex results in incomplete clay particles removal whereas overdosing of latex leaves behind excess free latex particles. The presence of sufficient metallic ions such as $CaCl_2$ is essential as electrostatic bridges especially at alkaline pH. Since only very small amount of NR latex (0.1% w/v) is required (to heterocoagulate completely a 8% w/v slurry) and is effective at neutral pH, the system is extremely attractive as a cheap, cost-effectve and environmentally friendly method of flocculating slurries (mainly clay particles) from mining source. Scaling-up of the method to on-site operation is in progress.

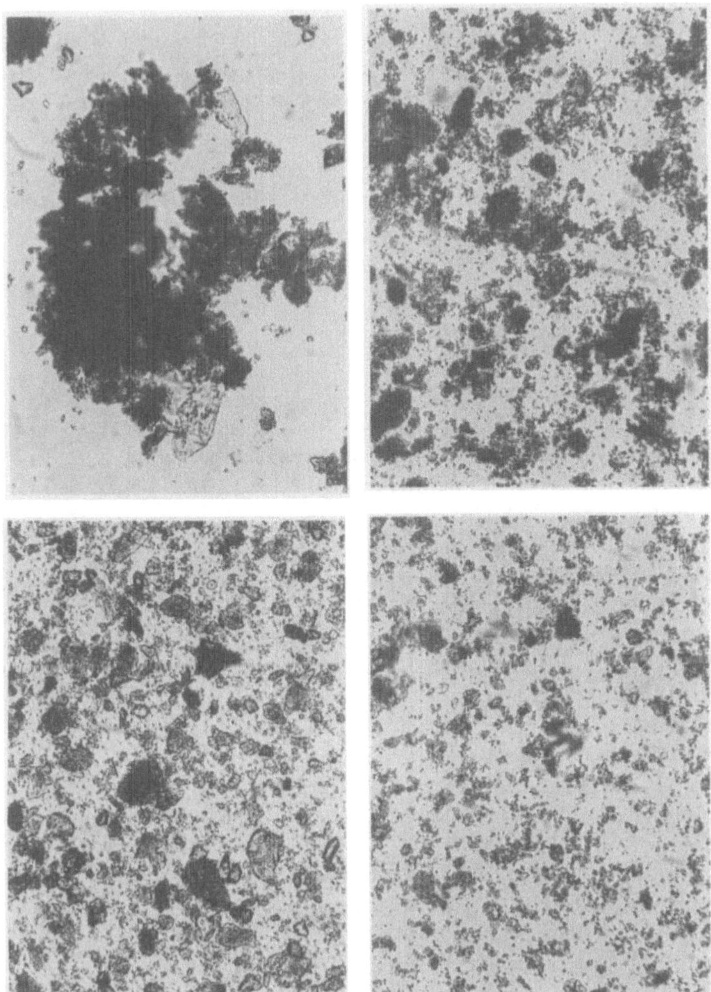

Fig. 10. Photomicrographs show the variation of floc size of the sediment of MN slime-latex dispersion (0.1% w/v latex and 90 ppm $CaCl_2$ at pH 7) with slime concentrations (a) 8% w/v; (b) 15% w/v; (c) 20% w/v; (d) 25% w/v.

ACKNOWLEDGEMENT

Financial support by the International Development Research Centre (IDRC), Canada and studentship to K.C.L by the University of Malaya are gratefully acknowledged.

REFERENCES

1 O'Gorman JV, Kitchener JA (1974) Int. J. Mineral Processing, 1:33.

2 Yong RN, Chen CK (1986) Second Annual Report IDRC Singapore Slurry Pond Reclamation Project, Vol. 1.

3 Yong RN (1981) U.S. Patent, No. 4,289,540

4 Chan YK (1993) M.Sc Thesis, University of Malaya

5 Ho CC (1989) Colloid & Polymer Sci. **267**: 643

6 Ho CC, Lee KC, Yeap EB (1992) Colloid & Surfaces **67** : 109

7 Ho CC, Lee, KC, Yeap EB (1992) Paper presented at 'Polymer Colloids' Conference, Bristol, England, March 1992

8 Ho CC (1988) Colloid & Polymer Sci. **266** : 70

9 Levy N, Bar-or Y, Magdassi S (1990) Colloid & Surfaces. 48:337

10 Poirier JE, Cases JM (1990) Colloid & Surfaces. **55**: 333

11 van Olphen H (1963) An Introduction to Clay Colloid Chemistry Interscience, New York

12 Michaels AS, Bolger JC (1962) I & E C Fundamentals **1**: 24.

13 Slater RW, Kitchener JA (1966) Discuss Far Soc. **42**: 267

Liquid Natural Rubber: Preparation and Application

Ibrahim Abdullah

Chemistry Department
Universiti Kebangsaan Malaysia
43600 UKM Bangi, Malaysia

ABSTRACT

Liquid natural rubber (LNR) is a modified form of natural rubber (NR) with a shorter polymeric chain. The preparation of LNR can either be done via mechano-oxidation or redox reaction or photosensitized oxidation. Mechanooxidation, being the earliest method to be developed, is still widely used with some modifications. Pilot plant scale has already been tested on two redox reaction methods using phenylhydrazine-air and sodium nitrite/chlorite - hydrogen peroxide as oxydo-reducing couples. Photosensitized oxidation of NR is made possible by using a photosensitizer which is active in visible light and able to induce degradative oxidation. Research in the field is quite active but so far no method has developed into a commercial scale. The short nature of the LNR polymeric chain allows many chemical modifications possible and thereby enlarging the field of applications of NR. The potential applications of LNR are in the preparation of thermoplastic NR, adhesives, binders, bitumen, low hardness vulcanizates, paints and varnishes.

INTRODUCTION

Chemical modifications of NR vary from very low mole percent in the incorporation of antioxidants and cross-linkings without affecting the physical properties, to a more than 50 mole percent in the epoxidation, hydrogenation, cyclization etc., and may result in changing the nature of the polymer from a rubber to a more plastic-like or resinous materials. Some of the modified forms of NR are hydrogenated, hydrohalogenated, chlorinated, polymer grafted, epoxidised, cyclised and depolymerised or liquid NR. Liquid NR (LNR) is a product of NR depolymerization and as such it is not in a true sense a modified NR. The LNR has received a lot of attention lately due to the high potential as a starting material for a wide range of products.

The physical properties of processed rubber is greatly influenced by the molecular weight and molecular weight distribution of the vulcanised NR. Uniformation of the molecular weight through mastication which result in the chain degradation is usually done prior to processing

K.P. Ghiggino (Ed.)
Progress in Pacific Polymer Science 3

with the aim of facilitating easy processing, addition or incorporation of fillers, vulcanizing additives and other chemicals. Molecular weight reduction mainly concerns NR only as in the case of synthetic rubber the molecular weight is well controlled during production, and only a specific molecular weight rubber is produced for certain application. A method of controlling the molecular weight of NR would be advantageous as it allows a greater flexibility on the production of various grades of NR.

Liquid rubber is defined in the UNIDO[1] report s a class of elastomers or elastomer precussors that can be poured or pumped at temperature below 100°C without the assistance of an external medium. Frequently liquid natural rubber (LNR) can be defined as a form of NR with a similar microstructure but shorter polymeric chain and with molecular weight, Mw, lower than 10^5. A much shorter polymeric chain LNR is more open to structural design such as addition of new functional groups or active groups onto the chain:[1,2,3]

PREPARATION OF LIQUID NATURAL RUBBER

According to RAPRA, most of the LNRs produced in recent years belong to the first generation liquid rubber which implies the lack of active terminal groups on the polymer chain. The nature of LNR is greatly influenced by the manner in which the NR is degraded leading to depolymerization. Mechanochemical oxidation, redox reaction and photosensitized oxidation are the three general techniques of molecular weight reduction of NR exploited thus far. Other methods like exposing NR to high energy radiation are found to be not very suitable on a large scale and thus are not exploited.

Mechanochemical Oxidation

Mechanical degradation, normally known as mastication, of NR in air is the earliest known method of molecular weight reduction process tried and still in use till the present. Although it is far from efficient but is the most convenient, easy and free from any kind of chemical complexities. The earliest reported patent on the depolymerization of NR to LNR is by Hardman et. al in 1944[4]. They described a process of making articles such as drums, cans, tanks and pipes from depolymerised natural rubber (DPNR) via casting-vulcanization process. The degradation apparently occurs during the extensive mastication in air at high temperature around 300°C for 2-4 hours.

In 1952, the British Rubber Producers' Research Association (BRPRA) had reported a detail method of NR depolymerisation through mastication[5]. The stages of the process are:

- mastication in a mill or internal mixer until the viscosity drops to about 20 Monney;
- further milling with a Z-blade mixer at 120 - 140°C for 6-8 hours until the viscosity drops to 50,000 -70,000 poises and finally;

–the semifluid material is heated at 250 to 300°C for about 2 - 2.5 hours until the product reaches a viscosity of the order 2500 poises.

Chloride Lorival of United Kingdom later adopted this method to produce depolymerised natural ruber in 1953. This method of depolymerization is still in use today with some improvements. Instead of milling at high temperature, the mastication is done in an intemal mixer at about 120°C in the presence of a peptizer over a long period[6]. Some of the commonly used peptizers are hydrazine derivatives and organic sulphur like thio-β-naphthol, pentachlorothiophenol and dibenzoyl disulphide. Being easy to process without any complex chemistry is an advantage but it is overshadowed by its high cost as the method consumes too much energy. However the method is largely employed to produce depolymerised NR for in-house use only.

Redox Reaction

The chain degradation in this method involves catalytic oxidation of rubber in an oxidising agent such as air and a catalyst or a reducing agent. The first patent on catalysts for the oxidation of rubber was by Stevens who patented the use of lead, cobalt and manganese linoleates[7]. However this process was not further developed and soon discarded in favour of the simpler mechanochemical oxidation method.

In 1984 UNIDO has set up a pilot plant to produce LNR in Ivory Coast. The aim of the project is to select the best and most promising technique of LNR generation and also to evaluate the industrial applications base on physical properties and economics. The technology adapted is based on the process developed by the Rubber Research Institute of Africa (IRCA) and patent registered by Institute Francais du Caoutchouc, France[8]. The reaction involves the use of an oxydo-reducing couple of oxygen-phenylhydrazine to depolymerise natural polyisoprene in latex phase. The depolymerization mechanism is as shown in Figure 1.

In the study, the nature of the latex, the phenylhydrazine concentrations, temperature and air flow were the parameters investigated in detail. As shown in Table 1, the degree of polyisoprene degradation depends on the concentration of phenylhydrazine relative to rubber. The other reaction conditions were found to be not very sensitive and thus the reaction conditions adapted were:

Latex concentration	- 30% rubber content
Phenylhydrazine	- 8×10^{-4} mole /g of rubber
Temperature	- 65°C
Effective reaction	- 24 hours

Although the project seems to be quite promising but it never proceed any further due to the highly toxic nature of the phenylhydrazine and the search for an alternative reducing agent is still very active.

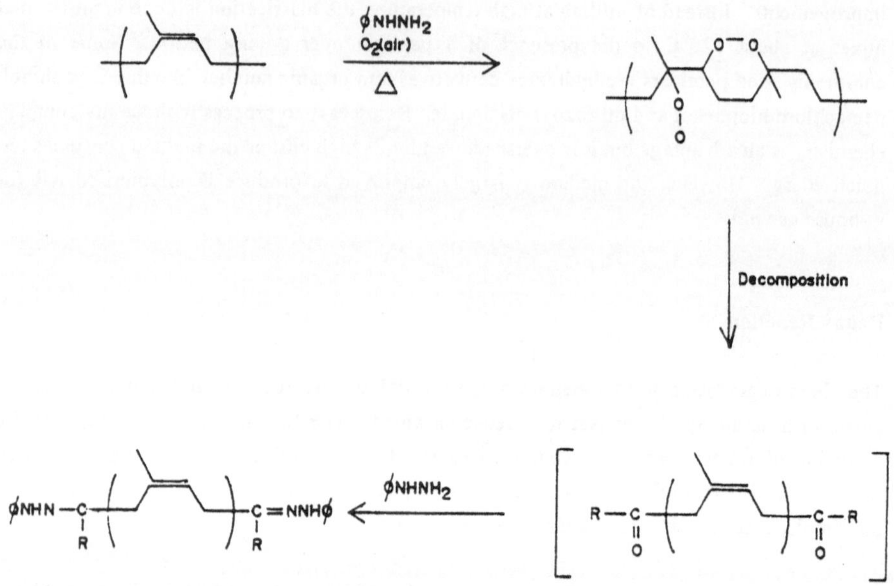

Figure 1 : Depolymerization mechanism of 1.4-cis-polyisoprene

Table 1: The influence of phenylhydrazine concentration on the molecular weight of LNR or degree of depolymerization

Phenyl hydrazine (mole/g rubber)	Mol. wt. (Mv) of LNR
4×10^{-4}	17,000
6×10^{-4}	12,000
8×10^{-4}	8,000
12×10^{-4}	8,000

The Malaysian Rubber Producers' Research Association (MRPRA) in 1985[9] has developed a method of generating LNR base on a redox reaction on NR in latex phase. The reducing agent is either sodium chlorite or sodium nitrite and the oxidising agent is hydrogen peroxide. The depolymerization seem to occur efficiently at around pH = 7 and the LNR produced has a wide spread of molecular weights, 20,000 to 150,000 depending on the peroxide concentration. The reaction is believed to involve singlet oxygen which attack the double bonds to form allylic/ hydroperoxides attached to the polymer chain[10]. Subsequent decomposition of these hydroperoxides leads to chain degradation.

$$-CH_2-\overset{\overset{CH_3}{|}}{C}=CH-CH_2- \xrightarrow{O_2} -CH_2-\overset{\overset{CH_3}{|}}{\underset{\underset{\underset{OH}{|}}{O}}{\underset{|}{C}}}-CH=CH- \; + \; -CH=\overset{\overset{CH_3}{|}}{C}-\underset{\underset{\underset{OH}{|}}{O}}{\underset{|}{CH}}-CH_2-$$

$\downarrow$

hydroperoxides decomposition and subsequent chain degradation

Photosensitized Oxidation

The photochemical reaction occurring in natural rubber on exposure to visible light is most likely to be sensitised by any reagent present in it as the polyisoprene molecule does not absorb any light at that frequencies. It was reported that the depolymerization of NR was found to be quite significant when rubber solution in the presence of nitrobenzene was exposed to sunlight[12]. As discussed by Rabek [11] the most probable reaction pathway is the abstraction of hydrogen by the excited nitrobenzene from the polyisoprene chain producing a macroradical which proceeds to oxidation and degradation.

Ravindran et al.[13] had reported the generation of hydroxyl- terminated LNR by exposing the rubber solution to UV light or sunlight in the presence of hydrogen peroxide. The degree of depolymerization of rubber was found to be about the same in UV- light and in sunlight and the rate was proportional to the exposure time. The suggested mechanism is as shown in Figure 2.

In our study on the photostability of NR towards visible light, we observe only a slight degradation of rubber on exposing the solution to light. However in the presence of a photosensitizer, chain degradation on a massive scale is recorded. One set of photosensitizer used are benzophenone, flavone and xanthone which have a common carbonyl functional group[14].

These molecules have an absorption shoulder at around 360 nm due to the $(n\pi)^*$ transition[15]. On absorption of light energy, excitation of the molecule into the singlet excited state, $^1(n\pi)^*$

occurs which then crossover into the triplet excited state, $^3(n\pi)^*$. Interaction of this state with the polyisoprene chain, leads to hydrogen abstraction and an isoprenyl macroradical is formed. Molecular rearrangement of the macroradical leading to chain scission subsequently occurs. However, in the presence of oxygen, the radical will be oxidised to form hydroperoxides which on decomposition results in further chain degradation. Thus an increase in the degree of depolymerization will be observed in the undegassed sample. The reaction mechanism concluded is summarised in Figure 3.

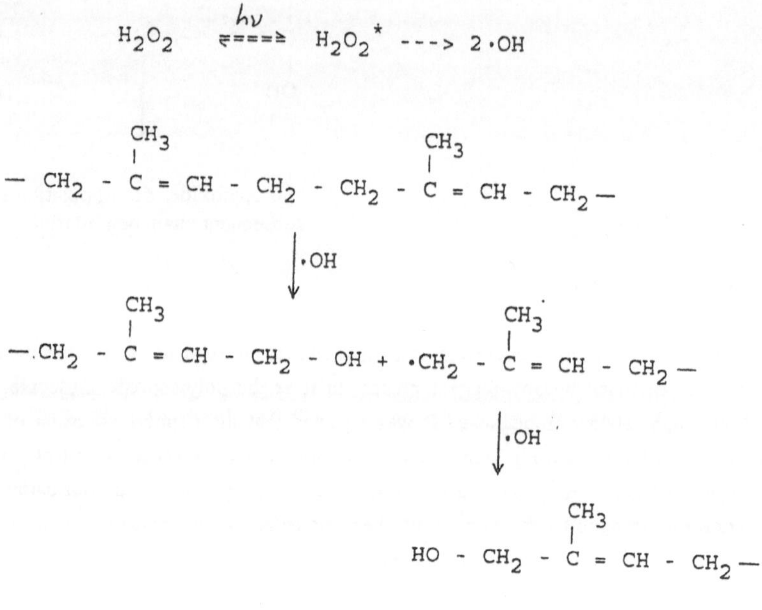

$$H_2O_2 \overset{h\nu}{\rightleftharpoons} H_2O_2^* \dashrightarrow 2\cdot OH$$

$$-CH_2-\underset{}{\overset{CH_3}{\overset{|}{C}}}=CH-CH_2-CH_2-\overset{CH_3}{\overset{|}{C}}=CH-CH_2-$$

$$\downarrow \cdot OH$$

$$-CH_2-\overset{CH_3}{\overset{|}{C}}=CH-CH_2-OH \;+\; \cdot CH_2-\overset{CH_3}{\overset{|}{C}}=CH-CH_2-$$

$$\downarrow \cdot OH$$

$$HO-CH_2-\overset{CH_3}{\overset{|}{C}}=CH-CH_2-$$

Figure 2 : Photodepolymerization of polyisoprene in the presence of H_2O_2

Recombination of radicals forms crosslinks among the chains and will give rise to gel. As expected the yield of gel increases with the increase in the concentration of rubber. The relative effectiveness of the photosensitizing ability of these molecules are of the order benzophenone > xanthone > flavone.

As discussed above, the main primary step is hydrogen abstraction and hence the different abilities of these molecules to sensitize degradation of the polyisoprene chain lie in their efficiency to promote hydrogen abstraction. Consequently the interaction between the $^3(n\pi)^*$ and $^3(\pi\pi)^*$ is the determining factor[15,16,17]. The degree of conjugation within the molecule which influence the mixing of the two excited triplet states is most extensive in flavone and least in benzophenone. The $^3(n\pi)^*$ state of benzophenone is thus least influence by the neighbouring $^3(\pi\pi)^*$ and thereby becomes the most active in the hydrogen abstraction. The activity of xanthone is between the two molecules as its conjugation state lies somewhere between them.

The effect of ferrocene and benzoylferrocene on the photodegradation of natural rubber has also been studied[18]. Ferrocene apparently causes a very small effect on rubber photolysis. However benzoylferrocene with an active carbonyl group induces a greater degree of depolymerization.

The photosensitised isomerization and dimerization of conjugated dienes by ferrocene had been reported[19,20]. The activity was rationalised as due to the ability of ferrocene to form a complex with the polymer. On excitation by light, the complex is excited and interact with the diene resulting in energy transfer to the diene to form triplet excited diene:

$$\text{Ferrocene + diene} \quad \equiv \quad \text{complex} \quad \text{---} \quad {}^1(\text{complex})^*$$

$$\downarrow$$

$${}^3(\text{diene})^* + \text{ferrocene} \longleftarrow {}^3(\text{complex})^*$$

$$\downarrow$$

Isomerization of diene

This reaction pathway is not very favourable with the polyisoprene of NR as the formation of the complex is expected to be incomplete and thus restricts the transfer of energy. However if a triplet excited isoprene is formed a small degree of molecular weight reduction will be observed according to the following pathway:

$${}^3(-CH_2-\overset{\overset{\large CH_3}{|}}{C}=CH-CH_2)^* \text{-----} (-CH_2-\overset{\overset{\large CH_3}{|}}{C}-CH-CH_2-)$$

$$\downarrow$$

Isomerization and chain scission

Benzoylferrocene, however, can also induce NR degradation through hydrogen abstraction reaction facilitated by the carbonyl group. The photosensitizing effect of ferrocene, benzoylferrocene and benzophenone is found to increase significantly on addition of carbon tetrachloride to the solvent. The rate of degradation increases with the increase in carbon

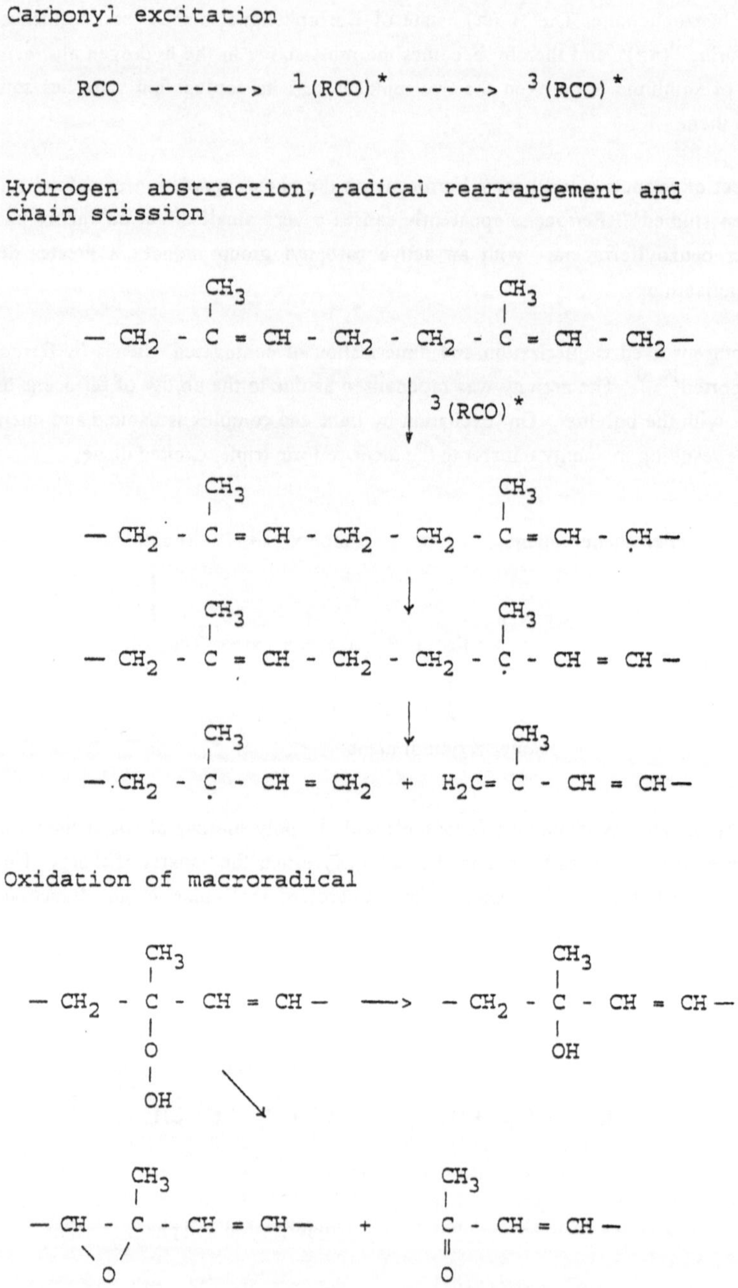

Carbonyl excitation

$$RCO \dashrightarrow {}^1(RCO)^* \dashrightarrow {}^3(RCO)^*$$

Hydrogen abstraction, radical rearrangement and chain scission

$$-CH_2-C(CH_3)=CH-CH_2-CH_2-C(CH_3)=CH-CH_2-$$

$$\downarrow {}^3(RCO)^*$$

$$-CH_2-C(CH_3)=CH-CH_2-CH_2-C(CH_3)=CH-\dot{C}H-$$

$$\downarrow$$

$$-CH_2-C(CH_3)=CH-CH_2-CH_2-\dot{C}(CH_3)-CH=CH-$$

$$\downarrow$$

$$-CH_2-\dot{C}(CH_3)-CH=CH_2 + H_2C=C(CH_3)-CH=CH-$$

Oxidation of macroradical

$$-CH_2-C(CH_3)(O-OH)-CH=CH- \longrightarrow -CH_2-C(CH_3)(OH)-CH=CH-$$

$$\searrow$$

$$-CH-C(CH_3)-CH=CH- \text{ (epoxide, } -CH-C- \text{ bridged by O)} + C(CH_3)(=O)-CH=CH-$$

Figure 3: Carbonyl photosensitised degradation of natural rubber

tetrachloride until the composition of the solvent becomes 10% in carbon tetrachloride. Further addition of carbon tetrachloride does not seem to enhance the rate but instead a steady drop is observed. A similar enhancing effect of carbon tetrachloride was also observed in other photodegradation studies on polymers[21] and high energy irradiation curing of natural rubber[22].

The role of carbon tetrachloride in the polymer degradation can be seen from it ability to enhance radical reactions. Figure 4 shows the generation and role of trichloromethyl radical in the photodegradation of polyisoprene chain. Ferrocene and carbon tetrachloride form a charge-transfer complex in solution which on irradiation decomposes to form a trichloromethyl radical. The triplet carbonyl regenerates the radical on reacting with the $CHCl_3$ produced. This extended reaction is very likely in the benzoylferrocene case where both functional groups are present in the same molecule. As shown in Table 2, however the degradation of NR in the presence of carbon tetrachloride is about the same in ferrocene and benzoylferrocene. The increase in the stearicity of the molecule from ferrocene to benzoylferrocene leads to a decrease in the stability of ferrocene - CCl_4 complex which does not favour the generation of the active CCl_3.

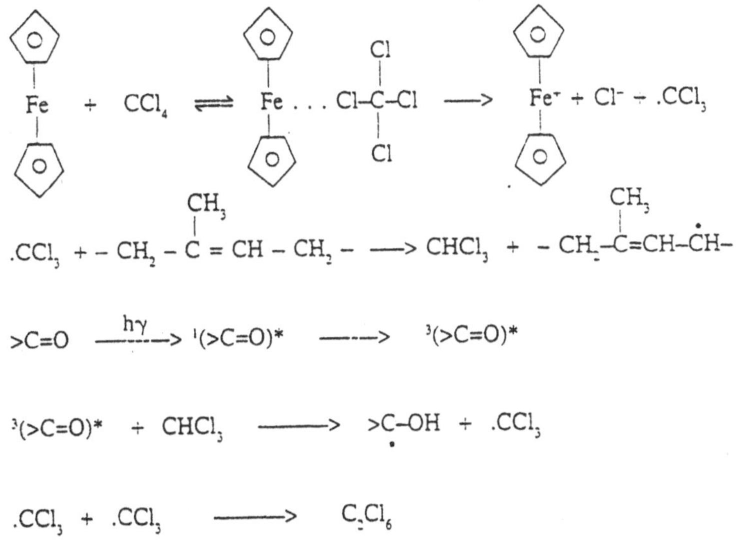

$$Fe(C_5H_5)_2 + CCl_4 \rightleftharpoons Fe(C_5H_5)_2 \ldots Cl\text{-}CCl_3 \longrightarrow Fe(C_5H_5)_2^{+} + Cl^- + .CCl_3$$

$$.CCl_3 + -CH_2-C(CH_3)=CH-CH_2- \longrightarrow CHCl_3 + -CH_2-C(CH_3)=CH-\dot{C}H-$$

$$>C=O \xrightarrow{h\gamma} {}^1(>C=O)^* \longrightarrow {}^3(>C=O)^*$$

$${}^3(>C=O)^* + CHCl_3 \longrightarrow >\dot{C}-OH + .CCl_3$$

$$.CCl_3 + .CCl_3 \longrightarrow C_2Cl_6$$

Figure 4 : Generation of trichloromethyl radical

Table 2: The molecular weight of degraded NR after 20 hours of exposure to visible light in the presence of ferrocene, benzoylferrocene and carbon tetrachloride

Irradiated system	Mv x 10^{-3}g mol
NR only	970
NR + Ferrocene	460
NR + Ferrocene + CCl_4	60
NR + Benzoylferrocene	380
NR + Benzoylferrocene + CCl_4	70

Infrared spectrum of the products i.e degraded rubber, shows all similar peaks to that of the original NR except for a broad peak at around 3450 cm^{-1} and two sharp peaks at 725 cm^{-1}, and 1250 cm^{-1}. The assignment as reported in the literature[23,24] are the 3450 cm^{-1} to -OH vibration, 1725 cm^{-1} to C = O vibration and 1250 cm^{-1} to C-O vibration of epoxide.

Figure 5 shows the proton nmr spectra of degraded rubber taken immediately after preparation of sample and retaken five days later. A summary of peaks with assignment is depicted in Table 3[25]. The proton nmr analysis thus comfirms the observation in infrared study.

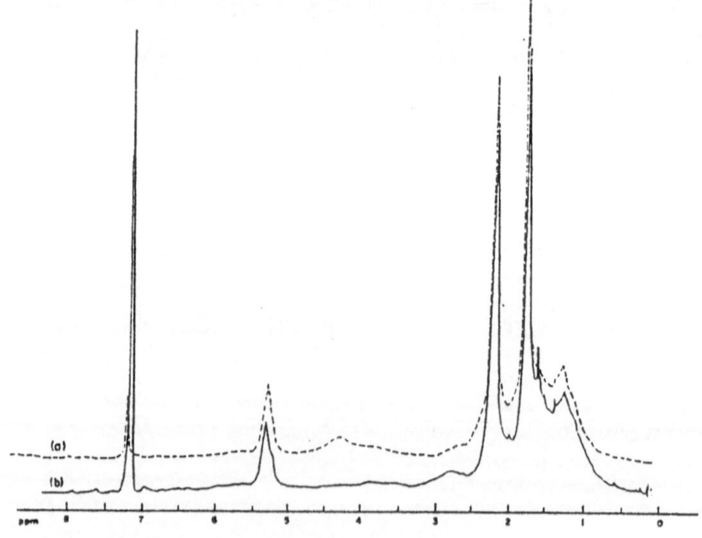

Figure 5: Proton nmr spectra of LNR for a fresh sample (a) and retaken five days later (b).

Table 3: The NMR spectral assignment of LNR in C6D6 solvent

Fresh Sample	5 days sample	Assignment
1.40 broad	1.40 broad	$-CH_3-C-C$
		O
	1.61	$-CH_3$
	1.62	trans $CH_3-CH=$
1.72	1.78	$-CH_3$
2.20	2.20	$-CH_2-$
2.70 broad	2.60 broad	-C-CH-
		O
4.30 broad	3.80 broad	-O-OH, -CH-OOR
5.30	5.30	-CH=

These new groups formed are the products resulting from the oxidation of rubber. The appearance of -OOH and -OOR groups in the freshly prepared sample implies the involvement of hydroperoxides in the intermediate stage which over a period decompose to more stable products like hydroxide, expoxide and carbonyl as shown in Figure 3. Although these groups are presence in very small amount as terminal active groups on the polyisoprene chain, they are very important in determining the reactivity of LNR.

APPLICATIONS OF LIQUID NR

The potential commercial applications of LNR have been exhaustively studied in a project sponsored by the United Nations Industrial Development Organisation (UNIDO) at the Rubber Research Institute of Africa (IRCA), Invory Coast, in 1985. The physical properties of LNR and related products, and the relevant economics are the main factors considered in evaluating the applicability of LNR. Table 4 lists some of the important applications explored and general comments about the products.

Low hardness vulcanisates can be prepared by vulcanisation of LNR at moderate temperatures; a few hours at 100°C or several days at ambient temperature. Coating of toxic chemicals used in the rubber processing has been shown to be feasible by mixing them with LNR but the coating technology is inadequate. In the processing of high molecular weight polymers, LNR is found to be effective in replacing aromatic oil where it also act as a reactive plasticiser.These vulcanisates seem not only able to maintain the physical properties but also to enhance in some aspects like green strength and improve oxidative ageing property. Incoporating LNR as a component in bitumen formulation helps to improve the waterproofness property of the product.

Due to the complex nature of NR, chain grafting reaction is either difficult or impossible but it is otherwise in LNR. Antioxidants[26,27], such as 4-nitrosodiphenylamine, can be grafted onto the polyisoprene chain of LNR quite easily and this provides a method of introducing rubber bonded antioxidants into any rubber network. Chlorination of LNR is possible by bubling chlorine gas into LNR solution in carbon tetrachloride. The chlorinated LNR has a high potential application in the manufacture of anticorrosive paints and adhesives. Addition of maleic anhydride to LNR chain is possible via moderate heating the mixture in the presence of free radical catalyst or strong heating without any catalyst. The grafted polymer finds application in adhesive industry.

Wider application of LNR is not possible due to the ill-defined nature of the molecular dimension of LNR. Reactive sites or active terminals which is a very important feature in facilitating chemical modifications is lacking in LNR[28] and thus restricts the field of applications. However the LNR produced by photodegradation carries active groups on the chain and the hydroxyl group is the most prominent. Appropriate active groups can also be implanted onto the polyisoprene chain of LNR to suite certain application.[29]

A new role of LNR as an NR-plastic compatibiliser in the preparation of thermoplastic based on natural rubber (TPNR) is being investigated in our laboratory. The ability of the photogenerated LNR to compatibilised NR-plastics has been tested on NR-Polypropylene (PP), NR-high density polyethylene (HDPE) and NR - Low linear density polyethylene (LLDPE) systems. The TPNR is prepared by melt-blending in a mixer at a temperature close to the melting point of the plastic component. For NR-PP, NR-HDPE and NR-LLDPE, the processing temperatures are 170°C, 150°C and 135°C respectively and all the mixes are processed between 12 to 15 min. Table 4 shows a typical set of mechanical and thermal properties of the material prepared.

The strength of the material as expressed by the stress and strain at break increases significantly with the addition of LNR for both systems, NR-PP and NR-HDPE. For the NR-PP, the optimum value of LNR is about 15% whereas in the NR-HDPE, 27% LNR yields the maximum physical properties. A single glass transition temperature, Tg, strongly indicates a high homogeneity of the blend in both thermoplastics NR.

A complete understanding on how the LNR function in compatibilising the NR-plastic phase is difficult. However for simplicity it can be rationalised in term of inducing interaction between the phases. As reported by Gelling and Mente[30], the role of LNR in influencing the physical properties of the vulcanizate depends on the chain length of polyisoprene. At molecular weights around 6000 LNR acts as bulk viscosity modifiers and this role decreases with increasing molecular weights. Covulcanization of LNR into the rubber network is optimum at molecular weight around 30,000 and yields vulcanizate with high hardness and modulus.

Table 4 : The potential commercial applications of Liquid Natural Rubber

Application	Remark
Low hardness vulcanizates	Hardness 15 to 30 degrees Shore
Ingredient coating for rubber chemicals	Safer handling of chemicals
	Inadequate technology
Processing aid or reactive plasticizer	Substitute for oil
	Improve tensile strength
	Formulating high hardness mixes
	Non-extractible plasticizer
Bitumen	Good waterproof properties
Flexible Moulds	Cheaper than silicon moulds
Paint and varnish	LNR needs modification
Adhesive and binder	High tackifying and filler acceptance
	Improve with chemical modification
	Restricted to coloured products

Our photogenerated LNR has moleculer weight around 50,000 and reactive terminals such as OH and C=O groups. The LNR is soluble in NR and as such all kind of interactions, chemical and physical are possible in the rubber network, but on the other hand, migration of LNR into the plastic matrix is very unlikely or impossible. Through these active terminals, interaction between the polyisoprene chain of LNR and the polymer chain of thermoplastic is believed to occur. Thus the overall role of LNR in the preparation of TPNR are as cross-linking agent within the NR phase and bonding agent between the NR and thermoplastic interphase. Excessive addition of LNR above the optimal concentration results in the overall increase in the low molecular weight rubber component in the mixture which causes the TPNR to become soft and have lower tensile properties.

Table 5 : The mechanical and thermal properties of NR-PP and NR-HDPE thermoplastics

Blend	Composition		Maximum		Hardness		Tg
	% NR	%LNR	Stress MPa	Strain %	Shore A	Shore D	oC
NR-PP	50	5	7.9	450	93	42	
		10	8.3	670	94	43	-45
		17	8.8	500	95	44	
NR-HDPE	60	7	4.3	580	90	33	
		13	5.3	650	93	35	-55
		20	6.2	700	94	37	
		27	7.1	760	94	38	

CONCLUSION

It is clear that LNR has unfolded a new era in natural rubber industry. With new forms of natural rubber, new applications are expected to develop. The technological requirements of the present industry are too demanding to be satisfied by the conventional natural rubber and in most cases it loses out to synthetic materials. The most versatile rubber would be the form that can easily be modified to meet the specification set by the product to be manufactured. Although LNR, to an extent is able to cope with this demand, it is far from satisfactory as the most important aspect of modification, that is, polymerisation of LNR to give NR of specific molecular weight with attached functional groups and thereby the physical properties required, is not possible at the moment. However LNR has created new avenues in rubber research long lost to the synthetic polymers. With our ever-increasing knowledge in polymer and related chemistry, the chemical modifications of LNR and thus new NR applications look very promising.

REFERENCES

1 UNIDO (1986). Development of Liquid Natural Rubber. Project No UF/GLO/81/059, IRCA
2 Ravindran T, Gopinathan Nayar MR, and Joseph Francis D(1991). J. Apple Poly. Sci. 42: 325
3 Capitoni Clito (1975) Annual Meet Proc Intern Inst Synth Rubb Prod. 15: 16
4 Hardman HV and Hardman KV. (1944). Method of making depolymerised rubber articles, US pat. 2,349,549
5 BRPRA (1954) Tech Bull No 1 "Heveaplus M", London

6 Kuipers RJ, Dulfer RV and Van Der Veen WJ (1972) Materials and Technology Vol 5, Ch. 14, Longman, London
7 Stevens HP (1934) British Patent 407,038
8 Pautrat R and Marteau J (1974) Institut Francais du Caoutchouc, French Patent No 2,259,110
9 Gazeley KF and Mente PG (1985) Malaysian Rubber Producers' Research Association, UK Patent No 8529685
10 Lemoine AJ (1988) Proc Intern Rubb Technol Conference, Penang, Malaysia
11 Ranby B and Rabek JF, (1975) Photodegradation, Photooxidation and Photostabilization of Polymers, Wiley, New York
12 Tillekeratne LMK, Perena PVAG, De Silve MSC and Scott G (1977) J Rubb Res Inst, Sri Lanka 54(2): 501
13 Ravindran T, Gopinthan Nayar MR and Joseph D (1988) J Appl Polym Sci. 35: 1227
14 Ibrahim A. (1983) Sains Malaysiana 12(3): 235
15 Turro NJ (1967) Molecular Photochemistry, Benjamin New York
16 Murov SL (1973) Handbook of Photochemistry, Marcel Dekker New York
17 Golub MA, Rosenberg ML and Germmer RV (1978) Applications of Polymer Spectroscopy, Academic New York
18 Ibrahim A, Ku Halim KB and Maimon A (1991) Sains Malaysiana 29(2): 47
19 Dannesberg JJ and Richard JH (1965) J Amer Chem Soc 87(7): 1626
20 Gullory JP, Cook CF and Scott DR (1967) J. Amer Chem Soc 89(25): 6776
21 Ranby B and Rabek JF (1979) J Appl Polym Sci: Appl Polym Symp 35: 243
22 Puig JR (1971) Atomic Energy Review 9: 373
23 Golub MA, Hsu MS and Wilson LA (1975) Rubb Chem Technol 48: 953
24 Rodriguew MA and De Paoli MA (1985) Eur Polym J 21: 15
25 Ibrahim A and Zuriati Z (1989) Sains Malaysiana 18(2): 99
26 Derouet D, Phinyocheep P, Boccaccio and Brosse JC (1991) J Nat Rubb Res 6(1): 39
27 Derouet D, Brosse JC and Tillekeratne LMK (1990) J Nat Rubb Res 5(4): 296
28 Xuan LH and Decker C (1993) J Polym Sci: Pt A: Polym Chem 31: 769
29 Ravindran T, Gopinathan Nayar MR and Joseph Francis D (1991) J Appl Polym Sci 42: 325
30 Gelling IR and Mente PG (1993) Intern Rubb Conference 93, Kuala Lumpur

A Nuclear Magnetic Resonance Study of the Microstructure of Poly (Vinyl Alcohol)

S. Amiya

Central Research Laboratories, Kuraray, 2045
Sakazu, Kurashiki, Okayama 710, Japan

ABSTRACT: The microstructures of Poly (vinyl alcohol) (PVA) and Poly (vinyl alcohol-acetate) copolymers were studied by 1D and 2D NMR techniques. The assignments of all CH proton resonances of PVA relating to 1,2-1,4-glycol and terminal 1,2-glycol linkages were made. The concentrations of 1,2-glycol linkages located at inner and terminal positions of a polymer chain were determined to be ca. 1.61 and 0.06mol%. End-groups of PVA, which were arising from chain-transfers to solvent, monomer and initiator or from the termination of hydrogen abstraction, could be readily determined by using $\alpha,\alpha,\beta-$ trideuterated PVA. The short chain branches consisted of two monomer units (butyl branches) resulting from intramolecular chain-transfer to methine carbons have been also identified. Other structural irregularities, polyene structures which arising from heat treatment, were analyzed by 2D COSY NMR spectroscopy. The sequence distributions of (PVA-Ac) copolymers were studied by ^{13}C-NMR spectroscopy. In the saponification reactions of poly(vinyl acetate), the presence of an irregular bonding could make the rate of saponification more slowly. The relationship between the sequence distribution and a saponification distribution was also discussed and the saponification distribution of (PVA-Ac) copolymers could be evaluated from a block character η which is obtained by ^{13}C-NMR spectroscopy. Combination of 1D and 2D NMR technique provides a powerful method for gaining information on the microstructure of these polymers.

INTRODUCTION: It has been clearly established that the structures of polymer chains are an important factor in determining their properties. In obtaining the structure-property relationship of a polymer, the most important work is to collect the high quality information about its structure. Various kinds of microstructural informations are present in polymers, such as tacticity and sequence distribution for major structures and end-groups, in addition to irregular bonds and short branching within minor structures. It is therefore very important to analyze these microstructures using advanced quantitative techniques. NMR spectroscopy is a powerful method for characterizing synthetic polymers. High-resolution ^{1}H and ^{13}C-NMR spectroscopy and more recently multi-dimensional homonuclear and heteronuclear correlation spectroscopy have revealed structural features of complex polymers in detail [1~9]. In particular, two-dimensional (2D) NMR has been proved as a useful technique to identify small amounts of irregular

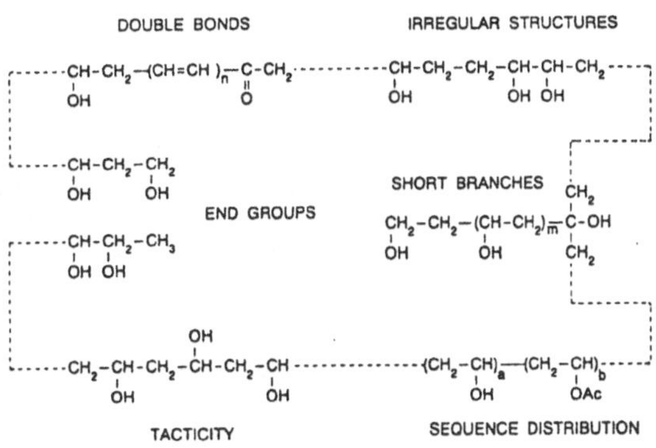

Figure 1 Microstructures of poly(vinyl alcohol)

K.P. Ghiggino (Ed.)
Progress in Pacific Polymer Science 3

structures in synthetic polymers.[10~13] This study focuses on the application of conventional 1D and 2D NMR to determination of the microstructures of poly (vinyl alcohol) and also describes the structure-property relationship. PVA is usually obtained by radical polymerization of vinyl acetate followed by saponification of poly(vinyl acetate). Physical and mechanical properties of PVA are influenced by the degree of polymerization as well as the degree of saponification. Moreover a small amount of irregular structures also affects properties such as durability, stability of the concentrated solution of PVA and so on. As shown in Figure 1, there are many kinds of microstructures in PVA. Each of the items has been studied using conventional 1D and a variety of 2D techniques.

EXPERIMENTAL

MATERIAL: PVA samples (Kuraray Co.Ltd.) are commercially available, which were derivatives of poly(vinyl acetate)(PVAc) synthesized from vinyl acetate(VAc) with α,α'–azobisisobutyronitrile (AIBN) as a initiator in methanol. ($\overline{Pn}$=1700, degree of saponification, DS=98.5mol% and $\overline{Pn}$=170, DS=99.9mol%)

An α,α,β-trideuterated PVA (d_3-PVA) was derived from a sample taken during polymerization of d_3-VAc by using AIBN in methanol at 60℃. ($\overline{Pn}$=1650, DS=98.0mol%)

The d_3-VAc was synthesized through the intermediate d_2-acetylene (CD≡CD) and d_1-acetic acid (CH_3COOD).

Poly(vinyl alcohol-acetate) copolymers with about the degree of saponification of 90mol% were prepared in three ways :

a) Reacetylation ; PVA was dissolved and acetylated in the mixture of acetic acid and water by HCl as catalyst at 30℃. The reaction was controlled by the relative amounts of acetic acid and water.

b) Acid-hydrolysis ; PVAc was dissolved in methanol and hydrolyzed by H_2SO_4 at 63℃ for 2hours. The concentration of H_2SO_4 was almost 0.17N.

c) Alkali-hydrolysis ; PVAc was dissolved in methanol and hydrolyzed by NaOH at 40℃. The degree of saponification was controlled by the molar ratio of PVAc and NaOH (0.02).

MEASUREMENTS: The ^{1}H-NMR spectra were obtained at 60℃ with a JEOL JNM-GX500 spectrometer at 500MHz. The sample concentration was 5%(w/v) in perdeuterated dimethyl sulfoxide (d_6-DMSO), which provides the internal lock signal. Approximately 256-10000 free induction decays were accumulated using a pulse width of 4μs, a pulse interval of 5.0s, and spectral width of 10kHz.

Homonuclear 2D correlation (COSY) spectra were acquired according to the method of Nagayama [14] et al, and Homonuclear Hartmann-Hahn experiment (HOHAHA) according to Bax.[15] The 2D data were zero filled and weighted prior to Fourier transformation as appropriate. A sine-bell filtering function was used in both dimensions.

The ^{13}C-NMR spectra were obtained with the JNM-GX500 at 125MHz. The sample concentration was 10%(w/v) in d_6-DMSO, D_2O and a mixture of D_2O and perdeuterated methyl alcohol (CD_3OD). The spectra were recorded at 80℃, using pulse width of 10μs and pulse intervals of 3.0s. The spectra were obtained after accumulation of 5000-20000scans. Tetramethylsilane, TMS was added as an internal standard to the d_6-DMSO solution and sodium 3-trimethylsilylpropionate, TSP, to the D_2O or the mixture solution. Concentrated solution (12% by weight) viscosities were determined with a Brookfield viscometer.

TACTICITY: Various 2D-NMR methods including ^{1}H J-resolved spectroscopy, ^{1}H-^{1}H COSY spectroscopy, ^{13}C-^{1}H COSY spectroscopy and INADEQUATE spectroscopy have been applied to determine absolute assignments of ^1H and ^{13}C-NMR spectra at the triad-tetrad and pentad-hexad level, respectively.[1,16)]

The configurational assignments have been almost finished through these 2D-NMR methods.

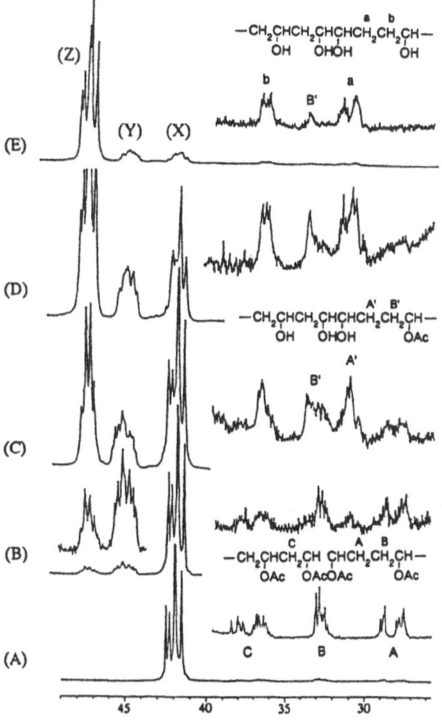

Figure 2 ^{13}C-NMR spectra of methylene carbon of (PVA-Ac) copolymers prepared by alkali hydrolysis ; degree of saponification (mol%) : (A) 0, (B) 6, (C) 45, (D) 63, (E) 88

SEQUENCE DISTRIBUTION: It is very important for understanding structure-property relationships to determine the sequence distributions of vinyl acetate and vinyl alcohol units as well as the degree of saponification. The former, in particular, strongly affects the solution and solid properties (e.g. the viscosity, the surface activity, the melting point [17)] and so on). The determination of the sequence distribution of (PVA-Ac) copolymers using ^1H and ^{13}C-NMR techniques have been reported by a number of researchers.[18~21)] Figure 2 shows ^{13}C-NMR spectra in the region of CH_2 carbons in (PVA-Ac) copolymers with various degree of saponification. Resonances (X) at 41~43ppm were assigned to the methylene carbons of poly(vinyl acetate) [see Figure 2-(A)]. With an increase of the degree of saponification, the peak (X) gradually decreases in intensity and conversely, the peak(Z) at around 47ppm increases. After the middle peak (Y) increases slightly, the intensity falls off again. These three peaks have been assigned to the three dyad sequences : acetate-acetate, alcohol-acetate (including acetate-alcohol) and alcohol-alcohol with decreasing the field strength. (The mole fractions of the dyad sequences will be abbreviated (OAc-OAc), (OAc-OH), (OH-OH), respectively.) Thus, the integrated intensities of the carbon resonance are directly proportional to the number of carbons contributing to that signal. A quantitative description of the sequence distribution and a chemical composition is possible by applying the following equations to the integrated areas of the three methylene carbons dyads :

$$(OH)=(OH\text{-}OH)+(OH\text{-}OAc)/2 \quad (1)$$

$$(OAc)=(OAc\text{-}OAc)+(OH\text{-}OAc)/2 \quad (2)$$

$$\eta=(OH\text{-}OAc)/2\cdot(OH)(OAc) \quad (3)$$

where (OH)=mole fraction of vinyl alcohol units
(OAc)=mole fraction of vinyl acetate units
η=block character

The η value is a useful guide to characterize a sequence distribution in binary copolymers ; it takes $0 \leq \eta < 1$ for blockier distributions, $\eta=1$ for completely random cases, and $1 < \eta \leq 2$ for alternate-like cases. The sequence distributions of each sample in the Figure 2 are blocky with $\eta \leq 0.45$. The samples prepared by reacetylation from PVA are close to random distributions with η=1.0. In the sample obtained by acid hydrolysis the sequence distributions are blocky, but their randomness increases compared with the copolymers prepared by alkali hydrolysis ; η=0.70~0.80

On the other hand, some small peaks are observed in the region between 27 and 39ppm in the spectra. As indicated in this Figure, these peaks are identified as the methylene carbons belonging to the 1,2- and 1,4-linkages. Peak assignments have been made by comparison with the spectra of polymers like PVAc, PVA, and (PVA-Ac) copolymers as well as the empirical additivity rules.[22,23] Although the acetyl groups 1, 2, 3 [see eq. (4)] could be readily converted to hydroxyl groups, the acetyl group 4 wouldn't be converted as far as the degree of saponification of 90mol%.

$$\begin{array}{c} \underset{OAc}{\overset{1}{-C}}-C-\underset{OAc}{\overset{2}{C}}-\underset{OAc}{\overset{3}{C}}-C-C-\underset{OAc}{\overset{4}{C}}- \xrightarrow{\text{saponification}} \underset{OH}{\overset{1}{-C}}-C-\underset{OH}{\overset{2}{C}}-\underset{OH}{\overset{3}{C}}-C-C-\underset{OAc}{\overset{4}{C}}- \end{array} \quad (4)$$

Consequently, the presence of the irregular bonding could make the rate of the saponification reaction more slowly. It is generally believed that the sequence distribution of PVA-Ac copolymers becomes random if the rate decreases. In the PVA-Ac copolymer with the irregular bonding of ca. 2.2mol%, the block character, η is about 0.50, which is larger than that of the commercial available copolymers (η=0.45), having the irregular bonding of ca. 1.7mol%. The block character increases, namely the randomness increases with increasing content of the irregular bonding.

The mechanism of saponification by the alkali catalyst has been reported [24] that the catalyst is adsorbed on the free hydroxyl groups and this induced successive hydrolysis in the acetyl groups of the direct neighbors. Therefore the rate of the reaction depends on the distance between the hydroxyl group and acetyl group. In order to accelerate the rate of saponification reaction, the structures of the 1,2-sequences are favorable, however, the 1,4-sequences are unfavorable structures. The relationship between the sequence distribution and the distribution of a degree of saponification (or composition distribution) has not been fully elucidated. As the sequence distribution is defined by the distribution of each monomer in a polymer chain, the distribution of a degree of saponification indicates the distribution of each polymer chains with various degrees of saponification.

The η value is substituted for the sequence distribution and Dc for the composition distribution. Dc indicates the index of deviation of a compositional uniformity and is defined as follows : [25]

$$Dc=\Sigma\,|W_i \Delta Ac_i|$$

Here W_i represents the weight fraction at the fraction number i, ΔAc_i represents the difference between the mean degree of saponification and the degree of saponification at the fraction

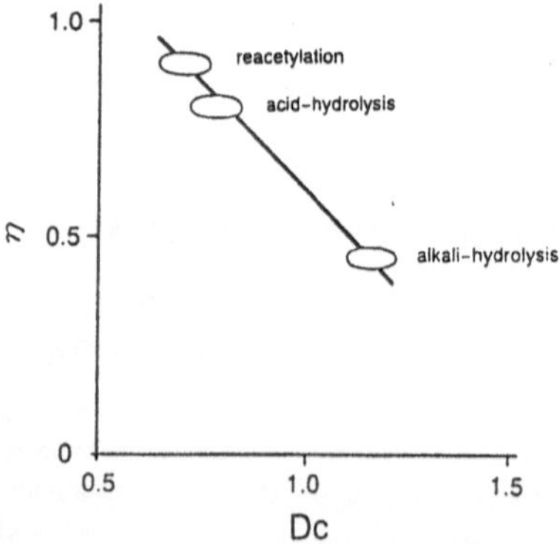

Figure 3 Relationship between η and Dc

number *i*. Dc is the sum of the absolute value of the product of W_i by ΔAc_i . As Dc decreases, the composition distributions become sharp. Dc's of (PVA-Ac) copolymers prepared in three ways were determined by the method of Spencer[26] as follows :

(A) reacetylated copolymer, 0.65~0.75
(B) acid-hydrolyzed copolymer, 0.75~0.80
(C) alkali-hydrolyzed copolymer, 1.15~1.20

Figure 3 shows the relationship between η versus Dc. This reveals the strong correlation between η and Dc. Dc decreases with increasing η, namely the composition distribution becomes sharp as the randomness of the sequence distribution increases. These results suggest the composition distributions of (PVA-Ac) copolymers can be readily evaluated from the η values. Figure 4 shows ^{13}C-NMR spectra of methylene carbons in (PVA-Ac) copolymers with and without heat treatment (135℃, 10 and 12hours). In the figure, the intensity of (OAc-OAc) lines slightly decreases. The intensity of the (OAc-OH) lines, on the other hand, increases. The η values of the copolymers before and after heat treatment could be determined to be 0.45 and 0.54 respectively. Namely, the randomness of the sequence distribution of copolymers increases by heat treatment. The degree of saponification of each copolymer was nearly equal. Therefore the heat treatment could give rise to change the sequence distribution, but not to change the degree of saponification. These results reveal that the ester-exchange reactions between acetyl groups and hydroxyl groups occur both inter and/or intra molecular chains. Heat treatment of (PVA-Ac) copolymers is able to improve the properties, in particular, the surface activity [27] and the ability of foam formation to some extent.

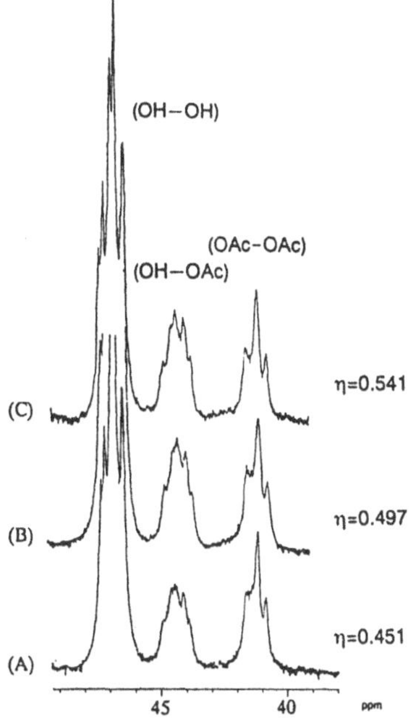

Figure 4 ^{13}C-NMR spectra of methylene carbon in (PVA-Ac) copolymers treated with heat : (A) original (B) 135℃ * 10hrs (C) 135℃ * 12hrs

1,2-GLYCOL LINKAGES : 1,2-glycol linkages as irregular bondings are present along the PVA chains. The determination of 1,2-glycol linkages is important because the irregular bonding produces serious undesirable effects on the properties of PVA, such as thermal deterioration and reduction of mechanical properties. Chemical assay methods [28] are usually adopted for 1,2-glycol bonding determinations, but there is some disagreement between methods for determining 1,2-glycol content of low molecular weight PVA (e.g. viscometry after 1,2-cleavage with

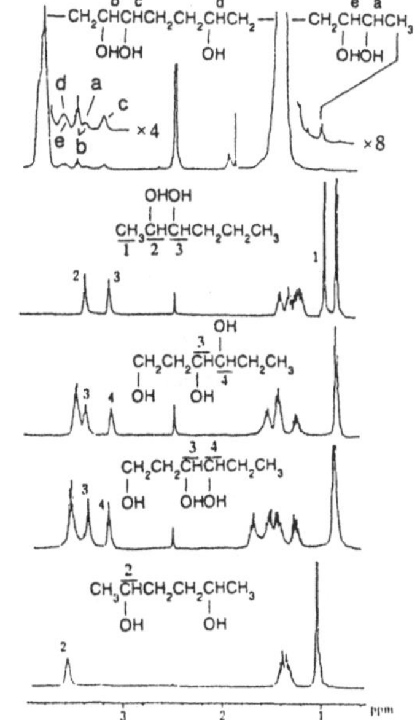

Figure 5 ^{1}H-NMR spectra of Poly(vinyl alcohol) and its model compounds of irregular bondings in PVA

$NaIO_4$ and $NaIO_4$ titrimetry). Therefore new techniques are continually being developed.

^{1}H-NMR spectra of PVA dissolved in DMSO-d_6 is shown in Figure 5, where hydroxyl proton signals were removed from the observable range by adding a few drops of trifluoroacetic acid to the solution. The expanded peaks resonated in the chemical shift region 3.0~3.8ppm are due to methine protons evaluated against the standardized chemical shifts.

Triplet resonance at 3.55ppm has been identified as the methyrol proton (-$CH_2$$\underline{CH_2}$OH). The asymmetrical triplet due to unresolved stereosequence effects at ca. 1.03ppm has been assigned to methyl group which belongs to terminal 1,2-glycol linkages.[29] In addition to these known peaks, undetermined ones were also appeared.

Specific sequence assignments can be made from the COSY spectrum shown in Figure 6. In the COSY spectrum, the normal spectrum is contained along the diagonal, and symmetric pairs of cross-peak connect peaks which correspond to J-coupled protons.

The assignment of the peaks in the downfield region from 3.1 to 3.7ppm are made as follows :

The highest field cross-peak (cross-peak 1) is assigned to CH_3 (1.03ppm) and methine proton H_B (3.43ppm). The methine proton H_B also couples to the methine proton H_C. Therefore the cross-peak comes from H_B and H_C coupling. Two cross peaks (cross-peak 2, 4) correlated with H_B were observed in the Figure.

The assignment of the H_C proton can be made from the HOHAHA spectrum. In the HOHAHA spectrum (not shows here), the cross-peak connecting peaks due to long range J-coupled protons can be observed. Then, the connectivities of more than 3 bonds through the long range coupling can be readily determined. The cross-peaks due to H_A and H_C as well as H_A and H_B were observed, however, the

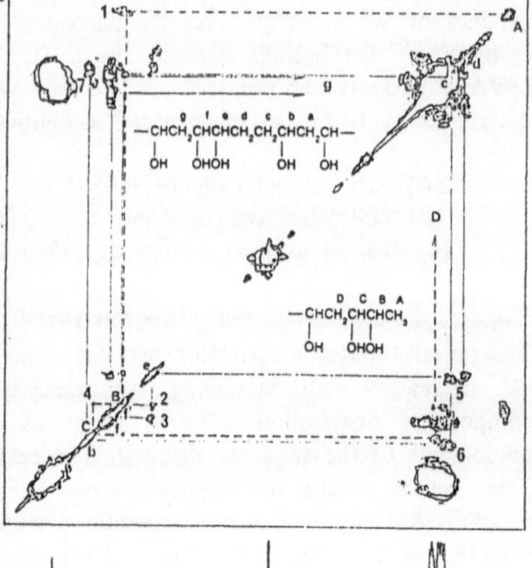

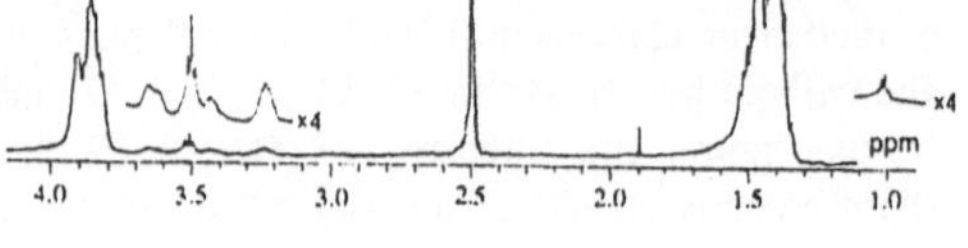

Figure 6 COSY spectrum of Poly(vinyl alcohol) in DMSO-d_6

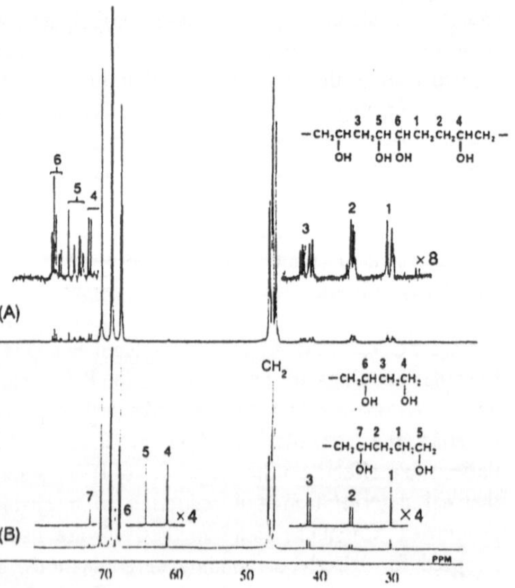

Figure 7 ^{13}C-NMR spectra of Poly(vinyl alcohol) in D_2O : (A) original (B) treated with $NaIO_4$ followed by $NaBH_4$

cross-peak due to H_A and CH(e) (3.23 ppm) was not observed. Consequently the peak at 3.62ppm can be assigned to the H_C proton.

The determination of 1,2-glycol linkages located at the inner position in a polymer chain is made as follows :

As shown in Figure 6, the peak at 3.23ppm has been assigned to the CH(e). The CH_2(d) couples to the CH(e) and CH_2(c). Therefore the cross-peak 5 comes from the CH_2(d) and CH(e) coupling. Then both cross-peaks 2 and 3 are due to CH(e) and CH(f) coupling. The CH(f) peak is probably decomposed two peaks (3.43 and 3.52ppm). This phenomenon may be due to threo and erythro stereo configurational isomers around 1,2-glycol linkages, while, the cross-peaks(6) and (7) are due to CH_2(a)/CH(b) and CH(b)/CH_2(c) coupling respectively. Because the CH(b) couples to both the CH_2(a) and CH_2(c). The CH(f) peaks at 3.43 and 3.52ppm can't be identified to erythro or threo isomers from the 2D spectra.

Using the proton chemical shifts of some model compounds (1,3-4(threo)-hexanetriol, 1,3-4(erythro)-hexanetriol), the peak at 3.43ppm could be assigned to the erythro isomer and the peak at 3.52ppm to the threo isomer as shown in Figure 5. Therefore we could accurately assign all the CH proton resonances of PVA relating to 1,2-1,4-glycol and terminal 1,2-glycol linkages. Using the signal intensity measurements, the concentration of 1,2-glycol linkages located at the terminal and inner positions of a polymer chain were determined to be about 0.06 and 1.61mol% respectively.

Figure 7-(A) shows ^{13}C-NMR spectrum of PVA in D_2O. The expanded signals have been assigned as illustrated in Figure according to the previous data.[30] In the Figure, ^{13}C peaks of terminal 1,2-glycol linkages were not observed because of its low concentration. ^{13}C signal of each carbon of 1,2-1,4-glycol linkages should be basically split into eight peaks which are arising from the stereochemical configuration of both of neighboring hydroxyl groups to 1,2-glycol linkages. 1,2-glycol linkages is cleaved to a long chain aldehyde by $NaIO_4$. After scission with $NaIO_4$, the aldehyde was reduced with $NaBH_4$ resulting in PVA, having both the end-groups of methyrols (-CH_2OH). Figure 7-(B) also shows ^{13}C-NMR spectrum of the PVA in D_2O.

The ^{13}C-peaks arising from 1,2-glycol linkages were completely undetectable. Some new peaks were observed instead. These signals are probably assigned to the methyrol end-groups on the basis of the ^{13}C additivity rules and the relative intensity as shown in the Figure.

$$-CH_2\underset{OAc}{CH}\cdot + CH_2{=}\underset{OAc}{CH} \xrightarrow{k_p} -CH_2\underset{OAc}{CH}-CH_2\underset{OAc}{CH}\cdot \quad (5)$$

$$-CH_2\underset{OAc}{CH}\cdot + \underset{OAc}{CH}{=}CH_2 \longrightarrow -CH_2\underset{OAc}{CH}-\underset{OAc}{CH}CH_2\cdot \quad (6)$$

$$-CH_2\underset{OAc}{CH}-\underset{OAc}{CH}CH_2\cdot + CH_2{=}\underset{OAc}{CH} \xrightarrow{k'_p} -CH_2\underset{OAc}{CH}-\underset{OAc}{CH}CH_2CH_2\underset{OAc}{CH}\cdot \quad (7)$$

$$-CH_2\underset{OAc}{CH}\cdot + RH \xrightarrow{k_{tr}} -CH_2\underset{OAc}{CH_2} + R\cdot \quad (8)$$

$$-CH_2\underset{OAc}{CH}-\underset{OAc}{CH}CH_2\cdot + RH \xrightarrow{k'_{tr}} -CH_2\underset{OAc}{CH}-\underset{OAc}{CH}CH_3 + R\cdot \quad (9)$$

It is generally believed that normal VAc chain growth occurs in a head-to-head addition as in eq. (5), but that occasional head-to-head followed by tail-to-tail addition polymerization occurs as in eq. (6) and (7). The growing radical chain mainly undergoes termination by chain-transfer in the VAc polymerization. The chain-transfer reaction occurring with the normal polymer chain radical leads to an end-group which is an acetate ester of a primary alcohol (8). The chain-transfer with abnormal polymer chain radicals would lead to a methyl end-group (9). where RH denotes a chain-transfer agent such as monomer, solvent, polymer and initiator, k_p and k'_p are the rate constant of normal and abnormal polymer propagation , k_{tr} and k'_{tr} are also the rate constant of the chain-transfer for normal and abnormal polymer chain radical, respectively.

The transfer constants are defined as follows :

$$C = k_{tr}/k_p \ , \ C' = k'_{tr}/k'_p$$

The values of these constants indicate the situation with which chain- transfer occurs.

The C value for VAc or methanol is small, in the order of 10^{-4} , which suggests that the chain-transfer reaction occurs every 10^4 propagation reactions. That is, the degree of polymerization maximizes at about 10^4 , while, the value of C' can be calculated to be 3×10^{-2} (=0.06/1.68), which is much larger than that of C. Since 0.06 of a -CH(OH)CH(OH)CH_3 group is formed every 1.68 vicinal glycol groups, about 3% of the steps following abnormal addition result in chain transfer as in eq.(9). The most important factor affecting the content of 1,2-glycol linkages is the temperature of polymerization, as was first pointed out by Flory. [31] With an increase of the temperature, the 1,2-glycol content goes up, as a result, the degree of polymerization decreases. The properties of PVA seem to be affected by 1,2-glycol linkages even if the amount is very small (1~2mol%). It has been reported that 1,2-glycol linkages are present in the non-crystal region but not in the crystal region.[32] As a result, 1,2-glycol linkages may be cited as a structure-confusing factor. Figure 8 shows the change in the viscosity of concentrated aqueous solution(12%) of two poly vinyl alcohol samples with the same molecular weights ($\overline{Pn}$=170). Although the solution viscosity of commercial PVA, which contains about 1.7mol% of the 1,2-glycol linkages, is almost stable with the elapse of time, the viscosity of PVA treated with $NaIO_4$ followed by $NaBH_4$, which has no 1,2-glycol linkage, increases gradually. The 1,2-glycol linkage in PVA is one of the most important factors to the stability of concentrated aqueous solutions. This phenomenon indicates that 1,2-glycol linkages might play an important role of interrupting the coagulation of polymer chains due to the inter-hydrogen bonding. In order to investigate the reason, we tried to determine the conformation of 1,2-glycol linkages from information of a J-coupling constant between both of the methine protons of 1,2-glycol linkages. The peak of each proton is too broad to determine the J-coupling constant. The broadness of these peaks probably arises from the configurational effect surrounding 1,2-glycol linkages and scalar couplings among neighboring methine and methylene protons. Then, combination of the conformation obtained from the J-coupling constants of model compounds and preferred conformations calculated from a molecular mechanics method(MM2) provided the determination of the conformations at 1,2-glycol linkages in PVA. As a result, the preferred conformation of 1,2-glycol linkages was a gauche

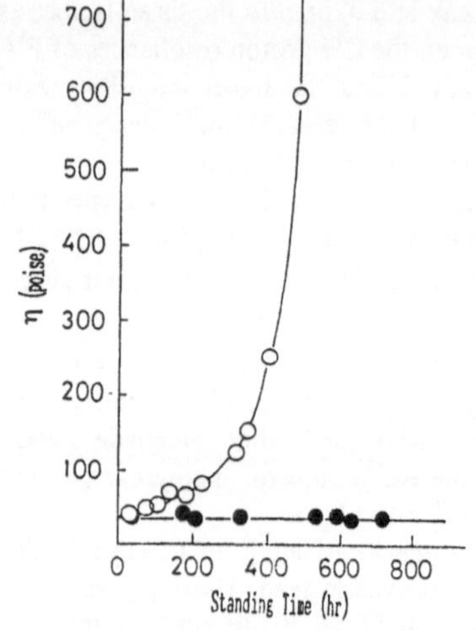

Figure 8 The change in the viscosity of aqueous solution (12%) of two Poly(vinyl alcohol) samples contained the different amount of 1,2-glycol linkages (○) 0%, (●) 1.7mol% ($\overline{Pn}$=170, 20℃)

conformer rather than a trans conformer. For a 1,3-glycol linkage as a regular structure, the trans zigzag conformation was preferable. Therefore the stereostructure in PVA polymer chains may be bent at the point of 1,2-glycol linkages.

END GROUPS: Chain-transfer predominates over termination under the usual conditions of polymerization of vinyl acetate. Consequently, major end-groups can arise from chain-transfer reactions to monomer, solvent, polymer, initiator and some kinds of additives. Then, various kinds of end-groups are present in PVA. The determination of end-groups in PVA reveals the condition of polymerization in details such as temperature of polymerization, concentration of monomer to solvent, a kind of initiator and additives used. Figure 9 shows ^{1}H-NMR spectrum of d_3-PVA in D_2O. In the figure, peaks at 1.50~1.80ppm and 3.95~4.15ppm are assigned to residual protonated CHD proton and CH proton of a main chain. The splitting of these peaks are due to only the configurational effect because of no scalar couplings between methylene and methine. Peaks at 2.05~2.15ppm are due to CH_3 proton in the unconverted residual acetate. In the polymerization of d_3-vinyl acetate, the termination of hydrogen abstraction, the initiation by AIBN and chain-transfer would lead to some kinds of end-groups having nondeuterated hydrogen as follows :

Decomposition of an initiator:

$$(CH_3)_2C(CN)N{=}NC(N)C(CH_3)_2 \longrightarrow (CH_3)_2C(CN)\cdot + N_2$$

Polymerization by the initiators fragment:

$$(CH_3)_2C(CN)\cdot + nCD_2{=}\underset{\substack{|\\ OAc}}{CD} \longrightarrow (CH_3)_2C(CN){-}\!\left[CD_2{-}\underset{\substack{|\\ OAc}}{CD}\right]_n\!\cdot$$

$$\xrightarrow{RH} \xrightarrow{NaOH} (CH_3)_2C(CN){-}\!\left[CD_2{-}\underset{\substack{|\\ OH}}{CD}\right]_n\!H \qquad (10)$$

Head-to-head addition followed by termination:

$$-CD_2\underset{\substack{|\\ OAc}}{CD}\cdot + \underset{\substack{|\\ OAc}}{CD}{=}CD_2 \longrightarrow -CD_2\underset{\substack{|\\ OAc}}{CD}{-}\underset{\substack{|\\ OAc}}{CD}CD_2\cdot$$

$$\xrightarrow{RH} \xrightarrow{NaOH} -CD_2\underset{\substack{|\\ OH}}{CD}{-}\underset{\substack{|\\ OH}}{CD}CD_2H \qquad (11)$$

Chain transfer to methanol:

$$-CD_2\underset{\substack{|\\ OAc}}{CD}\cdot + CH_3OH \longrightarrow -CD_2\underset{\substack{|\\ OAc}}{CD}H + HOCH_2\cdot$$

$$HOCH_2\cdot + nCD_2{=}\underset{\substack{|\\ OAc}}{CD} \longrightarrow HOCH_2\!\left[CD_2\underset{\substack{|\\ OAc}}{CD}\right]_n\!\cdot$$

$$\xrightarrow{RH} \xrightarrow{NaOH} \underset{\substack{|\\ OH}}{CH_2}\!\left[CD_2\underset{\substack{|\\ OH}}{CD}\right]\!{-}CD_2{-}\underset{\substack{|\\ OH}}{CD}H \qquad (12)$$

to monomer:

$$-CD_2\underset{\substack{|\\ OAc}}{CD}\cdot + CD_2{=}CDO\underset{\substack{\|\\ O}}{C}CH_3 \longrightarrow -CD\underset{\substack{|\\ OAc}}{C}DH + CD_2{=}CDO\underset{\substack{\|\\ O}}{C}CH_2\cdot$$

$$CD_2{=}CDO\underset{\substack{\|\\ O}}{C}CH_2\cdot + nCD_2{=}\underset{\substack{|\\ OAc}}{CD} \longrightarrow CD_2{=}CDO\underset{\substack{\|\\ O}}{C}CH_2\!\left[CD_2\underset{\substack{|\\ OAc}}{CD}\right]_n\!\cdot$$

$$\xrightarrow{RH} \xrightarrow{NaOH} NaO\underset{\substack{\|\\ O}}{C}CH_2\!\left[CD_2\underset{\substack{|\\ OH}}{CD}\right]_n\!H \qquad (13)$$

The peak at the highest field, 1.13ppm is assigned to one H of the methyl group in terminal 1,2-glycol linkages. [eqs. (11)] The splitting into two or three peaks may be due to the tactility of the 1,2-glycol sequence and monomer sequence at the chain end. Resonances at 1.25 and

1.27ppm are assigned to an initiation fragment, 1-cyano-1-methylethyl group located at the chain end of a polymer molecule, because these peaks were not observed when PVA was prepared without AIBN. The splitting into two peaks of equal intensities is due to the existence of an adjacent asymmetric center in the polymer chain. [eqs. (10)]

Peaks at 2.20~2.30 and 2.55~2.68ppm, which show a typical AB spin system, is probably assigned to methylene proton next to a carboxylic acid end group. When NaOH was added into the D_2O solution of PVA, the resonance centered at 2.63ppm disappeared and the resonance centered at 2.25ppm increased in intensity as shown in Figure 9-(b). An opposite phenomenon occurred on addition of hydrochloric acid into PVA solution. [See Figure 9-(c)].

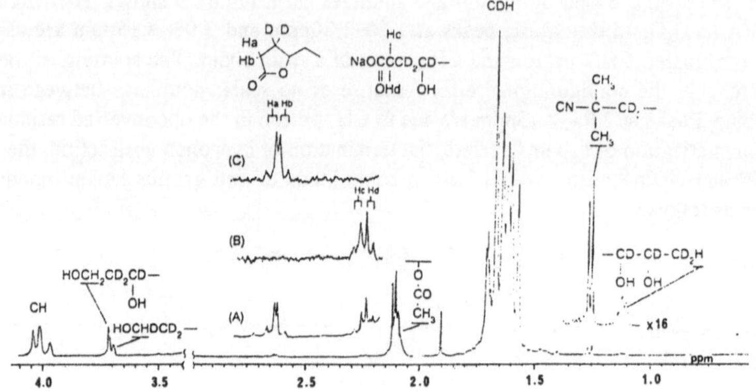

Figure 9 ^{1}H-NMR spectra of d_3-poly(vinyl alcohol) in D_2O at 80℃ : (A) normal spectrum (B) in the presence of excess sodium hydroxide and (C) in the presence of excess hydrochloric acid

These results and our previous paper [31] indicate that carboxylic acid sodium salts and neighboring hydroxyl groups in PVA undergo an intramolecular reaction to form the γ-lactone structures in the presence of hydrochloric acid and reversed in the presence of sodium hydroxide. Thus peaks Ha and Hb are assigned to methylene groups in the closed form and peaks Hc and Hd are also assigned to those in the open form as pointed in the figure. No carboxylic acid proton was observed in the spectrum of PVA in DMSO-d_6 at 60℃(not shows here). These spectra revealed a pH-dependent equilibrium as follows :

$$NaO-\overset{\displaystyle O}{\overset{\|}{C}}-CH_2-CD_2-\underset{\displaystyle OH}{\underset{|}{CD}}-CD_2-\underset{\displaystyle OH}{\underset{|}{CD}}- \underset{NaOH}{\overset{HCl}{\rightleftharpoons}} \underbrace{{}_2HC-CD_2-CD-(O=C-O)}_{\text{ring}}-CD_2-\underset{\displaystyle OH}{\underset{|}{CD}}-CD_2-$$

Among various kinds of end-groups formed by the chain transfer reactions to monomer or polymer, one of end-groups was found to be present in the γ-lactone form under the acidic condition, while in the carboxylic acid sodium salt form under the alkaline condition. No free carboxylic acid group was detected in ordinary PVA. The concentrations of these end-groups depend on the degree of saponification of PVA used. With increasing the degree of saponification, the concentration of the open form increases.

Singlet peaks at 3.69 and 3.72ppm may be assigned to $-CD_2\underline{CDH}OH$ and $-CD_2\underline{CH_2}OH$ groups respectively. The assignment was on the basis of the relative intensity. The latter group would occur as end-groups formed by chain-transfer with the methanol solvent, [see eqs. (12)] and the former group at 3.69ppm occurs as end-groups formed by the termination of hydrogen abstraction [see eqs. (12)-(13)]. The structures of these two end-groups would be same in an undeuterated polymer chain, therefore it is impossible for the commercially available PVA to distinguish between methanol fragment groups and methyrol end-groups formed by normal termination.

SHORT BRANCHES: Although a number of researchers[27,32,33] indicated that small amounts of short branches was presented in PVA, NMR signals of proton or carbon atoms at or near branch points or chain ends have never been directly observed. Figure 10 shows the expanded ^{13}C-NMR spectra of PVA in D_2O at 80 ℃, which showing only terminal methylene resonances in $-CH_2OH$. Peaks at 61.33 and 61.42ppm are assigned to a methylene carbon of $-CH_2OH$ group in the main-chain end. These resonances are split into two peaks which are attributable to the configurational effect of near terminal, that is random dyad tacticity. The peak at 61.42ppm is further split into two peaks due to triad tacticity. Visible in Figure 10 are two weak peaks at 61.30 and 61.24ppm which are observed at higher field than that of the main-chain end by about 0.1ppm. On the basis of a back-biting mechanism proposed by Roedel,[36] short branches in PVA may take either of the following model structures designated by B_1 or B_2, and ^{13}C chemical shifts of these structures are made by the ^{13}C additivity rules.

$$\overset{75.0}{CH(OH)}-\overset{44.4}{CH_2}-\overset{56.5}{CH_2}-OH$$

B1

$$\overset{75.6}{CH(OH)}-\overset{48.7}{CH_2}-\overset{66.4}{CH(OH)}-\overset{41.9}{CH_2}-\overset{59.0}{CH_2}-OH$$

B2

The structure of B_1 corresponds to a monomer unit of vinyl alcohol, while the structure B_2 is two monomer units. By comparison of observed chemical shifts with those of the model compounds, the signals at 61.30 and 61.24ppm are assigned to the ultimate carbon atom in the end-group in model B_2 and two peaks also coming from essentially dyad tacticity. The content of the PVA was about 0.027 B2 end-groups per 100 repeat units. When VAc was polymerized in bulk or methanol solution at the range of conversion 40-90%, the concentrations of short branches in the PVA's were almost same. The correlation between the concentration and the conversion was not found.

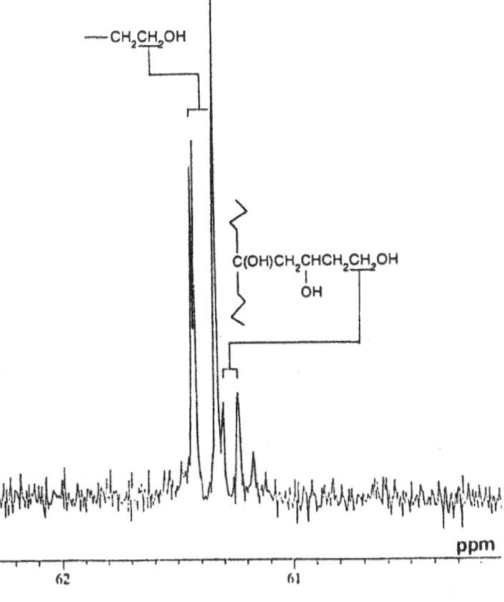

Figure 10 Expanded ^{13}C-NMR spectrum of Poly(vinyl alcohol) in D_2O at 80℃, showing only terminal CH_2 resonances

POLYENE STRUCTURE: Treatment of PVA with heat results in a formation of conjugated polyene structures. Consequently PVA is colored yellow. The conjugated polyene structure has usually been studied by UV spectrum.[37] 2D-NMR experiments could be performed on the sample with thermal treatment to determine the polyene structure in detail. Figure 11 shows COSY spectrum of PVA treated with heat. COSY gives correlation maps showing the connectivity of spin-spin coupled protons. The F2 axis (horizontal) represents ^{1}H-NMR spectrum of PVA as shown in Figure 11. In the case of triene (n=3), in which n indicates the conjugation sequence of $-(CH=CH)_3-$, starting from the H-1 signal at 6.08ppm, a unique transverse connectivity with H-2 at 7.21ppm is then possible. Hence H-2 is correlated with H-3. From H-3 on, the same mechanism reveals connectivities with H-4, H-5 and H-6. Figure 11 also showed other kinds of connectivities with monoene (n=1) and diene (n=2) structures. From analyzing 2D-COSY spectrum, we found that three kinds of conjugated sequences were contained in the polymer. ^{1}H-NMR spectrum was utilized in the quantitative analysis of PVA treated with heat. The concentration of n-polyenes, $-(CH=CH)_n-$, was as follows : monoene (n=1), 0.137mol% ; diene (n=2), 0.083mol% ; triene (n=3), less than 0.01mol%.

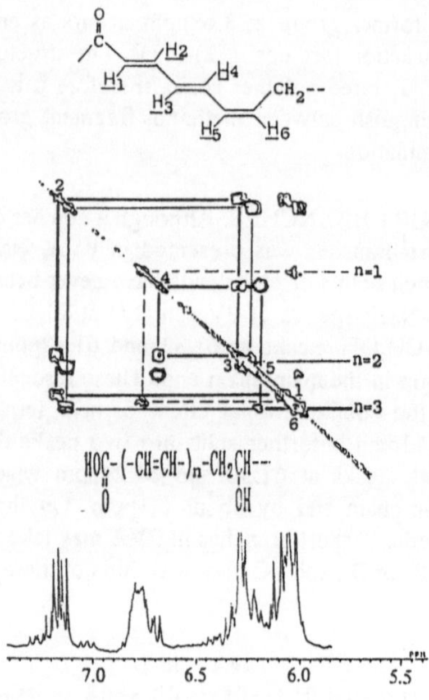

Figure 11 COSY spectrum of Poly(vinyl alcohol) treated with heat (150℃ * 14hrs) in vacuum atmospher in DMSO-d_6 at 80℃

We concluded that various kinds of microstructures in PVA were determined by 1D and 2D NMR spectroscopy. Moreover, the determinations of these microstructures could reveal a detailed property-structure relationship of PVA.

References:

1) Hikichi K, Yasuda M (1987) J Polymer 19:1003
2) Crowther MW, Szeverenyi NM, Levy GC (1986) Macromolecules 19:1333
3) Bruck MD, Bovey FA, Cais RE (1984) Macromolecules 17:2547
4) Asakura T, Demura M, Nishiyama Y (1991) Macromolecules 24:2334
5) Hikichi K, Ikuda M, Yasuda M (1988) J Polymer 20:851
6) Hikichi K, Hirai T, Ikura M, Higuchi K, Ohuchi M (1987) J Polymer 19:1317
7) Beshah K (1992) Macromolecules 25:5597
8) Ketels H, Beulen J, Velden GV (1988) Macromolecules 21:2032
9) Doremaela GHJ, German AL, Veries NK, Valden GPM (1990) Macromolecules 23:4206
10) Benedisk GM (1992) Pure appl. Chem. A29(2):85
11) Hatada K, Ute K, Tanaka K, Imanari M, Fujii N (1989) J Polymer 21:447
12) Asakura T, Nakayama N, Demura M, Asano A (1992) Macromolecules 25:4876
13) Amiya S, Uetsuki M (1985) Anal. Sci. 1:91
14) Nagayama K, Kumar A, Wurlblich K, Ernst PR (1980) J Mag Reson 40:321
15) Bax A, Davis DG (1985) J Mag Reson 65:355
16) Gippert GP, Brown LR (1984) Polymer Bull 11:585
17) Tubbs RK (1966) J Polymer Sci. PartA-1 4:623
18) Moritani T, Fujiwara Y (1977) Macromolecules 10:532

19) Velden G, Beulen J (1982) Macromolecules 15:1071
20) Toppet S, Lemstra P, Velden G (1983) Polymer 24:507
21) Bugada DC, Rudin A (1984) Polymer 25:1759
22) Roberts JD, Weigert FJ, Kroschwitz JI, Reich HJ (1970) J Am Chem. Soc 92:1388
23) Lindeman LP, Adams JQ (1971) Anal. Chem. 43:1245
24) Sakurada I (1968) Koubunshi 17:21
25) Sakurada I, Sakaguchi Y, Shima S (1956) Koubunshi Kakagu 13:348
26) Spencer RS (1948) J Polymer Sci. 4:606
27) Sakurada I, Osugi T (1944) Gosei Seni Kenkyu 2:192
28) Imoto S, Ukida J, Kominami T (1957) Koubunshi Kagaku 14:214
29) Adelman RL, Ferguson RC (1975) J Polymer Sci. Polymer Chem. Ed 13:891
30) Amiya S, Uetsuki M (1982) Macromolecules 15:166
31) Flory PJ, Leutner FS (1948) J Polymer Sci. 3:880
32) Imoto S (1959) Koubunshi Kagaku 16:125
33) Amiya S, Tsuchiya S, Qian R, Nakajima A (1990) Pure Appl. Chem. 62:2139
34) Morishima Y, Iimuro H, Nozakura S (1975) J Polymer Sci. Polymer Chem. Ed 13:891
35) Nozakura S, Morishima Y, Iimuro H, Irie Y (1976) J Polymer Sci. Polymer Chem. Ed 14:759
36) Roedel MT (1953) J Am Chem. Soc 75:6110
37) Maruyama K, Kuramoto Y, Yagi M, Tanizaki Y (1988) Polymer 29:24

Enhancing Reinforcement Effects of Polymers in Asphalt: An Applied Approach to Solid Waste Disposal

William H. Daly, Zhaoyao Qiu, Jeffrey Youngblood and Ioan Negulescu.

Macromolecular Studies Group, Department of Chemistry
Louisiana State University, Baton Rouge, LA 70803-1804

Abstract: Polyethylene, a recyclable waste which is abundantly available either as virgin off specification material or from solid waste classification processes, was slightly chlorinated to improve its compatibility with asphalt. High density polyethylene, HDPE, and chlorinated polyethylene (CPE) modified asphalts were evaluated using dynamic mechanic analysis (DMA), DSC, and epifluorescence microscopy. The results confirm that CPE's with low chlorine contents (<15 wt %) are more compatible with asphalt than HDPE. Blends with up to 20% ground vulcanized rubber (both crumb and 200 mesh powder particles) from recycled tires were prepared with asphalt cements of various grades (AC5 - AC30) and evaluated using DMA. Blends produced from powdered rubber particles exhibited Newtonian behavior at high temperatures; similar behavior was not observed with crumb rubber blends. The mechanical properties of asphalt-rubber mixtures depended upon the concentration of rubber additives, the particle dimensions, and the chemical composition of the asphalt. Some asphalts induced excessive swelling of the rubber particles; blends with high temperature viscosities substantially greater than the upper limit of 3 Pa·sec resulted. The dynamic mechanical characteristics of all blends are discussed in terms of $G^*/\sin \delta$ and G"; comparative data are presented according to the new SHRP binder specifications. Constant stress creep and creep recovery of the polymer or rubber asphalt blends proved to be a sensitive measure of the additive content.

INTRODUCTION

Since the advent of asphalt paving blocks in 1824, efforts to improve the properties of asphalt surfacing materials have centered on polymeric additives. In the 1840's patents describing the modification of bitumen with gutta percha or natural rubber appeared and as each new polymer was developed, its potential interaction with asphalt was evaluated. Since the addition of polymers increases the cost of the corresponding polymer modified asphalt cement (PMAC), the new material must increase the durability of the mixture and meet all climatic requirements without compromise. Indeed, addition of appropriate polymers to asphalt increases fatigue cracking resistance, reduces the extent of permanent

K.P. Ghiggino (Ed.)
Progress in Pacific Polymer Science 3

deformation, improves thermal cracking resistance, lessens moisture sensitivity and reduces age hardening. Polymers typically employed in PMAC's include natural rubber latexes, styrene-butadiene latexes, styrene-butadiene-styrene thermoplastic rubbers, polyethylene and other polyolefins, and ethylene-vinyl acetate copolymers [1]. Polyolefins and vulcanized rubber also represent major components of solid wastes, so incorporation of these materials into polymer modified asphalt cements would be an effective method of polymer recycling.

Addition of SBS block elastomers to asphalt increases resistance of the resultant PMAC to rutting at warm temperatures and improves the low temperature ductility, elasticity, and cyclic loading properties of the mixture [2]. The potential for realizing similar improvements with ground rubber additives, coupled with the growing disposal problem of waste rubber tires, have lead to extensive reprocessing of tire rubber for use in asphalt concrete mixtures. In the United States, the Intermodal Surface Transportation Efficiency Act of 1991 (ISTEA-91) mandates that all asphalt products purchased with federal aid dollars must incorporate discarded tires. The goal of the legislation is to add the equivalent of one tire carcass to each ton of aggregate asphalt mix employed in highway construction. Incorporation of ground rubber tires must begin in 1994 and full implementation of the statute must be completed in 1997. Thus, it is imperative that the properties and potential problems associated with rubber containing asphalt mixes be ascertained as soon as possible. In order to assess the potential benefits of rubberized asphalt mixtures, especially at hot temperatures characteristic of Louisiana, a laboratory research program has been initiated in which the rheological properties of binders containing ground tire rubber are compared to those of asphalt binders based upon other polymeric materials.

Blends of polymeric materials with asphalt are complex and characteristically unique paving material systems. For any specific asphalt cement, (AC), the physical properties of the asphalt-polymer blend are affected by the amount of polymeric material added, its composition, its physical properties, i.e., molecular weight, degree of branching or crosslinking, and, the most important variable, the compatibility of the AC with the admixed polymer. The ability to enhance polymer/asphalt compatibility is complicated further by source dependent variations in asphalt compositions within a given grade. Our research efforts are directed toward enhancing the compatibility of polymer additives with asphalt, assessing the phase structure of the asphalt-additive mixture, and comparing the properties of polymer/asphalt cements (PMAC) with those of the pure asphalt matrix.

An extensive study on the chemistry and physical properties of asphalts has just been completed under the auspices of the Strategic Highway Research Program (SHRP) and new performance based specifications based upon dynamic testing techniques before and after controlled aging have been proposed [3]. Molecular structures of asphalt are highly diverse, but three basic structures are generally recognized to coexist, i.e., straight or branched aliphatic chains, simple and complex naphthenic rings, and heteroaromatic systems [4]. Durable asphalts are comprised of these components interacting to form a balanced, compatible system. The asphaltenes are suspended in the oils by the resins, making asphalt a colloidal system. The introduction of any incompatible polymer under agitation into such a system at high temperature generally results in asphaltene flocculation and oil bleeding, leading to a binder having no cohesion [5]. Even if the phase separation of the asphalt components is not apparent, extended mixing times will be required to achieve acceptable mixtures [6]. Polymers must improve not only AC properties, but must improve the performance of binder-aggregate combination as well [7]. Studies on the application of SHRP procedures to PMAC's indicate that the long term behavior of aggregate-PMAC mixes can also be predicted by dynamic rheological tests [2].

Plastics give strength to the asphalt, but the ability to recover from extension is lost [8]. Thermoplastics which are partially crystalline such as polyethylene (PE) and polypropylene (PP), have attracted more and more attention since this class of polymer combines the advantages of rubber and fibers. The crystalline segments of polyolefins serve as high strength fillers in the asphalt-polymer blend and improve the blend properties over all service conditions. However, the polyolefins are only slightly compatible with asphalt; thus, the blends tend to separate at high temperatures. Ethylene-vinyl acetate copolymers, EVA, are more compatible and improve resistance to permanent deformation and increase modulus . The presence of vinyl acetate repeat units decreases the crystallinity of the PE blocks, leading to increased flexibility and toughness. In one study on surface dressings, the EVA modified asphalt ranked among the best of the asphalts tested in the areas of fatigue resistance and field performance [9]. Unfortunately, the properties of EVA blends may vary substantially depending upon the asphalt used to prepare the PMAC and unsatisfactory mixtures can be produced.

EXPERIMENTAL

Asphalt Cements: Tank AC-10 asphalts provided by Louisiana refineries were used for the preparation of polymer/asphalt mixtures. The distribution of components was determined by solution NMR in perdeuterotetrachloroethane. A relaxation agent, $Cr(acac)_3$, 12 mg/ml, was added to the ^{13}C NMR samples. Table 1 summarizes the observations on four samples of Louisiana asphalts; three different AC sources are represented in the samples. The relative crystallinity of a given asphalt as measured by DSC is also reported.

Table 1. NMR And DSC Characterization Of Asphalt Composition

Sample	AC10-1	AC10-2	AC10-3	AC5-1
Arom H%	6.22	4.53	6.90	5.88
Arom C%	37.3	33.4	37.3	
Benzylic H%	14.89	5.62	14.89	12.30
Linear Aliph H%	78.89	89.85	81.73	81.82
Methyl/Aliphatic	21.22	30.08	20.95	26.14
ΔH_f, (J/g)	7.7	5.8	8.5	
% cryst.[a]	3.9	2.9	4.3	
Glass Transition Tg,(°C)	-12	-20	-3	

[a]Average enthalpy for 100% crystallinity = 200 J/g [10]

Polymeric Materials: High density polyethylene (HDPE) was supplied by Paxon Polymer Co; a sample with a weight and number average molecular weight of 8.5×10^4 and 1.9×10^4, respectively, and a melt index of 25 was employed. Chlorination of HDPE was performed in tetrachloroethylene to a level of 10 wt% Cl [11]; finely divided chlorinated polyethylene (CPE) was obtained. Rubber from ground tires was used as received from the suppliers as crumbs (Baker Rubber Inc., Southbend, IN) or fine powder (Rouse Rubber Industries, Inc., Vicksburg, MS). The crumb rubber, CR, had a maximum particle size of 1 mm and contained residual polyester fiber particles. The powder rubber, PR, was a fine 200 mesh (0.075 mm) ground rubber.

<u>Polymer/Asphalt Cement Mixtures:</u> Except for NOVAPHALT AC-10 asphalt binder, which was tested as received, (NOVAPHALT AC-10 was considered as a reference polymer modified asphalt binder), all the other polymer/asphalt mixtures were prepared by mixing the polymeric materials (i. e., HDPE, CPE, CR and PR) with the asphalt cements. Mixing was done in aluminum containers normally used as a soft drink cans using a high speed lab stirrer. The cans were immersed in an oil bath maintained at T+5°C, where T is the mixing temperature. HDPE or CPE were mixed with AC-10 at T = 160°C for 40 min. CR was added to AC at T = 220°C, the temperature was allowed to drop to T = 170°C and the mixing was continued at this temperature for 40 min. PR was added to AC at T = 170°C and the mixture was stirred for 20 min. The cooling to ambient temperature was not monitored. The cans were also used for the storage of mixtures.

Methods.

The relative crystallinity of a given asphalt can be measured by DSC [12, 13]. A Seiko DSC 220C instrument calibrated for temperature and enthalpy with indium was employed to estimate the relative volume of the crystalline phase in each of the asphalt samples. The DSC analysis was conducted on ≅ 10 mg samples sealed in an aluminum sample pan using an empty aluminum sample pan with cap as a reference. Initially each sample was cooled at 3°C/min to -45 C and then heated at 3°C/min. The heats of fusion (ΔH_f) observed are listed in Table 1. The percent crystallinity was estimated from this data by assuming that completely crystallized hydrocarbons in an asphalt matrix exhibit an average enthalpy of 200 J/g [10]. The glass transition temperatures, Tg, of the neat asphalts, PMAC's and RAC's were estimated by, DSC at a heating rate of 10 °C/min. The glass transition was observed after the sample had been subjected to one heating/cooling cycle (RT→ 150°C→ -30°C).

A Brookfield viscometer (Spindle # 3) was used to evaluate the change in consistency of binders at high temperatures. Viscosity variations based upon both the amount of rubber added and the mixing/storage times were measured. A BOHLIN CS rheometer and a SEIKO DMS110 dynamic mechanical spectrometer were used to investigate the rheological behavior of neat, PMAC's and RAC's. Using the Bohlin rheometer, the measurements were made with a 4° cone and plate with a diameter of 20 mm; the minimum gap was set at 0.15 mm. The SEIKO DMS employs a vertical parallel square plate system; the contact area of the plates was 100 mm^2 and a gap varying from 0.7 to 1.5 mm. A constant stress creep test was run at 35°C with the BOHLIN CS rheometer; the stress applied was 590 Pa. Creep recovery was measured after stressing the sample for 10, 100 and 1000 sec.

In a master representation of DMA data, a viscoelastic function is plotted against frequency. The William-Landel-Ferry (WLF) equation [14] has been widely used to predict the temperature shift factors, a_T, for asphalt binders which allow the generation of rheological master curves based on the principle of time(frequency)-temperature superposition [15-17]. Multiple frequency determinations (f = 1, 2, 5, 10, 20, 50 and 100 Hz) were used in the present study for building the master representations. The reference temperature was taken as T = 64 °C to comply with the performance graded asphalt binder specification [4] particularized for the hot climate typical of Louisiana. Using the WLF equation to estimate the shift factor, the best grouping of data was obtained when the values of C_1 and C_2 were 17 and 60, respectively.

RESULTS AND DISCUSSION

Polymer additives should reduce the temperature sensitivity of asphalt cements by increasing the high temperature viscosity. Ideally, the resultant binder should have a very low thermal sensitivity throughout the utilization range, but a low viscosity at mixing temperatures [5]. Resistance to permanent deformation as well as tensile and fatigue strength should be enhanced relative to the properties of the unmodified asphalt. However, the binder must maintain good adhesive properties and exhibit high aging resistance. A suitably prepared PMAC should exhibit reduced plastic deformation at high temperatures, better low temperature resiliency and better performance in service under heavy oscillatory loads.

If price and availability of various polymers that have been proposed for asphalt modification are considered, it is obvious that polyethylene or waste polyolefins (PO) would be more economical than other polymeric candidates. Further, since polyolefins comprise approximately 60% of plastic solid wastes, a reliable source of polyolefins, either from virgin material or from recycled waste is assured. Polyethylene (PE) is a potentially useful modifier for increasing the low temperature fracture toughness of asphalt concrete [18], and it may confer additional pavement stability at elevated temperatures, which would minimize rutting and distortion due to creep. High density linear PE, HDPE, is highly crystalline but the amorphous domain exhibits a very low glass transition temperature. This allows PE to contribute additional toughness and ductility at low temperatures to

PE/asphalt blends, particularly those prepared from soft asphalts. However, it is known that asphalt-polyethylene mixtures have a tendency toward gross phase separation, i.e., gross incompatibility, when standing at elevated temperature for long periods 19]. Therefore modification of PE is needed to enhance its compatibility with asphalt.

Modification of PE by chlorination is a simple technique to change the polarity, to reduce the crystallinity, and to increase the elasticity of the polymer. Partially chlorinated polyolefin waxes were reported to improve stability of asphalt-polymer blends [20, 21], so we elected to prepare and characterize polyethylenes with various degrees of chlorination to improve the polymer interaction with polar components of asphalt. The extent of chlorination can be used to vary the crystallinity of the polymer additive. The crystalline domains of polyolefins contribute to high temperature reinforcement while their amorphous domains, which exhibit very low glass transition temperatures, contribute additional toughness and ductility at low temperatures to PO/asphalt blends, particularly those prepared from soft asphalts.

Viscosity Behavior of PMACs

The Brookfield viscosities of three representative neat asphalts and the corresponding PMACs produced by addition of HDPE, CPE, PR and CR are compiled in Table 2. Addition of linear polymers to asphalt at concentrations approximating 5% can produce a ten to twenty fold increase in the high temperature (135°C) viscosity. The rather undesirable effect can be attributed to partial solubility of the PE and CPE in the asphalt at high temperatures. The high viscosities of these mixtures indicate that more time and energy will be required to mix PMAC's with aggregate. SHRP specifications for neat asphalts suggest that the maximum high temperature viscosity for the binder should not exceed 3 Pa·sec. The viscosities of PMAC's containing 5% PE or CPE remain below the recommended limit, but the viscosities of mixtures with higher polymer contents will exceed the limit.

Since the scrap rubber was highly crosslinked during vulcanization, the rubber particles are not soluble in asphalt even at high temperatures. Thus, RAC's containing 5% rubber can be produced with very modest increases in the high temperature viscosity. However, the rubber particles are swollen as components of the asphalt mixture are imbibed and at the concentrations of rubber additive proposed by the ISTEA legislation, a substantial increase in the viscosities of the binders is observed (Table 3). Crumb rubber particles enhance the viscosity more than powder rubber at the same concentration. Vulcanized rubber is infusible and insoluble; the particle size can only expand during the mixing process unless some polymer degradation occurs. The effect is further complicated by the variable swelling capacity of the neat asphalts. These variations may contribute to problems in mixing the binders with aggregate. Mix designs based upon asphalt grade and rubber content will not adequately predict the increased viscosity observed with asphalts from different sources. The mix specification should be expanded to include some measure of the asphalt chemical composition.

Table 2. Influence of Asphalt Type and Polymer Type on SHRP Stiffness Parameter, Shear Loss Modulus and Dynamic Viscosity of Neat and Rubber Modified Paving Grade Asphalt Cements.

ASPHALT MIXTURE	POLYMER %	$G^*/\sin\delta$, Pa @64°C, 1Hz	G", Pa @64°C, 1Hz	$G^*(20)/G^*(64)$	η, Pa·s @ 135°C, 1Hz
NOVA.AC10	ca. 5.0-7.0	10,010	9,193	86	1.37
AC10-3(<64)	0.0	898	909	383	0.10
AC10-3/PE	5.0	2,008	2,008	400	1.86
AC10-3/CPE	5.0	3,365	3,210	167	1.80
AC10-3/PR	5.0	1,279	1,254	303	0.27
AC10-1(58)	0.0	592	568	492	0.21
AC10-1/CR	5.0	1,146	1,085	317	0.86

Epifluorescence Microscopy

The macrostructure of the asphalt polymer mixtures was examined using epifluorescence microscopy. Our observations confirm previous reports [6, 22, 23] that there are basically two phases existent in asphalt-polymer blends, a continuous asphalt rich phase and a polymer rich phase. The dispersed polymer rich phases are expected to improve the toughness of brittle asphalt at low temperatures and reinforce asphalt at high temperatures [24, 25]. The high temperature enhancement will be more pronounced if the polymer can form a network-like structure, i.e. the polymer concentration exceeds the overlap concentration, c* [6]. The CPE rich phase was larger than the HDPE rich phase in asphalt blends prepared under comparable conditions indicating that a higher percentage of the asphalt components have been absorbed in the polymer phase. Samples containing 10 wt% CPEC exhibited bicontinuous phase morphology; the phase transformation was accompanied by a marked increase in viscosity as expected at concentrations greater that c*. Blends with 10 wt% HDPE retained the polymer droplets in an asphalt continuous phase; the concentration of HDPE in the asphalt phase was minimal.

The enhanced compatibility of CPE in asphalt can be attributed to a change in the polymer polarity as well as changes in morphology stemming from the reduced crystallinity. In crystalline polymers like HDPE, interaction with solvents and reagents is limited to the readily accessible amorphous regions. Although chlorination of HDPE was conducted in solution, analysis of the chlorine distribution in the chains indicates that chlorination is not perfectly random. Runs of unreacted methylene groups that can crystallize remain. Chlorinated methylene groups do not enter the crystallites so the amorphous region contains a higher chlorine content than that measured in bulk samples. Thus, the amorphous regions are substantially more polar, and the presence of chlorine atoms on the polyolefin chain will improve the compatibility of the polymer with aromatic components and functional groups containing heteroatoms, such as N, S and O in asphalt. The polar components of asphalt would have a greater affinity for the amorphous regions of CPE and one would expect a corresponding increase in the compatibility of these polymers with asphalt. Introduction of chlorine adjusts the interaction parameters to reduce single component extraction. Thus the delicate equilibrium among the asphalt components is maintained. Examination of a RAC containing 17.5% powdered rubber revealed that a swollen interface approximately 8 μm wide is created during the high temperature mixing process. Detailed examination of the interface is limited by the low contrast, but its presence confirms the interaction between the asphalt matrix and rubber particles.

Dynamic Rheology

Rheological measurements under oscillating conditions yield the dynamic mechanical properties of polymers, i.e. the storage modulus, G', the loss modulus, G", and a mechanical damping or internal friction, tan δ. In studies of the response of a material to vibrational forces, stress, strain, frequency and temperature are the key variables. When a material is subjected to cyclical stress under conditions analogous to those encountered in the intended applications, the data reflect both short-term and long-term responses to the stress conditions. If time-temperature superposition can be applied, dynamic data obtained at short time intervals at high temperature can be transformed to yield long loading time data relevant to thermal cracking [26]. Although extensive data can be acquired using dynamic mechanical analysis, DMA, selection of the correct parameters in the data analysis for predicting service performance of asphalt binders is not immediately obvious. G^*, the ratio of the peak stress to the peak strain, reflects the total stiffness. The in-phase component of $|G^*|$ is the shear *storage modulus*, G', and represents the part of the input energy which is not lost to heat (the elastic portion). In the SHRP specifications, the resistance of the asphalt binder to fatigue cracking is considered by specifying a maximum value for the stiffness parameter $G^* sin\delta$ at the average pavement design temperature [17]. Since by definition $sin\delta = G''/G^*$, it follows that $G^* sin\delta = G''$, implying that the loss modulus is indicative of this pavement distress. This parameter relates the contribution of the asphalt binder to the dissipation of energy in a pavement during each loading cycle. Since the loss modulus is a measure of the viscous flow within a viscoelastic fluid, this parameter relates also to permanent deformation at high temperatures. King et al. [2] demonstrated that G'' does indeed correlate quite well with results from the rutting simulator. Below certain values at the test temperature, there was an approximately linear relationship of rut depth to $log(G'')$. A similar correlation between rutting resistance and $G^*/\sin\delta$ was anticipated by SHRP investigators.

An estimate of the temperature susceptibility can be obtained from a ratio of the one of the moduli at low vs. high temperature [6]. We have elected to report the characteristic ratio of the complex moduli, G*(20°C)/G*(64°C) as this indicator. Not surprisingly, changes in the temperature susceptibility relative to neat asphalt reflects the additive compatibility as well. The DMA data obtained on mixtures produced from several grades and sources of asphalt by blending either powdered or crumb rubber additives is compiled in Table 3.

The dimensions of rubber particles significantly affect the rheology of the rubberized asphalts as indicated by variation in the phase angle, δ, with increasing temperature. The isochronal sin δ curves of both neat asphalt and powdered rubber asphalt cements (PRAC) approach one asymptotically as the materials begin to flow (Figure 1). However, this is not the case for crumb rubber asphalt cements (CRAC) where $G^*/sin\delta < G''$ and the trend of sin δ at high temperatures is substantially less than 1, indicating that the flow is far from Newtonian. This may be due to limitations in the measuring technique since the rubber particles are relatively large relative to the gap between the rheometer plates; further work to define the magnitude of this problem will be required. Thus, data obtained from PRACs can be reasonably compared with other PMACs, but similar comparisons with CRAC's must be made with reservations.

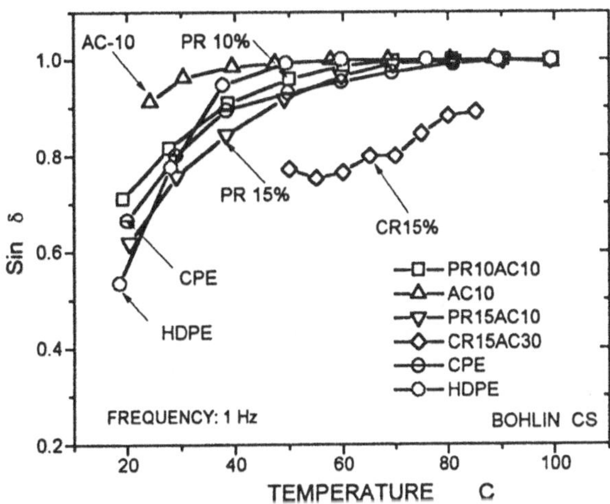

Figure 1. Plot of isochronal sin δ versus temperature for representative asphalt, powder rubber and crumb rubber asphalt mixtures.

The polymeric additives increase the storage modulus of the asphalt mixtures by an order of magnitude at lower frequencies and with the exception of HDPE mixtures the temperature susceptibility is decreased. This is apparent from the slope of the plots of G' vs. reduced frequency (Figure 2). The impact of polymeric additives on the loss modulus is illustrated by a plot of G" vs. reduced frequency shown in Figure 3. A commercial PMAC, Novaphalt, is also included in both plots. At low frequencies, corresponding to high temperatures, the higher viscosities of the PMAC's is apparent, particularly with the HDPE mixture. Chlorination of the HDPE reduced the high temperature G" and produced a mixture quite comparable to Novaphalt.

The rubber content of the RAC mixtures shown in Figure 3 was set at 10 wt% to allow more direct comparison with Novaphalt. The absolute values of *G"* of the PR modified AC-10 binder are comparable to that of the standard NOVAPHALT AC-10 and are superior to those of the higher viscosity grade AC-30 binder. At high frequencies, the properties of the CR/AC-10 mixture fall below those exhibited by

Table 3. Influence of Asphalt Type and Rubber Type and Concentration on SHRP Stiffness Parameter, Shear Loss Modulus and Dynamic Viscosity of Neat and Rubber Modified Paving Grade Asphalt Cements.

ASPHALT MIXTURE	POLYMER %	G*/sinδ, Pa @64°C, 1Hz	G", Pa @64°C, 1Hz	G*(20)/G*(64)	η, Pa·s @ 135°C, 1Hz
AC5-1(<58)*	0.0	396	393	166	0.08
AC5-1/PR	2.5	838	838	144	0.45
AC5-1/PR	5.0	2,283	2,166	103	0.70
AC5-1/PR	10.0	4,827	4,289	49	1.02
AC5-1/PR	17.5	9,940	7,712	36	5.00
AC5-1/PR	20.0	11,003	8,821	32	6.76
AC5-1/CR	15.0	14,235	8,883	30	2.31
AC10-1(58)*	0.0	592	568	492	0.21
AC10-1/PR	5.0	1,310	1,305	256	0.36
AC10-1/CR	5.0	1,146	1,085	317	0.86
AC10-1/PR	10.0	2,080	2.059	174	0.57
AC10-1/PR	15.0	3,811	3.650	132	2.12
AC10-1/PR	20.0	5,898	5,859	93	4.01
AC10-1/CR	20.0	20,308	11,301	36	6.34
AC10-2(58)*	0.0	520	520	300	0.08
AC10-2/PR	5.0	2,104	2,036	103	0.18
AC10-2/PR	10.0	5,842	5,184	63	0.89
AC10-2/CR	15.0	23,318	12,893	30	4.23
AC10-3(<64)*	0.0	898	909	383	0.10
AC10-3/PR	2.5	1,165	1,139	308	0.12
AC10-3/PR	5.0	1,270	1,254	303	0.36
AC10-3/PR	10.0	1,701	1,683	216	0.42
AC10-3/PR	15.0	4,343	4,165	127	1.21
AC10-3/PR	20.0	5,462	5,023	87	3.20
AC10-3/CR	15.0	9,964	6,659	69	2.00
AC30-1(64)*	0.0	1195	1177	364	0.33
AC30-1/CR	15.0	31,025	17,149	37	6.68
AC30-5(64)*	0.0	1,521	1,501	796	0.29
AC30-5/PR	2.5	3949	3892	224	0.98
AC30-5/CR	15.0	18,835	12,890	57	3.42

* The AC grade according to SHRP specifications is shown in parentheses.

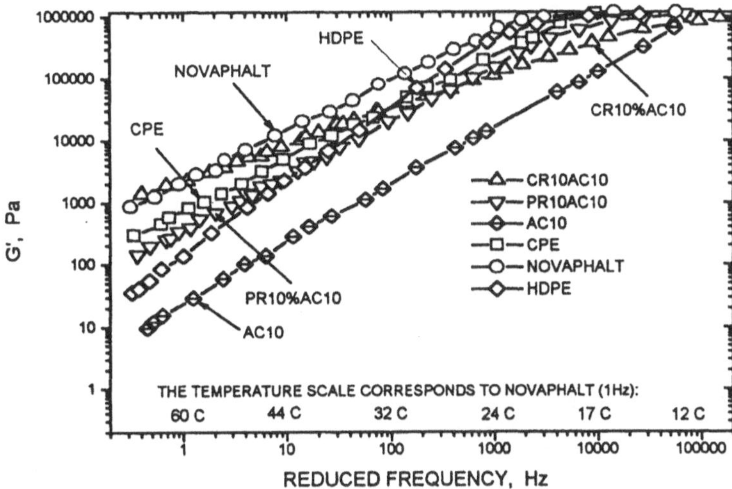

Figure 2. Plot of storage modulus, G', versus reduced frequency for neat, polymer and rubber modified paving grade AC10-3 binders. Reference material: NOVAPHALT AC10.

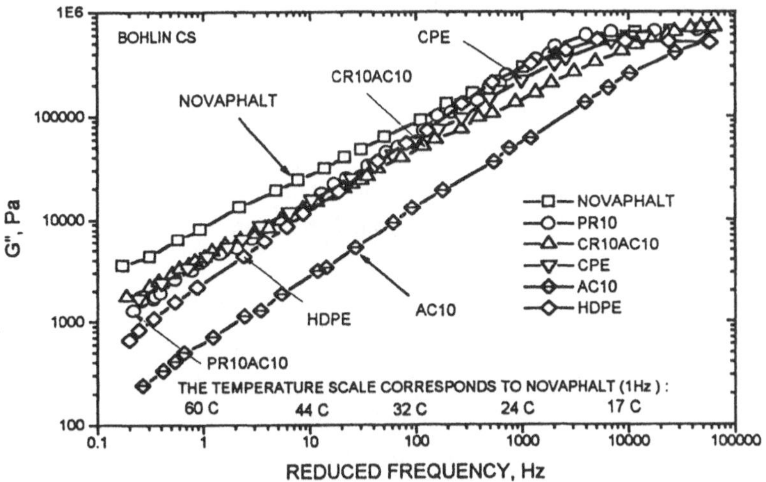

Figure 3. Plots of shear loss modulus, G", versus reduced frequency neat, polymer and rubber modified paving grade AC10-3 binders. Reference material: NOVAPHALT AC10.

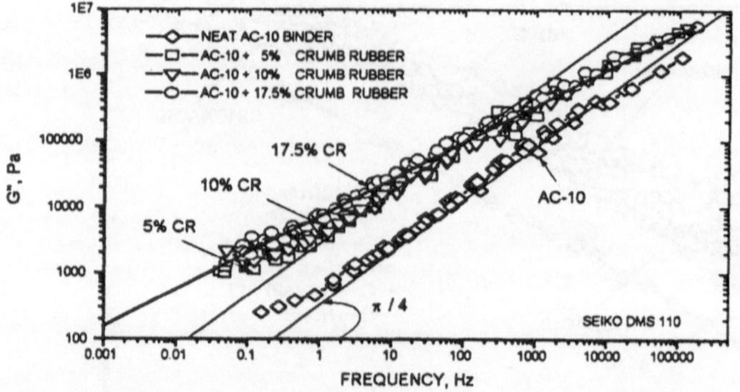

Figure 4. Plots of shear loss modulus versus reduced frequency for neat and rubberized paving grade AC-10 binders containing different amounts of crumb rubber.

AC-30. It is expected that the field performances of mixes prepared with blended CR binders will be inferior to those based on PR/AC-10 mixtures.

The variation of the shear loss modulus of rubberized binders with content of crumb rubber is shown in Figure 4. Note that the slope of the curve for the neat AC-10 and for the mixture containing 5% CR approaches 1 for a large domain of frequencies. This is the region of frequency (or temperature) where the flow properties of the binders exhibit Newtonian behavior, i. e. the fluid exhibits stress in proportion to the rate of the strain but not to the amount of strain. It is in this region that simple shear (e. g., cone and plate, sliding plate) and dynamic tests have been found to coincide [15]. Increasing the content of large rubber particles in the asphalt mixture produces both a deviation from Newtonian flow and an increase of the viscous component.

The contribution of the asphalt binder to permanent deformation of asphalt mixtures is considered in terms of the minimum value for the stiffness parameter, $G^*/\sin\delta$, at a maximum pavement design temperature (i. e., 1000 Pa @ 10 rad/s). This parameter correlates to that portion of the accumulated, non-recoverable deformation occurring in a pavement that is attributable to the asphalt binder. The higher the value, the less deformable is the pavement. Isochronal plots of $G^*/\sin\delta$ reveal distinct differences in the contribution of the additives to the mixture stiffness. In Figure 5, it is clear that the base asphalt would qualify for a maximum use temperature of 58°C; addition of any polymer additive raises the qualification to > 64 °C. Using CPE at 5 wt% or powder rubber at 10 wt% enhances the high temperature properties without changing the low temperature properties extensively, i.e. the temperature susceptibility of the mixtures are lowered. The values of this stiffness parameter (read at 64°C from isochronal plots at 1Hz of G*/sinδ versus temperature) and the characteristic ratios are listed in Table 2 for neat asphalt cements and mixtures prepared from the respective AC and polyolefins and or rubber additives. The relative softness of the rubber is apparent; addition of a small amount of rubber (5%) induces a much lower increase of the stiffness parameter and G" compared to the corresponding polyolefin modified asphalt cements, particularly the AC/CPE mixture. Moreover, the rheological behavior of all materials listed in this table is Newtonian at the reference temperature, 64°C, as shown by the value of the ratio (G*/sinδ)/G" ≅ 1. The change in the characteristic ratio, G*(20° C)/G*(64°C), is quite revealing; the mixture containing HDPE exhibits a ratio comparable to the neat asphalt. Powdered and crumb rubber reduce the ratio by approximately 25% and CPE cuts the ratio in half.

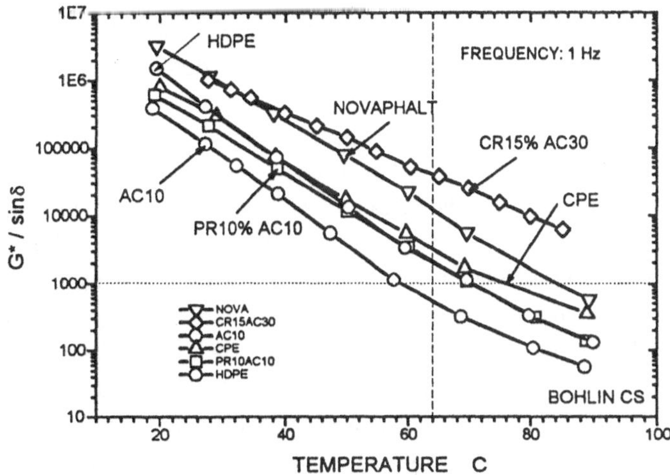

Figure 5. Isochronal plots of G*/sin δ for neat, polymer and rubber modified paving grade AC10-3 binders. Reference material: NOVAPHALT AC10.

As one might expect, vulcanized rubber can also improve elasticity (increase the storage modulus) of the asphalt binder and strengthen the asphalt - aggregate bond at high temperatures. A comparison of G*/sin δ, G", a characteristic ratio, and high temperature viscosity for a series of rubber mixtures prepared from AC-10 and

AC-30 from different sources is presented in Table 3. Addition of 5% rubber is sufficient to produce markedly enhanced properties without raising the high temperature viscosity excessively. Depending upon the source of the asphalt, addition of 15-20% rubber can produce very stiff mixtures. Since a swelling-reaction takes place when the asphalt cements are mixed with vulcanized rubber [27], the degree of the interaction - and therefore the storage modulus of the mixture - is mainly dependent upon the chemical composition of the AC and less on its grade. There is little correlation between the AC grade and the variation of the storage modulus with rubber content. Clearly, one can not predict the elastic behavior of a rubberized asphalt based on the AC grade. Further, since the rubber particles absorb the lighter fractions of the asphalt cement, this interaction could result in an increased binder viscosity and hence higher mixture stiffness. The low characteristic ratios observed for mixtures containing high rubber contents are a result of the very high values for G^* at high temperatures. Note that the $G^*/\sin\delta$ plot for the crumb rubber mixture in figure 5 exhibits a much lower slope than the curves for the other mixtures.

Creep analysis.

A constant stress creep/creep recovery test (Figure 6) illustrates a much more pronounced difference between different polymer/AC mixtures than the master or isochronal curves discussed above. If the complex compliance $J^* = 1/G$ is considered, then the significant difference between the compliance of PRAC10 and CRAC10 mixtures might be an indication of their rutting resistance, i. e., the lower the 1/J value, which is directly proportional to the extent of deformation, the better the binder response to rutting. The Novaphalt AC 10 mixture was completely resistant to creep under the conditions employed and thus would be expected to yield a cement with the maximum rutting resistance that can be imparted by the binder. Obviously, the aggregate plays the dominant role in controlling rutting. After long term loading (1000 sec), none of the RAC's exhibited significant recovery. The asphalt matrix allowed some mode of slow stress relaxation that dominates over the expected stress recovery imparted by the presence of the more elastic fillers.

Elastic creep recovery can be observed after short term loading (10 sec) as illustrated in Figure 7. Since this recovery is primarily induced by the filler, one can observe distinct differences in the properties of the RAC's as the filler content is increased. Addition of 10% powdered rubber to an AC10 produced a mixture with a creep recovery better than that of an AC30 from the same source. The stress relaxation process is rather slow; creep recovery after medium term loading (100 sec) is almost comparable to that observed after 10 sec loading considering that the sample deformation is greater. As expected, AC30/rubber mixtures exhibited higher creep recovery than AC10/rubber mixtures. Crumb rubber mixtures always displayed higher compliance and creep recovery than the corresponding mixtures with finely divided PR particles, regardless of the rubber content or the loading time.

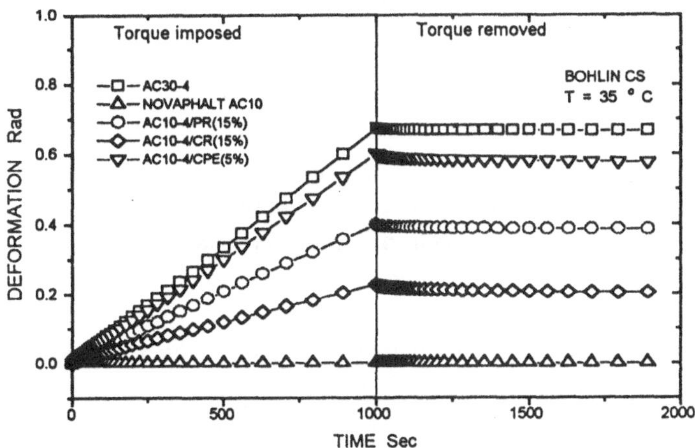

Figure 6. Constant stress creep/creep recovery curves at 35°C for neat, polymer and rubber modified paving grade AC10-3 binders. Reference material: NOVAPHALT AC10.

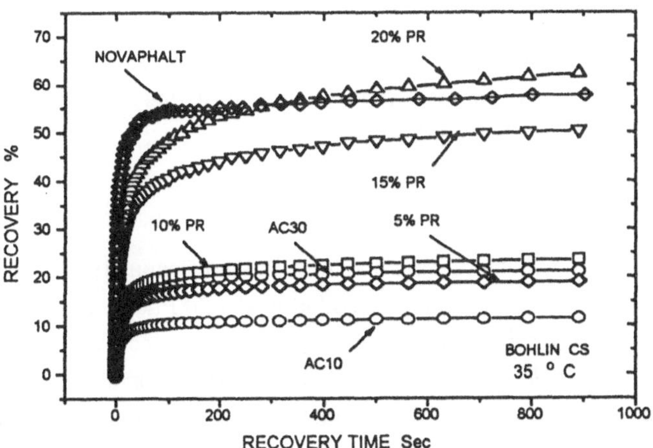

Figure 7. Creep recovery after 10 sec loading of powdered rubber AC10-3 mixtures. Reference binder: AC30-4.

CONCLUSIONS

1. Low level chlorination of polyethylene can be controlled to produce semicrystalline polymeric additives. Chlorinated polyethylenes containing less than 15 wt% chlorine interact more extensively with an asphalt matrix that polyethylene.

2. The chemical composition of the asphalt is a very important parameter in determining the physical properties of RACs.

3. Blends of vulcanized rubber particles may be analyzed using dynamic mechanical rheological techniques if the particle size is less than 1/10 of the gap between the plates of the rheometer. If the average particle size approaches the width of the gap, anomalous results may be obtained. More research is needed to establish the optimum conditions for analyzing crumb rubber mixtures.

4. Addition of ground rubber particles to asphalt affects the performance related properties of the rubber-asphalt cement as follows:

a. The stiffness of the mixture as reflected by $G^*/\sin\delta$ can be enhanced several fold depending upon the concentration of rubber added. This should improve the rutting resistance of the RACs.

b. The increase in the loss modulus is parallel to the increase in stiffness. If the maximum value for G" is considered 3000 Pa @ 64°C for example, the concentration of rubber which may be added to a given asphalt is limited to approximately 10%.

c. The high temperature viscosity increases significantly when the rubber content exceeds 10%. The ability to prepare RACs with 15% rubber that exhibit η @ 135°C less than 3.0 Pa·s depends upon the nature of the rubber additive and the chemical composition of the asphalt.

5. Creep and creep recovery (resiliency) of RAC's are directly dependent upon the rubber content. These tests may be the most effective method for assessing the rubber/asphalt interactions.

ACKNOWLEDGMENTS

This work is supported by a contract from the Louisiana Transportation Research Center. Special appreciation is extended to Harold Paul who served as the contract technical representative and provided helpful and incisive comments and suggestions. We wish to thank Seiko Instruments, USA, for their generous support to enable us to purchase Seiko Thermoanalysis Stations and LEQSF for the funds to purchase a Bohlin Rheometer.

REFERENCES

1. Stewart L (1989) Highway & Heavy Construction 132:48
2. King GN, King HW, Harders O, Chavenot P, Planche J-P (1992) Proc, Asso Asph Pav Tech 61:29
3. AASHTO (1993) Provisional Standard Specification for Performance Graded Asphalt Binder based on Strategic Highway Research Program SHRP Product 1001, AASHTO Designations MPI and TP5
4. Goodrich JL, Goodrich JE, Kari WJ (1986) Trans Res Rec 1096:146
5. Brule B, Brion Y, Tanguy A (1988) Proc, Asso Asph Pav Tech 57:41
6. Collins JH, Bouldin MG, Gelles R, Berker A (1991) Proc, Asso Asph Pav Tech 60:43
7. King GN, Muncy HW, Prudhomme JB (1986) Proc, Asso Asph Pav Tech 55:519
8. Collins JH , Mikols WJ (1985) Proc, Asso Asph Pav Tech 54:1
9. Coyne LD (1988) Proc, Asso Asph Pav Tech 57:545
10. Elliot RP , Herrin M (1985) Proceedings, AAPT 54:209
11. Daly WH, Qiu(Chiu) Z-Y, Negulescu II (1993) Trans Res Rec 1391:56
12. Noel F , Corbett LW (1970) J Inst Petrol 56:261
13. Brule B, Planche JP, King GP, Claudy P, Letoffe JM (1990) Prepr Am Chem Soc, Div Pet Chem 35:330
14. Williams ML, Landel RF, Ferry JD (1955) J Am Chem Soc 77:3701
15. Goodrich JL (1991) Proc, Asso Asph Pav Tech 60:80
16. Anderson DA, Christensen DW, Bahia H (1992) Proc, Asso Asph Pav Tech 61:437
17. Christensen DW, Anderson J, Anderson DA (1992) Proc, Asso Asph Pav Tech 61:67
18. Jew P , Woodhams RT (1986) Proc, Asso Asph Pav Tech 55:541
19. Hesp SA , Woodhams RT (1991) Polyolefin-Asphalt Emulsions, In: K.R. Wardlaw and S. Schuler (ed). ASTM STP 1108 Philadelphia, PA
20. Pitchford AC , Sarret HJ (1967) U S Pat 3,312,649
21. Fogg SG , Westerman PH (1977) British Pat 1,475,924
22. Brule B, Ramond G, Such C (1986) Trans Res Rec 1096:22
23. Lenoble C (1992) Fuel Science and Technology Int'l 10:549
24. Bucknall CB , Stevens WW (1980) J Material Sci 15:2950
25. Bucknall CB , Page CJ (1982) J Material Sci 17:808
26. Pink HS, Merz RE, Bosniack DS (1980) Proc, Asso Asph Pav Tech 49:64
27. Ensley K , Tauer J (1993) Petersen Asphalt Research Conference, Laramie, Wyoming

Rigid-Rod Polyimides with n-Alkoxymethyl Side Chains

S.-B. Park, Y.-J. Choi, W.-C. Zin, J. C. Jung*

Department of Materials Science & Engineering, Pohang University of Science & Technology
P. O. Box 125, Pohang, Kyungbuk, 790-600 KOREA

Abstract: Three series of new polyimides were prepared by condensation of imidazole-blocked 2,5-bis(n-alkoxymethyl)-1,4-benzene diisocyanates with pyromellitic dianhydride (PMDA), benzophenone tetracarboxylic dianhydride (BPDA), and naphthalene tetracarboxylic dianhydride (NTDA), respectively. After the polymers obtained were spectroscopically characterized, their solubilities, thermal properties and crystalline structures were measured and discussed. It was found that structures and properties of the polyimides having regularly substituted n-alkoxymethyl(n-$CH_2OC_mH_{2m+1}$, m=4, 6, 8) side branches are governed not only by side chain length, but also by main chain rigidity.

INTRODUCTION

In recent years rigid-rod polymers with regularly substituted linear and flexible side chains have drawn special attention due to their improved processibility-thermal stability balances and unique layered cystalline structures[1 - 3]. Based on this new molecular design a large number of rigid-rod polymers of this type has been synthesized, such as polyesters[4 - 7], polyamides[8, 9], polyphenylenes[10, 11], polyimides[12 - 14] and polyazomethines[15, 16]. Many of the rigid-rod polymers with flexible side branches have been prepared from monomers substituted laterally by linear n-alkoxy[4, 5, 8, 9, 12-15], or n-alkyl[10,11], or methylpoly(ethyleneoxy)[13], or n-alkoxycarbonyl[6].

Recently, we have designed a new n-alkoxymethyl side branch and synthesized various kinds of rigid-rod polymers containing this linear pendant group, such as polyazomethines[16] or poly(1,4-phenylene-1,2,4-oxadiazole)s[17-18]. In this report three series of polyimides were prepared by solution condensation of PMDA, BPDA and NTDA, respectively, with imidazole-blocked 2,5-bis(n-alkoxymethyl)-1,-4-phenylene diisocyanates (DIS-IBDI), in which the side chain length was varied from n-butoxy via n-hexoxy to n-octoxy. After spectroscopic and viscometric characterization of the polymers we measured their thermal properties and crystalline structures and discussed the effect of side chain length and main chain rigidity on properties and crystalline sturctures. To get more clear understanding on the polymerization reaction and polymer characterizations we also prepared corresponding model compounds by reacting imidazole-blocked phenyl isocyanate with PMDA and BTDA and by condensing aniline with NTDA.

K.P. Ghiggino (Ed.)
Progress in Pacific Polymer Science 3

MONOMERS SYNTHESIS

The monomeric imidazole-blocked 1,4-phenylenediisocyanates with n-alkoxymethyl substituents were prepared from 2,5-bis(n-alkoxymethyl) terephthalic acids by chlorination with oxalyl chloride in DMAc at 0 °C[19 - 21] followed by reaction with aqueous NaN_3, as shown in Scheme 1. The isolated azido products were rearranged to isocyanates in the presence of excess imidazole at 85 °C in toluene. On cooling were precipitated the DIS-IBDIs, which were thoroughly purified by recrystallization from acetone. The substituted terephthalic acids were prepared by basic hydroysis of 2,5-bis(n-alkoxymethyl)-terephthalonitriles[17 - 18].

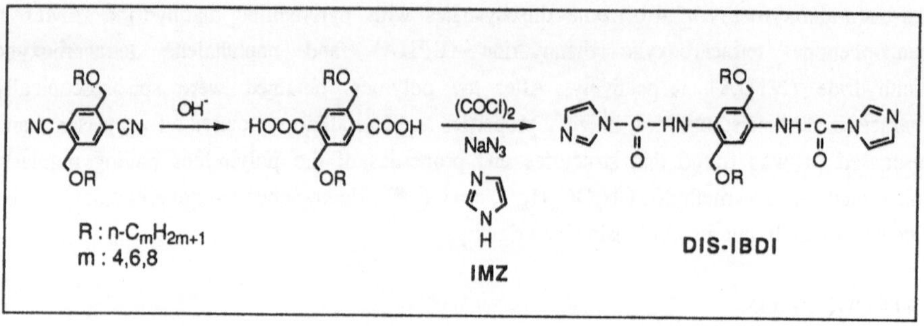

OH^- ; $(COCl)_2$, NaN_3, IMZ

R : n-C_mH_{2m+1}
m : 4,6,8

DIS-IBDI

Scheme 1. Synthesis of Monomers

The DIS-IBDIs prepared were characterized by IR, mass spectrometry(MS) and elemental anaysis and these results are summarized in Table 1. This shows that the three phenylene diisocyanates in imidazole-blocked form are obtained in polymerizable purity and with increasing side chain length melting point of the monomers drops. The latter fact indicates that at high temperature near melting there exists an attractive interaction between the flexible side chain moiety and the rigid main chain moiety, leading to an internal plasticization[3].

Table 1. Synthesis of monomeric DIS-IBDIs.

m	yield (%)	mp (°C)	IR (cm^{-1})	MS (m/e)	elemental analysis	%C	%H	%N	%O
4	47	162	1720, 1556, 1074	332, 333, 334	obsv.	61.80	6.60	14.52	13.66
					calc.	61.52	6.88	17.94	13.66
6	43	152	1720, 1560, 1076	388, 389, 390	obsv.	64.65	7.21	15.38	13.18
					calc.	64.10	7.64	16.02	12.20
8	44	146	1720, 1560, 1072	444, 445, 446	obsv.	66.85	8.02	13.47	12.27
					calc.	66.18	8.33	14.47	11.02

Imidazole-deblocking behavior was examined for the solid samples by DSC scanned up to 220°C at a scan rate of 20°C/min. All the three DIS-IBDIs began to deblock slowly at 200°C, meaning that the deblocking temperatures of the monomers lie at higher temperatures than their melting points in solid state under the measured conditions.

SYNTHESIS OF MODEL COMPOUNDS

To prepare model compounds representing the polymers structures two model compounds were prepared by reacting imidazole-blocked phenyl isocyanate (T_m 81°C) with PMDA and BPDA (coded PMDA-M and BPDA-M, respectively) under the same conditions as polymerization reaction. However, from the same reaction with NTDA we could not obtain pure compound NTDA-M and the yield was low(27 %). The product obtained always revealed some incomplete cyclization, and it needed be additionally cyclized by reflux in acetic anhydride for prolonged time. More conveniently and in higher yield, it could be prepared by direct condensation with aniline in cyclohexane/DMAc mixture at reflux temperature.

When PMDA or BPDA is reacted with imidazole-blocked phenyl isocyanate, the reaction proceeds, as known[22], via a 7-membered ring intermediate. However, in the same reaction with NTDA, the intermediate must be an 8-membered ring which should be so unstable that it can only hardly be formed, and the yield and the degree of imidization should be low.

POLYMERIZATION AND CHARACTERIZATION

Ar	Dianh.	Polym. Code
	PMDA	C_m-PY-PI
	BTDA	C_m-BP-PI
	NTDA	C_m-NA-PI

Scheme 2. Polymerization

For polymerization the DIS-IBDIs with varying alkyl side chain lengths were condensed with PMDA, BTDA and NTDA, respectively, in N,N'-dimethylpropylene urea (DMPU) at 120°C in the presence of 4-N,N-dimethylaminopyridine (DMAP) as catalyst, as shown in Scheme 2. In this polymerization reaction imidazole is deblocked from the DIS-IBDIs to form free diisocyanates which attack the anhydride ring to form a 7-membered cyclic intermediate[22], from which one molecule of CO_2 splits off. Resultantly the polymerization does not give rise to any precursors, but proceeds directly to form cyclized polyimides. To thoroughly purify the polyimides obtained the imidazole was removed by repeated washing with acetone.

The polyimide samples thus obtained showed typical imide ring absorptions in IR spectroscopy, indicating that the polymerization led directly to cyclized imide polymers. The degree of cyclization was investigated by FT-IR and in C_6-BP-PI and C_8-BP-PI it could be done also by ^{1}H-NMR spectroscopy in $CDCl_3$. In C_m-PY-PIs and C_m-BP-PIs the cyclization was confirmed to be complete, but in C_m-NA-PIs it was not complete. This lower reactivity toward imidization of NA-PIs must be caused from lower thermodynamic stability of 8-membered ring intermediate to be inevitably formed from the 6-membered NA ring, whereas the 7-membered intermediates formed from 5-membered anhydride rings of PY and BP are more preferable. Therefore, C_m-NA-PIs had to be additionally cyclized by reflux in acetic anhydride, but C_4-NA-PI, as will be discussed later in greater details, could not be completely cyclized even under prolonged reflux with acetic anhydride.

In Table 2 are summarized the solubility behaviors and inherent viscosities of the polyimides. It shows that the polymers generally do not have very high molecular weights and especially the polymers having the shortest C_4 side chain reveal the lowest inherent viscosities. This might result from the lowest solubility of the C_4-PIs in the polymerization solvent DMPU. Table 2 also shows that BP-PIs having one kinked C=O bond on the main chain exhibit the highest solubilities in DMAc, DMPU and chloroform, and C_6-BP-PI and C_8-BP-PI are completely soluble even in chloroform, while all the PIs are completely soluble only in very strong protonic acid solvents such as concentrated H_2SO_4 and MSA.

Table 2. Solubility behavior and inherent viscosity of the polyimides.

polymer code	inh. vis[a)] (dL/g)	solubility[b)]				
		DMAc	DMPU	H_2SO_4	MSA[c)]	$CHCl_3$
C_4-PY-PI	0.14	-	++	+++	+++	+
C_6	0.20	-	++	+++	+++	+
C_8	0.19	+	++	+++	+++	+
C_4-BP-PI	0.23	+	+++	+++	+++	+
C_6	0.33	+++	+++	+++	+++	+++
C_8	0.34	+++	+++	+++	+++	+++
C_4-NA-PI	0.15	+	++	+++	+++	-
C_6	0.21	+	+++	+++	+++	-
C_8	0.23	+++	+++	+++	+++	-

a) determined from 0.2 g/dL 97% H_2SO_4 at 25°C.

b) notation: +++, soluble at room temperature; ++, partially soluble; +, slightly soluble to swellable; -, insoluble

c) methanesulfonic acid

Structures of C_6-BP-PI and C_8-BP-PI could be elucidated by ^{1}H-NMR spectrum taken from $CDCl_3$ at room temperature. From the spectrum of C_6-BP-PI reproduced in Figure 1 could be observed a triplet at δ =0.82 ppm ascribable to CH_3, a broad multiplet at δ =1.17 ppm absorbed by $(CH_2)_3$ linked to CH_3, a sharp multiplet at δ =3.34 ppm arising from OCH_2 bonded with trimethylene, and a singlet at δ =4.51 ppm assignable to benzylic CH_2. Aromatic protons revealed absorptions at δ=7.49 ppm as singlet, 8.16 ppm as doublet, 8.29 ppm as doublet and 8.34 ppm as singlet.

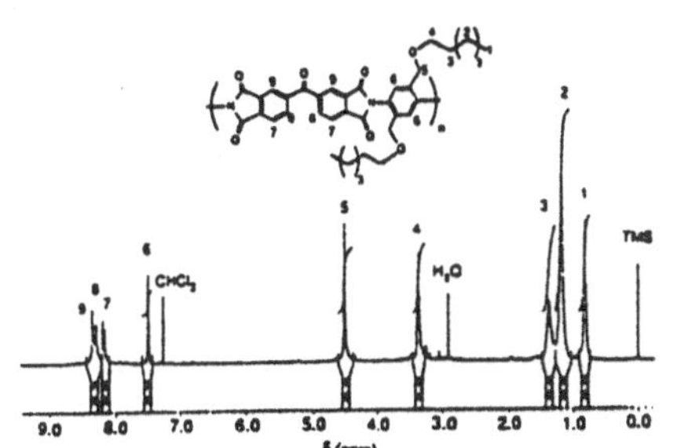

Figure 1. ^{1}H-NMR spectrum of C_6-BP-PI.

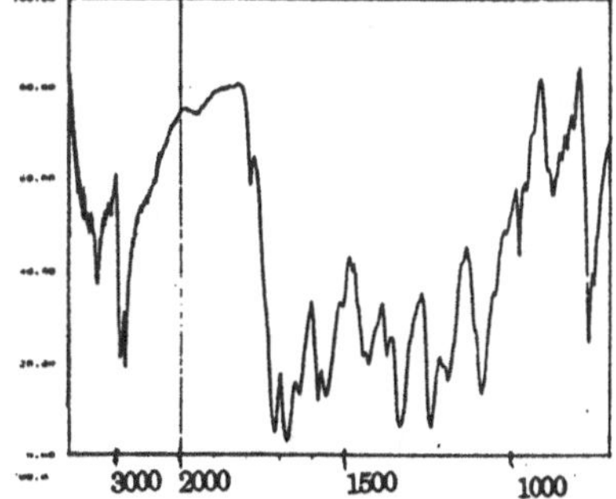

Figure 2. FT-IR spectrum of C_4-NA-PI.

Structures of all other samples were characterized by FT-IR spectra, because they were not soluble in $CDCl_3$. The FT-IR peaks could be well interpreted by comparison with the spectra of the corresponding model compounds. From every spectrum could be confirmed specific absorptions arising from imide I, II, III and IV vibrations at 1717, 1339, 1096 and 766 cm^{-1}, respectively. However, in the spectrum of C_4-NA-PI which is reproduced in Figure 2, there still remains a small peak at 1788 cm^{-1}. This peak should be ascribed to an amide bond which remains uncyclized. It did not disappear even after treatment with acetic anhydride for prolonged time. The reason of this greater resistance against imidization of this PI can not be clearly understood at present.

The model compounds were also useful in UV-spectrosopic characterization of the PIs. When the UV spectra of the PIs taken from concentrated H_2SO_4 were compared with those of the model compounds, it could readily be confirmed that all π-π* absorption maxima of PY-PIs and BP-PIs revealed a bathochromic shift, whereas those of NA-PIs exhibited no one. This fact indicates that π-electron conjugation length along the main chains of the PY-PIs and BP-PIs has become increased by polymerization, whereas the polymerization does not bring about any change in π-electron conjugation of NA-PIs at all. The π-electrons of NA-PIs seem not conjugatable along the main chain but localized. This localization of the π-electrons in NA-PIs led us to an assumption that the 6-membered imide ring might not exist coplanar with the phenylene ring but tilted, as drawn in Figure 3. This assumption is based on the trigonal pyramid structure of p^3 hybrid orbital of threefold-substitued ammonia. If this bent structure may be assumed to be true, there must be a cis and trans isomerism in the chain structure, as shown in Figure 3. This cis/trans isomerism should then lead to an amorphous morphology of the solid NA-PIs, which will be discussed later in greater details.

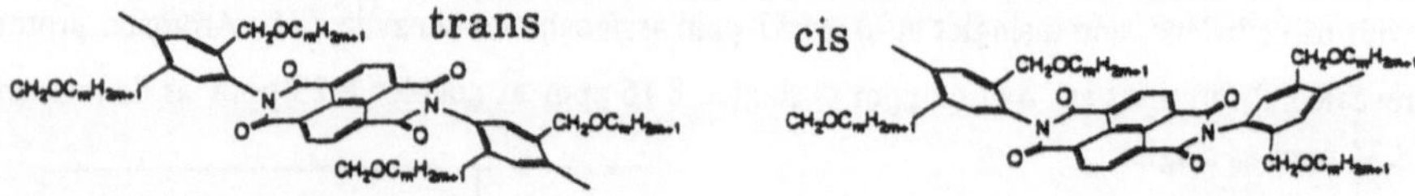

trans

cis

Figure 3. Cis/trans isomerism of C_m-NA-PIs.

Every attempt to detect the cis/trans isomerism of the model compound NTDA-M went to failure, presumably due to the umbrella effect, in which the phenyl group bonded to the nitrogen atom with the unshared electron pair rapidly oscillates from one side to the other. In NA-PIs the phenylene ring attched to the nitrogen atom is enormously heavy, because it is linked with a long chain at 4-position, so that it can not rapidly oscillate any more. Therefore, the cis/trans isomerism should be observed in NA-PIs.

THERMAL STABILITY

Thermal stabilities of the polyimides were studied by TGA scanned at 10 °C/min in N_2. In Figure 4 are reproduced the pyrograms of PY-PIs, BP-PIs and NA-PIs.They clearly show that all the samples of PY-PIs and BP-PIs as well as C_4-NA-PI are pyrolyzed in two separate steps, while two samples of C_6- and C_8-NA-PI reveal a smooth one-step pyrogram. The two-step pyrolysis behaviors could easily be understood when we simply assume that in PY-PIs and BP-PIs the flexible side chains first split away in the low-temperature step and the rigid main chains then in the high-temperature step. This assumption may not be quite unreasonable, because in both PY-PIs and BP-PIs the rigidity of main chains has been greatly increased by polymerization accompanying the extension of π-electron conjugation length, whereas the flexibility of side chains can not be changed by polymerization.

Based on this assumption we determined the weight losses of the low-temperature pyrolysis steps from Figure 4 and compared these loss values with side chain contents calculated theoretically for PY-PIs and BP-PIs. As to see from Table 3, both the values are well coincident with each other within experimental error. This coincidency indicates that the assumption stated above is reasonably established. Weight fractions degraded in the second step is constant, independent of side chain length. This two-step degradation behavior leads to a conclusion that the bond linking the main chain with the side chain is the weakest site for thermal degradation of C_m-PY-PIs and C_m-BP-PIs.

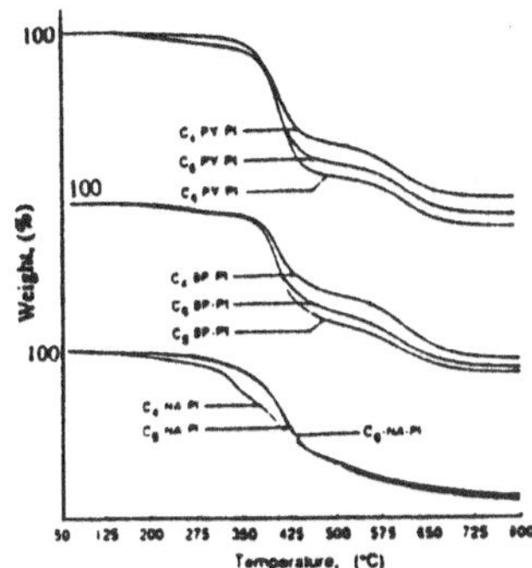

Figure 4. Thermogravimetric pyrograms of the polyimides in N_2.

In Figure 4 the TGA curve of C_4-NA-PI shows also a two-step degradation behavior. As already mentioned, the sample of C_4-NA-PI turned out incompletely imidized, and thus it is presumable that a thermal post-imidization in solid state takes place in the low-temperature degradation step. The smooth one-step pyrolysis of the C_6- and C_8-NA-PIs indicates that in these PIs the weakest bond for degradation does not exclusively lie on the bond linking the side chain with the main chain.

Table 3. Weight fractions of side chains calculated and determined in PY- and BP-PIs.

polymer code	weight fractions of n-alkoxymethyl group in %		
	calculated(W_c)	determined(W_d)	W_d - W_c
C_4-PY-PI	38	40	2
C_6	44	45	1
C_8	50	49	1
C_4-BP-PI	38	40	2
C_6	44	45	1
C_8	50	49	1

PHASE TRANSITION AND CRYSTALLINE STRUCTURE

C_m-PY-PIs: Independent of side chain length, thermocurves obtained from DSC measurements showed no detecable transitions, and therefore, crystalline structures were investigated by WAXS. In Figure 5 are reproduced the diffractograms, from which it is readily to see that the PY-PIs have layered crystalline structures. The detailed layered crystalline structures of the PY-PIs could be somewhat quantitatively elucidated. The layer spacing values measured from the smallest-angle peaks proved that the side chains emanating from neighboring backbones are interdigitated with each other. When the layer spacing values are plotted against number of side chain atoms, a good linearity is obtained, whose slope is 0.983 Å/atom and intercept is 3.8 Å. From the slope value we could estimate that the side chains emanate not perpendicular but 85° tilted from the backbone. The side chains are once more 30° tilted from the backbone plane. Lateral layer spacing was measured to be 4.4 Å and this spacing value was found to be independent of side chain length of the PY-PIs.

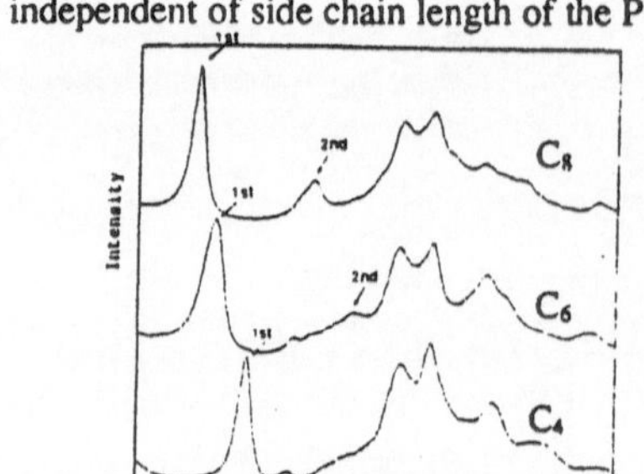

Figure 5. X-ray diffractograms of C_m-PY-PIs.

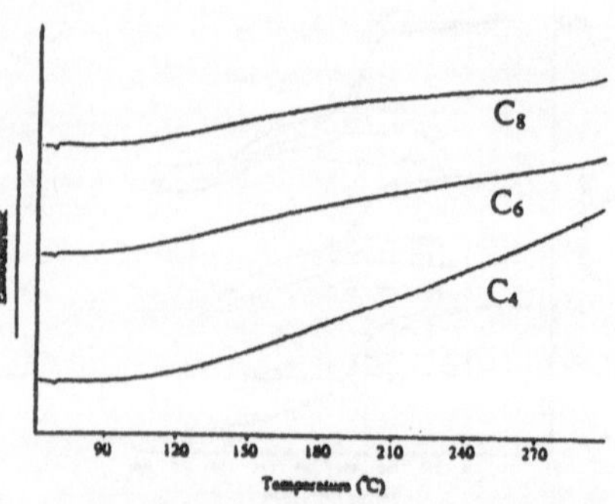

Figure 6. DSC thermograms of C_m-PY-PIs.

The experimental observation that, in spite of presence of the crystalline structures, no transitions were observable from DSC, was reconfirmed by taking X-ray diffractograms of the PY-PIs at varying temperatures. In these X-ray experiments undertaken from room temperature to 250°C any noticeable changes in diffractograms were not observed, either, although the samples had been previously annealed at 150°C for prolonged time. Normally, rigid-rod polymers with regularly substituted flexible side chains exhibit three transitions, from order to disorder of side chains, from solid to mesophase and from mesophase to isotropic melt. Because the main chains of PY-PIs have a symmetric structure and a very strong bckbone-backbone attraction, it is reasonable that they can pack very tightly with each other to form such tight crystallites that the highest transition from mesophase to isotropic melt can not be observable in the measurable temperature range. We measured also layer spacing values in the presence of eicosane at elevated temperatures, but it was found that no eicosane penetrates into the crystal layers. This indicates that the side chain-side chain interaction is much stronger than the side chains-eicosane interaction.This strong attractive interaction beween the side chains themselves may be induced from the far much stronger backbone-backbone attration. This very strong backbone-backbone attraction leads to the well ordered crystalline structures, among which the side chains are so tightly packed that they are not mobile and stay frozen even at temperatures near decomposition.

<u>C_m-BP-PIs:</u> Phase transitions were first studied by DSC. In Figure 7 are reproduced the thermocurves, which show the presence of glass transitions. As already shown in Figure 6, X-ray diffractograms of these samples show only broad halos, indicating no presence of crystallinity. The amorphous structure of these PIs might be resulted from the presence of kinked C=O group in their repeating units. If the T_g values are plotted as a function of side chain length, it was found that the T_g decreases logarithmically with increasing side chain length. This decrease caused by an attractive interaction between backbone domain and side chain domain is a typical behavior observed in many other rigid-rod polymers of this type.[1-3, 16-18]

When the samples were annealed at temperatures exceeding T_g, C_6- and C_8-BP-PI showed an endotherm in DSC at 212°C and 227°C, respectively. In X-ray diffractograms taken at elevating tamperatures for C_8-BP-PI, a small but distinctive peak could be observed from the curve obtained at 180°C and this peak disappeared on further heating. If a cross-polarized microscopic picture is taken at 180°C, presence of a nematic texture could be confirmed, as reproduced in Figure 8, indicating that C_8-BP-PI forms a nematic mesophase at 180°C. If eicosane was put to this mesophase, the interrod distance was observed to increase, demonstrating that eicosane molecules had penetrated into the amorphous region of the mesophase. In C_4-BP-PI no presence of this mesophase could be confirmed at temperatures below 300°C, perhaps because the side chain is too short, and hence the side chain content is too low, to make the system mobile and to reorient the molecules to form a mesophase.

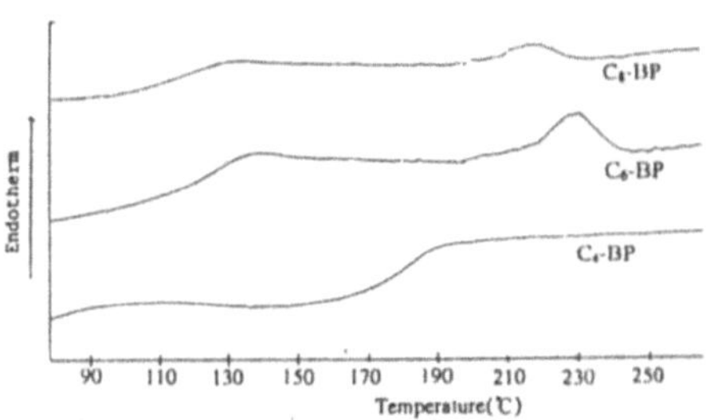

Figure 7. DSC thermocurves of C_m-BP-PIs.

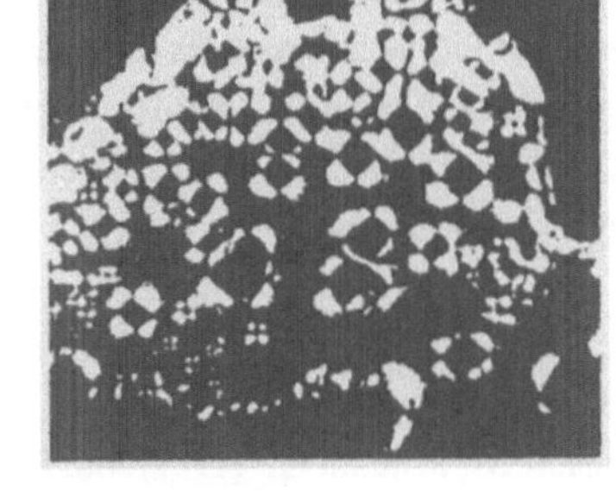

Figure 8. Microscopic picture of C_8-BP-PI.

C_m-NA-PIs: DSC thermograms to study phase transitions of these PIs did were measured to 330°C at 20°C/min, but any noticeable peaks assignable to distinctive transitions could not be detected in the temperature range. These PIs do not contain any crystalline structures, presumably due to the presence of the bent structures resulting from the cis/trans isomerism described above. The amorphous structures can also be confirmed from the X-ray diffractograms taken at room temperature, as reproduced in Figure 9. This shows a complete amorphous structure in C_6- and C_8-NA-PIs, while two broad peaks at $2\theta = 6°$ and 8° are observed in C_4-NA-PI. These two peaks are believed to result from the incomplete imidization of the sample. If it is heated to 330°C, the peak of $2\theta = 6°$ disappears completely, indicating that the cyclization of the C_4-NA-PI can take place at temperatures near 330°C. This difficulty in cyclization has also been observed in TGA pyrogram of the PI.

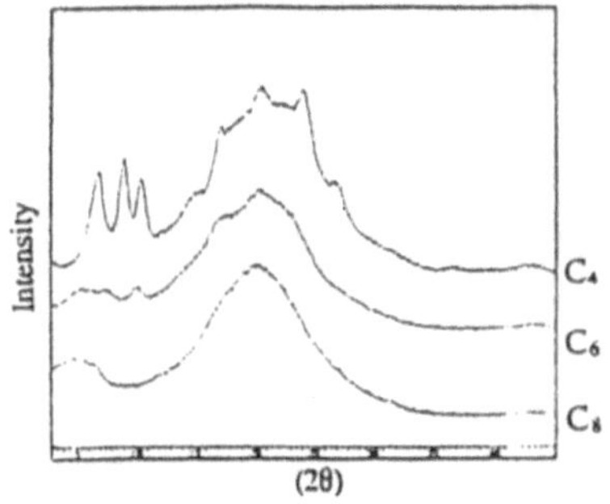

Figure 9. X-ray diffractograms of C_m-NA-PIs.

The fact that no T_gs were observable in C_m-NA-PIs at temperatures below 300°C suggests that the T_gs must lie at higher temperatures than 330°C. A repeating unit of these PIs contains a fused ring consisting of four rings, and this big-sized fused ring should reveal a very rigid disclike bahavior, bringing about a very strong interdisc attraction. This strong attraction makes the polymers so rigid that their T_gs are not observable from DSC's scanned to 330°C.

CONCLUSIONS

By one-step condensation of DIS-IBDIs having n-alkoxymethyl side chains with PMDA, BPDA and NTDA three series of new rigid-rod polyimides are synthesized and characterized. X-ray studies on their structures revealed that the C_m-PY-PIs have well-ordered layered crystalline structure and the C_m-BP-PIs and the C_m-NA-PIs are amorphous, but the former with C_6 and C_8 side chains show a nematic mesophase at temperatures near 180°C and the latter gave no observable phase transitions.

References :

1 Ballauff M, Schmidt GF (1987) Mol Cryst Liq Cryst 147:163
2 Ballauff M, Berger K (1988) Mol Cryst Liq Cryst 157:109
3 Ballauff M (1989) Angew Chem Int Ed Engl 28:353
4 Rodriguez-Parada JM, Duran R, Wegner G (1989) Macromol 22:2507
5 Stern R, Ballauff M, Lieser G, Wegner G (1991) Polymer 32:2096
6 Harkness BR, Watanabe J (1991) Macromol 24:6759
7 Kallitsis JK, Wegner G, Pakula T (1992) Makromol Chem 193:1031
8 Herrmann-Schoenher O, Wendorff JH, Ringsdorf H, Tischirner P (1986) Makromol Chem Rapid Commun 7:791
9 Ballauff M, Schmidt GF (1987) Makromol Chem Rapid Commun 8:93
10 Rehahn M, Schlueter A-D (1989) Polymer 30:1045
11 Rehahn M, Schlueter A-D, Wegner G (1990) Makromol Chem 191:1991
12 Wenzel M, Ballauff M, Wegner G (1987) Makromol Chem 188:2865
13 Helmer-Metzmann F, Ballauff M, Schulz RC, Wegner G (1989) Makromol Chem 190:985
14 Helmer-Metzmann F, Rehahn M, Schmitz L, Ballauff M, Wegner G (1992) Makromol Chem 193:1847
15 Lee K-S, Won JC, Jung JC (1989) Makromol Chem 190:1547
16 Park S-B, Kim H, Zin W-C, Jung JC (1993) Macromol 26:1627
17 Park S-B, Jung JC (1991) Makromol Chem 193:2519
18 Park S-B, Jung JC (1993) Angew Makromol Chem 208:97
19 Johnson DA (1953) J Am Chem Soc 75:3636
20 Weinstock J (1961) J Org Chem 26:3511
21 Chow RCL, Tarbell DS (1967) J Org Chem 32:2188
22 Meyers RA (1969) J Polym Sci Part A-1 7:2757

New Developments in Perfluoroelastomers: Chemistry, Properties and Applications

Anestis L. Logothetis

DuPont Central Research and Development, Experimental Station, P.O. Box 80328, Wilmington, DE 19880-0328, USA

Abstract: Two new perfluoroelastomer crosslinking systems have been developed, one based on the carbon-hydrogen (C-H) functionality and the other on high energy irradiation. The hydrogen atoms located between fluoro-substituted methylene groups are acidic and react with strong nucleophiles. A dinucleophile, which is also a strong base, first abstracts the acidic hydrogen, giving an anion on the backbone which, in turn, eliminates a fluoride anion to generate double bonds across which the dinucleophiles then add to make a crosslinked network. Tensile properties, compression set resistance and thermal resistance are excellent and compare favorably with previously developed crosslinking systems. Crosslinking by high energy irradiation, e-beam or γ-irradiation, is carried out on preformed articles and requires no chemical reagents or high temperatures. Doses of 10-15 Mrads are adequate and there is little selectivity in terms of efficiency among polymers containing various functionalities. Some chain scission also occurs. The crosslinks appear to be carbon-carbon bonds.

I. INTRODUCTION

Perfluoroelastomers are specialty products in the much larger fluoroelastomer area dominated by vinylidine fluoride based copolymers like Viton® [1-5]. Although discovered more than 20 years ago, perfluoroelastomers represent production volumes of only a few thousand kilograms a year. However, they are growing at a fast pace in terms of applications and introduction of new compositions to meet industrial needs. Perfluoroelastomers contain only carbon, fluorine and oxygen atoms and are characterized by having all-carbon backbones. Small amounts of nitrile, hydrogen or halogen functionality may be present to be used for crosslinking. Their presence has only minor effects on the overall polymer backbone stability.

Perfluoroelastomers are designed to maintain elasticity in severe environments, i.e., at high temperatures and in the presence of aggressive chemicals. Primary use of these materials is in sealing applications to prevent leakage after prolonged compression. These materials are very rarely used under tension and then under only modest elongations. They are copolymers of TFE and perfluoro(alkylvinyl ethers) the most readily copolymerizable homolog and commercially most important being

K.P. Ghiggino (Ed.)
Progress in Pacific Polymer Science 3

perfluoro(methyl vinyl ether) (PMVE) [6-9]. There are also commercial perfluoroelastomers consisting of TFE and perfluoroalkoxy(alkyl vinyl ethers) [10,11]. A third monomer is usually added in small amounts (0.5-1.0 mol%) to make these polymers crosslinkable. It is imperative that this third monomer has the right functionality and copolymerizes with TFE and perfluoro(alkyl vinyl ether) in a random fashion to give high molecular weight polymers.

Perfluoro(alkyl vinyl ethers) carrying reactive groups are the most useful and versatile functional monomers. One such functional group is the perfluorophenyl [7] which is susceptible to nucleophilic attack and is crosslinked with bisphenols. Another is the nitrile [12] functional group which crosslinks by forming 1,3,5-triazines. Substituted-fluoroolefins carrying the bromine functionality have also been used [1,14]. A novel approach is to utilize the chain transfer reaction to introduce functionality to the polymers. Thus perfluorodiiodo-alkanes [10,11,16] provide iodo groups to the two ends of the macromolecule while perfluoro(α,ω–iodobromo-alkanes) introduce an iodo group on the one end and a bromo group on the other [15]. All these bromo-, iodo- and bromoiodo-containing polymers are crosslinked by peroxides in the presence of a coagent like triallyl isocyanurate. In this paper two new curing systems are described which compare favorably with those mentioned above.

II. EXPERIMENTAL DETAILS

A. Polymerizations

Polymerizations were conducted in aqueous emulsion systems, under pressure with free-radical initiation behind protective barricades. Typical pressures are 5-7 MPa and temperatures 50-100 °C. A persulfate/sodium sulfite redox initiator was used at 50-70 °C or thermally induced persulfate initiation at 70-120 °C. All these aqueous polymerizations require a surfactant, usually a salt of a perfluorinated carboxylic acid (C_8-C_{12}), and a buffer to maintain the pH close to neutral. The polymer is isolated by coagulation, either by freezing or adding an electrolyte to the emulsion.

B. Crosslinking

A rubber mill was used to do the compounding and the parts were cured in a press at 190 °C under pressure. The parts were post-cured at 260 °C for 48 h in an oven under an atmosphere of nitrogen. High energy irradiation was performed with a high energy electron-beam accelerator on preformed parts at room temperature. The samples were irradiated stepwise to doses ranging from 5 to 200 KGy (0.5 to 20 Mrads). No particular effort was made to completely exclude air from these samples. Chemical reagents or fillers were absent in these polymers. In some experiments samples were placed in pyrex tubes, evacuated for 6 h at 25 °C, sealed and then irradiated. The tubes were opened and the gases passed over an infrared cell and identified by their spectra. Weight loss was determined by weighing the samples before and after irradiation. The samples were also evaluated by ^{19}F nmr, solubility in fluorosolvents like FC-75 [perfluoro(n-butyltetrahydrofuran)] and physical properties.

III. RESULTS AND DISCUSSION

A. C-H Bond as Crosslinking Functionality

Hydrogen atoms are introduced in the polymer by selecting the proper fluoro-olefin from among those shown in Fig. 1 (13). The C-H bond is relatively acidic

$$R^1CH=CR^2R^3 \quad \text{where} \quad R^1, R^2, R^3 = \begin{array}{l} H \\ F \\ R_F X \end{array}$$

Fig. 1. Monomers with C-H functionality that can be used to make perfluoroelastomers

because it is situated between fluoro-substituted methylene groups and is susceptible to attack by strong bases. A dinucleophile, which is also a strong base, first attacks the acidic hydrogen, giving initially an anion on the backbone. The anion, in turn, eliminates a fluoride anion or a perfluoroether functionality to generate a double bond across which the dinucleophile then adds. The various steps of this process are outlined in Fig. 2. Selection of the proper amount of C-H bond concentration is important

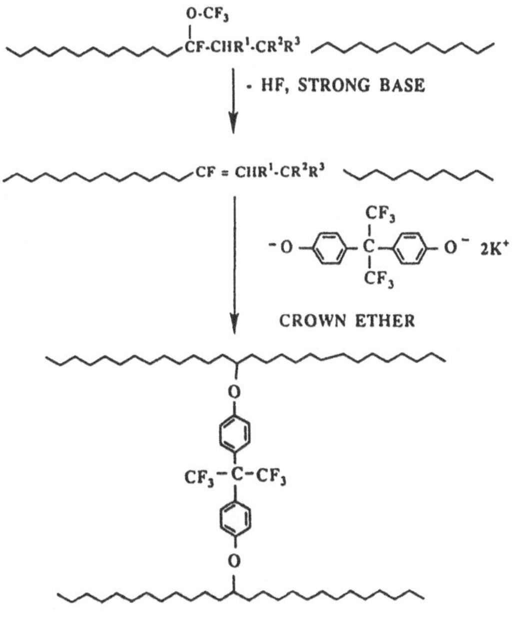

O-CF₃
CF-CHR¹-CR²R³

- HF, STRONG BASE

CF = CHR¹-CR²R³

$^-O-C_6H_4-C(CF_3)_2-C_6H_4-O^-$ 2K⁺

CROWN ETHER

$-O-C_6H_4-C(CF_3)_2-C_6H_4-O-$

Fig. 2. Proposed pathway for crosslinking C-H containing perfluoroelastomers

(0.5 to 1.0 mol %) in order to give the optimum properties for cured parts. If too many C-H bonds are present, some thermal and chemical resistance properties are sacrificed. If not enough C-H bonds are present, then the compression set resistance and tensile properties are compromised. The dinucleophile used is the dipotassium salt of Bisphenol AF. A small amount of a hydrocarbon polyether like "crown ether" facilitates the reaction because it acts as a complexing agent to the metal cation, thus making the dinucleophilic anion more reactive. The physical properties of cured parts (Table 1) compare favorably with those of the previously described systems. The low

TABLE 1 PHYSICAL PROPERTIES OF C-H CONTAINING PERFLUOROELASTOMER

Specific gravity, g/cm^3	2.05
Tensile at 100% elongation, MPa (psi)	7.6 (1100)
Tensile strength, MPa (psi)	15.3 (2225)
Elongation at break, %	160
Durometer hardness, Shore A	80
Glass transition temperature (T_g), °C	-28
TR_{10} retraction temperature, °C	-20
Compression set, % (O-rings)	
204 °C / 70 hrs	33
230 °C / 70 hrs	42
280 °C / 70 hrs	81

temperature properties are particularly good because a special nonfugitive compatible plasticizer is incorporated. Compression set resistance is excellent and even after 1000 h the values remain in the 50-60% range (Fig. 3).

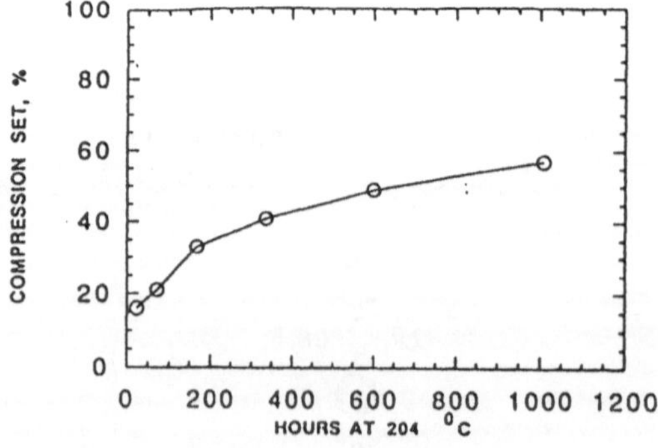

Fig. 3. Compression set of pellets after long exposures

The chemical resistance towards aggressive fluids is shown in Table 2. O-rings were immersed in these fluids at the conditions indicated and the percent swell was measured. These values are compared with the results obtained from similar O-rings containing the nitrile functionality which was chemically crosslinked by conversion to 1,3,5-triazine and with others crosslinked by peroxides. There are small differences from one curing system to the other depending on the fluid exposed but overall the swell resistance is excellent considering the aggressive nature of these fluids. Thermal and oxidative resistance was evaluated by exposing samples to air at 260 °C for a number of days. Very little or no change in the tensile and compression set properties was observed as compared to the unexposed samples.

TABLE 2 CHEMICAL RESISTANCE OF PERFLUOROELASTOMERIC PARTS CROSSLINKED BY DIFFERENT METHODS

Chemical	Conditions (°C / 70 hrs)	Crosslinking System (% swell)		
		Bisphenol	Triazine	Peroxides
Nitric acid (70%)	85	16	2	3
Ethylene diamine	90	21	28	25
Skydrol 500B	125	4	10	1
Water	225	28	27	13
Butyraldehyde	70	16	18	3
Methyl ethyl ketone	70	6	5	4
Ethylene oxide	50	10	34	3
Acetic acid	100	17	4	4
MTBE[1]	40	3	3	6

[1] Methyl t-butyl ether.

B. Crosslinking by High Energy Irradiation

The previously described crosslinking systems are based on a chemical reaction between a functionality on the macromolecule and a reagent. The reagent can be a dinucleophile or a peroxide or a nitrile trimerization catalyst. In addition fillers such as carbon black, titanium dioxide or barium sulphate may also be added for reinforcement and improvement of some properties. In high energy irradiation no chemical reagents nor any special functionalities in the polymer are required [17]. All perfluoroelastomers crosslinked when exposed to either electron beam or γ-irradiation. There is, however, a difference in properties of crosslinked parts depending on the polymer molecular weight and functionality. Fillers can also be used to give reinforcement and do not interfere with the irradiation process. In the work described in this paper, electron-beam irradiation was used exclusively. However, γ-irradiation on the same polymers gives similar results [18].

<u>1. Changes in composition</u>: Samples exposed to various doses were found to lose a small amount of their weight upon irradiation. Infrared spectra indicate that there is formation of carboxylic acid groups but no other major compositional change takes place. The higher the irradiation dose, the higher the weight loss and the carboxylic acid concentration. The weight loss which increases with increasing irradiation dose is associated with the evolution of gases identified by infrared spectroscopy as being mainly carbon tetrafluoride, carbonyl fluoride, carbon dioxide and perfluoro(methyl ether). The crosslinked perfluoroelastomer shows a strong peak at 1775 cm^{-1}, indicating the presence of carboxylic acid groups.

<u>2. Crosslinking</u>: The crosslinking process was followed by gel formation and by the development of physical properties. The starting polymers are soluble in fluorinated solvents like FC-75. As the irradiation dose proceeds the parts become less and less soluble and they remain swollen rather than dissolve, indicating formation of a network. In Fig. 4 we show the formation of gel as a function of radiation dose. Even after only 2 Mrads irradiation, there is considerable amount of gel formed.

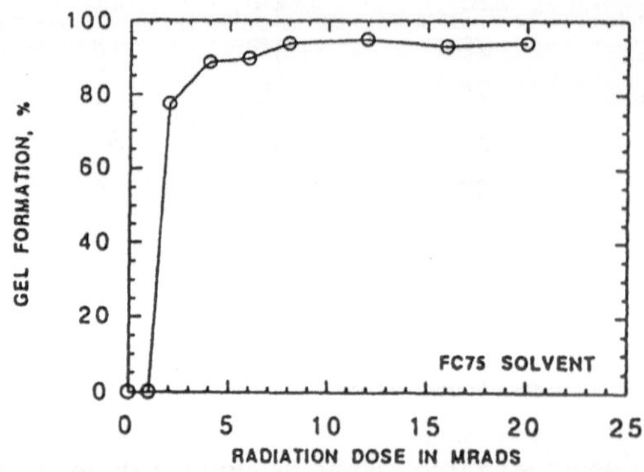

Fig. 4. Gel formation upon e-beam irradiation of perfluoroelastomer.

In Fig. 5, we show the tensile strength and elongation at break as a function of irradiation dose. There is a minimum for the tensile strength and a corresponding maximum for the elongation at break at around 1-2 Mrad dose. As the irradiation dose increases the tensile strength values rise above the minimum and plateau at about 8-10 Mrads. Elongation-at-break values, on the other hand decrease below the maximum as the dose increases. These results indicate that both chain scission and crosslinking take place. At low radiation doses the effects of chain scission dominate the properties, while as the radiation dose increases and crosslinking takes place, the properties improve substantially. Carboxylic acid formation is the result of chain scission.

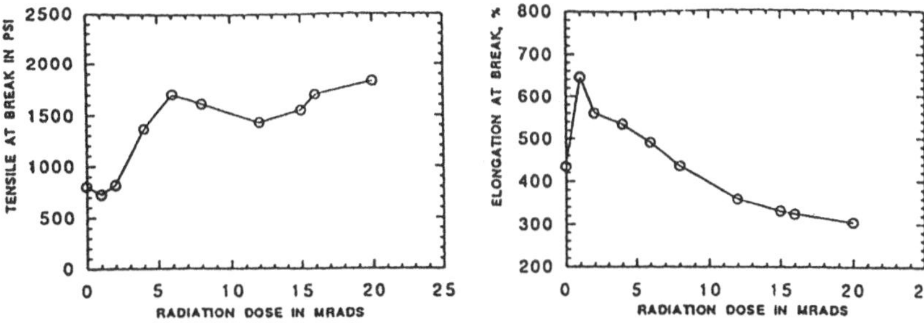

Fig. 5. Tensile strength and elongation at break of perfluoroelastomer after e-beam irradiation.

3. Crosslinking Process: In electron and γ-irradiation the energy is high enough to cause ionization of the substrate molecules through ejection of electrons, resulting in excitation of the macromolecules. The reactive intermediate radicals and ions produced by the ionization and excitation steps undergo chemical reactions which cause changes in the molecular structure of the polymer. The chemical changes observed in hydrocarbon type polymers are [19]: a) scission of the polymeric chains in the backbone, b) crosslinking of the macromolecules, and c) formation of small molecules. Similar results are observed with perfluoroelastomers. There seems to be little selectivity in the irradiation attack on the macromolecule. Small amounts of functionalities (< 1 mol%) such as nitrile, bromine, iodine and perfluorobenzene remain essentially intact in the polymer. When the chain scission occurs next to an ether functionality, it leads to an acid fluoride end group. Hydrolysis of these acid fluoride groups by moist air leads to the carboxylic acids observed by infrared. Crosslinking takes place by reactions of radicals and ionic intermediates. As the concentration of these intermediates increases with the higher irradiation doses so does the probability of crosslinking. Similar observations, scission and crosslinking, were made with vinylidene fluoride-hexafluoropropene based fluoroelastomers [20]. All these polymers are flexible, being amorphous and having a *Tg* below room temperature. Thus the macromolecular chains are highly mobile and allow intermolecular interactions which lead to crosslinking. On the other hand stiff, crystalline perfluorinated polymers like PTFE and TFE copolymers, whose macromolecular motions at room temperature are limited, mostly degrade upon exposure to high energy irradiation [21,22]. A proposed pathway of the process is shown in Fig. 6.

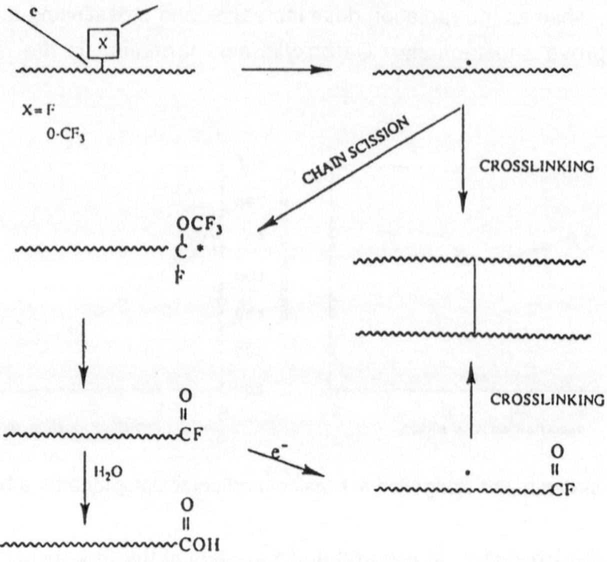

Fig. 6. Proposed pathway for crosslinking of perfluoroelastomers with high energy irradiation.

4. Properties: The best physical properties are obtained when the materials are exposed to 12-15 Mrads irradiation. In Table 3 we show the compression set results at various temperatures and the tensile properties at room temperature. The

TABLE 3 PHYSICAL PROPERTIES OF IRRADIATION CROSSLINKED PERFLUOROELASTOMER

Property	Crosslinking System (% swell)	
	Irradiation[1]	Triazine[2]
Compression Set		
336 h at 23 °C	35	23
70 h at 175 °C	22	25
336 h at 175 °C	28	25
70 h at 204 °C	72	32
Tensile Properties		
M_{100} , psi	305	000
T_b , psi	1130	2000
E_b , %	285	160
Hardness, Shore A	70	78

[1] Clear.
[2] Filled and chemically compounded.

results are compared with those obtained from the same polymer which has been filled with carbon black and chemically cured to give triazine crosslinks. The compression set resistance values at 25 and 175 °C temperatures are comparable, however, at higher temperatures (204 °C), the irradiation crosslinked O-rings showed poorer values. This is probably related to the detrimental effects of the chain scission during irradiation and the generation of ionic end groups. The chemically cured O-rings have considerably higher tensile strength associated with the filler reinforcement while the irradiation crosslinked O-rings contain no filler; therefore, the values are lower. In Table 4 the chemical resistance to various aggressive fluids of radiation crosslinked O-rings is compared to that of chemically cured O-rings. In spite of the fact that the chemically cured O-rings contain 10 phr carbon black the results are almost identical, indicating that the chemicall resistance is determined mainly by the polymer backbone structure and the structure of the crosslinked network.

TABLE 4 CHEMICAL RESISTANCE OF IRRADIATION CROSSLINKED PERFLUOROELASTOMER

Chemical	Conditions (°C / 70 h)	Crosslinking System (% swell)	
		Irradiation[1]	Triazine[2]
Ethylene diamine	90	36	28
Nitric acid (70%)	85	2	2
Trifluoroacetic acid	50	28	26
Acetic acid	100	3	3
HF (5%)	100	35	31
HF (60%)	100	30	28
HF gas	100	25	23
Freon 113	50	25	26

[1] Clear.
[2] Filler and chemically compounded.

IV. SUMMARY

Each crosslinking system provides its own advantages. The C-H functionality is another way to chemically crosslink perfluoroelastomers and the properties obtained compare favorably with the curing systems which utilize the perfluorobenzene, nitrile, bromine and iodine functionalities. The crosslinks are based on Bisphenol AF, and are thermally stable and chemically inert. Many inexpensive and commercially available fluorolefins can be used to introduce C-H functionality in the polymer. This gives an economic advantage over the more complicated and custom-made cure site processing, aerospace and oil exploration industries. Crosslinking by high energy irradiation, on the other hand, is aimed for uses in the semiconductor industry where cleaniness, lack of contamination and plasma resistance is of paramount importance. It is the only crosslinking system which requires no chemical agents or accelerators and at the same time it gives very good properties.

V. REFERENCES

1 Logothetis AL, (1989) *Prog Polym Sci* 14:251

2 Arnold RG, Barney AL, Thompson DC (1973) *Rubber Chem Technol* 46:619

3 Carlson DP, Schmiegel WW (1988) Ullman's Encyclopedia of Industrial Chemistry, VCH Verlagsgesellschaft, Weinheim A11:393

4 Schroeder HE (1986) In: Seymour RB, Kirshenbaum GS (eds) High Performance Polymers: Their Origin and Development, Elsevier, Amsterdam, p 389

5 Logothetis AL (1986) In: Encyclopedia of Materials Science and Engineering, Pergamon Press, New York, p 1809

6 Barney AL, Keller WS, Van Gulick NM (1970) *J Polym Sci* A-1, 8:1091

7 Kalb GH, Khan AA, Quarles RW, Barney AL (1973) ACS Advances in Chemistry Series 129:13

8 Barney AL, Kalb GH, Khan AA (1971) *Rubber Chem Technol* 44:660

9 Kalb GH, Quarles RW, Graff RS (1973) Applied Polymer Symposium No 22 p 127

10 Tatemoto M, Amano T (1984) US Pat 4,487,903 to Daikin [CA 99:39639j]

11 Nakagawa S, Nakagawa T, Ihara K, Amano T, Omori M, Asano K (1985) US Pat 4,499,249 to Daikin [CA 100:52217d]

12. Breazeale AF (1981) US Pat 4,281,092 to DuPont [CA 93:187543g]

13 Finlay JB (1985) US Pat 4,529,784 to DuPont [CA 102:205254g]

14 Apotheker D, Finlay JB, Krusic PJ, and Logothetis AL (1982) *Rubber Chem Technol* 55:1004

15 Masatoshi A (1992) US Pat 5,159,026 ; Masatoshi A, Haruyoshi T (1992) US Pat 5,151,492 ; Haruyoshi T (1988) US Pat 4,748,223 to Nippon Mektron

16 Oka M, Tatemoto M (1984) In: Contemporary Topics in Polymer Science, Plenum Press, New York, Vol 4:763

17 Logothetis AL (1991) Preprints, Second Pacific Polymer Conference, Otsu, Japan p 374

18 O'Donnell JH (1993) personal communication

19 Reichmanis E, O'Donnell JH (eds) (1989) The Effect of Radiation on High-Technology Polymer, ACS Symposium Series No 381, Washington DC, USA

20 Carenza M, Lora S, Pezzin G, Faucitano A, Buttafava A (1990) *Radiat Phys Chem* 35:172

21 Florin RE (1972) In: Wall LA (ed) Fluoropolymers, Wiley-Interscience, New York p 317

22 Burger W, Lunkwitz K, Pompe G, Petr A, Jehnichen D (1993) *J Appl Sci* 48:1973

Oriental Lacquer

Otto Vogl(a), Jan Bartus(a), Meifang Qin(a) and John D. Mitchell (b)

(a) Polytechnic University, Six MetroTech Center, Brooklyn, NY 11201-2990;
(b) The New York Botanical Garden, Bronx, New York, NY 10458.

ABSTRACT

The oriental lacquer is prepared from the sap of the "varnish tree" *Rhus verniciflua,* which is an emulsion of an aqueous phase and an organic phase called urushiol. The urushi fraction consists of a mixture of catechol derivatives substituted in 3-position with aliphatic C_{15} or C_{17} side chains. About 60% of these side chains are trienes. The structure of the trienes is similar to that in linseed or tung oil and is essential for the effectiveness of the oxidative curing of oriental lacquer. We have utilized new techniques and used combinations of modern techniques to analyze the urushiol mixtures and to characterize and identify each individual compound. We have also developed ultraviolet stabilizers for oriental lacquer, stabilizers that could be incorporated into the polymerizing mixture during the curing process to result in ultraviolet stabilized oriental lacquer.

INTRODUCTION

Toxicodendron vernicifluum, also called *Rhus verniciflua* is the source of Japanese lacquer. The correct name of the 'varnish tree' (Japanese lacquer tree) is *Toxicodendron vernicifluum (Stokes) F. Barkley*, the names *Rhus verniciflua Stokes,* and *Rhus vernicifera D.C.* are synonyms [1,2].

The Japanese lacquer tree belongs to the family *Anacardiaceae* which consists of about 600 species. A number of *Anacardiaceus* species produce poisonous saps, oleoresin mixtures of phenolic compounds that cause contact dermatitis. These compounds are similar in chemical structure to the well known dermatologically toxic components of the sap of *poison ivy* or *poison sumac.*

The temperate genus *Toxicodendron* [Figure 1] grows in North America, East and South East Asia, and has been extensively studied. In Asia *Rhus verniciflua* trees grow in China, primarily in the Sichuan region and in Japan; similar species also grow in Vietnam, in Thailand and in Taiwan [Figure 2][3].

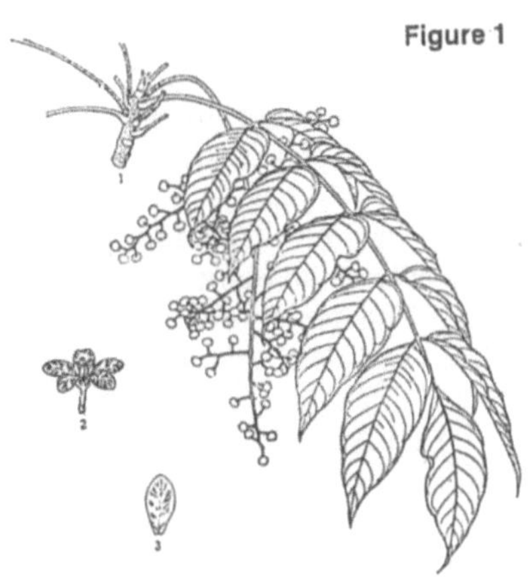

Figure 1

漆. Toxicodendron vernicifluum (Stokes) F. A. Barkl.:
1.果枝; 2.花; 3.花瓣。

K.P. Ghiggino (Ed.)
Progress in Pacific Polymer Science 3

Oriental lacquer is an emulsion, the sap is obtained by tapping lacquer trees in the same way that natural rubber is harvested from rubber trees. A *Rhus verniciflua* tree can grow to a height of 30 feet, its leaves, blossoms and seeds are shown in Figure 3. After ten years of growth, the stem of the lacquer tree is mature enough, it can be cut and a sap oozes out of the tree which is similar to the sap from the rubber tree.

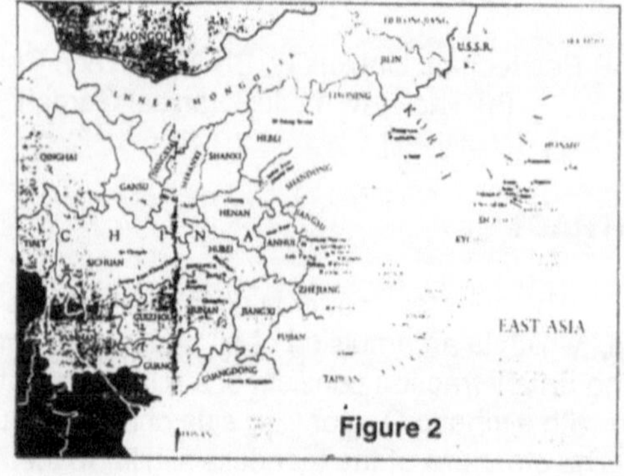

Figure 2

The difference between the oriental lacquer trees of the *Rhus* variety and the rubber tree is that the rubber trees have polyisoprene as the desirable ingredient while the urushi components are mixtures of substituted catechol derivatives.

漆树 Rhus verniciflua

Figure 3

The sap of the Japanese lacquer tree consists of a water/oil emulsion, the "oily" or urushiol fraction (about 65%) and an aqueous fraction (35%) which contains water-soluble plant gums and polysaccharides as well as small amounts of enzymes. Urushiol is a mixture of catechol derivatives substituted in 3-position with unsaturated (trienes, dienes, monoenes) and saturated hydrocarbon chains of C15 and C17 chain lengths (Scheme). The quality of the Japanese lacquer depends to a great extent on the quantity of the triene component and on the amount and activity of the enzyme laccase. The composition of the urushiol varies depending on growing conditions of the *Rhus vericiflua* tree and on the season of harvesting.

To prepare the lacquer sap for the actual coating material, the sap is stirred in an open vessel at room temperature for about 1.5 hours, and then at a temperature increasing from 20 to 45°C where it is kept for 2 to 4 hours until the water content is reduced to 2-4% ("sugurome" process). The temperature cycle must be carefully controlled to retain the activity of the enzymes. At this point the sap has become clear, has changed in color and has increased in viscosity. This liquid, known as "raw lacquer", consists of urushiol and oligo-urushiol, small amounts of water and other components and is ready for application. Urushiol is cured by oxidation of the

Side chain R (catechol ring with OH, HO)	%
$-(CH_2)_7-(CH{=}CH)_{cis}-CH_2-(CH{=}CH)_{trans}-(CH{=}CH)_{cis}-CH_3$	55
$-(CH_2)_7-(CH{=}CH)_{cis}-CH_2-(CH{=}CH)_{cis}-CH_2-CH{=}CH_2$	2
$-(CH_2)_7-(CH{=}CH)_{cis}-CH_2-(CH{=}CH)_{trans}-(CH_2)_2-CH_3$	6
$-(CH_2)_7-(CH{=}CH)_{cis}-CH_2-(CH{=}CH)_{cis}-(CH_2)_2-CH_3$	5
$-(CH_2)_7-(CH{=}CH)_{cis}-(CH_2)_5-CH_3$	15

Scheme

urushiol, particularly the triene components with other ingredients of the raw lacquer by the enzyme laccase, present in the sap. It is a copper-glycoprotein and functions as the polymerization (primarly dimerization) catalyst of urushiol, by assisting the oxidation of the catechol derivatives of urushiol. The mechanism of urushiol curing and especially the oxidative dimerization of the urushiol compounds has been extensively investigated.

For the preparation of lacquer ware, the oriental lacquer is usually applied on cedar wood, particularly from the Japanese cedar tree because it is the most warp resistant wood. One coat or two coats of primer, unprocessed urushi sap is applied followed by subsequent very thin layers (from twenty to forty layers) of processed "raw urushi" lacquer, one layer at the time every one or two days per application and curing. High relative humidity at the time of application and during curing is very critical; it must be in the range from 65 to about 80% .

The number of layers depends on the intended quality of the final product, each layer is cured and sanded to guarantee the smoothness of the layers. Specially prepared urushi samples are used for the preparation of high quality Japanese lacquer ware.

The Japanese lacquer used for individual applications contains several ingredients in addition to the urushiol base depending on what special effects and what coloring is desired. As the final layer, a clear raw lacquer coat is applied The final product of the oriental lacquer is a crosslinked polymer and is resistant to water, alcohol and oils.

EXPERIMENTAL PART

We have investigated [3-7] several samples of the urushi fraction of Japanese lacquer, obtained from the Urushi Co., Ltd., Kyoto, Japan: by a combination of a.) capillary gas chromatography, b.) supercritical fluid chromatography, c.) high resolution NMR spectroscopy and d.) two techniques of mass spectrometry. The composition of the urushi fraction of these samples was found to consist of most of the components that have been identified by earlier analyses of urushi samples [8].

RESULTS AND DISCUSSION

Up to the present time, a number of techniques have been used for the analysis and characterization of the ingredients of urushiol. They include traditional separation techniques, but more recently such techniques as high performance liquid chromatography (HPLC), and gas chromatography (GC) have been used. As mentioned earlier (Scheme), the sap of the lacquer tree is a mixture of components with distinctively different molecular weight (degree of unsaturation and geometric isomers of the olefinic groups of the hydrocarbon side chain of the urushiol components) and polarity, like the water soluble gums and polysaccharides.

We have investigated several samples of oriental lacquer by a combination of capillary gas chromatography, supercritical fluid chromatography, high resolution NMR spectroscopy and two techniques of mass spectrometry, especially potassium ionization of desorbed species (K^+IDS) mass spectrometry for the isolation and characterization of the components of urushiol. The composition of the urushi fraction of these samples was found to consist in general of the components that had been found by former analyses of urushi samples. Our new understanding allows us to recognize quickly and more accurately individual components of oriental lacquers.

K^+IDS mass spectrometry of the several samples of urushi showed the typical peaks that were expected for the trienes at [M]K^+=353 daltons, the dienes at [M]K^+=355 daltons and the monoenes at [M]K^+=357 daltons. There was also a peak noticed at [M]K^+=371 daltons that could not be idetified and seemed to be a satellite peak of the [M]K^+=353 dalton peak. The peak showed up at different intensities in the various samples and seemed to be triene peak [M]K^+=353 daltons with an additional mole of water.

In addition, especially in the processed urushi samples, peaks with masses [M]K^+=665 daltons and [M]K^+=667 daltons which are the expected products of the dimerization reaction by oxidation of the phenol part of the catechol derivatives of urushiol. Supercritical fluid chromatography in combination with electron bombardment mass spectrometry was then used for the identification of some of the major components of the urushiol samples. Most prominently were found the trienes with [M]=314 daltons, and [M]=317 daltons and the monoenes. Again the compound with the mass of [M]=332 daltons was noticed which was the water adduct of the triene [M]=314 daltons.

The urushi samples were now subjected to capillary gas chromatography. A number of components were found in the samples but most compounds were present in amounts less than 1-2%. Only two large peaks were found in all samples, one with a retention time of 11.2 minutes and the other with a retention time of 12.2 minutes. The two main components were present in a ratio of 6:2. This result together with the mass spectrometry result and later the NMR result indicated that the peak with the retention time of 11.2 minutes was the triene and the peak with the retention time of 12.2 minutes was the monoene.

A most careful study was now carried out by NMR spectroscopy, not only by regular ^1H 500 MHz and ^{13}C but also by ^1H COSY NMR spectroscopy [4,5]. Individual compounds were isolated by preparative HPLC of the mixture in urushiol, particularly of kuro-urushi and all the isomers, including the cis-trans isomers were identified and characterized [Figure 4] [5]. Particular care was taken to identify the water addition product $[M]K^+$=371 daltons. It was found that water had indeed been added to the triene to form a C_{16} hydroxylated diene [Formula] [5]. This water addition is catalyzed by some additives such as ferric chloride which are used in the *sugurome* process to optimize the lacquer properties. The details of the process is a closely guarded secret and is slightly different from manufacturer to manufacturer.

CH_3—CH=CH—CH=CH—CH_2—CH_2—CH(OH)—$(CH_2)_7$—C_6H_3(OH)(OH)

Formula

The composition and the content of the triene fraction is very essential because the amount of triene determines the reactivity of the oriental lacquer. Very important is the location of the methylene groups between two double bonds which is similar to that in linseed or in tung oil. Analyses of urushiol samples from Chinese and Japanese sources are shown in the Table.

Table. Urushi Compositions from Different Growing Regions

	China					Japan		
	Hubei	Shanxi	Huxian	Fendju	Songya	Yamagata	Nakuro	Ibaraki
Trienes	65.9	56.5	57.7	61.5	67.7	54.6	63.5	59.5
Dienes	5.1	11.9	26.1	14.9	15.1	16.2	12.3	7.0
Monoenes	23.2	27.9	13.2	19.6	13.3	25.3	19.3	27.1
Saturated	3.1	3.6	2.0	2.5	2.0	3.1	3.3	4.4

The urushiols, harvested from varnish trees other than those grown in China or Japan may have different amounts and compositions of the ingredients. Analyses from Vietnamese, Thaiwanese and Thai lacquer trees, have higher amounts of monoene components in addition to the trienes. Although the aliphatic side chain is still basically a C_{15} side chain the group might be attached in the 4-position of the catechol group or just subtituted in the 3 or 4-position of phenol.

As has been indicated earlier, the curing of the lacquer is a mixture of enzymatic oxidation and the radical curing of the highly unsaturated systems initiated by the hydroperoxides that are formed on the methylene groups between the double bonds. The curing of the lacquer does not work in a dry atmosphere and the exact nature of the curing mechanism is not known. However, it is also well known that even after a month the curing is not complete and some post curing still continues; the lacquered sample is still slightly soft even after one month!

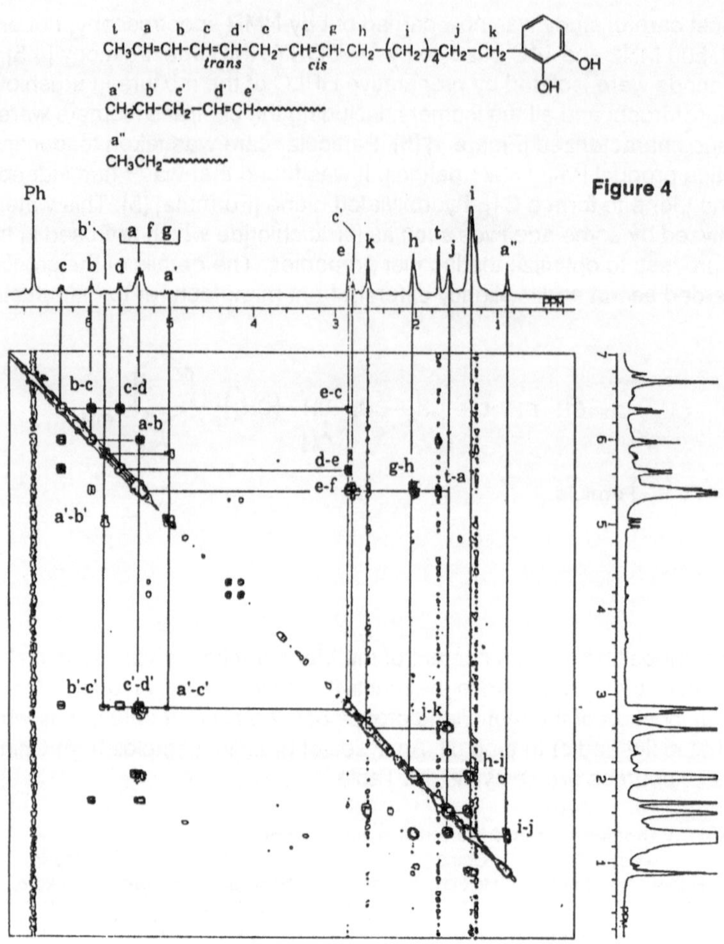

Figure 4

Oriental lacquers have two weak points. Firstly, the coating is not very stable to light, especially to UV light. On exposure to light, further oxidative photo-cross-linking occurs which causes cracking and sometimes chalking. As a consequence, oriental lacquer is mostly utilized for indoor uses and lacquered boxes of museum quality are kept under dim lights. Fugure 5 shows the interior of the Matayama mortuary in the temple Kodaiji, in Kyoto [9] for Hideyoshi which was built in 1606 upon his wife's commission. The lacquering technique used for this temple is called *makie* and was done by artists of the Kaomi school.

Secondly, processed oriental lacquers cannot be produced and are consequently not available in colorless form and artists who are using oriental lacquer would like to have colorless oriental lacquers.

Figure 5

We have made it our objective to develop UV stabilizers that could be added to urushi and "cocured" to obtain a photo-stabilized oriental lacquer. We have synthesized several 2(2-hydroxyphenyl)2H-benzotriazole UV stabilizers, stearic, oleic, linoleic and linolenic acid esters of 2[2-hydroxy-3-tert-butyl-5(3'-hydroxypropyl)-phenyl]2H-benzotriazole [Equation] [7]. The compounds, pale yellow oils were characterized by their UV and K^+IDS mass spectra [Figure 6] [7]. All four compounds showed nearly identical UV spectra with two maxima at 305 and 340 nm. Those UV stabilizers have aliphatic side chains that are by six carbon atoms longer than the carbon side chains of the catechol derivatives of the urushi components . The unsaturated esters, especially the linolenic ester could readily be incorporated into urushiol or drying oil compositions and the unsaturated esters can be "co-cured" into UV stabilized oriental lacquer films. They could also be added to linseed oil and cured into films of drying oils.

OH, $C(CH_3)_3$, N, N, N, $(CH_2)_2$, CH_2OH — RCOOH, esterification → OH, $C(CH_3)_3$, N, N, N, $(CH_2)_2$, CH_2OCOR

Equation

R = $(CH_2)_{16}{-}CH_3$
$(CH_2)_7CH{=}CH(CH_2)_7CH_3$
$(CH_2)_7CH{=}CH{-}CH_2{-}CH{=}CH(CH_2)_4CH_3$
$(CH_2)_7(CH{=}CH{-}CH_2)_3CH_3$

Specifically, the linolates and linolenates of 2[2-hydroxy-3-tert-butyl-5(3'-hydroxypropyl)phenyl]2H-benzotriazole were blended into several samples of urushi lacquers in about 3% per weight quantities and this composition was used to paint cedar wood and polyethylene surfaces. After curing the oriental lacquer composition appropriately, the coatings were exposed to the light of a Sylvania Sunlamp 078 in an attempt to mimic accelerated weathering. The coatings showed increased UV stability by an estimated factor of 2 as compared to coatings made with urushi samples without UV stabilizer added.

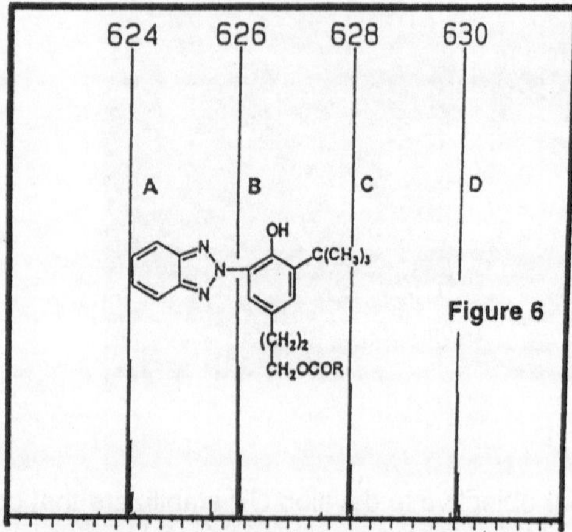

Figure 6

K^+IDS mass spectrum (M/Z in Daltons) of the linolenic [A], linoleic [B], oleic [C] and stearic [D] acid esters of 2[2-hydroxy-3-tert-butyl-5-(3'-hydroxy-propyl)-phenyl)2H-benzotriazole

Oriental Lacquer Ware

The Japanese oriental lacquer has been used for several thousands of years as a coating material for boxes and other articles. The first lacquered box of good quality dates from about 200 B.C., the Qin period in China. A Chinese lacquer box from the 14th century is shown in Figure 7 [9]. These types of boxes are well known for their thick layers of lacquers which are then carved. Two types of colors have been used to pigment oriental lacquer boxes even in early periods. One is the vermilion color caused by an iron oxide pigment and the black color which is caused by either magnetite, elemental iron or carbon black. Heavy pigmentation is needed because processed raw urushi used for painting has a deep blueish/black to black color.

Figure 7

Japanese lacquer trays from the Momoyama period of Japan, show the influence of the Zen culture, the simplicity of the neguro technique of applying the lacquer. Numerous items of these periods, which include trays all sorts of containers and boxes, even tea pots, sake containers, musical instruments like drums and stirrups were coated with Japanese lacquer. Figure 8 [9] shows a box from the late18th century and early19th century. For the decorations of such items as boxes, gold and silver flakes were incorporated when the last few layers of lacquers, but before the final topcoat, were applied. As indicated earlier, fine lacquer ware may have as many as 30-40 layers of thin coating of oriental lacquer.

Lacquer ware originating from Okinawa and Korea, but also some Chinese lacquer ware uses a considerable amount of mother of pearl or of shells as inlay for their decoration. The regional styles may differ considerably, but are quite easily distinguishable from each other.

Figure 8

In the Meiji period, the end of the 19th century, Japanese lacquer products became very fashionable world wide and the European market demanded considerable amounts of Japanese lacquer ware. A new art form was developed to satisfy the European market.

Figure 9

Japanese lacquer ware is also currently very popular. It is possible to purchase very nice lacquered boxes [Figure 9] or trays [Figure 10] in department stores. Very valuable items of high artistic value can be obtained in special art stores where artists sell very high quality lacquer ware, of traditional but also of novel and modern styles.

Acknowledgements. This is our 6th communication on Oriental Lacquers. The work was supported by the Mark Chair of the Polytechnic University and was inspired by the Pacific Polymer Federation's series of Symposia sessions on "Polymer Science and the Arts". We are greatly indebted to Koichi Hatada, Tatsuki Kitayama, William J. Simonsick Jr. and Ju Kumanotani for their help in our work on oriental lacquers.

Figure 10

References.

1. "The New Encyclopedia Britannica," Vol.10, 15th ed., Chicago, p. 575 (1984)
2. J.D. Mitchell, Advances in Economic Botany, 8, 103 (1990)
3. O. Vogl and J. Bartus, Polymer Preprints, ACS Division of Polymer Chemistry, 14 (1), 582 (1993)
4. J. Bartus, W.J. Simonsick, Jr., C. Garner, T. Nishiura, T. Kitayama, K. Hatada and O. Vogl, Polymer J., 26, 67 (1994)

5. K. Hatada, T. Kitayama, T. Nishiura, A. Nishimoto, W.J. Simonsick, Jr. and O. Vogl, Makromol. Chem, 195, 1865 (1994)
6. O. Vogl, J. Bartus, M.F. Qin, W.J. Simonsick, Jr., J.D. Mitchell, T. Kitayama and K. Hatada, Pacific Polymer Preprints, 3, 272 (1993)
7. O. Vogl, J. Bartus and W.J. Simonsick Jr., Polymer Preprints, ACS Division of Polymer Chemistry 14(1), 584 (1993)
8. Y.M. Du, in "Urushi", N.S. Bromelle and P. Smith, Ed., The Getty Conservation Institute, Marina de Rey, CA, 1985.
9. J.C.Y. Watt and B.B. Ford, in "East Asian Lacquer," J.P.O. Neill, Ed., The Metropolitan Museum of Art of New York, 1992.

Author Index

Abdullah, I. 351
Ajji, A. 25
Amiya, S. 367

Bartus, J. 423
Bolto, B.A. 325
Bottle, S.E. 85
Bousmina, M. 25
Bowmaker, G.A. 123
Bu, H. 313
Busfield, W.K. 85

Carlsson, D.J. 171
Carreau, P.J. 25
Chan, H.S.O. 237
Chen, E. 313
Chen, X. 63
Choi, Y.-J. 401
Christie, D.I. 203
Chujo, Y. 41
Clay, P.A. 203
Cooke, J.M. 171
Cooney, R.P. 123

Daly, W.H. 381

Economy, J. 271

Falicki, S. 171
Fujishige, S. 191

Gardella Jr., J.A. 63
Gilbert, R.G. 203
Gosciniak, D.J. 171
Grice, I.D. 85

Han, J.H. 307
Heiland, K. 85
Heuts, J.P.A 203
Ho, C.C. 337
Ho, T. 63
Honda, M. 159
Hozumi, S. 159
Huang, H.H. 237
Hutovic, J. 203

Imanishi, Y. 181

Jenkins, I.D. 85
Jung, J.C. 401

Kable, S.H. 203
Kang, T.J. 307
Kawaguchi, S. 247
Kerr, J.C.H. 99
Kim, C.Y. 225
Kim, D.H. 225
Kim, D.Y. 225
Kim, H.-B. 247
Kimura, S. 181
Kitayama, S. 159
Kuramoto, M. 53

Lee, H.B. 13
Lee, J.H. 13
Lee, J.Y. 225
Lee, K.C. 337
Lim, J.C. 271
Logothetis, A.L. 411

McCann, G.F. 123
Meutermans, W. 85
Millar, G.J. 123
Mitchell, J.D. 423
Miyaki, Y. 307
Miyata, S. 307
Monteiro, M. 85
Motobe, T. 307

Nakano, T. 111
Natansohn, A. 295
Negulescu, I. 381
Ng, S.C. 237
Nissan, R.A. 63
Nose, T. 1

Okamoto, Y. 111

Park, S.-B. 401
Penfold, J. 99
Piton, M.C. 203
Potter, C.W. 271
Pulpoka, B. 137

Qin, M. 423 Qiu, Z. 381

Rochon, P. 295

Saville, P.M. 99
Schué, F. 137
Seow, S.H. 237

Serpe, G. 271
Shi, G. 259

Tantayanon, S. 137
Ternai, P.C. 137
Thomas, R.K. 99
Tsuchimoto, T. 181

Van Luyen, D. 147
Vogl, O. 423

Wagener, K.B. 217
Wallace, G.G. 283
Whang, Y.E. 307
White, J.W. 99
Winnik, M.A. 247
Wynne, K.J. 63

Xu, S. 313

Yeap, E.B. 337
Youngblood, J. 381

Zin, W.-C. 401

Subject Index

ab initio quantum mechanics 203
acyclic diene metathesis polymerization 217
acylphosphine oxide 85
adsorbents 325
albumin 13
aminoxyl 85
amphiphilic properties 99
asphalt 381
asymmetric polymerization 111
azo polymers 295

birefringence 295
Bragg diffraction 159
tert-butoxyl radicals 85

cell adhesion 13
charge-transfer complex 85
Chinese hamster ovary cells 13
chitin 147
chitosan 147
chitosan applications 147
chlorinated polyethylene 381
clay mineral particles 337
coexistence curve 1
coil-globule transition 191
concentration fluctuation 1
conducting polymer colloids 283
conducting polymer-hydrogel composites 283
conductive electroactive polymers 283
contact angle 13
copolyesters 271
copolymerisation mechanism 85
corona discharge 13
COSY NMR 367
creep 381
crosslinking 411
β-crystalline form of polypropylene 259
crystallization 259
cyclic-peptide template 181

diethyl fumarate 85
diethyl maleate 85
differential scanning calorimetry 191
diffusion coefficient 247
diphenyl-3-pyridylmethyl methacrylate 111
diphenylphosphinoyl radicals 85
DNA binding 181
dynamic shear moduli 381

electrically conducting polymers 123
electrically conductive polymer composites 307
electrochemical polymerization 225,237
electron diffraction 313
energy transfer 247
epifluorescence microscopy 381
ESCA 13,63

flocculants 325
flotation agents 325
fluorescence probes 181
free-radical polymerizations 203
FTAR-ATR 13

Gamma 171

helix 111
heterocoagulation 337
heterogeneous titanium catalyst 53
high energy irradiation 411
hindered amines 171
homogeneous titanium catalyst 53
hydroboration polymerization 41
2,6-hydroxynaphthoic acid coplyester 271

impact modification 259
In-situ Raman spectra 225
interchain transesterification reactions 271
ion exchanges 325
ionomer 247

isotactic polystyrene 313

kinetics 1

latex films 247
light control film 159
light scattering 191
liquid natural rubber 351
lower critical solution temperature 191

mechanical properties 259
mechanism of initiation 85
methacrylate 111
microlithography 137
microporous film 259
microstructure of poly(vinyl alcohol) 367
minimally adhesive surfaces 63
molecular weight fractionation 191
morphology 1

natural rubber 351
natural rubber latex 337
negative neighbouring group effect 217
nitroxide 85
NMR 367
nucleating agent 259

optical storage 295
optically active polymer 111
organoboron polymer 41
oriental lacquer 423
oxidation 171

p-hydroxybenzoic acid 271
perfluoroelastomers 411
performance on demand 283
phase separation 1
phase transition 191
phenyl[bis(2pyridyl)]methyl 111
1-phenyldibenzosuberyl methacrylate 111
photoacoustic infrared spectroscopy 123
photoelectron spectroscopy 123
photoisomerization 295
photopolymerization 159
photoresists 137
PMMA/rubber blends 25
polarized light 295
poly(3,3'-dibutoxy-2,2'-bithiophene) 237
poly(3,3'-dichloro-2,2'-bithiophene) 237
poly(cyclodiborazane) 41
poly(ether-urethane) 63
poly(ethylene oxide) 313
poly(ethylene oxide) macromonomer core-shell latex 247
poly(Lewis acid) 41
poly(organoboron halide) 41
poly(p-epoxystyrene-co-N-phenylmaleimide-co-styrene) 137
poly(siloxane-urea-urethane) 63
poly(vinylacetate)-polypyrrole composite 307
poly(vinyl alcohol) 367
poly(vinylchloride)-polypyrrole composite 307
polybithiophene 237
polyelectrolytes 325
polyethylene 13
polyimides 401
1,4-polyisoprene-polyacetylene copolymers 99
polymer blends 1, 25, 259
polymer diffusion 247
polymer modified asphalt 381
polymer toughening 259
polypeptide tertiary structure 181
polypropylene 171
polypyrrole 123,225,283
propagation rate coefficients 203
protein adsorption 13
PS/PMMA blends 25
PS/PVME blends 25
PS/rubber blends 25
pulsed-laser polymerization 203

radical-trapping 85
Raman spectroscopy 225
recycled tire rubber 381
redox reaction 225
refractive index modulation 159
resonance Raman 123
rheological properties 25
rigid-rod polyimides 401

SHRP specifications 381
single-chain single crystals 313
SIROFLOC 325
SIROTHERM 325
specular neutron reflection 99
spherical diffraction 159
stabilization phenols 171
stereospecific polymerization 53
stress whitening 259
structure reactivity relationships 217
styrene 53
surface excess structure 99
syndiotactic polystyrene 53

template-assisted synthesis 181
thermal degradation 271
thermal stability 401
tin tailings slurry 337
transfer-rate coefficient 203
triphenylmethyl methacrylate 111

ultraviolet stabilizers 423
urushiol 423

Vietnam 147
viscometry 191

wettability gradient 13

x-ray 123

yellowing 171

zeolite channels 123